D1292951

DEMCO

Mathematics in Historical Context

Images on Cover from Left to Right:
Issac Newton
Felix Klein
James Joseph Sylvester
Joseph-Louis Lagrange
Solomon Lefschetz
Hypatia
Brahmagupta

Mathematics in Historical Context

Jeff Suzuki
Brooklyn College

Published and Distributed by
The Mathematical Association of America

SPECTRUM SERIES

The Spectrum Series of the Mathematical Association of America was so named to reflect its purpose: to publish a broad range of books including biographies, accessible expositions of old or new mathematical ideas, reprints and revisions of excellent out-of-print books, popular works, and other monographs of high interest that will appeal to a broad range of readers, including students and teachers of mathematics, mathematical amateurs, and researchers.

Mathematical Carnival, by Martin Gardner

Mathematical Circles Vol I: In Mathematical Circles Quadrants I, II, III, IV, by Howard W. Eves

Mathematical Circles Vol II: Mathematical Circles Revisited and Mathematical Circles Squared, by Howard W. Eves

Mathematical Circles Vol III: Mathematical Circles Adieu and Return to Mathematical Circles, by Howard W. Eves

Mathematical Circus, by Martin Gardner

Mathematical Cranks, by Underwood Dudley

Mathematical Evolutions, edited by Abe Shenitzer and John Stillwell

Mathematical Fallacies, Flaws, and Flimflam, by Edward J. Barbeau

Mathematical Magic Show, by Martin Gardner

Mathematical Reminiscences, by Howard Eves

Mathematical Treks: From Surreal Numbers to Magic Circles, by Ivars Peterson

Mathematics: Queen and Servant of Science, by E.T. Bell

Mathematics in Historical Context,, by Jeff Suzuki

Memorabilia Mathematica, by Robert Edouard Moritz

Musings of the Masters: An Anthology of Mathematical Reflections, edited by Raymond G. Ayoub

New Mathematical Diversions, by Martin Gardner

Non-Euclidean Geometry, by H. S. M. Coxeter

Numerical Methods That Work, by Forman Acton

Numerology or What Pythagoras Wrought, by Underwood Dudley

Out of the Mouths of Mathematicians, by Rosemary Schmalz

Penrose Tiles to Trapdoor Ciphers . . . and the Return of Dr. Matrix, by Martin Gardner

Polyominoes, by George Martin

Power Play, by Edward J. Barbeau

Proof and Other Dilemmas: Mathematics and Philosophy, edited by Bonnie Gold and Roger Simons

The Random Walks of George Pólya, by Gerald L. Alexanderson

Remarkable Mathematicians, from Euler to von Neumann, by Ioan James

The Search for E.T. Bell, also known as John Taine, by Constance Reid

Shaping Space, edited by Marjorie Senechal and George Fleck

Sherlock Holmes in Babylon and Other Tales of Mathematical History, edited by Marlow Anderson, Victor Katz, and Robin Wilson

Student Research Projects in Calculus, by Marcus Cohen, Arthur Knoebel, Edward D. Gaughan, Douglas S. Kurtz, and David Pengelley

Symmetry, by Hans Walser. Translated from the original German by Peter Hilton, with the assistance of Jean Pedersen.

The Trisectors, by Underwood Dudley

Twenty Years Before the Blackboard, by Michael Stueben with Diane Sandford

Who Gave You the Epsilon? and Other Tales of Mathematical History, edited by Marlow Anderson, Victor Katz, and Robin Wilson

The Words of Mathematics, by Steven Schwartzman

MAA Service Center
P.O. Box 91112
Washington, DC 20090-1112
800-331-1622   FAX 301-206-9789

Introduction

This book emerged from a discussion I had with Don Albers in the Spring of 2002, when he suggested the idea of a book that would describe the world of the great mathematicians: "What would Newton see, if he looked out his window?" I really liked the idea, and planned to pick out a dozen or so mathematicians and write a hundred or so pages of history discussing how mathematics, society, and mathematicians interacted with one another. Before I knew it, I had written six hundred pages. I've cut those down to the present text, which is part mathematics, part mathematical biography, and part history.

My hope is to convey some of the fascinating and complex relationships between mathematicians, mathematics, and society. In these pages, you will find how world events shaped the lives of mathematicians like Archimedes and de Moivre; how artistic conventions inspired some of the mathematical investigations of Abu'l-Wafā and al-Khayyāmī; how Newton and Poincaré affected the political events of their time; and how mathematical concepts like the irrationality of $\sqrt[7]{2}$ drove cultural development.

Finally, some thanks and dedication. First, to Rob Bradley of Adelphi University and Ross Gingrich of Southern Connecticut State University for suggesting the title of the book (my title was a completely uninspired "Mathematics and History"). Next, to my wife Jacqui, who listened with great forbearance as I tried to reconcile information from different sources, which often spelled and indexed the same name in different ways, or gave different dates for the same historical event, or completely different interpretations of the significance of an event. Last but not least, I'd like to dedicate this book to my children, William and Dorothy: May the past be the guide to the future.

Contents

1

The Ancient World

1.1 Prehistory

When anthropologists speak of a group's culture, they mean the sum of all the human activity of the group: how they talk, think, eat, play, and do mathematics. Every culture in the world creates some sort of mathematics, though it may be on a very basic level. The most ancient evidence of mathematical activity comes from a wolf's bone on which fifty-five notches are carved, grouped in sets of five. The bone was found in the Czech Republic and is about 35,000 years old. A similar artifact was found at Ishango, on the shores of Lake Edward in Zaire. The Ishango bone (from a baboon) dates to around 18,000 B.C., and is particularly intriguing since along one side, the notches are grouped into sets of 11, 13, 17, and 19, suggesting an interest in prime numbers.

At that time, mankind obtained food by hunting animals and gathering plants. But about ten thousand years ago, an unknown group of people invented agriculture. This forced people to establish permanent settlements so the plants could be cared for until harvest: the first cities.

The culture of city-dwellers is called *civilization* (from *civilis*, "city-dweller" in Latin). Since cities were originally farming communities, they were established in regions that provided the two fundamental needs of agriculture: fertile soil and a reliable water supply. River flood-plains provide both, and the most ancient civilizations developed around them.

Figure 1.1. The Ishango Bone. Photograph courtesy of Science Museum of Brussels.

1.2 Egypt

In Egypt, civilization developed around the Nile river, whose annual flooding deposited silt that fertilized the fields of the Egyptian farmers. As early as 4000 B.C. the Egyptians may have noticed that it took approximately 365 days from one Nile flood to the next. The Egyptian calendar divided the year into 12 months of 30 days, with 5 extra days celebrated as the birthdays of the main gods of the Egyptian pantheon. These days of religious festivities were the original *holidays* ("Holy days").

Egypt is divided into two main parts, Lower Egypt and Upper Egypt. Lower Egypt consists of the marshy areas where the Nile empties into the Mediterranean, known as the Delta (because of its resemblance to the Greek letter Δ, pointing southwards); the term would later be applied to *any* river's outlet into a sea or lake. Upper Egypt is formed by the region between the Delta and the First Cataract, or set of waterfalls, at Aswan. According to tradition, Upper and Lower Egypt were united around 3100 B.C. by Narmer, whom the Greeks called Menes. Archaeological evidence points to an even earlier King Scorpion who ruled over a united kingdom, and recently the tombs of King Scorpion and his successors have been identified.

The kings of Egypt were known as *pharaohs* ("Great House," referring to the Royal Palace). The pharaohs were revered as gods, though this did not spare them from being criticized, plotted against, and deposed. Though the rulers changed, the fundamental in-

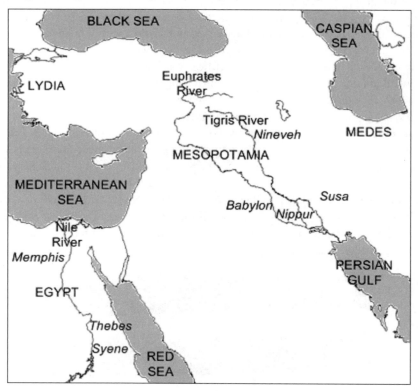

Figure 1.2. Egypt and the Fertile Crescent

stitutions of pharaonic Egypt would last for nearly three thousand years, a period of time usually referred to as Dynastic Egypt.

Dynastic Egypt is traditionally divided into thirty dynasties, according to a division first described by the Greek-Egyptian scholar Manetho around 280 B.C. These thirty dynasties are grouped into five periods: the Old Kingdom (roughly 3100 B.C. to 2200 B.C.); a First Interregnum (2200 B.C. to 2100 B.C.); a Middle Kingdom (2100 B.C. to 1788 B.C.); a Second Interregnum (1788 B.C. to 1580 B.C.); and a New Kingdom (1580 B.C. to 1090 B.C.).[1] Manetho's division of the reigns has been retained, though his dates are no longer considered accurate; indeed, the dates of ancient Egypt are very uncertain and Egyptologists themselves differ by up to a hundred years on the dates of the reigns of early pharaohs.

The best-known features of Egyptian civilization are the pyramids. The step pyramid at Sakkara, near the Old Kingdom capital of Memphis, is the most ancient; it was completed under the direction of the architect Imhotep around 2650 B.C. The largest and best known pyramid is the Great Pyramid at Gizeh, finished around 2500 B.C. and the oldest of the Seven Wonders of the World. Moreover, it is the *only* one of the Seven Wonders still standing, a tribute to the skill of the Egyptian builders and the dry climate of Egypt.

By 2700 B.C., a form of writing had been invented. Since many examples were found adorning the walls of Egyptian temples, it was erroneously believed that the writings were religious in nature. Hence this form of Egyptian writing became known as *hieroglyphic* ("sacred writing" in Greek).

Hieroglyphs may have originally been pictograms (small pictures of the object being represented, such as a set of wavy lines ≈ to represent a river) or ideograms (stylized figures that represent an abstract concept, such as the emoticon ; -) to indicate "just kidding"). However, this form of writing would require the knowledge of hundreds, if not thousands, of symbols to represent the common words in a language, so hieroglyphs soon took on new meanings as *sounds*. For example, we might draw ≈ to represent the word "river" or to refer to a man named Rivers, but in time ≈ might come to represent the sound "riv-" or even the initial sound "r-."

1.2.1   Egyptian Mathematics

To write numbers, we could spell them out: two hundred forty-one. However, writing out number words is tedious and all civilizations have developed special symbols for numbers. The use of a vertical or horizontal stroke to represent "one" is nearly universal, as is the use of multiple strokes to represent larger numbers: thus, "three" would be | | |. Of course, such notation rapidly gets out of hand: try to distinguish | | | | | | | | | | | | | from | | | | | | | | | | | | |. The next logical step would be to make a new symbol for a larger unit. In hieroglyphic, ∩ represents ten, ৩ represents one hundred, ⌇ represents one thousand, and other symbols were used for ten thousand, one hundred thousand, and one million. To indicate a large number, the symbols would be repeated as many times as necessary: two hundred forty-one would be | ∩∩∩∩৩৩ (note that the Egyptians wrote from right to left, and placed the greatest values first). Since the value of the number is found by adding the values of the symbols, this type of numeration is called additive notation.

---

[1]Menes is the first pharaoh of the First Dynasty. King Scorpion and his immediate successors, who predate Menes, are collectively grouped into Dynasty Zero.

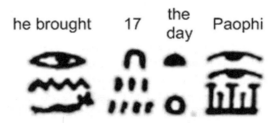

Figure 1.3. From the Rosetta Stone, line 10. Inscription reads, "On the 17th day of Paophi, he [Ptolemy V] brought...."

Fractions were generally expressed as sums of unit fractions. Thus $\frac{2}{5}$ would be written as $\frac{1}{15}$ and $\frac{1}{3}$. These unit fractions were written by placing a symbol, $\bigcirc$, read as *ro* (an open mouth), above the hieroglyphic representation of the denominator. Thus $\frac{2}{5}$ would be

Some common (non-unit) fractions had special symbols, though the only one used with any consistency was a symbol for two-thirds. The immense length of Egyptian history can be illustrated by the two thousand year history of this symbol: in the Old Kingdom, two-thirds was written as ⟨ but a thousand years later, this form would change into ⟨ . It would be another thousand years before this form was written ⟨.

By 2600 B.C., a cursive form of hieroglyphic, suitable for writing on softer materials, came into existence: hieratic. All Egyptian mathematical treatises are written in hieratic, on papyrus, cloth, or leather. Unfortunately, these materials disintegrate rapidly when exposed to water, insects, and sunlight, so our knowledge of Old Kingdom mathematics is limited to indirect evidence (like the existence of the pyramids or an accurate calendar) and hieroglyphic numerals.

The oldest mathematical texts date to the Middle Kingdom, six hundred years after the construction of the Great Pyramid of Gizeh. The Reisner papyrus, named after its discoverer, dates to the reign of the Twelfth Dynasty pharaoh Sesostris (around 1900 B.C.) who established a chain of forts south of Aswan to keep Egypt's borders secure against the Nubians. Sesostris also sponsored a number of building projects, including additions to the great temple complex at Karnak (near Thebes, the capital of the Middle Kingdom).

The Reisner papyrus is a set of four worm-eaten rolls, and appears to be a set of books for a construction site. The readable portions include a list of employees, as well as calculations of volumes and areas. A typical computation is determining the number of workmen needed to excavate a tomb, given the dimensions of the pit and the expected volume of dirt to be removed by a workman each day (apparently 10 cubic cubits, or roughly 30 cubic feet).

Sesostris's dynasty ended with the reign of Sebeknefru (around 1760 B.C.), the earliest female ruler whose existence is definitively established. Her reign was peaceful, but her death touched off a civil war and over the next century, nearly seventy pharaohs would rule in quick succession.

The Moscow or Golenishchev papyrus (named after its current location or discoverer) was written during this era. It includes a number of geometric problems. For example, Problem 7 asks to find the dimensions of a triangle with area 20 *setat*, where the ratio of the height to the base is 2 1/2 to 1. The scribe's solution is to double the area (making a rectangle with sides in ratio 2 1/2 to 1) then multiply by 2 1/2: this produces a square (of area 100), whose side can be found by taking the square root; this is also the triangle's height. Dividing by 2 1/2 gives the base. Problem 9 finds the surface area of a basket (possibly a hemisphere or a half-cylinder).

Problem 14 contains what is perhaps the greatest mathematical discovery of the ancient Egyptians: the computation of the exact volume of the frustum of a pyramid. The idea of a mathematical formula was as yet non-existent; the student was expected to generalize from the given example. Thus the Moscow papyrus gives:

Problem 1.1. *Find the volume for a frustum 6 cubits high with [square] base 4 cubits and [square] top 2 cubits. Square the base to get 16; multiply [top side] 2 by [bottom side] 4 to get 8; square the top 2 to get 4; add these together to get 28. Multiply the height 6 by 1/3, to get 2; multiply 2 by 28 to get 56, the volume.*

Thus if the pyramid has a square base of side a, square top of side b, and height h then its volume will be given by $V = \frac{1}{3}h\left(a^2 + ab + b^2\right)$.

Shortly after the Moscow papyrus was written, Egypt was invaded by a mixed group of tribes, consisting primarily of Semites from Palestine and Hurrians from Asia Minor (modern-day Turkey). These invaders swept away all opposition by using a new weapon of war: the horse and chariot. They conquered Egypt but kept its political and religious institutions intact, establishing themselves around 1680 B.C. as the Fifteenth Dynasty. The Egyptians named them the *Hyksos* ("Rulers of Foreign Lands").

Our first complete text of Egyptian mathematics dates to this period, a thousand years after the building of the first pyramids. The Rhind papyrus (named after its discoverer) was, according to the preface, written down in the fourth month of the Inundation Season in the thirty-third year of the reign of A-user-Re by the scribe A'H-MOSÈ (fl. 1650 B.C.?). The introduction to the Rhind papyrus highlights an important problem of history: precise dating of an event. Often we have a better idea of the day and month than for the year itself: because of the reference to the fourth month of the Inundation Season, we know that the Rhind was copied around September. But we do not know A-user-Re's dates with any certainty, so the best guess for the thirty-third year of the reign of A-user-Re is around 1650 B.C.

A'h-mosè claimed the Rhind was based on an older work, dating back to the Thirteenth Dynasty pharaoh Ne-ma'et-Re and thus contemporaneous with the Moscow papyrus, a claim that might or might not be true. Why would someone deny that their work was original? In the days before printing, all texts had to be copied by hand: they were literally *manuscripts* ("hand written" in Latin). Since it takes just as much effort to copy a worthless text as a worthwhile one, this meant that the only texts that were copied and recopied were the most important. One way to give your work a veneer of importance is to attribute it to some great author of the past or some time period that people looked back to with nostalgia: hence, the first five books of the Old Testament are attributed to Moses, even though they were not written down until a thousand years after the events they claim to describe.

The first part of the Rhind papyrus consists of the quotient of 2 and the odd numbers from 3 to 101. From this, we can discern many details of how the Egyptians performed calculations. A typical computation is 2 divided by 13, whose quotient is $\frac{1}{8} + \frac{1}{52} + \frac{1}{104}$.

The computation proceeds as follows:

1	13		
1/2	6	1/2	
1/4	3	1/4	
\ 1/8	1	1/2	1/8
\ 4	52	1/4	
\ 8	104	1/8	

The second, third, and fourth lines seem to represent the product of 13 and $1/2$, $1/4$, and $1/8$, respectively. The fifth line seems to refer to the fact that $13 \cdot 4 = 52$ implies $13 \cdot \frac{1}{52} = \frac{1}{4}$. Likewise, the last line obtains $13 \cdot \frac{1}{104} = \frac{1}{8}$ from $13 \cdot 8 = 104$. The lines indicated with the \ can then be interpreted to read:

$$13 \cdot \left(\frac{1}{8} + \frac{1}{52} + \frac{1}{104} \right) = 1 + \frac{1}{2} + \frac{1}{8} + \frac{1}{4} + \frac{1}{8}$$

Hence $13 \cdot \left(\frac{1}{8} + \frac{1}{52} + \frac{1}{104} \right) = 2$, and thus 2 divided by 13 is $\frac{1}{8} + \frac{1}{52} + \frac{1}{104}$.

In the Rhind papyrus, A'h-mosè shows how to solve simple linear problems using a variety of methods, including the method of false position. The student was expected to generalize from the examples, which began with:

Problem 1.2. *A number and its seventh make* 19 *[i.e.,* $x + \frac{1}{7}x = 19$*]. Suppose* 7*; its seventh is* 1*, and together* 8*. Divide* 19 *by* 8*, then multiply by* 7*. Solution:* 16 *and* $\frac{1}{2}$ *and* $\frac{1}{8}$*.*

To find the area of a circle, A'h-mosè gave the example:

Problem 1.3. *Find the area of a circular plot of land with a diameter of* 9 khet. *Take away one-ninth of the diameter, leaving* 8*; multiply* 8 *by itself to get the area,* 64 setat.

The volume of a cylinder was computed by multiplying the area of the base (using the above method) by the height.

The problems in the Rhind are all linear, though the contemporary Kahun papyrus seems to suggest the Egyptians could also solve non-linear equations. Unfortunately the problem statement is missing, and must be reconstructed from the method of solution.

1.2.2 The New Kingdom

The Hyksos conquest of Egypt depended on advanced technology: the horse, chariot, and compound bow. But all of these were soon duplicated by native Egyptians, who made ready to throw out their masters. A revolt began in Thebes, and by 1521 B.C., the Hyksos were expelled and a new dynasty, the eighteenth, was formed. Mathematicians might smile at the name of the first pharaoh of the new dynasty: Ahmose.[2] Ahmose went on to conquer Nubia, to the south, and his successors established a vast Egyptian Empire that ultimately extended well into Palestine. The Kingdom of Egypt underwent a great cultural renaissance.

One of the critical events of the Eighteenth Dynasty concerned religion. For millennia, the Egyptians were polytheists, revering many gods, though chief among them was Amon. Around 1350 B.C., the pharaoh Amenhotep IV ("Amon is satisfied") became a monotheist,

---

[2]The name was a popular one during and immediately after the Hyksos era.

worshipping a single god. In Amenhotep's case, the single god was the sun god Aton, and he renamed himself Akhenaton ("One Useful to Aton"). Akhenaton established a new capital at Tell-el-Amarna, where Egyptian art and sculpture flourished.

Akhenaton attempted to eliminate the worship of other gods besides Aton, and in this he failed. His appointed successor Smenkhkare had a very short reign, and was succeeded by an eight-year-old boy named Tutankhaton. With the accession of a minor, the polytheists saw the chance to return Egypt to the old religion. They forced the young pharaoh to restore Amon to the position of chief among many gods, and to rename himself Tutankhamun ("Living Image of Amon"). Tutankhamun's unplundered tomb was discovered in 1922 by Howard Carter and the fifth Earl of Carnarvon, George Edward Stanhope. A few months later Stanhope died suddenly, giving rise to a belief in a "curse of the pharaohs," though Carter, the first to actually enter the tomb, survived another seventeen years.

The internal religious warfare weakened the Eighteenth Dynasty and led to the rise of the most famous Egyptian dynasty of all, thanks to its pharaoh Rameses the Great. Rameses, the third pharaoh of the Nineteenth Dynasty, built enormous monuments to himself all over Egypt. The most spectacular was at Abu Simbel, where four 67-foot-high statues of a seated Rameses loom beside the entrance to a temple. In the 1960s, the damming of the Nile at Aswan formed a lake that threatened to submerge the temples, so they were cut apart and reassembled on higher ground.

The Berlin papyrus was written during the Nineteenth Dynasty and gives the first clear example of a non-linear problem solved by the ancient Egyptians. The problem is:

Problem 1.4. *A square with an area of* 100 *square cubits is to be divided into two squares whose sides are in ratio* 1 *to* $\frac{1}{2}$ $\frac{1}{4}$ *[i.e.,* 1 *to* $\frac{1}{2} + \frac{1}{4}$*].*

The scribe's solution is by means of false position: supposing the side of the bigger square is 1, making the side of the smaller square $\frac{3}{4}$, the total area will be 1 $\frac{9}{16}$, which is the area of a square of side 1 $\frac{1}{4}$. Divide 10 by 1 $\frac{1}{4}$ to get 8, then multiply the initial guess (1) by 8 to yield the actual side of the larger square; the side of the smaller square is thus $\frac{3}{4}$ of 8 or 6.

After Rameses, the Egyptian Empire underwent a long and final decline. The reasons are varied, but one factor may have contributed more than anything else. The Egyptian Empire was conquered and controlled by warriors using bronze. But around the reign of Rameses the Great and the writing of the Berlin papyrus, the Hittites, an obscure tribe living in the foothills of the Caucasus mountains, learned how to refine iron from its ores. An iron equipped army could easily destroy one outfitted with bronze. The Egyptians adopted the horses, chariots, and bows of the Hyksos, and might have adopted the iron technology of the Hittites as well—but there are few sources of iron ore in Egypt.

1.3 Mesopotamia

Egypt is the westernmost of the ancient river civilizations, and is just south of one end of the Fertile Crescent, a region that includes modern Iraq, southeastern Turkey, Syria, Israel, Lebanon and Jordan. A second great river civilization arose at the eastern end of the Fertile Crescent, in modern Iraq. Located between the Tigris and Euphrates rivers, it became known as Mesopotamia ("between the rivers" in Greek). Today, the Tigris and

Euphrates merge about a hundred miles inland of the Persian Gulf, though in ancient times, the two rivers flowed separately into the sea.

Most historians believe that agriculture was invented in Mesopotamia, and thus civilization began in this land between the rivers. The most ancient civilization in this region was the Sumerian, and by 3000 B.C., a group of Sumerian cities existed in Lower Mesopotamia, the region closer to the Persian Gulf. Unlike the contemporary Egyptians, the Sumerians never united into a single kingdom, and each city was a separate political entity with its own customs, army, and foreign policy; hence they are generally referred to as city-states.

Like the Nile, the Tigris and Euphrates flood periodically. But while the flooding of the Nile is a gentle, welcome event that signals the beginning of the year and a renewal of life, Mesopotamian floods are often catastrophic events. One flood tale was included in the story of Gilgamesh, the King of Uruk (a city along the Euphrates). When a close friend dies, Gilgamesh searches the world for the secret of immortality, and comes across the one man who has found it: Utnapishtim. He tells Gilgamesh the story of how the world became so populated that the gods, tired of hearing the racket made by so many people, decided to exterminate mankind with a great flood. One of the gods, Ea, warned Utnapishtim of what was to happen, and had him build a boat in which to save himself and his family. To find when the waters had receded enough to land, Utnapishtim released a dove, which could find no resting place and thus returned to the boat, and later a raven, which found a roost and never returned. The legend of Gilgamesh is almost certainly one of the inspirations of the Biblical story of Noah. As for immortality, Utnapishtim explained to Gilgamesh how to find the plant that confers immortality, but the plant is eaten by a snake while Gilgamesh sleeps. Again, the parallels to the Biblical story of how mankind was deprived of immortality through the actions of a serpent suggest that the ancient Hebrews drew heavily upon these legends.

1.3.1   Positional Notation

The Egyptians wrote on papyrus, leather, cloth, or stone. The Mesopotamians had none of these in abundance; instead, they had mud. Fortunately, mud could be written on with a stylus while still wet, and then baked into a brick-hard clay tablet. Such tablets can last for millennia, virtually unchanged. In contrast to a handful of Egyptian papyri, we have thousands of clay tablets.

Mesopotamian script consists of *cuneiform* ("wedge-shaped" in Greek) symbols, distinguished by the number of marks and their orientation. Deciphering an arbitrary text with no knowledge of the content would be virtually impossible, but if the text is mathematical, we can make considerable headway. Consider the table text reproduced in Figure 1.4. As you read down the rows of the first and second vertical columns, the number of ⅄ symbols increases from one to nine, after which a new symbol, ⟨, appears. It seems reasonable to suppose that ⅄ indicates a unit, ⅄⅄ indicates two units, and so on, up until ⟨ indicates ten units; then ⟨⅄ indicates eleven and so on. Mesopotamian numeration *appears* to be additive.

However, consider the third column. The first entry is obliterated, but if our assumption that ⅄ represents one and ⟨ represents ten is correct, then the second through seventh rows read: four, nine, sixteen, twenty-five, thirty-six, and forty-nine. These are the squares of

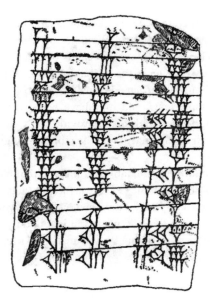

Figure 1.4. A Table Text from Nippur.

the numbers in the second column, so the eighth entry ought to represent sixty-four. So how shall we interpret the fact that the number is written using five ⌐s? If there is any consistency in the table text, then one of the ⌐s must represent sixty and the remaining four must represent four. This suggests that Mesopotamian notation was positional, base 60.

There are some drawbacks in the positional notation used in Mesopotamia. First, there was no way to indicate the order of magnitude: ⌐ by itself might mean one unit, or it might mean one sixty, or it might even mean one sixtieth. Context was the only way to decide between the possible interpretations.

Another problem was that there was no way to indicate the *lack* of an order of magnitude. For example, consider the eleventh line of the table text. The number in the third column should be the square of eleven, or one hundred twenty-one. This is two sixties and one unit, and the scribe has properly recorded the number, leaving a wide space between the ⌐s representing the sixties and the ⌐ representing the unit. But a careless or inexperienced scribe might not leave enough space between the orders of magnitude, making it possible to read this as three ⌐s: three units or possibly three sixties.

Our own system uses a number of devices to solve these problems. First, we have nine different quantity symbols (1 through 9), so that adjacent symbols actually represent different orders of magnitude: 123 represents 3 units, 2 tens, and 1 hundred. Moreover, we use a decimal point to separate units from fractions of a unit, while a tenth symbol (0) indicates the lack of an order of magnitude. Thus 10.1 is 1 ten, no units, and 1 tenth. In modern transcriptions of Babylonian sexagesimals, the comma is used to separate the orders of magnitude and the semicolon (;) separates the units from the fractions; the zero is used freely. Thus 1 sixty and 1 sixtieth would be written 1, 0; 1. However (and this is important to emphasize), the Mesopotamians *never* indicated the sexagesimal point, and it was not until very late in their history that a symbol for an empty space was used.

In 1898 an American expedition began excavating the city of Nippur, the center of worship of the Sumerian god Enlil ("Lord Wind"), and thus one of the main religious centers in Mesopotamia. There they found thousands of mathematical cuneiform texts in the eastern section of the city (known as the scribal quarter). Originally attributed to being from a "temple library", though later expeditions cast doubt on this claim, these tablets are of the type known as table texts, and may have been written as early as 2200 B.C. These are essentially multiplication tables, and provide many examples of Mesopotamian numeration; there are enough careless errors on the tablets to suggest they were written by apprentice scribes.

One might expect the multiplication tables to include the multiples of 1, 2, 3, and so on, and most of the tablets do. But tables exist for the products of (in sexagesimal) 3, 20 or even 44, 26, 40. Such tables might be explained as "make work" assigned by teachers to keep students busy, but a more likely explanation is found by noting that dividing by a number like 18 is the same as multiplying by $\frac{1}{18}$, and in sexagesimal, $\frac{1}{18} = 0; 3, 20$. For 44, 26, 40 we note that $0; 0, 44, 26, 40 = \frac{1}{81}$. Thus these tablets were probably used for division problems.

1.3.2   Babylon

The disunity of the Sumerican city-states made them easy prey for any unified conqueror. Under Sargon of Agade (or Akkad), a group of tribes speaking a Semitic language conquered the Sumerians of Mesopotamia around 2300 B.C. Since Sargon and the Akkadians ruled over a diverse population of which they themselves were only a small minority, we speak of the Akkadian *Empire*. It was the first empire in history. It lasted less than a century, and by 2200 B.C., the empire was destroyed and the capital city of Akkad was so devastated that even today its exact location is unknown. But the brief existence of the Akkadian Empire had some lasting consequences. One of the main results was that Sumerian (a language unrelated to any others) gradually disappeared as a spoken language, to be replaced with Semitic languages (such as Akkadian). This linguistic shift affected place names, and an obscure Sumerian city, *Ka-dingr* ("The Gate of God" in Sumerian) was translated literally into Akkadian as *Bab-ilu*, a name that eventually became Babylon.

Around 1800 B.C. the Babylonians conquered all of Mesopotamia. Since most of the problem texts seem to date to around this time, Mesopotamian mathematics is frequently referred to as Babylonian mathematics.

The most famous Babylonian king was Hammurabi, who reigned around 1750 B.C. Hammurabi achieved lasting fame for having inscribed on a *stele* (stone pillar) a complex and sophisticated legal code. This was not the first written code of laws, though it is one of the earliest that we have in its entirety.

The stele was originally in Sippar, at the temple of the sun god Shamash. At the top is a depiction of Hammurabi receiving the laws from Shamash. Flames emerge from the god's shoulders, which is suggestive of the much later Biblical myth of Moses receiving the ten commandments from a flaming god. The code of Hammurabi divides people into several classes, including noblemen, commoners, and slaves. The law is primarily punitive and consists of prescriptions: if X occurs, the penalty is Y. Many crimes are punishable by death or mutilation, but monetary fines are also common. The legal theory of Hammurabi's empire gave the classes distinct responsibilities and punishments: in general, harm to a noble warranted the greatest punishment, and harm to a slave the least. But nobles were also held to a higher standard of behavior, and misconduct by a noble was more severely punished: if a noble stole livestock (as opposed to other types of property), they had to pay 30 times the value, but a commoner only 10 times. The code also regulates prices and wages. The law code must have made a great impression on the people of the time, for shortly after Hammurabi's death, an invading army carried the stele away to Susa, where it was discovered in 1901 by French archaeologists.

We cannot compare the legal codes of Babylon and Egypt, for the surviving Egyptian law codes date to a time much later than Hammurabi. We may, however, compare their mathematics, for the Rhind papyrus and the Babylonian problem texts are contemporaneous. Many Babylonian problems dealt with canals, a necessity of life in Mesopotamia, required both for irrigation and flood control. A typical problem was:

Problem 1.5. *A canal 5 GAR long, $1\frac{1}{2}$ GAR wide, and $\frac{1}{2}$ GAR deep is to be dug. Each worker is assigned to dig 10 GIN, and is paid 6 SE. Find the area, volume, number of workers, and total cost.*

where GAR, GIN, and SE are units of quantity (the GIN is equal to a cubic GAR, for example).

The Babylonians frequently ventured beyond simple linear equations. Many canal problems resulted in quadratic equations, though a few "pure math" problems were posed:

Problem 1.6. *The* igibum *exceeded the* igum *by 7, and the product of the two is* 1, 0. *Find the* igibum *and the* igum.

The problem is equivalent to solving $x(x + 7) = 60$, and the scribe's solution relied on an identity we would write as

$$\left(\frac{a + b}{2}\right)^2 - \left(\frac{a - b}{2}\right)^2 = ab$$

Thus given the product ab and $a \pm b$, we can find $a \mp b$. Finally, the *igibum* and the *igum* can be found using

$$\frac{a + b}{2} + \frac{a - b}{2} = a$$
$$\frac{a + b}{2} - \frac{a - b}{2} = b$$

There is some evidence that degenerate cubics with rational solutions were solved by transformation into a standard form and referring to tables of values. Even approximate solutions to exponential equations were found using linear interpolation (also known as the method of double false position or the method of secants).

Much of our knowledge of Babylonian geometry comes from a French expedition to Susa, about 200 miles east of Babylon. In 1936, the expedition uncovered a horde of cuneiform texts, including some on geometrical subjects, which seem to be contemporaneous with the arrival of Hammurabi's stele. One tablet shows a circle with the cuneiform numbers 3 and 9 on the circumference and 45 on the inside. This suggests that the area of a circle was found by taking the circumference, 3, squaring it to get 9, then dividing by 12 to get 0; 45. This gives an even worse approximation for the area of a circle than the one used by the Egyptian scribes. This might suggest the Mesopotamians were poor geometers, though other tablets of the Babylonian era show they knew of and used the Pythagorean Theorem; the most remarkable is a tablet that seems to indicate the diagonal of a square is 1; 24, 51, 10 of its side which, if converted to its decimal equivalent, gives $\sqrt{2} \approx 1.414213\ldots$.

In addition, Plimpton 322, believed to have been written around Hammurabi's time, gives a list of 15 rational numbers that satisfy $a^2 + b^2 = c^2$, which suggests an early interest in number theory, and might point to the knowledge of the Pythagorean Theorem as well. Another tablet gives relationships between the area A_n of a regular n-gon with a side length of S_n as:

$$A_5 = 1; 40S_5^2 \qquad\qquad A_6 = 2; 37, 30S_6^2 \qquad\qquad A_7 = 3; 41S_7^2$$

These values are accurate to within about 3%. Moreover, if C is the circumference of the circle circumscribed about a hexagon, then the tablet also suggests $S_6 = 0; 57, 36C$, accurate to within 1%.

1.3.3 The Iron Empires

Around 1400 B.C., the Hittites of Asia Minor learned how to smelt iron. Before they could turn this to their advantage, they were conquered by the Assyrians, who were the first to field an army of iron.

In the ancient world, a conquered population could expect death or enslavement. The Assyrians added a new horror to war, and routinely tortured and mutilated their captives in what can only be called a calculated attempt to terrify potential adversaries into submission. The end result of this policy was that the Assyrian empire, established by force, had to be maintained by force and was ultimately destroyed by force.

In 612 B.C., the Assyrian capital of Nineveh was utterly destroyed by an alliance that included the Medes under Cyaxeres and the Chaldeans under Nabu-apal-usur ("Nabu guards the prince"), usually known as Nabopolassar. Under him the Chaldeans rose to become a mighty empire. His son, Nabu-kudurri-usur ("Nabu protects my boundaries") rebuilt Babylon and turned it into the capital of a new empire, properly called the Chaldean Empire but also referred to as the Babylonian or Neo-Babylonian Empire.

Nabu-kudurri-usur is better known as the Nebuchadnezzar of the Bible (though we will use the more correct spelling Nebuchadrezzar). According to tradition, Nebuchadrezzar married a Medean princess, but she became so homesick for her homeland in the foothills of Asia Minor that he had built for her a fabulous tiered garden. The "Hanging Gardens of Babylon" became one of the Seven Wonders of the World.

In 598 B.C., Judea revolted against Babylonian rule. Nebuchadrezzar suppressed the revolt, deported some of the leading citizens to Babylonia, but left the kingdom, temple, and local government intact. A second revolt began. Nebuchadrezzar captured Jerusalem in 586 B.C., destroyed the Temple of Solomon, and arranged for a second deportation. This began the era of Jewish history known as the Babylonian Captivity.

Meanwhile, the Medes were moving into Asia Minor, where they met the expanding Lydian Empire. Five years of inconclusive warfare between the Lydians and the Medes were brought to an abrupt end when, during a battle, "day turned into night:" a solar eclipse occurred. The opposing commanders took this as a sign of the displeasure of the gods, and signed a hasty peace. We can calculate that a total eclipse was visible from Asia Minor on May 28, 585 B.C., so this battle is the earliest historical event that can be given an exact date.

According to one story, the eclipse was predicted by the first mathematician we know by name: Thales of Miletus.

For Further Reading

For the history of Egypt and the Fertile Crescent, see [1, 18, 60, 68, 106, 130]. For the mathematics and science of the region, see [22, 41, 52, 84, 85].

2

The Classical World

2.1 The Greeks

Greece is a mountainous land of limited fertility, so settlements tended to cluster in the narrow valleys and on the various peninsulas that jut out into the Mediterranean. The relative isolation of the settlements encouraged them to form independent city-states. Like Sumeria, the individual Greek city-states were no match for a united conqueror, so by the time of A'h-mosè, mainland Greece was part of the Minoan Empire, centered on the island of Crete. The Minoan royal palace at Knossos was an impressive structure. Spread over six acres of land, it contained hundreds of rooms—and flush plumbing. Around 1500 B.C., the volcanic island of Thera exploded, causing a tsunami so devastating that the Minoans never recovered; this was probably the origin of the legend of Atlantis, retold by Plato a thousand years later.

The tsunami weakened the Minoans and the Greeks revolted. They invaded Crete, burned the palace and other centers of civilization, and lost the secret of flush plumbing for thousands of years. The Greeks established the Mycenaean Empire, named after one of their main cities. Around 1250 B.C., the Mycenaens besieged and destroyed Troy, a city on the coast of Asia Minor. But shortly after, the bronze-equipped conquerors of Troy were overcome by iron-wielding Dorians from the north. The Dorian dialect of Greek was difficult for the Mycenaeans to understand; they lampooned Dorian speech as "bar-bar" (much as we might describe someone's ramblings as "yadda-yadda") and called them *barbarians*, a term that has since been applied to cultures that have no permanent cities. Since it is easy (and usually incorrect) to associate "uncivilized" with "simple-minded," a particularly plain type of architectural column was later called *Doric* in the (mistaken) belief that they were products of Dorian builders.

By 700 B.C., Greece had recovered from the Dorian invasions, and the population was booming. To relieve crowding, a city-state would establish colonies around the Mediterranean. The "mother city," or *metropolis* in Greek, would equip a colony with people, ships, and government. As often as not, the colonies became independent of the mother city within a few generations, though some kept close ties.

These colonies went out in two directions, which can lead to confusion among students of classical geography. The Greek settlers who went west, towards the Italian peninsula and the island of Sicily, were the first Greeks encountered by the Romans. Thus the Romans referred to southern Italy as *Magna Graecia*, Latin for "Greater Greece" (where in this context, "Greater" means "Larger").

Other Greek colonists went east, to Asia Minor. Many who settled there came from the shores of the Ionian Sea, on the west coast of modern Greece. In commemoration of their original homeland, the Greek colonies in Asia Minor were collectively known as Ionia, even though they nowhere border the Ionian Sea.

2.1.1 Thales and Pythagoras

THALES (fl. 6th cent. B.C.) lived in Miletus, in Ionia. He is credited with having "discovered" five geometrical propositions (which suggests that he did not prove them):

1. A circle is bisected by its diameter.

2. The base angles of an isosceles triangle are equal.

3. Vertical angles are equal.

4. If, in a triangle, a side and its two adjacent angles are equal to the side and two adjacent angles of another triangle, then the two triangles are equal.[1]

5. An angle inscribed in a semicircle is a right angle.

There is a story that Thales sacrificed an ox upon his discovery of the last theorem.

The life of Thales is well documented in the *Histories* of Herodotus, who lived in the fifth century B.C. Herodotus has been called the "father of history," because he was the first to apply recognizably modern historical methods to the problem of the past, as well as the "father of falsehoods," because so much of what he concludes was fantastical and hard to believe. However, the more fantastical statements were usually Herodotus quoting what someone else said, so a proper title for Herodotus might be the "father of journalism."

According to Herodotus, Thales predicted the eclipse that ended the wars between the Lydians and the Medes. The peace was sealed by a marriage between Astyages, the King of the Medes, and Aryenis, the sister of Croesus, the King of the Lydians. Croesus then turned his attention south, towards the Greeks, so by Thales's time, most of Ionia had been absorbed by Lydia. Miletus, recognizing the futility of resistance, allied itself with Lydia and retained a measure of independence.

Part of the secret of the success of the Lydian Empire was a remarkable invention made around 650 B.C.: coinage. The value of money is often underrated and frequently misunderstood. Economic activity relies on trade, but trade requires two people, each having what the other person wants. Money serves as a universally desirable commodity. The problem is making theory and practice coincide. One way is to make money out of silver or gold, both rare, durable metals.

However, silver or gold by themselves are merely valuable; they are not money. Imagine the problems that would arise from trying to pay for a purchase using silver or gold. First,

---

[1]Greek geometers used the term "equal" to mean one of three things. First, two figures are equal if one can be superimposed on the other through a sequence of rigid transformations; today we use the term "congruent." Alternatively, two figures were equal if one could be dissected and reconstituted to form the other; today we would say the two figures have the same area. Finally, two figures were equal if their lengths, areas, or volumes had a ratio of one to one.

the purchaser would have to cut off a sliver of metal and measure its size (weighing would be the easiest method). But before accepting it the prudent seller would verify its weight and purity. Discrepancies over perceived weight or purity would lead to endless argument, and trade would grind to a halt. In the seventh century B.C., Croesus's predecessors had a marvelous idea: a trusted authority could stamp a disk of metal guaranteeing its weight and purity. Thus, coinage was invented. The commerce flowing through Lydia made her kings immensely wealthy, so the Greeks used "rich as Croesus" to describe anyone of great wealth.

Around 553 B.C., a usurper named Kurush from Fars rebelled against Astyages, and deposed him by 550 B.C. in a relatively bloodless civil war. Kurush, his homeland Fars (east of Mesopotamia), and his family, the Hakhamani, are better known to us by the Latinization of the Greek forms of their names: Cyrus, Persia, and the Achaemenids. Croesus saw an opportunity to invade and conquer the Medes (which he could do under the guise of liberating them from a foreign conqueror). Before invading, he consulted his advisors, and got an outside opinion from the Oracle of Delphi, supposedly possessed of the gift of prophecy. She told him that if he invaded, he would bring down a mighty empire. This sounded like good news to Croesus, and he sent in his army.

The border of the Lydian Empire was the river Halys. Herodotus noted (with doubt) a belief among the Greeks that Thales diverted the river so a bridge could be built across it. If this story is true, then Thales, like so many later mathematicians, was a military engineer. Croesus's army invaded Persia, and promptly fulfilled the Oracle's prophecy. Unfortunately the mighty empire that Croesus brought down was his own, which Cyrus conquered in 547 B.C. Ionia, as part of the Lydian Empire, became part of the Persian Empire. Again, Miletus remained independent.

Cyrus employed a novel strategy: tolerance. For example, the worship of Marduk had been suppressed by the Chaldean king Nabonidus; Cyrus promised to restore the worship of Marduk. With the worshippers and priests of Marduk on his side, Cyrus conquered the Chaldean Empire easily in 538 B.C. After the conquest, Cyrus kept his promise.

Another group disaffected by the Chaldeans were the Jews, many of whom had been living in Babylon since the time of Nebuchadrezzar. Cyrus allowed the Jews to return to Palestine to rebuild the temple. Few bothered: most had made comfortable lives for themselves in Babylon, though they did send monetary gifts with the returnees.

While in Babylon, the Jews maintained their cultural identity by a rigid adherence to the old customs. The Palestinian Jews, on the other hand, let their customs evolve naturally. As a result, the Babylonian Jews, returning to Palestine, felt the indigenous Jews were practicing the "wrong" form of Judaism. Since the chief city of the Palestinian Jews was Samaria, this was the beginning of a great schism in Judaism and as a result, the Babylonian Jews (who became dominant) learned to hate and detest the Samaritans. Half a millennium later, this hatred would be used to good effect in a story that argued for the fundamental unity of all mankind.

Egypt was added to the Persian Empire in 525 B.C. by Cyrus's son Cambyses, and by the time of his death in 521 B.C., the Persian Empire included Egypt and most of the Fertile Crescent. It was the largest empire the West had yet seen.

During the conquest of Egypt, a number of Greeks were captured by the Persians and transported to Babylon. There is a tradition that PYTHAGORAS of Samos (580–500 B.C.)

was among this group. Pythagoras had been in Egypt since 535 B.C., and spent five years in Babylon before making his way back to Greece. In 518 B.C., Pythagoras left the eastern Mediterranean forever and settled in Croton, at the heel of the "boot" of modern Italy.

Croton was the home of Milon (or Milo) the wrestler, the most famous athlete in antiquity. Milon's strength was legendary. During an Olympic procession around 540 B.C., he is said to have carried a fully grown ox across the stadium, a distance of about 600 feet. Milon, as the local sports hero, was greatly honored in Croton and in 510 B.C. led an expedition that destroyed the neighboring town of Sybaris, whose inhabitants were well known for their rich lifestyle (hence the word *sybaritic*).

At Croton, Pythagoras established a secretive, mystical school that lasted about a century. At least one story suggests that Milon was Pythagoras's patron, and that Milon's daughter became one of the first Pythagoreans. None of Pythagoras's own work has survived; we have only what his followers claim he said. Some of the rules of the school, such as always wearing white and not eating beans, have parallels with ancient Egyptian practices. Pythagoras and his followers were struck by the many relationships among numbers; they sought to analyze the physical world in terms of these number relations. For example, they apparently discovered that the sum of the first n odd numbers is n^2. Since the Greek word for number is *arithmos*, this study of number properties became known as arithmetic.

Perhaps the best evidence of the beauty of mathematics comes from the Pythagorean study of music. According to one rather dubious story, Pythagoras happened to be passing by a smithy, and noticed that the noise of the falling hammers sounded pleasant. Upon investigation, he discovered that the weights of the hammers had a whole number ratio to one another. This began a study of music from a mathematical perspective.

Rather than using hammers, the Pythagoreans used a monochord: a one stringed instrument with a movable bridge. If the bridge divided the string into two equal parts, the two parts could be plucked one after another (melodically) or simultaneously (harmonically). In both cases, the sounds went together well: the string ratio of 1 to 1 produced a consonance (now known as unison, from the Latin "one sound").

Suppose the bridge was moved so the string was divided in a 2 to 1 ratio. The shorter part would produce a higher pitched version of the tone produced by the longer part, and the 2 : 1 ratio produces another consonance.[2]

Another consonant ratio corresponds to string lengths in a ratio of 3 : 2. Finally, an "inversion" takes the lower note of a set and replaces it with a higher pitched version of the same note (in this case, by halving the string length). Inverting the 3 : 2 ratio produces a 2 : 3/2 ratio, which we can simplify to 4 : 3. The evidence of the senses, as well as the elegance of the numerical ratios, make us regard this ratio as consonant. Moreover, the three lengths together give us a 6 : 4 : 3 ratio, which has a remarkable property: the ratio of the difference between the first and second to the difference between the second and third is equal to the ratio between the first and third. In this case, the ratio of $6 - 4$ to $4 - 3$ is equal to the ratio of 6 to 3. This type of ratio is now called a harmonic ratio (and the numbers are said to form a harmonic progression). On the other hand, if we began with the 4 : 3 ratio and inerted it, we would obtain a 3 : 2 ratio, giving us three numbers in an arithmetic progression.

---

[2]Changing the tension or thickness of the string also affects the tone. This fact makes the hammer story extremely improbable, since several factors would influence the fundamental tones produced by dropping hammers.

Suppose we tuned an instrument so the strings had a $4 : 3 : 2$ or a $6 : 4 : 3$ ratio. The instrument would produce three notes, with the property that any combination of the notes would form a consonance. However, a three note repertoire is rather limited, so additional notes were added. Various schemes were tried, but eventually the division of the $2 : 1$ interval into eight notes became standardized (Euclid was the first to note this division, though it certainly predated him). Consequently, the $2 : 1$ ratio is said to correspond to an interval of an octave. From lowest to highest, these notes are now designated as C-D-E-F-G-A-B, with the eighth note also called C and beginning the pattern again.[3] The C-G interval, which spans five notes, corresponds to the $3 : 2$ ratio; hence this ratio is referred to as a fifth. Likewise the C-F ratio, which spans four notes, corresponds to the $4 : 3$ ratio, and is designated a fourth. This means that the F-G interval must correspond to a ratio of $9 : 8$. We can designate this ratio as defining a tone. But the internote ratios cannot all be $9 : 8$, since $9^7 : 8^7 \neq 2 : 1$. In fact, the problem is worse than that: there are *no* whole numbers p, q for which $p^7 : q^7 = 2 : 1$. Thus it is impossible to divide the octave into equal intervals. This is the root of what is called the tuning problem: how can we make a product of the power of one rational number equal another rational number? If the two rational numbers have different prime factors, then the problem is unsolvable; the best we can hope for is an approximation.

Any solution to the tuning problem must sacrifice some of the intervals. In Pythagorean tuning, the octave and fifths are retained, while the fourths and tones are sacrificed where necessary. For example, if C-D and D-E both correspond to a $9 : 8$ ratio, then to make C-F correspond to a fourth (and thus the ratio $4 : 3$), then E-F must correspond to the ratio $256 : 243$. Since $256^2 : 243^2$ is approximately equal to the ratio $9 : 8$ that defines a tone, this new and inelegant ratio is designated a semitone. In order for all fourths to be perfect, it is necessary that any sequence of four notes consist of two tones and a semitone. Likewise, for all fifths to be perfect, any sequence of five notes must contain three tones and a semitone. The octave, which consists of a fourth and a fifth together, thus consists of five tones and two semitones, with the semitones five notes apart. Reconciling all these factors produces the Pythagorean tuning:

C to D	D to E	E to F	F to G	G to A	A to B	B to c
$9 : 8$	$9 : 8$	$256 : 243$	$9 : 8$	$9 : 8$	$9 : 8$	$256 : 243$

The F-B fourth is slightly sharp (the upper note is slightly high in pitch), but the remaining fourths and fifths are perfect. This solution is uniquely determined up to the starting point. In the above, the lowest note of the scale is C (so we have two tones followed by a semitone, followed by three tones and a semitone), but if the lowest note is E (and thus we begin with a semitone, followed by three tones and a semitone, then two tones), you are playing in the "Dorian mode."

In addition to studying music, Pythagoras began a tradition of examining mathematical results in an "immaterial" and "intellectual" manner, thereby turning it into a "liberal art." This probably means that Pythagoras introduced the deductive method to mathematics, while restricting its domain to the theoretical properties of abstract objects. We can

---

[3]To distinguish the two Cs, the German physicist Hermann von Helmholtz used a system of marks. Thus if our first note is C, then the same note one octave higher is designated c, and the same note one octave higher still is designated c/. This last note corresponds to "middle C."

draw an interesting parallel: the ancient Olympic games highlighted the skills of the warrior, presented in an abstract form divorced from an actual siege or battle. Although the term itself is Latin, the idea of liberal arts originated with the Greeks—and slavery. Slavery was a key element of almost every culture before the present day, though generally men (and women, and children) were slaves because of military conquest, rather than race. Pythagoras himself was probably a slave during his time in Babylon.

Slaves worked with their hands: hence, the work done by slaves came to be known as manual labor, from *manus* ("hand" in Latin). Free men, on the other hand, were expected to pursue the liberal arts, from the Latin *libera* ("free man" in Latin). Free women, incidentally, were expected to bear children; *raising* the children was a task for slaves.

The mathematics of the Egyptians and the Babylonians concerned itself with practical matters, whether it was computing the height of a pyramid or the cost of digging a canal. Because of its association with manual labor, this type of mathematics was deemed fit only for slaves. The Greeks gave the name logistics to this practical, computational mathematics. The difference between logistics and a liberal art like geometry can be illustrated in the following way. The procedure for finding the area of a parallelogram:

Rule 2.1. *The area of a parallelogram is the base times the height.*

is logistics, while the theorem:

Theorem 2.1. *If two parallelograms have equal bases, and equal heights, then either can be dissected and rearranged to form the other.*

is geometry.

The Pythagoreans seem to have been the first to make proof an essential part of mathematics. They are credited with discovering and proving at least five theorems. Four of these are:

1. The angles in a triangle are together equal to two right angles.

2. The angles in an n-gon are together equal to the angles in $n - 2$ triangles.

3. The exterior angles in a polygon are together equal to four right angles.

4. The space about a point can be filled with regular triangles, squares, or hexagons.

The last may have led to Pythagorean discovery of three of the five regular solids: the tetrahedron, formed by equilateral triangles; the cube, formed by squares; and the dodecahedron, formed by regular pentagons.

The fifth Pythagorean discovery was the Pythagorean Theorem. A few myths about the Pythagorean Theorem ought to be discussed. The most often repeated myth comes to us from Proclus (writing in the fifth century A.D.): "Those who like to record antiquities" claim Pythagoras sacrificed an ox (a hundred oxen in some accounts) upon the discovery of the relationship between the sides of a right triangle. Proclus sounds dubious: the Pythagoreans believed in transmigration of souls, and were adamantly opposed to animal sacrifices. The story is also suspiciously like one told about Thales. Another story is that Pythagoras learned the theorem in Egypt, from "rope stretchers" who routinely formed 3-4-5 right triangles using a knotted chord. There is no evidence of any Egyptian knowledge of the Pythagorean Theorem in any form.

If Pythagoras did not discover the theorem independently, then he may have learned about it in Babylon. More concretely, Pythagoras may have learned how to construct what are now called Pythagorean triplets: three numbers, a, b, and c, that satisfy $a^2 + b^2 = c^2$. Pythagoras constructed triplets in the following way: if a is any odd number, b half of one less than the square of a, and c one more than b, then $a^2 + b^2 = c^2$. For example, if $a = 5$, $b = \frac{1}{2}\left(5^2 - 1\right) = 12$, and $c = 12 + 1 = 13$, and $5^2 + 12^2 = 13^2$.

Both the Pythagorean Theorem and the tuning problem lead to incommensurable quantities, discovered by the Pythagoreans during the fifth century B.C. Today we would say that two quantities are incommensurable if the ratio between them corresponds to an irrational number. We do not know how incommensurable quantities were discovered or who discovered them, and even the identity of the first pair of incommensurable quantities is unknown. One of the better candidates is the side and diagonal of a regular pentagon inscribed in a circle. This suggests that the discoverer was HIPPASUS (fl. ca. 430 B.C.) from Metapontum, who was apparently expelled from the order for revealing to outsiders the Pythagorean methods of inscribing a regular pentagon in a circle and a regular dodecahedron in a sphere.

The school did not long survive the discovery of incommensurable quantities, for it began to interfere in local politics. By the middle of the fifth century B.C., it was suppressed by the authorities. Real progress in mathematics came from the Greek heartland—which was at that point in a struggle for its very existence.

2.1.2 The Wars of Greece

The greatest of the Persian kings was Cambyses's successor Darius, from a collateral branch of the Achamaenids. Darius became king in 521 B.C., and concentrated on internal improvements. He established a vast system of roads, complete with military patrols to deter bandits, and an efficient postal system. Regarding the latter, Herodotus wrote: "Neither rain, nor sleet, nor dark of night stays [prevents] these couriers from the swift completion of their appointed rounds."[4]

In 500 B.C. the Ionian city-states, led by independent Miletus, revolted against Persian rule. They pled for help from the Greek mainland, but only Athens and Eretria sent more than token assistance. Miletus finally lost its independence when the Persians conquered it in 494 B.C., and Darius organized a punitive expedition to deal with the interfering Greeks. Darius, a shrewd diplomat, encouraged the neutrality of the other Greek city-states by announcing that his battle was with Athens and Eretria alone.

In 490 B.C., twenty thousand Persian troops laid siege to Eretria. After six days, a traitor opened the city gates to the Persians, who destroyed the city and carried its population off as slaves. With Eretria successfully reduced, the Persians crossed the narrow strait separating Euboea from mainland Greece, and made ready to crush Athens. The Athenians sent a runner, Pheidippides, to seek help from Sparta.

The Spartans conquered their neighbors the Messinians around 600 B.C., and reduced the inhabitants to the status of serfs, or *helots*. The helots outnumbered the Spartans ten to one, so to keep the helots subjugated, the Spartans made their society increasingly militaristic. Children were subject to a rigorous examination upon birth, and any who seemed

---

[4]The architect William Mitchell Kendall thought the phrase appropriate for the New York General Post Office, and when the building was completed in 1912, the inscription appeared along its façade. It is not the motto of the U.S. Postal Service.

weak or sickly were exposed and left to die. Education emphasized physical fitness, poetry, and music.

At the age of 20, a Spartan male joined one of several military units. The members of the unit ate together, lived together, fought together—and if necessary, died together. One Greek visitor, after eating at the communal barracks, found the food so unpalatable that he is said to have remarked, "Now I know why the Spartans are unafraid of death." Silence was encouraged, and if one had to speak, one should be brief about it and issue short statements that cut right to the point. Since Sparta is located in the region of the Peloponnesus known as Laconia, the other Greeks commented on their *laconic* way of speaking.

The Spartans agreed to help Athens, but for political reasons, they could not send assistance until after the full Moon. Pheidippides ran back to Athens to report the bad news. Boldly, the Athenian commanders made the decision to attack before the Spartans arrived. This turned out to be the correct decision, for the lightly armed Persians were no match for the heavily armed and armored Greek hoplite (which referred to their armor). The Persians were forced back to their ships, losing over six thousand men in the process. The Athenians lost less than two hundred. Pheidippides ran the 26 miles from the battle site to the marketplace in Athens, where he announced victory (*nike* in Greek) over the Persians on the beaches of Marathon. Then he collapsed and died of exhaustion, having run about 150 miles in two days.

Egypt revolted against the Persians at the same time, and keeping the rich province of Egypt was far more important than punishing the Greeks. Thus the next great invasion of Greece did not occur until 480 B.C., under Darius's successor Xerxes. This time it was clear that the intent was not to punish Athens, but instead to conquer all of Greece.

At the pass of Thermopylae, 1400 Greeks, including 300 Spartans, faced the entire Persian army. The exact size of the Persian army is unknown, but it may have been around 100,000 soldiers, which included 10,000 elite troops, known as "Immortals." According to Herodotus, the Spartans were warned by a native of Trachis (a village near Thermopylae) that when the Persians fired their arrows, they would be so numerous they would blot out the sun. A Spartan named Dienices remarked, "Good! We will have shade to fight in." Defeat was inevitable, but the Spartans stopped the Persian advance for three days (dying to the last man), buying enough time for other parts of Greece to improvise hasty defenses. It was not enough: Xerxes laid siege to Athens and burned it to the ground. But at the naval battle of Salamis, the Persians lost nearly half their fleet; Greek losses were insignificant. Xerxes withdrew—for one year. The next year, Xerxes returned with another army, suffered another defeat (at Plataea), and withdrew again.

To fight Persia, Athens organized an alliance, now known as the Delian League, because its headquarters and the treasury were located on the island of Delos. Member states of the alliance could either contribute ships and men, or the equivalent in money. Most members chose to supply money, letting Athens build and man the ships of the fleet. Eventually the Delian League included all of Ionia, and the Persian threat began to recede.

In theory the alliance, with no enemy to fight, ought to have been disbanded, but Athens argued that it was still necessary to defend Greece from Persia. Member states grudgingly agreed to maintain what was rapidly becoming the Athenian Navy, but after a while, the threat of Persian invasion seemed remote. Tired of paying for a navy that did it no good, the inhabitants of the island of Naxos attempted to withdraw from the alliance. The Athenians

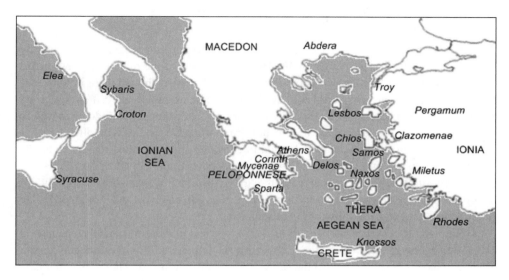

Figure 2.1. The Athenian Empire

refused to accept the withdrawal, laid siege to Naxos, and in 470 B.C., captured the city, destroyed its fortifications, and sold the inhabitants into slavery. Other member states were cowed into submission, their annual dues being converted into tribute, and the Athenian Empire was established.

Pericles was an ambitious aristocrat who came to power in 461 B.C. and presided over the Golden Age of Athens. The great playwrights Aeschylus and Sophocles lived and worked in Athens in this period; indeed, one of Pericles's first recorded actions was paying for the production of one of Aeschylus's plays in 472 B.C. The Parthenon, considered one of the most elegant buildings in the world, was also constructed in the time of Pericles, using funds technically belonging to the Delian League. When Thucydides of Melesias objected to this misappropriation of funds, Pericles arranged to have him ostracized (a political maneuver that forced Thucydides into exile for ten years).[5]

The Golden Age of Athens was a great era of building, democracy, literature—and slavery. At the height of the Golden Age, there were perhaps fifty thousand (male) citizens with the franchise—and over a hundred thousand slaves. Many worked at the state-owned silver mines of Laurion, which were so profitable that by 483 B.C. each Athenian citizen received a share of the revenues. Even today, mining is an extremely dangerous profession; in the fifth century B.C., the death toll among the slaves must have been horrific, and some estimate that slaves only survived an average of two years in the mines.

Only three states of the Delian League maintained a measure of independence, contributing ships and men to the fleet rather than money to the treasury. Geography may have played a role: all were large islands off the shore of Asia Minor, so of necessity they were forced to have strong navies. From north to the south, the islands were Lesbos, Chios, and Samos.

---

[5] The term comes from *ostraka* ("pottery shard") onto which the name of an individual might be written: those who received more than a specified percentage of the votes were sent into exile, with the understanding that their property and person were to remain undisturbed in their absence.

2.1.3   Mathematicians of the Golden Age

Chios joined the league in 479 B.C. There may have been a school there established by
Pythagoras, for OENOPIDES of Chios (fl. 450 B.C.) was one of the more celebrated mathematicians of the era. We know nothing about Oenopides's life, though there is some evidence he visited Athens. He would not have been the first: as the vibrant center of a growing empire, Athens attracted the best minds of the age. For example, Herodotus moved to
Athens to write his *Histories* about the recent Persian wars. Oenopides is credited with being the first to construct a perpendicular and to construct an angle equal to a given angle.
As the Parthenon and other large buildings were under construction by the time of Oenopides, it seems unlikely that he was actually the first to do either of these things. More likely,
Oenopides was the first to construct the figures using *only* compass and straightedge, the
tools of the liberal arts, not those of the manual ones.

Around 462 B.C., ANAXAGORAS (500–428 B.C.) came to Athens from Clazomenae, in
Ionia. Anaxagoras may have been the first to bring the Ionian tradition of rational inquiry to
Athens. Anaxagoras suggested that the sun was not a god, but instead a hot rock larger than
the Peloponnese. This suggestion might have been based on the following logic: First, the
Moon eclipses the sun, which implies the sun is more distant. Second, the sun and Moon
appear to be the same size, so the sun must be larger than the Moon. Hence the region of
totality during an eclipse must be smaller than the Moon. From accounts of the total eclipse
of April 30, 463 B.C., Anaxagoras might have learned that the shadow of the Moon covered
most of the Peloponnese; this would allow him to give a lower bound for the size of the
Moon and sun.

The other Athenians were not as impressed with the Ionian traditions, and Anaxagoras was charged with heresy. In an account written five hundred years later, Plutarch said
Anaxagoras "wrote on" the squaring of the circle while awaiting trial, though Plutarch gave
no details. The events surrounding the trial itself are somewhat hazy; Anaxagoras was apparently condemned to death, though in fact he lived out the remaining years of his life in
Lampsacus in Ionia. Pericles may have secured Anaxagoras's acquittal (there is a tradition
that Pericles was a student of Anaxagoras), but other accounts suggest that Anaxagoras had
already left Athens by the time charges were brought, and that the trial and condemnation
occurred *in absentia*.

If Plutarch's account can be taken at face value, Anaxagoras may have been the first to
examine one of what became known as the Three Classical Problems of antiquity:

1. **Trisection of an angle**: given an angle, divide it into three equal angles.

2. **Duplication of the cube**: given a cube, to construct another with twice the volume.

3. **Squaring the circle**: given a circle, to construct a square equal to it in area.

The greatest mathematician of the era was Oenopides's countryman, HIPPOCRATES
(470–410 B.C.). Hippocrates of *Chios* should not be confused with his contemporary, Hippocrates of *Cos*, the physician. Hippocrates of Chios was originally a merchant who came
to Athens to retrieve a cargo lost to piracy. The legal proceedings against the pirates took so
long that, to support himself, Hippocrates became the first known professional teacher of
mathematics. Moreover, he wrote a textbook, called the *Elements of Geometry* (now lost),
for his students.

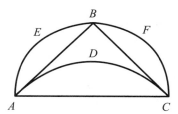

Figure 2.2. Hippocrates Lunes

Hippocrates was apparently aware that:

Theorem 2.2. *The area of a circle is proportional to the area of the square on its diameter.*

However, he probably did not have a proof.

While attempting to square the circle, Hippocrates was able to show that certain lunes (regions bound by circular arcs) were equal in area to certain rectilineal figures (see Figure 2.2). If ABC is an isosceles right triangle inscribed in a semicircle, and ADC is a segment of a circle similar to the segments AEB, BFC (two segments are similar when their central angles are equal), then the lune $ABCD$ is equal to the triangle ABC. Hippocrates found several other results, all equating the area of a lune with the area of a certain rectilineal figure.

Hippocrates also took steps towards duplicating the cube. Given two quantities, a and d, then the problem of inserting one mean proportional between them is the problem of finding b so $a : b = b : d$. We could insert two mean proportionals, b and c, if $a : b = b : c = c : d$. Hippocrates showed that in this case, $a^3 : b^3 = a : d$. Thus if $d = 2a$, then the cube with a side of b will have twice the volume of a cube with a side of a. Hippocrates himself was unable to solve the problem of finding two mean proportionals.

The most enigmatic mathematician of the era was DEMOCRITUS (ca. 460–370 B.C.), who came from Abdera at the northern edge of the Athenian Empire. Democritus traveled around the Mediterranean, and made at least one visit to Athens, where he was snubbed by Anaxagoras. Although none of Democritus's writings have survived, references to his work in other sources indicate that he was the first to give correct formulas for the volume of a pyramid and cone, though he apparently did not give a proof. A work of his known only by its title, *Two Books on Irrational Lines and Solids*, suggests that the existence of incommensurable figures was well known by his time.

HIPPIAS (b. ca. 460 B.C.) also came to Athens, from Elis, an independent city-state on the western shore of the Peloponnesus. He was a member of a new school of philosophy, the Sophists. Unlike the Pythagoreans, who were secretive and would teach only those who wished to become Pythagoreans, the Sophists would teach anyone who paid them. Teaching seemed too much like manual labor to other philosophers, who reviled the Sophists. It might seem that there was a small market for philosophy lessons, but one of the skills taught by the Sophists was the skill of debate, critical for success in the Athenian legal system.

Hippias might have taught in Sparta briefly, but found the Spartans uninterested in his primary subjects (astronomy, geometry, and logistics). Thus he made his way to Athens, where his skills were in demand. He was the first to invent a curve that could *not* be constructed using compass and straightedge. Since the curve could be used to square the circle,

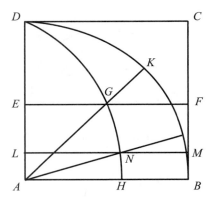

Figure 2.3. Trisectrix or Quadratrix

it is known as the quadratrix, from the Latin *quadratus*, "square" (see Figure 2.3); it is not known if Hippias knew that the curve could be used in this fashion.

To construct the quadratrix, take the square $ABCD$, and let DKB be the quadrant of a circle. Let the side DC drop parallel to itself towards AB while at the same time the radius AD rotates until it coincides with AB. The intersection of the radius and the side of the square at points G, N, etc. form the quadratrix. If angle KAB is to be trisected, then draw EF parallel to AB and divide EA into thirds at L (which can be done using compass and straightedge); draw ML which intersects the curve at N. Then $\angle NAB$ is one-third $\angle KAB$. Squaring the circle relies on the theorem that arc DB is to DA as DA is to AH; thus locating the point H on the quadratrix allows us to square the circle.

ANTIPHON (480 B.C.–411 B.C.) and BRYSON (b. ca. 450 B.C.) were two other important Sophists. Antiphon was Athenian, and Bryson may have come to Athens to study under Socrates. Antiphon worked on the problem of squaring the circle, and apparently suggested that its area could be found by considering the area of an inscribed polygon; this suggests an early use of the method of exhaustion. Bryson took Antiphon's method further and bounded the area between the area of inscribed and circumscribed polygons.

The problem of squaring the circle attracted enough popular attention by this time to warrant a reference in *The Birds* (414 B.C.) by Aristophanes. The main characters, Euelpides and Pisthetairos, leave Athens to found the utopian "Cloud Cuckoo Land," and are subsequently bombarded by unsolicited and impractical advice. The astronomer METON (b. ca. 440 B.C.), whose observation that 235 lunar months very nearly equals 19 solar years led to the development of a very accurate luni-solar calendar, appears as a character with a plan to design the city: "With this straight ruler here I measure this, so that your circle here becomes a square." Pisthetairos replies, "This man's a Thales," then drives him away with a beating.

Beating geometers played only a minor role in Aristophanes's work. A more constant theme was the stupidity and pointlessness of war. Athens's quest to build an empire led to the Great Peloponnesian War, which began in 431 B.C. In response, Aristophanes wrote *The Acharnians* (425 B.C.), *The Knights* (424 B.C.), and *The Peace* (421 B.C.), criticizing the leadership and conduct of the war, and calling for its end. The plays were well-received, but war proved more popular. The Athenians also passed a law limiting political satire, though this did not prevent Aristophanes from writing his greatest work, *Lysistrata* (411 B.C.). In it, Lysistrata convinces the wives and mistresses of the Athenian and Spartan soldiers to withhold sex until they cease their endless war.

Lysistrata ends happily. The Great Peloponnesian War ended in 404 B.C. with the destruction of the Athenian Empire. Sparta emerged the dominant power in the Greek world. Athens, having lost her worldly empire, was about to establish one far greater, and far more important. Built on ideas, not force, it would prove far more lasting, and to this day we are very much a part of the Athenian intellectual empire.

2.1.4 Mathematicians and the Academy

Plato, a student of Socrates, fought in the last five years of the Peloponnesian War, but was born to an aristocratic family and had political ambitions. But in 399 B.C. Socrates was ordered to commit suicide on charges of having "corrupted youth." This convinced Plato that politics was no place for a man with a conscience, and he left Athens. Plato spent the next dozen years traveling about the Mediterranean. He visited Egypt and southern Italy, where he met Pythagoreans and taught Dion, the brother-in-law of the Tyrant of Syracuse, Dionysius.[6]

According to one story, Dionysius grew angry at Plato and arranged to have the philosopher sold into slavery. Plato was saved from this unhappy fate by ARCHYTAS (428–347 B.C.), the leading citizen of Taras. Taras (Tarentum in Latin, and now Taranto, Italy) was the only colony ever successfully founded by Sparta, populated by bastard children born during the Messinian Wars. Spartan culture valued military prowess; Archytas had been chosen as *strategos* (general) seven times, and was never defeated.[7] Spartan culture also valued music; Archytas is believed to have invented the $\pi\lambda\alpha\tau\alpha\gamma\eta$, which Aristotle derided as a rattle, useful for keeping children occupied so they do not break things (although it has been suggested that the $\pi\lambda\alpha\tau\alpha\gamma\eta$ is the same instrument that adorns a number of vases of the fourth century B.C. alongside a number of other religious symbols). Archytas is also credited with associating the arithmetic, geometric, and harmonic means with musical intervals, and was the first to solve the problem of duplicating the cube, using a complex method involving the intersection of two space curves.

When Plato returned to Athens in 387 B.C., he founded the most famous school in history: the Academy. Plato may have been the first to use the term "mathematics", which comes from the Greek *mathema* ("that which is learned"). He emphasized mathematics as a way to train the mind in deductive thinking, so the Academy became a center for mathematical teaching and research. Archytas joined the faculty almost immediately.

The name of the school commemorates a rather unsavory event in the history of Athens. According to one story Theseus, a king of Athens and great hero, abducted Helen, a princess of Sparta (later to become Helen of Troy). An Athenian named Academus revealed where Theseus hid Helen, and she was rescued by her brothers Castor and Polydeuces (Latinized as Pollux). Tradition placed Academus's estate just outside the city walls of Athens; the site was purchased by a wealthy admirer of Plato and donated to him.

---

[6]In Greek political theory, a "tyrant" is the term used for any non-hereditary ruler.

[7]It is not clear which campaigns he fought in, but southern Italy was a constant battleground during the time period, and since, by tradition, a *strategos* could not succeed himself, the continued election of Archytas had to have some basis.

Other stories are told about the school. Johannes Tzetzes, a Byzantine author, claimed that a plaque above its entrance read, "Let No One Unversed in Geometry Come Under My Roof." Since Tzetzes wrote six hundred years after the Academy closed its doors forever, the claim is dubious. Another story highlights the difference between the manual and the liberal arts: a student asked Plato the value of knowledge, at which point Plato told a servant to give the student a coin, "since he must have value for what he learns." Then the student was expelled from the Academy.

The greatest mathematician associated with the Academy was EUDOXUS of Cnidus (408–355 B.C.), a student of Archytas in Tarentum. The life of Eudoxus coincided with a resurgence of Persian interest in Greece. Athens, Thebes, Corinth, and Argo all had historic grudges against Sparta, and the Persians were well aware of this. Late in 396 B.C., Pharnabazus, a *satrap* (provincial governor) of the Persian Empire, let these cities know that if they attacked Sparta, they would receive his support. Thus, less than a century after the Greeks united to fend off the Persian Empire, they united again—this time at the urging of the very same Persian Empire against one of their former allies. In 395 B.C., the Corinthian War began.

A Spartan fleet was destroyed just offshore Cnidus in 394 B.C., when Eudoxus was fourteen. Since the destruction of Sparta would leave expansionist Athens supreme among the Greek city-states, Persian policy reversed itself and in 387 B.C., the Spartans and Persians negotiated the "King's Peace," binding on all of Greece, though no Greek city-state besides Sparta was consulted.

Eudoxus visited Athens shortly afterwards. He was so poor he had to stay at the Piraeus (the port section), and walked ten kilometers uphill each day to the Academy. Eudoxus only stayed in Athens a few months, before going to Egypt. There a sacred bull licked his cloak, which meant (according to the priests) that he would die young, but famous. After a year in Egypt, he went to Cyzicus, where he founded a school before returning to Athens around 368 B.C. While Plato was away in Sicily, attempting to tutor King Dionysius the Younger of Syracuse, he left the Academy in the hands of Eudoxus. Dionysius, like his father, grew angry with the philosopher and eventually Plato returned to Athens. This allowed Eudoxus to return to Cnidus where he stayed the rest of his life.

Eudoxus was responsible for developing the theory of proportion and ratio, a crucial step in the development of the idea of a real number. Two magnitudes were said to have a ratio if either could be multiplied to exceed the other (thus, a circle and a square could have a ratio, but not a circle and a line). To compare two ratios, Eudoxus used the definition:

Definition. *Two ratios are equal if, given any equimultiples of the first and third, and any equimultiples of the second and fourth, the equimultiples of the first and second, and the equimultiples of the third and fourth, alike equal, exceed, or fall short of the latter equimultiples.*

What this complicated definition means is that given any two ratios, $\frac{A}{B}$ and $\frac{C}{D}$, and any whole numbers m and n, then if $mA = nB$ whenever $mC = nD$, and $mA > nB$ whenever $mC > nD$, and $mA < nB$ whenever $mC < nD$, then the ratios are equal. Another way to interpret Eudoxus's complex definition is that if two ratios $\frac{A}{B}$ and $\frac{C}{D}$ are equal, then given any rational number $\frac{n}{m}$, the ratios are both greater than, both less than, or both equal to the rational number.

With the theory of proportions (which Euclid preserved as Book V of the *Elements*), Eudoxus was apparently able to prove the propositions of Democritus and Hippocrates regarding the volumes of cones and pyramids, and the area of a circle. However, we have no copies of his proofs.

MENAECHMUS (fl. 350 B.C.) was a student of Eudoxus in Cyzicus, and was one of the first to study the conic sections in a systematic fashion.[8] He discovered specific properties, or symptoms of the conic sections, that translate into the modern algebraic equations $ky = x^2$ or $kx = y^2$ for the parabola and $xy = k^2$ for a hyperbola. Using these symptoms, Menaechmus presented a new and simple method of duplicating the cube. Hippocrates had shown that this problem reduced to inserting two mean proportionals between two quantities; Menaechmus showed that these two mean proportionals could be found using the intersection of a parabola and a hyperbola. In particular, suppose we wish to insert two mean proportionals between a and b. We seek to find x, y that satisfy the proportionality $a : y = y : x = x : b$. From the first and second terms of the proportionality, we have $a : y = y : x$; hence $ax = y^2$, so x, y, are on a parabola. From the first and third terms, we have $a : y = x : b$; hence $ab = xy$ and x, y are on a hyperbola. The intersection of the parabola $ax = y^2$ and $ab = xy$ can be used to find the desired quantities x and y.

One of the stories told about Menaechmus is that he came into the service of the household of Philip, the king of Macedon as a tutor for the young Prince Alexander. After a particularly difficult lesson, Alexander asked whether there was an easier way to learn mathematics. "Sire," Menaechmus replied, "there is no royal road to geometry."

2.1.5 The Hellenistic Kingdoms

By the time of Philip of Macedon, it was obvious that the Persian Empire was nothing but an empty shell, ready to topple at the slightest push. Philip intended to supply that push. After uniting the Macedonian tribes and modernizing their army, he conquered the Greeks and prepared to march against the Persian Empire. Unfortunately he was assassinated in 336 B.C.by a conspiracy that included his ex-wife Olympias.

The conspiracy placed their son Alexander on the throne. Over the next ten years, he embarked on a career of conquest never before seen in the ancient world, conquering an empire that stretched from Greece to the borders of India. He would have gone further, but his soldiers, who had faithfully followed him across the width of what was once the Persian Empire, rebelled at the thought of getting even farther from home. Thus, Alexander turned back. There is a story that he wept at the thought that there were "no more worlds to conquer," though this is probably apocryphal: Alexander knew that India lay just beyond the borders of his empire.

It cannot be denied that Alexander was a brilliant tactician, but his Persian campaign offered few opportunities to show it. The Persians never really learned how to fight the heavily armed Greek hoplites that defeated them at Marathon and Plataea, so the conquest of the Persian Empire was a relatively simple strategic exercise. Alexander's true greatness appeared in his domestic policy after the conquest. The empires of the past were conquered by force, maintained by force, and ultimately destroyed by force. Alexander intended to break the pattern.

---

[8]It may be significant that neither Eudoxus nor Menaechmus learned from their teachers while they were *at* the Academy.

First, he had to foster unity among his subjects. One way to do so was to give them a common heritage. In the ordinary course of events, the civilization of the conqueror inevitably percolated downward to the conquered, but Alexander hoped to expedite the process by establishing Greek colonies throughout the empire. According to his biographer, he founded seventy towns, though many of these were built around existing cities. At Issus, where he defeated the Persians, he established a new city and named it Alexandria (now Iskenderun, Turkey). At the mouth of the Nile in Egypt, he established another city and named it Alexandria (now al-Iskandarīyah, Egypt). At the eastern edge of the empire, he established yet another city and named it Alexandria (now Kandahar, Afghanistan).

By itself such a policy invited disaster, for the colonists would be viewed as foreign conquerors and, of course, the very name of the city would be a constant reminder. Thus Alexander encouraged his subjects to intermarry with the native population so that, in time, there would be no distinction between the Greek and non-Greek members of the empire. At the ancient Elamite capital of Susa, Alexander and eighty of his officers took Persian wives, and for the roughly ten thousand common soldiers who took Persian brides, he gave generous dowries. He even located the capital of his empire at Babylon, a symbolic choice for it was midway between Persia and Greece. Finally, he recruited Persians for key positions in the government and army.

All his plans for an empire unified by common interest and common heritage came to nothing when he died on June 11, 323 B.C. (probably from malaria, exacerbated by excessive drinking). His generals, who had followed him from Greece to the borders of India and back again, clustered around him, demanding to know who would inherit his empire. Who, they demanded, would be Alexander's heir? Who would the empire go to? They leaned close and heard his reply: "To the strongest." Thus for nearly twenty years, Alexander's generals fought for control of the empire during the Wars of the Diadochoi ("successors" in Greek).[9]

By 305 B.C., the surviving generals made an uneasy truce that divided the empire into five Hellenistic kingdoms. The two most important were those ruled by Seleucus and Ptolemy. The Seleucid Empire included most of what was once the Persian Empire, with the exception of the satrapy of Egypt; Seleucus himself was the only one of Alexander's officers to keep the Persian wife he married at Susa. Egypt was the domain of Ptolemy, and Ptolemaic Egypt would be the longest-lived of the successor states.

Ptolemy established his capital at Alexandria at the mouth of the Nile. There he established one of the most famous research institutions in the world: the Museum of Alexandria, named after a statue dedicated to the nine Muses (the patron goddesses of poetry, dance, music, history, and astronomy) that dominated its entrance. The Museum combined the teaching and research aspects of a university with the displays we associate with modern museums and zoos. It also contained a library which, in time, came to be the best-known part of the museum, and received funds directly from the king, making it one of the first state-supported research institutes.

A goal of the library was to collect every written Greek work: every tragedy, every comedy, every treatise on science and on mathematics. To augment its collection, royal decree

---

[9]There is an intriguing suggestion: One of Alexander's most honored generals was named Craterus, and in Greek, "To Craterus" is *kraTERoi*, while "To the strong*er*" is *KRATeroi*. Craterus was away at the time, building a fleet in Cilicia.

ordained that ships entering the harbor at Alexandria had to surrender their papyrus to the museum, which would make copies—and the *copies* would be returned to their owners. As a result of this, Alexandria became the publishing center of the Mediterranean. At its peak, the Library contained some 300, 000 Greek manuscripts, while another 200, 000 were kept at the nearby Temple of Sarapis, converted into a library annex around 235 B.C.

Ptolemy was more than a mere patron of the arts: he was himself interested in learning. The following suspiciously familiar story is told: one day, while taking lessons in geometry, he came across a very difficult proposition. Unable to understand its proof, he asked his teacher whether there was an easier way to learn mathematics. "Sire," the teacher replied, "there is no royal road to mathematics." The teacher was supposedly EUCLID (fl. 300 B.C.).

In Euclid's lifetime, Alexandria became the busy port of a prosperous Empire, and Euclid himself may have watched the construction of a new lighthouse (completed around 280 B.C.) on the island of Pharos, just offshore and perhaps a mile from the library itself. Teams of donkeys ascended to the top, bringing wood to burn in a fire that could be seen twenty miles out to sea; the lighthouse became one of the Seven Wonders of the World.

We know almost nothing about Euclid's life. He may have taught for a while at the Museum, and then may have established his own school nearby. An apocryphal story (again, suspiciously familiar) is that a student asked the value of a theorem; Euclid is said to have given the student a coin and dismissed him, chiding him for requiring that knowledge have value.

At the library, Euclid had access to every important mathematical work written by his predecessors. Euclid was not a brilliant mathematician: he discovered no great theorems, created no new methods. What he did create was a vast system for the elementary geometry of the time: a way that theorems could be deduced in a straightforward and logical manner, beginning with ten simple assumptions. Five of these relate to geometrical objects:

1. A straight line may be drawn between any two points.

2. A straight line may be extended in a straight line.

3. About any point a circle may be drawn with any given radius.

4. All right angles are equal.

5. If a line falling on two lines makes the interior angles on one side of the line less than two right angles, the two lines, if extended, will meet on that side.

The last is the so-called parallel postulate.

Notice that the first three postulates assume the existence of straight lines of arbitrary lengths and positions, and circles of arbitrary centers and radii: thus, Euclid's geometry is the geometry of straight lines and circles, and the figures that can be derived from them.

The five "common notions" that relate to logical deduction are:

1. Things equal to the same thing are equal to each other.

2. If equals are added to equals, the results are equal.

3. If equals are subtracted from equals, the results are equal.

4. Things which coincide are equal.

5. The whole is greater than any part of it.

From these ten simple assumptions, Euclid derived over a thousand theorems, which he divided into the thirteen "books" (equivalent to a modern day chapter) of the *Elements*. For the next two thousand years, the *Elements* dominated mathematics: one was not a mathematician, but a *geometer*. Indeed, a striking feature of the *Elements* is how little of it is geometry in our sense: the properties of plane figures are dealt with in Books 1, 3, and 4, and solid geometry is dealt with in Books 11–13.

The remaining books of the *Elements* show how the Greeks used geometry as the basis for all mathematics. Thus Book 2 concerns itself with algebraic propositions (like the expansion $(a + b)^2 = a^2 + b^2 + 2ab$, viewed from a geometric perspective); Books 5 and 6 discuss the theory of proportions and a geometric theory of the real numbers; and Books 7–9 are on number theory. Book 9 includes a proof that the number of primes is infinite, and concludes with a remarkable theorem on perfect numbers (numbers which are the sum of their proper divisors; this definition is believed original with Euclid): If $2^m - 1$ is prime, then $2^{m-1}(2^m - 1)$ is perfect. The longest book in the *Elements* is Book 10, with 115 propositions; in it, Euclid began classifying incommensurable magnitudes.

The Ptolemaic Kingdom reached its greatest cultural heights during the reign of the third Ptolemy, who was such a patron of the arts and sciences that he received the nickname *Euergetes*: benefactor. His reign saw the beginning of a project to translate the Hebrew Bible into Greek. According to tradition, 72 translators (6 from each of the 12 tribes of Israel) were placed in separate cells to translate the Hebrew Bible into Greek; when they completed their work, all 72 translations were identical. Since "duoseptuaginta" (Greek for seventy-two) is hard to pronounce, the version they produced is known as the Septuagint. Textual analysis of the surviving copies (which date to the 4th century A.D.) suggest that, while the translation might have begun during the reign of Ptolemy Euergetes, it was not finished until about a century later.

Ptolemy Euergetes invited ERATOSTHENES (276–196 B.C.) to Alexandria to tutor his son (the future Ptolemy IV). Around 235 B.C., Eratosthenes became head of the library of Alexandria, a post he retained until his death. Very little of Eratosthenes's original work survives; of the surviving work, the most remarkable is his use of simple geometry to measure the circumference of the Earth. The basis of Eratosthenes's calculation was that at Syene, 5000 *stadia* south of Alexandria, the sun was directly overhead at noon on the date of the summer solstice, while in Alexandria, the sun made an angle of about 7° to the vertical. Eratosthenes assumed a spherical Earth and concluded the difference in the angle was caused by the Earth's curvature. Since 7° is approximately $\frac{1}{50}$ of a circle, then the Earth's circumference was about $50 \times 5000 = 250,000$ stadia (see Figure 2.4). A stadia is believed to be about 0.1 miles, so the figure obtained for Earth's circumference by Eratosthenes is within a few percent of the correct value.

Eratosthenes's contemporaries thought highly of his work, and nicknamed him *beta*, after the second letter of the Greek alphabet (which itself comes from the names of the first two letters of the Greek alphabet, α and β). He was given this name because he was always "second best." He was the second best geographer in Alexandria. He was also the second

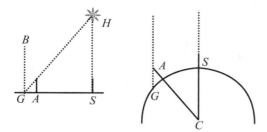

Figure 2.4. Two Explanations for the Angle of the Sun (greatly exaggerated for clarity)

best geometer, historian, astronomer, poet, and literary critic. One can only hope to be as good a "second" as Eratosthenes!

The "alpha" geometer in Alexandria was APOLLONIUS (262–190 B.C.), who came from Perga in Asia Minor. Apollonius himself received the nickname *epsilon* for his work on lunar theory, since the Greek letter ϵ resembles the Moon.

The year before Apollonius was born Eumenes, the governor of Pergamum, revolted against the Seleucids and established an independent kingdom of Pergamum. Apollonius went to Pergamum and stayed a while, though he eventually made his way to Alexandria, probably to study at the museum. His best-known work is his *Conics*, which originally included eight books, though only seven survive. Books I and II were addressed to Eudemus, another geometer in Pergamum, and Books IV through VII of *Conics* were addressed to Attalus, who became king of Pergamum in 241 B.C.

Apollonius investigated the conic sections in such a general and thorough manner that earlier works by Menaechmus and even Euclid were displaced, forgotten, and eventually lost. Euclid defined a cone as the surface of revolution generated by rotating a right triangle about one of its legs; this produces a single-napped, right circular cone. Based on scattered references in other authors, it seems that a conic section was formed by intersecting the cone with a plane perpendicular to the hypotenuse of the triangle. The conic section could be classified, depending on whether the vertex of the cone was acute, right, or obtuse; in modern terms, these sections would be the ellipse, parabola, or hyperbola. Note that since the cone is single-napped, the hyperbola only has one branch.

Instead of revolving a triangle about one of its sides to produce a cone, Apollonius considered instead a circle and a point P not in the same plane. The conic surface would be generated as a line through the point traced along the circle. The point would become the vertex of a double-napped circular cone. Moreover, he considered the intersection of this cone with an arbitrary plane. Besides introducing the conic sections in a new and more general way, Apollonius described and proved many of their properties, including those relating to their tangent lines and propositions concerning the greatest or least distance between a point C and a point on the conic section.

Apollonius completed *Conics* some time after he had moved to Alexandria. By then, Ptolemaic Egypt was in decline. Meanwhile the Seleucids began an expansionist policy under Antiochus the Great and his successors. In 168 B.C. Antiochus Epiphanes ("God manifest," ruled 175–163 B.C.) laid siege to Alexandria. But Egypt had one important ally: Rome. Antiochus was about to take Alexandria when he met an old friend, the Roman ambassador, who told him to leave Egypt or face war with Rome. Antiochus asked for time

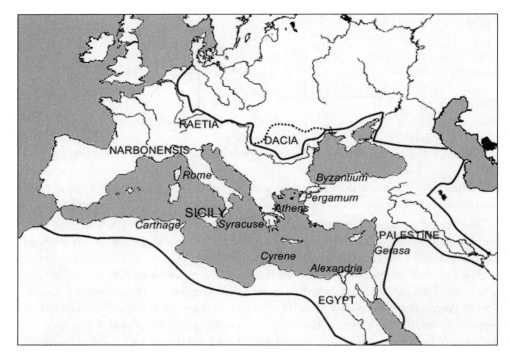

Figure 2.5. The Roman World

to decide, whereupon the Roman ambassador drew a circle in the sand around him and said, "Decide before you leave the circle." Antiochus, with his army around him and a string of victories behind him, was forced to back down before a single, unarmed Roman.

2.2   The Romans

According to tradition, Rome was founded by two brothers, Romulus and Remus, who were raised by a she-wolf. The brothers founded a city on the Tiber River, and shortly thereafter, Romulus killed Remus in a dispute over who got to name the new city. Early Rome was dominated by the Etruscans, a people about whom very little is known. By 509 B.C., the Romans expelled the Etruscan kings and founded the Roman Republic. Over the next few centuries, Rome gradually absorbed its neighbors.

In 282 B.C., Rome invaded Magna Graecia, and the inhabitants of Tarentum called for help from Pyrrhus of Epirus. At Heraclea in 280 B.C., Pyrrhus faced 35,000 Romans with 30,000 of his own soldiers, and a secret weapon, brought from the farthest borders of Alexander's former empire: elephants. The Romans had never before seen elephants, and were forced from the field, losing 7000 men. But Pyrrhus was struck—and concerned—by the bravery of the Roman legionnaires who stood their ground and died, rather than running away. It suggested that the Romans would prove a tough adversary.

At Ausculum in 279 B.C., Pyrrhus defeated the Romans a second time. This time the Romans fought even harder, and Pyrrhus lost so many of his own men that he reputedly said, "Another such victory and I am lost" (the original Pyrrhic victory). When Pyrrhus met Roman soldiers for the third time at Beneventum in 275 B.C., the Romans were ready, and dealt Pyrrhus a severe defeat, elephants and all. Pyrrhus saw nothing to be gained by staying in Italy, so he severed his ties with the Tarentines, who were conquered by Rome in 272 B.C. As he left Sicily, Pyrrhus is said to have remarked, "What a battleground I am leaving for Rome and Carthage."

Carthage was a city on the coast of North Africa (now in Tunisia) founded by the Phoenicians, a seafaring civilization as great as the Greeks and more daring as explorers. According to legend, around 800 B.C. Dido, the daughter of the king of Tyre (in Palestine), had to flee when her brother seized the throne. She made her way to North Africa, where the natives sold her as much land as she could enclose with a bull's hide: this is known as the isoperimetry problem, where one tries to determine the curve that will enclose the greatest area, and is one of the problems examined in the calculus of variations. Dido cut the hide into strips, tied them together, and marked out a semicircle along the coast.

Both Carthage and Rome had designs on Sicily, and an opportunity arose when the Greeks of Syracuse went to war against their old enemies, a group of Italian mercenaries known as the Mamertines ("Sons of Mars"), operating out of Messana. The Syracusans, led by a general named Hieron, drove the Mamertines back to Messana in 270 B.C.; the grateful Syracusans made Hieron their king. In 265 B.C., Hieron made ready to eliminate the Mamertine threat once and for all. Syracuse made an alliance with Carthage, and in response the Mamertines sought help from Rome. What began as a war between Syracuse and Messana turned into a war between Rome and Carthage, the First Punic War (*poeni*, "Phoenician" in Latin).

The war went badly for Syracuse, and in 263 B.C., Hieron made peace with Rome. Thereafter Hieron was one of Rome's staunchest supporters, and to the end of his life, he remained a faithful ally. When the First Punic War ended in 241 B.C., the peace treaty gave control of Sicily to Rome, who gave the eastern half of the island to her faithful ally Hieron and organized the western half of the island as the first province of Rome.

2.2.1   Archimedes

One of Hieron's relatives was ARCHIMEDES (287–212 B.C.), the greatest mathematician of antiquity and possibly the greatest mathematician of all time. As a youth, Archimedes visited Alexandria, probably to study at the museum, and he maintained contact and friendly professional rivalry with some of the geometers and astronomers he met there, notably Eratosthenes and Apollonius. However, he returned to Syracuse where he spent the rest of his life.

Archimedes pursued many different investigations throughout his life. In mathematical physics, he was one of the founding figures of statics and mechanics. In pure mathematics, he was the first to prove the surface and volume formulas for a number of figures, including the sphere and cone. In *The Sand Reckoner*, addressed to Hieron's son Gelon, he calculates the number of particles of sand (dust, really) it would take to fill up the Greek universe; to express the number, he invented octad notation, which has a curious echo more than two thousand years later in the researches of Georg Cantor. In *Quadrature of a Parabola*, he found the area of a segment of a parabola using a geometric sum. In *On Spirals*, he described the construction of a spiral, and proved theorems regarding the tangent to a spiral and the area enclosed by the first turn of a spiral.

The Method of Archimedes is perhaps his most interesting treatise. Archimedes proved many volume relationships, but he wrote *The Method* to provide some insight into how the relationships were discovered in the first place. He gave the details of a means by which problems in mathematics could be investigated using a technique very similar to integral calculus, though he emphasized the results obtained had to be proven "by geometry" (in other words, by a rigorous mathematical method). Essentially, Archimedes considered an n-dimensional figure to consist of an infinite number of $n - 1$-dimensional cross-sections; for example, a three-dimensional sphere could be viewed as consisting of an infinite number of two-dimensional disks. These cross sections could then be "weighed" on a balance beam, and in this way a second figure of known properties could be obtained.

Sometime during the Middle Ages, an unknown scribe in Constantinople copied *The Method* onto a piece of vellum. Vellum, made from sheepskins, is very expensive, so the scribe clearly thought very highly of this work. However, vellum has another useful property: it is durable, so that any writing upon it can be washed off and the vellum re-used. In 1229 a copyist named Johannes Myronas erased the text so he could reuse it to write a prayer book; such a recycled piece of vellum is called a *palimpsest*. Generally the erasure is incomplete, and some of the original text can be made out. In 1906, the Danish classicist J. Heiberg heard of a palimpsest in Istanbul that seemed to have originally been a mathematical treatise. Careful examination led to the recovery of the lost work of Archimedes, as well as an accurate date of when the erasure was made and the name of the copyist; in fairness to Myronas, his actions probably saved the text, since as a prayer book, it was kept in relatively good condition.

In addition to his work on quadrature and cubature, Archimedes was the first to consider π in its modern sense as the ratio of the circumference of a circle to its diameter; he also pointed out the correct relationship between the circumference, radius, and area of a circle. Archimedes estimated the value of π by considering the perimeters of circumscribed and inscribed 96-sided polygons and obtained the bounds $3\frac{10}{71} < \pi < 3\frac{1}{7}$. What is most remarkable about his work is his series of rational approximations to irrational numbers. For example, he estimated that $\frac{265}{153} < \sqrt{3} < \frac{1351}{780}$. He presents these approximations without comment, suggesting they (or at least the methods by which they were obtained) were well-known to his correspondents; we know nothing of how they were obtained.

One of the more famous stories about Archimedes concerns his discovery of the principles of hydrostatics. The story of how Archimedes, after trying to determine the volume of Hieron's crown, went to the public baths and had a flash of inspiration that led him to run through the streets crying "Eureka!" ("I have found it!" in Greek) is well known, as is the detail that he neglected to dress on the way out. The last detail ought to be clarified. The Greeks exercised in the nude (the contestants of the Olympic Games, for example, were nude, which is possibly one reason women were not allowed at the games), and the very word *gymnasium* ("to train naked") makes reference to this fact.

Perhaps the most famous story about Archimedes concerns the last few years of his life, when his home city of Syracuse was under siege by the Roman Republic. The siege came about as a result of a new conflict between Rome and Carthage.

Carthage recovered quickly from the First Punic War and built a new empire in Spain; some traces of Carthaginian influence remain to this day, such as the town of Cartagena, named after Carthage. In 218 B.C., after a series of incidents, the Second Punic War began.

The Carthaginian general Hannibal left Spain and crossed the Alps to take the war to Italy, the heartland of the Roman Republic. Hannibal's army, like Pyrrhus's, included elephants, though few survived the crossing of the Alps. Hannibal realized that Rome's greatest advantage was her alliance with the other Italian states. If he could divide Rome from her allies, he could conquer her easily. Thus, he repeatedly stated that his quarrel was with Rome, and with Rome alone. To make this claim convincing, he gave orders that non-Roman soldiers were to be released unharmed, while Romans were shown little mercy.

Hannibal first defeated the Romans at Lake Trasimene, in 217 B.C. The Consul, who led the Roman armies in wartime, was killed. His replacement was Quintus Fabius, who recognized a key weakness in Hannibal's army: it consisted mainly of mercenaries, interested in loot which they could take from a defeated army (which included the captured soldiers themselves, who could be sold as slaves). But if there were no battles, there would be no loot, and the mercenaries would abandon Hannibal for more lucrative ventures. Fabius used the legions to contain Hannibal while patiently waiting for his mercenary army to wither away.

To the Romans, however, the original Fabian policy smacked of cowardice, so Fabius was replaced with Gaius Terentius Varro. In 216 B.C., Varro's 85,000 troops met Hannibal's 50,000 at Cannae. But Hannibal set a brilliant trap for the Roman army, and by day's end, more than 50,000 Roman soldiers were killed, at a cost of 2000 Carthaginians. In all the centuries since, no general has inflicted a greater defeat on his opponent in a single day than Hannibal inflicted on the Romans at Cannae.

Since Hannibal's primary goal was to shatter the Roman confederation, a great victory would be essential: no one would abandon Rome to join the losing side. Cannae was exactly what Hannibal needed, and the Roman confederation began to waver. A key loss would be Sicily, because much of Rome's grain was grown on that island. Hieron stood by Rome, but when the aged king died in 215 B.C. his successor and grandson Hieronymus, abandoned Rome and allied Syracuse with Carthage.

Unfortunately, the Roman confederation proved more resilient than Hannibal imagined, and the core allies of Rome never deserted her. Moreover, at a time when money and equipment from Carthage would have helped Hannibal inflict yet another defeat on Rome, the rulers of Carthage feared what a successful general might do afterwards, so time after time they refused Hannibal's urgent requests for aid. Finally, the Roman people realized that Fabius had been right all along, and put together a new army that would try and contain Hannibal, but not confront him directly. Soon the tide of war turned against Hannibal, and he had to return to North Africa to defend Carthage itself.

Rome turned to deal with her rebellious allies. The pro-Roman faction in Syracuse assassinated Hieronymus and butchered most of the royal family, but when Marcus Claudius Marcellus, the Roman commander in Sicily, stormed the town of Leontini and beheaded 2000 of the troops there for desertion, the Syracusans realized that Rome would never forgive them for joining Carthage. Their choices were victory for Carthage—or annihilation.

In 214 B.C., Marcellus laid siege to Syracuse itself. According to Plutarch, Archimedes made several war machines for Hieron, whose peaceful reign never had need of them. But with the Romans besieging Syracuse, the weapons of Archimedes were put to use. Plutarch recounts tales of ships being sunk by enormous boulders, or lifted out of the water to be dashed on the rocks below, or arrows and darts flung in enormous numbers at Roman soldiers in the field.[10] After a while, the Roman soldiers would flee in terror if they caught sight of a length of rope or a piece of wood coming out of the wall, fearing it might be the precursor to some more horrible attack.

Most of these tales are hard to credit, but it is a matter of record that the siege of Syracuse took two years—somewhat longer than a typical Roman siege. The Romans finally stormed the city in 212 B.C. after the gates were opened by a member of the pro-Roman party. Archimedes was killed during the looting of the city.

According to Plutarch, Marcellus gave orders that Archimedes was to be taken alive. One should not read too much into this order. It was a Roman tradition that a victorious general, such as Marcellus, would be allowed to parade his troops and their trophies of war through the streets of Rome in a procession called a "Triumph." Archimedes, as part of the spoils of war, would probably have been a major display in such a spectacle.

In any case, Plutarch gives three versions of the death of Archimedes. The most likely is that Archimedes was on his way to meeting Marcellus, carrying mathematical and astronomical instruments. Some soldiers, thinking he was a rich citizen fleeing with something valuable, killed and robbed him. Another version is that a soldier came upon Archimedes in the chaos of the sack of the city, and killed him.

The most dramatic and best known version is also the least likely. According to this version, Archimedes was working on a problem when a Roman soldier came up to him and demanded that he accompany him to meet Marcellus. Archimedes refused to leave until he had finished the problem he was working on; the soldier, having no patience for insolence from a civilian, drew his sword and killed him. In the Roman army, insubordination was punishable by a gruesome death, and no Roman soldier would have knowingly disobeyed Marcellus's command. Plutarch reports that the killers of Archimedes were treated as murderers (they were probably executed), and that Marcellus treated Archimedes's surviving relatives with great honor.

The Second Punic War dragged on until 201 B.C., when Rome won a complete victory. The terms of the peace treaty were harsh: Carthage lost Spain and was restricted to North Africa; she was forced to pay an enormous indemnity; and was not to wage war without Rome's permission. Within a generation, she built a new commercial empire, paid off the indemnity, and achieved such success that Rome felt threatened again, this time economically. The anti-Carthaginian party was led by Cato the Elder. After every public speech, regardless of the topic, Cato would add, "And I am also of the opinion that Carthage must be destroyed." Through her allies, Rome provoked the Third Punic War, which resulted in the destruction of Carthage in 146 B.C.

Greece was added to the expanding Roman Empire about the same time. One wonders how the Greeks felt about the conquest. For 350 years, the Greeks fought amongst themselves. The Roman conquest ended centuries of warfare, and for the first time in a dozen generations, Greece was at peace.

2.2.2 The Fall of the Republic

The rest of the Mediterranean would undergo two more centuries of chaos and disruption. Not surprisingly, there was only one mathematician of note in the centuries following Arch-

---

[10]The infamous "burning mirror" of Archimedes is nowhere mentioned in Plutarch, and does not appear in sources until about four hundred years after the siege of Syracuse.

imedes and Apollonius: HIPPARCHUS of Nicaea (d. ca. 120 B.C.). Hipparchus was unusual for he did *not* work in Alexandria, but spent his life on the Greek island of Rhodes. He was apparently the first person to write a work on chords in a circle, though his work has been lost. The book included a table of chord lengths, the precursor to the modern sine table. By a clever application of geometry and careful observation, Hipparchus correctly determined that the distance to the Moon was thirty times the diameter of the Earth.

In 134 B.C., Hipparchus noticed a star that he had never seen before in the constellation Scorpius (incorrectly called Scorpio by astrologers). Unsure whether this was a new star or one that he simply hadn't noticed, he began to compile the first star catalog, showing the positions of the stars in the sky. He also divided the stars according to their apparent brightness: the twenty brightest stars in the sky were of the "first magnitude"; somewhat dimmer stars were of the "second magnitude", and so on down to the barely visible stars of the "sixth magnitude." The system, with a slight modification to make it more quantitative, is still used today.

While compiling his star chart, Hipparchus also noticed that the position of the sun on the date of the vernal equinox was undergoing a very slow shift that would take about 25,000 years to move it once around the sky. Nearly two thousand years later, the precession of the equinoxes would prove to be one of the key tests of Newton's theory of universal gravitation. One of the consequences of the precession of the equinoxes is that the astrological "signs of the Zodiac," which once corresponded to the constellations in which the sun was located, are now completely unrelated to any physical phenomena.

By the time of Hipparchus the Roman Republic had acquired a vast empire that included most of the Mediterranean, with the exception of the Ptolemaic Kingdom of Egypt. This was a disaster for the middle and lower class. Imported grain was cheaper than domestic grain, which bankrupted the Roman farmer; and slaves acquired as war booty provided labor more cheaply than any Roman citizen could, which wiped out the working class. As a result, the bankrupt farmers were forced to sell their land to the wealthy, who thus acquired huge estates, called *latifundia*, which were worked with slave labor. The working class was reduced to poverty.

In 133 B.C., when Hipparchus was in the midst of compiling his star catalog, Tiberius Sempronius Gracchus was elected to high office. Gracchus promised reform. He was murdered by reactionary nobles in June of that year, establishing a deadly precedent in Roman politics. His brother Gaius tried to continue his program, and met a similar fate (committing suicide before his enemies could murder him). For the next century, Roman fought Roman in a terrible civil war that pitted the aristocratic *optimates* ("best people") against the plebeian *populares* ("rabble rousers").

The leader of the military forces of the *optimates* was Gaius Pompey, a general with an impressive string of victories to his credit. In 67 B.C., pirates made it nearly impossible for Rome to get food shipments from overseas; it took Pompey just six months to wipe out the pirates. Between 66 and 64 B.C., Pompey swept through Asia Minor, and even conquered the troublesome kingdom of Judaea, adding more territory to Rome's vast empire. With proven military leadership and the resources to buy the best arms and armor, the defeat of the *populares* seemed inevitable.

But leading the armed forces of the *populares* was a wealthy aristocrat: Gaius Julius Caesar, who gained notoriety earlier by defeating the Gauls and the Britons. At Pharsalus,

Greece, in 48 B.C., Caesar's army, consisting mainly of poorly armed footsoldiers, met Pompey's army, consisting mainly of well-armed cavalry. Caesar demonstrated the true nature of military genius: understanding his opponents. Thus he instructed his footsoldiers to point their spears not at the cavalry horses, which was the standard tactic, but at the faces of the riders. The riders, all aristocrats, had a choice: they could win the battle, but be disfigured; or they could lose. They chose to lose.

Pompey fled to Egypt, hoping to recruit a new army. Egypt was in the midst of a civil war between the twelfth Ptolemy and his older sister. The arrival of Pompey put the regent, Pothinus, in a difficult position. The *optimates* were far from finished, so Pothinus could not afford to antagonize Pompey. But the *populares* had the upper hand, so Pothinus could not afford to antagonize Caesar, either. Pothinus solved his dilemma by having Pompey killed as he disembarked from his ship.

Pothinus showed the body to Caesar and suggested that Caesar return the favor by killing Ptolemy's sister: Cleopatra. The plan backfired, and Caesar threw his forces into the fight on the side of Cleopatra. The alliance of Caesar and Cleopatra has often been romanticized, but Caesar's previous and subsequent behavior showed that, above all else, the pursuit and maintenance of power was his one true love.

It is possible that the Library of Alexandria suffered great destruction during this time. Plutarch claims that Caesar set fire to his fleet as a diversionary tactic; the fire spread and destroyed the Great Library. Tellingly, the geographer Strabo, writing just a few years later, seems to indicate that the works available to him at the Great Library were only a fraction of those available to his predecessors.

Caesar eventually defeated his enemies and emerged the undisputed master of the Roman world. This allowed him to make some necessary changes, including a reform of the calendar. Creating a calendar is a task that lies at the intersection of aesthetics, astronomy, and mathematics. First, one must decide what the calendar is meant to track; a solar calendar tracks the seasons (so, for example, December is always a winter month in the Northern Hemisphere), while a lunar calendar tracks the phases of the Moon (so, for example, the first of the month is always the day of the new Moon), and a luni-solar calendar tracks both (so that the first day of first month of the year is always in the same season and has the same phase of the Moon).

However, the solar year (which determines the length of the seasonal cycle) is 365.2422 days in length, while the synodic month (the time between new Moons) is 29.531 days. Thus the basic mathematical problem of creating a calendar consists of selecting a small number of whole numbers whose mean approximates one of these periods. Most lunar calendars alternate 29 and 30 day months (with a mean of 29.5).

The traditional Roman calendar was a lunar calendar. The year began in March (hence, October was actually the eighth month, as its name suggests) and ended in February. However, twelve lunar months (alternating 29 and 30 days) contain only 354 days, more than 11 days short of the solar year. Without correction, the seasons would drift relative to the calendar: if the beginning of spring occurred in March one year, then three years later it would occur in April. To prevent this, the actual beginning of the year was announced by priests (in fact, the word "calendar" comes from the Greek word meaning "proclamation"). Unfortunately, in Rome these priests were elected officials. This meant they could (and would) delay or accelerate the start of the calendar year, depending on whether or not their party was in power. As a result, by the time of Caesar, the Roman calendar had no relationship to the seasons.

Based on the recommendations of an Egyptian astronomer named Sosigenes, Caesar made several important changes to produce what is now called the Julian calendar. First, to ensure the beginning of the calendar year was actually in the springtime, Caesar ordained 45 B.C. would last 445 days: 45 B.C. came to be known as "The Year of Confusion." Next, Caesar made the lengths of the months alternate, beginning in March with 31 days and April with 30 days (thus abandoning the lunar calendar entirely). The only exception was February, which was an unpopular month: all bills came due in February. Thus, February was shortened to 29 days. Finally, to make the mean calendar year approximate the solar year, every fourth year thereafter would be a leap year, in which February would have 30 days. Note that in the system originally established by Caesar, the months of September, October, November, and December had 31, 30, 31, and 30 days.

Shortly after Caesar's proclamation of the calendar in 45 B.C., he was assassinated on the floor of the Senate. Marc Antony, Caesar's friend, suggested the fifth month be named July, in honor of Caesar's family, the Julians. Caesar's death was followed by another civil war. The major contenders were Caesar's nephew Octavian, and Marc Antony. Marc Antony went to Egypt, to seek help from Cleopatra, while Octavian stayed in Rome and turned the Roman public against Marc Antony. It was not too difficult: Marc Antony had abandoned his wife and was having an affair with Cleopatra.[11]

Marc Antony and Cleopatra's forces suffered a disastrous naval defeat at Actium in 31 B.C. Marc Antony committed suicide. Cleopatra held out hope that she could entrance Octavian as she did Marc Antony and Caesar, but Octavian (something of a stern moralist) refused, offering her only a position in his triumph at Rome—as a captive. She killed herself instead. Octavian swept up the remnants of Ptolemaic Egypt and emerged undisputed master of the Mediterranean. The legions acclaimed Octavian as *Imperator* ("Supreme military commander" in Latin), which became the word Emperor in English.

Octavian was well aware that the last master of the Roman world was murdered on the floor of the Senate. To allay the fears of the Senate, he styled himself "First among equals" and gave himself the title of *Princeps* ("First citizen" in Latin). This became the word prince, and the Roman world became known as the Principate. This pleasant fiction was enough for the Senators, who wanted nothing more than to be treated well. Despite the existence of an elected Senate and Octavian's claim of being merely another citizen, the Republic had ceased to exist, and the Empire had been established.

Octavian took on the name Augustus ("exalted one") and added the name Caesar (to emphasize that he was Caesar's nephew): thus he was the Emperor Augustus Caesar. In time, "Caesar" came to be a title for many ruling monarchs: kaiser, tsar, czar, and (possibly) shah all stem from the word Caesar.

It was during the reign of Augustus that two changes were made to the calendar, which put it almost into its modern form. First, the month following July was named August in honor of Augustus. Then, since August would only have 30 days under the original Julian scheme, a day was taken from February (reducing it to its current 28 days, 29 in leap years). But since the month of September had 31 days, this would leave three 31-day months in

---

[11] Further circumstantial evidence for the destruction of the Great Library by Caesar's time is that Marc Antony is credited with giving the contents of the Royal Library of Pergamum to Cleopatra as a gift.

a row, so the lengths of the months from September to December were switched to their current amounts. The resulting calendar is now called the Julian calendar.

Augustus was succeeded by his stepson Tiberius, who reigned until 37 A.D. After Tiberius, Roman emperors tended to be weak, incompetent, insane, or all three. Most were murdered or committed suicide before they could be killed by rivals. Nero's suicide in 68 A.D. began the "The Year of Four Emperors," where Galba, Otho, and Vitellius reigned in quick succession, alternately succumbing to assassination or suicide. The "Year of the Four Emperors" ended when Vespasian became Emperor in 69; he ruled for ten years and re-formed the Empire's finances, army, and senate. His survival can be traced to a key factor: he was a popular general whose troops had been successfully fighting a revolt in Judea.

Since their conquest by Pompey in 64 B.C., the Jews of Palestine sought independence and eagerly awaited a *messiah* ("deliverer" in Hebrew). One sect believed the delivery would be a religious and spiritual one, while another waited for a great military hero. In either case it was traditional to anoint the deliverer, and in Greek, "the anointed one" is *christos*.

There is little doubt of the existence of a historical Jesus (the Latinized form of the very common Jewish name "Joshua"), though evidence outside the Bible is virtually non-existent and the best evidence is negative: at no point did contemporary Jewish critics of Christianity ever imply that Jesus was a fictional character. "Joshua the Anointed One" became Latinized to *Jesus Christ*. But because he promised a spiritual and religious salvation, and *not* a military solution, he found little support among the radicals. His supporters came to be known as Christians, and the Romans at first considered Christianity to be another sect of Judaism.

The most outspoken supporter of Christianity was Paul of Tarsus, a Roman citizen who lived around 60. Paul made Christianity more palatable to non-Jews by eliminating circumcision, dietary restrictions, and most of the complex rituals associated with Judaism. By the time of Paul's execution in Rome, Christianity was thriving. Ignatius of Antioch, later canonized as Saint Ignatius, gave the religion a new name: since it was not for Jews or Gentiles, but for everyone, it was *catholic*, which is Greek for "universal."

Meanwhile, the radicals continued to look for a military solution. In 66, the Roman governor of Judea infuriated the Jews by seizing the treasury in the Temple of Solomon to pay back taxes. Protests turned into riots, and riots turned into a full-scale revolt. The radicals, known as Zealots, saw this as an opportunity: why wait for a messiah if, by direct action, they could free themselves? The most fanatical of the Zealots were called Sicariots, (from *sicae*, Greek for "little knife"). The Sicariots believed the best way to deal with Rome was to kill Romans—any Romans, whether they were soldiers, politicians, or innocent bystanders. It has been suggested that Judas *Iscariot*, whom the Bible credits for turning Jesus over to the Roman authorities, is a Hellenization of Judas *Sicariot*. It took three Roman legions under Vespasian to end the revolt.

Vespasian's son Titus captured Jerusalem on September 7, 70. The Temple was destroyed (the "wailing wall" in Jerusalem is all that remains), and the surviving Jews were sold into slavery, beginning the dispersal (*diaspora* in Greek) of the Jews that spread them throughout the Empire. One last band of Zealots held out at the mountain fortress of Masada until 73. The Romans had to build a miles-long ramp to approach the fortress, and before the Romans could break in, the defenders killed themselves.

Clearly military force was not a solution, so there was an increasing turn towards the idea of a spiritual deliverer. Thus it is not surprising that the gospels, which emphasized the spiritual nature of the salvation of the Jews (and, thanks to Paul's work, the non-Jews as well), were written in this period. The gospels (Anglo-Saxon for "good news") were attributed to the disciples of Jesus. However even the earliest to be written down, the Gospel of Mark, was written a generation after the events it claims to describe.

2.2.3 The Return to Order

The Romans built roads, bridges, aqueducts, and buildings that outlived their empire, but theoretical mathematics made little progress during Roman times. This has led to accusations that the only Roman "contribution" to mathematics was killing Archimedes. Even the Roman numeration system is derided as clumsy and inefficient: it is additive, with symbols representing one (I), five (V), ten (X), fifty (L), and so on, so that twenty-three would be XXIII. If a smaller symbol preceded the next larger symbol, this indicated that the latter was to be reduced by the former: thus XL represented ten from fifty, or forty. A more charitable assessment is that Roman numerals were used primarily for record-keeping, where clarity and ease of interpretation are of paramount importance, and subtractive notation was used primarily when the writer ran out of space.

As for Roman contributions to theoretical mathematics, it would be reasonable to claim that theoretical geometry had exhausted the available tools, and progress became nearly impossible. Thus, the mathematically inclined turned to other fields, some of which included purely practical mathematics.

For example, HERON of Alexandria (10–75?) was probably an engineer who taught mathematics: his *Pneumatica* includes descriptions of many clever devices, like a vending machine (5 drachmas for an amount of sanctified water) and a primitive steam engine. Heron's *Metrica* appeared around the same time as the Gospel of Mark, and the subsequent fate of Heron's works mirrors the history of the time. The Gospel of Mark and the Catholic Church grew to become a dominant force in a western civilization that was growing increasingly more interested in philosophy and religion. Meanwhile Greek geometry declined into obscurity, and Heron's *Metrica* disappeared during the Middle Ages, not to be rediscovered until the late nineteenth century.

Metrica concerned itself with the measurement of figures, and in many ways reads like a modern textbook: some results are proven, others are justified non-rigorously, and others are simply attributed. For example, *Metrica* states and proves what is now known as Heron's Theorem: if the sides of a triangle have lengths a, b, and c, and s is half the perimeter, then the area is $A = \sqrt{s(s-a)(s-b)(s-c)}$. To justify the result that the area of a rectangle 5 by 3 units is 15 square units, he notes that the figure can be decomposed into 15 unit squares. The rule:

Rule 2.2. *To find the area of a regular enneagon [9-sided polygon]: Square the side and multiply by 51, then divide by 8.*

is derived using approximations from a table of chords, and for:

Rule 2.3. *To find the circumference of a circle, multiply the diameter by 22 and divide by 7. To find the area, take half the product of the circumference and the radius.*

Heron directed the reader to Archimedes.

Heron lived during the relative stability of the reign of Vespasian and his older son Titus. Titus oversaw several important projects, like the building of the *Colosseum* (named after its proximity to an enormous statue of Nero) and the relief effort following the eruption of Vesuvius in 79. Unfortunately he died suddenly in 81 and his younger brother Domitian became emperor. Unlike his father and brother, who coddled the Senators (though allowed them no real power worth speaking of), Domitian treated them like the impotent figures they were. His reign grew increasingly tyrannical, and in 96, he was assassinated in a palace coup whose conspirators included the Empress herself.

The conspirators chose Nerva, a respected Senator, as the new Emperor. He died two years later (of old age), but before his death he chose a conscientious and competent Spanish general, Trajan, to be his successor. This started a trend, and the "Five Good Emperors" (Nerva, Trajan, Hadrian, Antoninus Pius, and Marcus Aurelius) reigned in peaceful succession until 180, each appointing as his successor a man he felt was competent and conscientious.

NICOMACHUS (fl. 100) lived in Gerasa, Palestine (now Jerash, Jordan) around this time, though he probably studied in Alexandria. His *Introduction to Arithmetic* did for number theory what Heron's *Metrica* did for geometry: it presented results and theorems without proof. An important consequence of this is that number theory re-emerged as a subject independent of geometry.

The Pythagoreans developed an elaborate system of number classifications, which Nicomachus gave in detail. The polygonal numbers (from triangular to octagonal) are presented, with an observation equivalent to noting that the second differences of the k-gonal numbers is equal to $k - 2$. For example, the triangular numbers are 1, 3, 6, 10, 15, ...; the differences between successive triangular numbers form the sequence 2, 3, 4, 5, ... (the first differences), and the differences between successive terms of these numbers form the sequence 1, 1, 1, 1, ... (the second differences). We can reverse the process to form the n-gonal numbers. For example, the pentagonal numbers would have second differences 3, 3, 3, ..., so their first differences would form the sequence 1, $1 + 3 = 4$, $4 + 3 = 7$, $7 + 3 = 10$, ..., and the pentagonal numbers themselves would form the sequence 1, $1 + 4 = 5, 5 + 7 = 12, 12 + 10 = 22, \ldots$. Pyramidal numbers (with triangular and square bases) and other types are presented: the heteromecic numbers, for example, are numbers that are the product of two consecutive whole numbers.

In addition to classifying numbers individually, Nicomachus also classified relationships between numbers. We retain a vestige of this when we refer to one number as a multiple of another, but Nicomachus's classifications go much further. For example, 20 is the triple sesquitertian of 6 (since $20 = 3 \cdot 6 + \frac{1}{3} \cdot 6$). As with the polygonal numbers, Nicomachus makes observations on how to form the triple sesquitertians: in this case, the multiples of 10 are the triple sesquitertians of the corresponding multiples of 3 (hence 40, which is 4 times 10, is the triple sesquitertian of 4 times 3).

Another mathematician of the peaceful era of the Five Good Emperors was MENELAUS of Alexandria (fl. 98). His *Sphaerica* is the oldest surviving work to discuss the geometry of triangles on the surface of a sphere. In Book I, Menelaus develops the theory of spherical

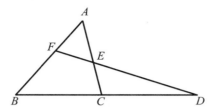

Figure 2.6. Menelaus's Theorem

triangles (defined as regions enclosed by three great circle arcs on the surface of a sphere), and in Book III, he develops some important propositions of spherical trigonometry, including what is known as Menelaus's Theorem. In the plane, let triangles ABC, FDB be as shown (see Figure 2.6). Then $BD \cdot CE \cdot AF = CD \cdot AE \cdot BF$. Menelaus showed that the same would be true if ABC, FDB were spherical triangles (where BD is the chord joining the points B, D).

Even greater contributions to trigonometry were made by CLAUDIUS PTOLEMY (fl. 125–151), who lived during the reign of Hadrian. About all we know for certain about Ptolemy's life is that he made astronomical observations from Alexandria between the years 125 and 151. He would, however, write a book that, like Euclid's *Elements*, incorporated the work of his predecessors in so successful a fashion that tracing the history of trigonometry before Ptolemy is quite difficult. Ptolemy called his work *The Mathematical Collection*, though later commentators called it *The Great Collection*, where "great" is used in the sense of "large." In Greek, this is *Megale Syntaxis*. When Islamic scientists began referring to it, they combined their definite article *al* with the Greek superlative *Megistos*, so the work came to be known as the *Almagest*.

The *Almagest* is mainly known as a work of astronomy, for in it Ptolemy describes a geocentric system that astronomers would use for fifteen hundred years. It highlights a fundamental difference between mathematics and science, for the science in the *Almagest* has been invalidated through the passage of time—but the trigonometry is just as valid and just as usable today as it was during the reign of Hadrian.

The trigonometry of the time was based on the lengths of chords in a circle of standard radius. Ptolemy began the *Almagest* by establishing the necessary theorems, complete with proofs, that allowed him to construct a table of the lengths of the chords in a circle of radius 60. Ptolemy's theorems include those equivalent to the half-angle, angle sum, angle difference, and double angle identities in trigonometry, most of which were derived from what is now known as Ptolemy's Theorem: Let $ABCD$ be a quadrilateral inscribed in a semicircle, with AD the diameter. Then $AC \cdot BD = BC \cdot AD + AB \cdot CD$.

If the radius of the circle is 60, then the chord with a central angle of 60° will have a length of 60 (or 60^P to distinguish it from the angular measure); likewise, we can find the exact length of the chords of 90° and 120°. Thus a combination of the Ptolemaic theorems will give us the lengths of chords of 30°, 45°, and many others. Inconveniently, all of these central angles are multiples of 15°. Fortunately, it is possible to construct a decagon in a circle, and thereby obtain an exact value for the chord with a central angle of 36°. Euclid included a construction, albeit a complex one; Ptolemy gives a very simple and elegant one as follows.

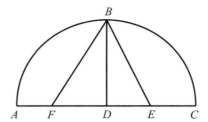

Figure 2.7. Ptolemy's Pentagon

Let AC be a diameter of a circle with center D; let DB be perpendicular to AC. Bisect DC at E and extend BE; let $EF = BE$, and join BF. Then the side of a regular decagon inscribed in the circle will equal FD, and BF will be the side of the regular pentagon inscribed in the same circle (see Figure 2.7).

This will give us the chord with a central angle of $36°$, and by applying the various theorems, we can eventually obtain the chords of $3°$, $1 \frac{1}{2}°$, and $\frac{3}{4}°$. By using linear interpolation on the last two, Ptolemy found an approximation for the chord of $1°$, and from there produced a table of chord values.

Ptolemy also produced another solution to the tuning problem. The difficulty with Pythagorean tuning is that, while most of the the fourths and all of the fifths are perfect, the intervals are inconveniently large. For example, no note between C and F will form a consonance with either note. However, the evidence of the senses points to a peculiarity. Consider an interval spanning three notes (and two whole tones), say C–E. This would be an interval of a third. Since it corresponds to the inelegant ratio $81 : 64$, thirds are considered dissonant in Pythagorean music theory. The problem is that if the notes are played melodically (one after the other), they *seem* to be consonant; it is only when they are played simultaneously that their dissonance is obvious.

While $81 : 64$ is not a ratio of small whole numbers, it is very nearly equal to the ratio $5 : 4$. Thus if we *define* the third to be $5 : 4$, we can have another set of consonant intervals. To determine the other ratios, let us begin with C–D, and again let this equal the $9 : 8$ ratio that defines an interval. In order for C–E to make a third, D–E must have a ratio of $10 : 9$. In order for C–F to make a fourth, E–F must have a ratio of $16 : 15$. The F–G interval must be $9 : 8$, and so on. Unfortunately, there is a problem; this makes the E–G correspond to $6 : 5$ not $5 : 4$: it is not a perfect third, but in fact represents a new ratio (again of reasonably small whole numbers). To resolve this problem, we note that in the Pythagorean scheme, the C–E interval includes two whole tones, while E–F includes a tone and a semitone; hence there is no reason to require that they have the same ratio. Unfortunately, this means that we must distinguish between the two types of thirds. A third that includes two tones is a major third, while a third that includes a tone and a semitone is a minor third.

Although a pure Pythagorean would deny the consonance of thirds of any type, they can actually be generated by number mysticism. Consider: the octave can be divided into a fourth and a fifth, either arithmetically $(4 : 3 : 2)$ or harmonically $(6 : 4 : 3)$. The fifth can be divided either arithmetically $(6 : 5 : 4)$ or harmonically $(15 : 12 : 10)$. In both cases, the ratios of $6 : 5$ and $5 : 4$ are generated, corresponding to the minor and major thirds. Combined with the evidence of the senses that suggests the $6 : 5$ and $5 : 4$ ratios correspond

Symbol	Diophantus's Term	English	Modern Notation
ς	arithmos	number	x
Δ^Y	dynamos	power	x^2
K^Y	cubos	cube	x^3
$\Delta^Y \Delta$	dynamodynamos	square square	x^4
ΔK^Y	dynamocubos	square cube	x^5
$K^Y K$	cubocubos	cube cube	x^6

Table 1. Diophantus's Notation for Positive Powers

to consonances, the temptation to incorporate thirds is overwhelming. Balancing all of these factors leads to Ptolemaic tuning:

C to D	D to E	E to F	F to G	G to A	A to B	B to c
$9:8$	$10:9$	$16:15$	$9:8$	$10:9$	$9:8$	$16:15$

There are many more consonances here than with Pythagorean tuning, which helped popularize Ptolemaic tuning. Unfortunately, we have sacrificed some of the consonances within the octave. Moreover, if we extend the scale to include two octaves (so, for example, we can consider a fourth whose lower note is A), we discover more dissonances. Consequently Pythagorean tuning also remained popular.

2.2.4   Decline and Fall

The most enigmatic figure in the history of mathematics is DIOPHANTUS of Alexandria, whose *Arithmetic* is the only surviving Greek algebra. The *Arithmetic* consisted of thirteen books, though only ten survive today, six in Greek, and four in Arabic. Most of the problems deal with finding rational solutions to indeterminate equations of the first or second degree, a branch of number theory now called Diophantine analysis.

Diophantus created a type of algebraic notation called syncopated notation, which essentially replaces the words describing an algebraic operation with an abbreviation (see Table 1). He indicated that a number represented a pure number (as opposed to the coefficient) by using the symbol $\overset{\circ}{M}$. Subtraction was indicated by using a truncated Ψ turned upside down, $\wedge$.

Diophantus's most famous problem appears in Book II, where he posed the problem of expressing a given square as the sum of two other squares:

Example 2.1. *Let* 16 *be the given square, and suppose it is divided into x^2 and $16 - x^2$, both assumed squares. Let $16 - x^2$ be the square of any number of x minus 4, the square root of* 16. *Thus:*

$$16 - x^2 = (2x - 4)^2$$
$$16 - x^2 = 4x^2 - 16x + 16$$

$$16x = 5x^2$$
$$x = \frac{16}{5}.$$

Hence the squares are $\frac{256}{25}$ *and* $\frac{144}{25}$.

The dates of Diophantus are very uncertain, though he probably lived during the third century and the dates 200–284 have been suggested for him (though there is some evidence that he predated Heron). Diophantus's life story is given in an epigram:

Problem 2.1. *For one-sixth of his life, Diophantus was a boy. A twelfth part later, he grew a beard, and married after a seventh. Five years later he had a son. The son lived only half the father's life, and Diophantus died four years later.*

It was not unusual for sons to predecease their fathers, but this tragic event was more likely during a time of chaos. This was precisely the state of the Mediterranean world during the third century A.D.

By then a dark change had come over the Empire. Marcus Aurelius, the last of the "Five Good Emperors," chose his son Commodus to be his successor. Not only was Commodus incompetent, he was delusional and believed himself to be the god Hercules. He entered gladiatorial games, killing wild animals in the arena, and even engaged in combats with opponents (the combats were, presumably, rigged). Tired of his excesses, a group of conspirators (including his mistress, Marcia) arranged to have Commodus strangled in 192 by a professional wrestler.

Commodus's successor, Pertinax, was a good man who sought to replenish the treasury by cutting expenditures: in particular, he refused to bribe the Praetorian Guard, the personal bodyguard of the emperor and the most powerful military force in the city of Rome. They killed him, and sold the empire to the highest bidder, a Senator named Marcus Didius Julianus. But Lucius Septimius Severus, a general commanding the Danubian legions, marched on Rome with the intent of declaring himself emperor; the Praetorians, knowing they stood no chance against real soldiers, deposed Julianus as Severus entered Rome. Julianus was executed after holding the throne for only two months, protesting to the end that he had done nothing wrong. If the Praetorians thought their support of Severus would save them, they were wrong: Severus disbanded the guard and established a precedent of military rule. Peace broke out: the peace of the sword.

If Diophantus lived in this era, then he was probably a Roman citizen, thanks to the Edict of Caracalla, proclaimed by Severus's son in 212. The Edict granted Roman citizenship to all freeborn inhabitants of the Roman Empire. By then, Roman citizenship mainly meant you were liable for an inheritance tax; indeed, this has been suggested as Caracalla's motive.

Severus's line came to an end in 235, when a military coup assassinated Alexander Severus (Septimius's grand-nephew), and the legions defending the Rhine acclaimed Maximinus the Emperor. The Senate had their own candidate, who was killed; Maximinus was killed by his own troops in 238. The next forty years saw this cycle repeated many times, with minor variations.

Finally a dour general named Diocletian became emperor in 284. Diocletian made two lasting changes to the empire. To avoid the fate of his predecessors, who reigned an average of two years before being murdered, Diocletian took on the trappings of an Oriental monarch, where the ruler was more god than man. One consequence was that if the emperor was a god, it was treason not to worship him. In particular, this meant the Christians, who refused to recognize any god but their own, were deemed a threat to the safety of the empire. Diocletian began a massive persecution of Christians—the last the Roman world would see.

The other reform of Diocletian was to divide the empire into two halves, an eastern and a western half, to be ruled by *Augusti*. The *Augusti* were equal in theory, though Diocletian made sure that *he* was the "first among equals." The co-emperors would each have co-rulers, called *Caesars*, and together they formed the *tetrarchy* (from the Greek word for "four rulers"). In theory, the Caesars would become emperors, after having gained experience governing. Diocletian's reward for his careful planning: he reigned as absolute master of the Mediterranean for twenty-one years, and retired in 305 to spend the remaining eight years of his life raising vegetables.

To further stabilize the empire, Diocletian mandated that all occupations were to be hereditary: sons had to follow in the professions of their fathers. Certain professions were forbidden entirely; in one edict, Diocletian forbade the practice of mathematics. By this, Diocletian meant the practice of astrology, numerology, and other forms of useless superstition.[12] Indeed, Diocletian's edict goes on to praise the work of *geometers*, who provide useful service to the state. The value placed on geometers can be gauged by their wages, also fixed by edict: an agricultural laborer could earn a maximum of 25 denarii per day, or around 600 denarii per month. For teachers, the pay ranged from 50 denarii per student per month for elementary teachers, 75 for teachers of arithmetic, 100 for teachers of architecture, and 200 for teachers of geometry.

PAPPUS (fl. 300–350) probably lived during the reign of Diocletian. Pappus's main work, his *Mathematical Collection*, was intended for use alongside the great books of the ancient Greek geometers, explaining the more difficult parts to students. There is evidence that the *Mathematical Collections* was meant to contain twelve books, but all that remains is half of Book 2, and Books 3 through 9.

It is possible that Pappus never wrote the last three books. Egypt revolted several times in the 290s. During one revolt, the legions besieged and eventually captured Alexandria but during the siege, a fire was started (no one knows by whom) that consumed half the city, possibly destroying the Great Library (though the annex at the Temple of Sarapis would have been untouched).

Thus the fire may have destroyed the source material Pappus was going to use for the last section of the *Mathematical Collection*. A glimpse of what Pappus had available to him appears in Book 7, where he listed the domain of analysis: those works beyond Euclid's *Elements* that were necessary to learn how to solve geometrical problems. These might be considered the "required reading" for the graduate study of classical mathematics. Table 2 lists the works, their length, and their availability: nine out of the twelve no longer exist. Note that the most advanced work was written by the "beta" mathematician Eratosthenes.

Pappus's intent in writing the *Mathematical Collections* was to revitalize the study of geometry, for he felt geometry had fallen to a very low level. He is extremely critical of

---

[12]Otho, whose reign was second in the "Year of Four Emperors," also banned astrologers and, according to Suetonius, ordered them to leave Italy by October 1, 69.

Author and Title	Length (Books)	Availability (Books)
Euclid, *Data*	1	1
Apollonius, *Cutting off of a Ratio*	2	2
Apollonius, *Cutting off of an Area*	2	LOST
Apollonius, *Determinate Sections*	2	LOST
Apollonius, *Tangencies*	2	LOST
Euclid, *Porisms*	3	LOST
Apollonius, *Neuses*	2	LOST
Apollonius, *Plane Loci*	2	LOST
Apollonius, *Conics*	8	7
Aristaeus, *Solid Loci*	5	LOST
Euclid, *Loci on Surfaces*	2	LOST
Eratosthenes, *On Means*	2	LOST
Total	33	10

Table 2. The domain of analysis

his fellow geometers, chiding them for doing nothing more than pursuing trivialities and making no new contributions to geometry. At one point, Pappus refers to Pandrosion, who is sometimes credited with being the first female mathematician in history. However, aside from this mention of her in Pappus, where he chides her students for their lack of knowledge, nothing else is known about her. It may be that Pappus invented the name as a way of criticizing a group of Athenian teachers, as Pandrosius was the daughter of Cecrops, the legendary king of Athens.

One of Pappus's original contributions, though he provided no proof, is now called the Pappus-Guldin theorem: the volume of a solid of revolution can be found by multiplying the area of the figure by the distance traveled by its center of mass. In the fifth book of the *Collection* Pappus solved the isoperimetry problem by showing that of all figures of a given perimeter, the circle has the greatest area.

If Pappus really lived in Alexandria during the reign of Diocletian, he would have witnessed the beginning of a crucial battle within the Church. An Alexandrian priest, Arius claimed that Jesus Christ was not God, but was instead created by God, a doctrine now called Arianism. Arius's main opponent was the Bishop of Alexandria, Athanasius, and as the fourth century opened, the conflict between Arians and Athanasians loomed.

The conflict between Arian and Athanasian might have been inconsequential had Diocletian's tetrarchy kept the peace. Unfortunately, the tetrarchy only lasted one generation before a new round of civil wars began. It ended when Constantine became sole emperor in 313. Before the Battle of Milvian Bridge (312), Constantine was said to have seen a flaming cross, with the words, "With This Sign, Conquer." Constantine had his troops paint the symbol on their shields, and won the day. Though it is claimed that Constantine also converted to Christianity after the battle, he was not baptized (if he ever was) until many years later; however, he did soften the official stand on Christianity. In 313, he proclaimed the Edict of Milan, which officially ended the persecution of Christianity. Constantine also sponsored the Council of Nicaea in 325 to mediate some of the doctrinal disputes within the increasingly influential Catholic church. The Council of Nicaea declared Arianism to be heretical, but in spite of this, the Emperors of Rome were alternately Arians (who censured and exiled Athanasius) and Orthodox (who allowed Athanasius to return).

Constantine moved the capital of the Empire to Byzantium. Surrounded on three sides by water, it was virtually impossible to capture, and was one of the few cities that withstood Philip of Macedon. After much new construction to make it a proper capital city, it was dedicated in 330 and given a new name: Constantinople ("Constantine's City", in Greek).

In 379, the Orthodox Theodosius became Emperor, and the Christian Church began to play an increasingly pivotal role in the governance of the Empire. Key among those who extended the power of the Church over the state was Ambrose, Bishop of Milan. In 384, Ambrose persuaded the Emperor Theodosius to reject an appeal for tolerance by the pagans. Four years later, the Emperor punished a Bishop who had burned down a Jewish synagogue, and Ambrose rebuked him for it. Intolerance for non-Christians was rapidly gaining a foothold in the Empire, and pagan learning became increasingly suspect.

For example Jerome, who had traveled about the Mediterranean studying Greek and Latin classics, had a vision that he was brought before God and accused of being a follower of Cicero, not Christ. As a result, he turned his enormous intellectual talents to religious pursuits. At the time, the version of the Bible in use was the Greek Septuagint. But few Christians could read Greek, so from 391 to 407, Jerome prepared a Latin version of the Bible. Since it was meant for the people (*vulgis* in Latin), it became known as the Vulgate and is the basis for the Catholic Bible.

Not all Christians were implacably opposed to pagan learning. Ambrose studied the Greek classics extensively, incorporating elements of Greek philosophy to make Christianity more palatable to philosophically minded pagans. This helped Ambrose convert Augustine, a Neo-Platonist, around 387; Augustine went on to become Bishop of Hippo in North Africa in 396, and is considered one of the major Christian philosophers. In his longest work, the *Ennarations* (sermons based on the *Psalms*), Augustine lists those who consult mathematicians among God's enemies. Since he lists mathematicians alongside sorcerers and pagan oracles, it is clear that he is referring to mathematicians in the Diocletian sense.[13]

The case of Synesius of Cyrene (in North Africa) is particularly interesting. Synesius was married and had three children; this did not prevent him from becoming the Bishop of Ptolemais. Earlier, he had studied at a Neo-Platonic Academy in Alexandria, and maintained contact with his old teacher: HYPATIA (ca. 370–March 415), the daughter of THEON of Alexandria (ca. 335–405).

Hypatia, as the first known female mathematician in history (with the possible exception of Pandrosion), has been so mythologized that it is difficult to separate fact from fantasy. The definite facts are that she helped Theon with his commentary on Ptolemy, and became the head of the Neo-Platonist school in Alexandria around 400, teaching mathematics and philosophy. She also wrote a commentary on Diophantus's work, which is believed to be the source of the six surviving Greek books of his *Arithmetic*.

In 391, the Bishop of Alexandria, Theophilus, with the support of the Emperor Theodosius, incited Christian mobs to destroy the remaining pagan temples. These included

---

[13] Interestingly enough, he does not attack mathematicians as such: only those who use them, because they are placing worldly concerns over spiritual ones.

the Temple of Sarapis, where the last surviving manuscripts of the once great Library of Alexandria were kept. Theophilus's nephew and successor Cyril continued his uncle's destructive work. Cyril encouraged his followers to massacre the Jews, which destroyed a population that had been living peacefully and prosperously in Egypt since the time of the first Ptolemy.

Such flagrant violations of the peace had to be addressed by Orestes, the governor of Egypt. Though Orestes was Christian himself, he could not, as governor, allow one segment of the population to terrorize the rest. However, Orestes dared not move directly against Cyril, so he appealed to Pulcheria, the regent for the Emperor Theodosius II. She declined to move directly against the Christians, so this appeal was useless. Indeed, even the attempt to move against Cyril caused Orestes to be assaulted by a group of fanatics. The chief attacker, Ammonius, was executed for the assault, but Cyril had his body exhumed and treated as if Ammonius were a martyr.

It was suggested to Cyril that the death of Hypatia, Orestes's friend and supporter, would lead to a reconciliation between Cyril and Orestes. Thus in March 415, she was set upon by a fanatical mob of Christian *parabalani* ("lay brethren") led by a lector named Peter who dragged her into the street, stripped her naked, then flayed the flesh from her body with shells (or tiles) before burning her body.

Clearly the intellectual environment in fifth century Alexandria had turned against pagan philosophy. Despite this, PROCLUS (410–485) came to Alexandria to study philosophy. Not surprisingly, he found the instruction poor, and moved to Athens around 430 to continue his study at Plato's Academy with Syrianus of Alexandria. When Syranius died in 437, Proclus became the head of the Academy, earning him the nickname *Diadochus* ("successor"). He wrote a number of works on philosophy and composed music, but his greatest contribution was his commentary on the first book of Euclid's *Elements*. Proclus gives us invaluable information on the development of Greek mathematics, partly because he had access to works now lost, including a history of geometry written by EUDEMUS (fl. 300 B.C.) and a comprehensive encyclopedia on mathematics written by GEMINUS (fl. 60 B.C.).

Proclus never married, though he had been engaged to Aedesia of Alexandria. However, an oracle foretold disaster if the wedding occurred, so she married Hermeias, another student of Syrianus. Hermeias taught Platonism at a Neoplatonic academy in Alexandria, indicating that pagan learning had not been completely suppressed, but died around 445. After his death, Aedesia moved to Athens with her sons, Ammonius and Heliodorus; subsequently, she sent them to study with Proclus.

Aedesia and her sons returned to Alexandria around 475, where Ammonius took up his father's former post after reaching some sort of arrangement with the Bishop of Alexandria, Peter III Mongus. The nature of the deal with the Bishop is unknown: suggestions range from modifying Neoplatonic doctrine so it supported Christianity to betraying the hiding places of other pagans. Most likely, Ammonius agreed to teach only non-offensive subjects.

Ammonius's continued teaching proved crucial to the history of mathematics, for one of his students was EUTOCIUS (ca. 480–ca. 540). Eutocius wrote commentaries on Archimedes's *On the Sphere and Cylinder*, *Measurement of a Circle*, *On Plane Equilibria*, as well on the first four books of Apollonius's *Conics*. Without these commentaries, our knowledge of the early history of Greek geometry would be much poorer; and in the case of Apollonius's *Conics*, we might not even have the Greek text.

Rome was even more backwards than Alexandria or Athens. By 500, the Italian peninsula was part of the Kingdom of the Ostrogoths under Theodoric the Great. The Kingdom of the Ostrogoths preserved some of the forms of Roman government, such as the consulship, though by then the position was largely symbolic. One consul was ANICIUS MANLIUS SEVERINUS BOETHIUS (ca. 480–524). After a disagreement with Theodoric, Boethius was imprisoned and then executed. While in prison, Boethius wrote *The Consolation of Philosophy*. Though Boethius himself was not a Christian, this work became one of the cornerstones of medieval Christian philosophy.

Boethius also translated many Greek works into Latin, including mathematical texts. However, his translations show the depths to which classical learning had fallen. His version of the *Elements* has been lost, but references to it suggest that it included only the simplest results, and omitted all proofs. Boethius also wrote an account of Nicomachus's *Arithmetic*, giving an account of Pythagorean number theory. These were meant to be part of a series of handbooks on the four mathematical sciences: arithmetic, geometry, music, and astronomy. These four subjects formed the *quadrivium* ("four roads" in Latin), and are roughly analogous to the course of study for a modern baccalaureate degree. Study of the quadrivium assumed a study of the *trivium* ("three roads" in Latin): grammar (Latin), rhetoric, and logic. Because the trivium was perceived to be easier than the quadrivium, the word "trivial" came to be applied to any simple study.[14]

Five years after Boethius's death, the last light of classical learning was put out when the Emperor Justinian closed the schools of pagan learning in Athens and elsewhere in 529. He also passed an edict that banned pagans from teaching, public office, and military service. Learning in the West declined precipitously, which entered a "Dark Age." Fortunately, learning and scholarship existed and thrived elsewhere.

For Further Reading

For the history of Greece and Rome, see [1, 11, 34, 60, 63]. For mathematics and science, see [48, 46, 49, 47, 84].

---

[14]Augustine began a similar project before his conversion to Christianity.

3

China and India

3.1 China

The easternmost of the great river civilizations of ancient times grew up along the banks of the Huang Ho (Yellow River) in China. Like the Nile, the Yellow River is rich in fertilizing silt (hence its name). Like the Tigris and Euphrates rivers, the Yellow River gives rise to disastrous floods, and as a result has received another name: China's sorrow. According to tradition the Emperor Yao employed the engineer Kung Kung to build dikes along the Yellow River to control flooding. Unfortunately the dams merely collected water until they burst, causing even more severe flooding. One tale tells of a flood around 2300 B.C. that lasted for an improbable thirteen years. Since dams failed, the Emperor Yu resorted to the use of building side channels and dredging. This controlled the floods more effectively, and Yu established the Hsia dynasty, which ruled from 2205 B.C. to 1766 B.C.

The very existence of the Hsia dynasty is debatable, though it is certain that, by this time, the Chinese had invented the arts of bronze casting, silk spinning, and writing. Written Chinese consists of thousands of ideograms, which probably originated as pictograms: "month" is a stylized Moon, "to shoot" is an arrow laid across a bow, "west" is indicated by a nesting bird.[1]

The Chinese believed their land to be in between Heaven and Hell; hence they called China the Middle Kingdom. The Emperor ruled with the consent of the gods (particularly Ti, the Supreme Emperor of Heaven), and was said to possess the Mandate of Heaven. So long as he was in heaven's favor, life was good, but earthquakes, floods, and other natural disasters indicated that the Mandate of Heaven had been withdrawn, a signal that the emperor had to be replaced. The inevitability of natural disasters means that rebellions were frequent in Chinese history, though revolutions, where the very nature of the government was changed, were virtually non-existent: it is said that no people are more rebellious and less revolutionary than the Chinese. Indeed, the imperial system was maintained for nearly four thousand years, and it was not abandoned until 1912. If Egypt and Mesopotamia offer antiquity, China offers continuity.

The possibly mythical Hsia were succeeded by the historical Shang (1766–1122 B.C.), who were in turn conquered by the Chou (1122–256 B.C.), barbarians from the north. But

---

[1] There is, by the way, no truth to the story that "trouble" is indicated by a pictogram of two women under one roof.

55

I	II	III	IIII	IIIII	T̄	T̄T̄	T̄T̄T̄	T̄T̄T̄T̄
One	Two	Three	Four	Five	Six	Seven	Eight	Nine

Table 1. Rod Numbers

by the fifth century B.C., the Chou realm was disintegrating, and China entered the Era of
Warring States. The Mandate of Heaven had clearly been withdrawn—but why, and how
might it be restored? Not surprisingly the "Hundred Schools" of Chinese philosophy came
into existence in this time period, all proposing a way out of the chaos. Three important
schools emerged.

The sixth century B.C. philosopher K'ung fu-tzu ("Master K'ung"), better known as
Confucius, held that order could only be restored when people began emulating the great
heroes of the past, such as the Duke of Chou (who helped establish the Chou dynasty
through sage advice to his brother, Wu-wang). Confucianism emphasized family piety and
a return to traditional ways. Legalists held that human beings are inherently selfish and will
never, without compulsion, sacrifice self-interest for the greater good; hence they advocated
strict laws and merciless punishment for transgressors. Taoism, promoted by Lao Tzu, a
contemporary of Confucius, held that happiness could be achieved by aligning one's desires
with the universe's.

3.1.1 Positional Numeration

By the Era of the Warring States, the form of written Chinese numerals had been established. There were two main types. Written Chinese has characters for the numbers one
through nine, as well as other characters representing "ten", "hundred", "thousand", and
so on, as well as "tenth", "hundredth", and so on. Thus 395.25 would be written in the
Chinese characters that meant "3 hundreds 9 tens 5, 2 tenths 5 hundredths." Examples of
this decimal system of numeration can be traced back as far as the later years of the Shang
dynasty.

It was a simple step to letting the position of the symbols alone determine their value.
This step was taken with the rod numerals (see Table 1). These were used in computation on
a counting board (essentially a flat board with ruled lines). Numbers in adjacent positions
were rotated or flipped: thus thirty-three would be ≡ |||, and three hundred three was
||| |||. The purely decimal and purely positional nature of Chinese numeration has given
rise to plausible suggestions that our own system of numeration originated in China, not
India. Indeed the earliest appearance of the zero symbol is from Cambodia, where the
date is inscribed on a monument; the date corresponds to the year 683. Unfortunately our
knowledge of ancient Chinese mathematics is limited because of a deliberate attempt to
wipe out the past.

3.1.2 The Rise of Imperial China

The man responsible was King Cheng of Ch'in, who conquered all of China by 221 B.C.
Cheng took the name Shih Huang Ti ("First Emperor Ti"), and on the advice of an astrologer, changed the name of his empire to Ch'in-a: China. Shih Huang Ti and his chief
minister, Li Ssu were both Legalists, which gave them the philosophical justification for

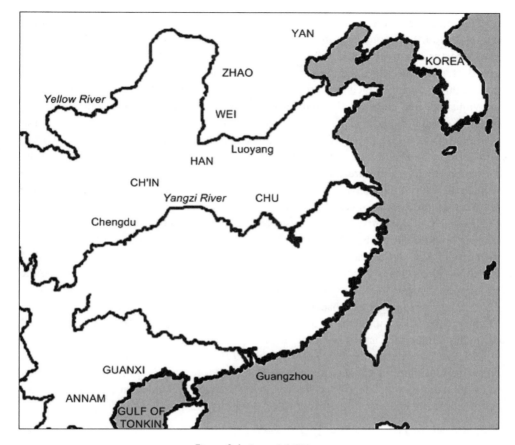

Figure 3.1. Imperial China

imposing some much-needed reforms. The various forms of written Chinese were standardized into a single, official style; forced labor was used to join several large walls across the northern part of China into the Great Wall of China; and two thousand years before the invention of the metric system, Shih Huang Ti instituted a system of weights and measures with each unit ten times the size of the next smaller unit.

The dark side of this passion for order was that the old ways of life had to be swept away. Shih Huang Ti accomplished this by ordering the burning of every book from the past: "History begins with Ch'in." There were a few exceptions: books on medicine, farming, and the chronicles of the Ch'in family were spared. Everything else was to be consigned to the flames.

The empire was to last for ten thousand years, though in fact it lasted only fourteen. Shih Huang Ti died of a stroke on a military campaign in 210 B.C. To keep the provinces from revolting, his death was kept a secret and his body was returned to the capital in a closed palanquin, and buried in great secrecy beneath a mountain with six thousand terra cotta statues of soldiers and a model of his imperial domains.[2]

---

[2]The discovery of the tomb in 1974 plays as central a role in Chinese archaeology as the discovery of Tutankhamen's tomb in Egyptian archaeology.

No amount of secrecy could keep the word of Shih Huang Ti's death from getting out, and soon the empire was in revolt. In 202 B.C., Liu Bang, a former Ch'in official, overcame the last rival to the throne and reunited China. Chinese naming practice is to give the family name first, followed by the personal name, though frequently the individual picks a third name by which they are known. This last practice was almost universal if the individual became emperor; the founder of the dynasty usually created (with the advice of astrologers) a dynastic name as well. Thus Liu Bang took the reign name of Kao Tsu ("Exalted Ancestor") and named his dynasty the Han. Except for a brief interregnum from 9 to 25 A.D., the Han dynasty would last until 221. The Han played such a key role in the formation of Chinese culture that for centuries after the fall of the Han dynasty, the Chinese would still refer to themselves as the "men of Han," and even today the Chinese language is known as the "speech of Han."

One of Kao Tsu's first acts was to seek out scholars who had memorized the ancient classics ordered burned by Shih Huang Ti, and have them recreate the lost classics from memory; in this way, much of Chinese culture was recovered. The subsequent survival of these and other texts is largely due to another Chinese invention that appeared during the Han dynasty: paper. According to tradition, the eunuch Tsai Lun invented paper around 105, though wrapping paper had evidently been used in China in the first century B.C. Writing paper would allow the creation of an inexpensive and relatively permanent record of anything important, and many things trivial.

Scholars also reconstituted the *Computational Prescriptions in Nine Chapters*, a work whose central place in Chinese mathematics is similar to the place of Euclid's *Elements* in Greek mathematics. The *Nine Chapters* covers a variety of mathematical problems with solutions clearly expressed in terms of rules. For example:

Rule 3.1. *To find the area of a circular field: multiply half the circumference by half the diameter. Or multiply the diameter by itself; multiply the result by 3 and divide by 4. Or multiply the circumference by itself, and divide the result by 12.*

Clearly the geometry of the Chinese leaves much to be desired, but in algebra, ancient Chinese mathematicians showed their genius and creativity. In Chapter Seven of the *Nine Chapters* we find the first systematic appearance of the method of double false position:

Problem 3.1. *There is a wall* 90 cun *high. A melon grows from the base of the wall, creeping upward and becoming* 7 cun *taller each day. A gourd grows from the top downward, becoming* 10 cun *longer each day. When do they meet?*

After 5 days, the two are 5 cun *apart, after 6 days, they are 12* cun *past each other. Write down the excess and shortfall, with their corresponding trial numbers:*

$$5 \quad 5$$
$$6 \quad 12$$

Cross multiply and add the products: hence $5 \cdot 12 + 5 \cdot 6 = 90$. *Use this sum as the dividend. Add the excess and shortfall; use it as the divisor:* $5 + 12 = 17$. *Divide the dividend by the divisor; this will be the number of days:* $\frac{90}{17} = 5\frac{5}{17}$.

The principle later known as Cramer's rule appeared in Chapter Seven in the problem:

Problem 3.2. *Some men buy an item together. If each pays 8 coins, there are 3 coins too many; if each pays 7, there are 4 coins too few. How many men are there, and how much is the item?*

The amounts assumed were 8 and 7, which gave an excess of 3 and a shortfall of 4, respectively. Writing these down, we have

$$8 \quad 3$$
$$7 \quad 4$$

Cross multiplying and adding, we have $8 \cdot 4 + 7 \cdot 3 = 53$, which is the dividend. The excess plus the shortfall is $3 + 4 = 7$, the divisor. The difference between the assumed amounts is $8 - 7 = 1$; divide this into the dividend, 53, to get the total price, 53; and divide 1 into the divisor, to get the number of people, 7.

The author of the *Nine Chapters* gives the general rule as:

Rule 3.2. *Put down the assumed amounts, and under them the corresponding excess and shortfall. Cross multiply; add the products and use the sum as the dividend. Add excess and shortfall; it is the divisor.*

From the assumed amounts, find the difference, and divide the dividend by the difference to find the price; divide the divisor by the difference to find the number of people.

Corresponding rules were also given if both assumed amounts resulted in an excess, a shortfall, or if one was exact and the other was an excess or a shortfall.

A precursor of Gaussian elimination appears in Chapter Eight of the *Nine Chapters* under the name "Rectangular Tabulation" (in Chinese, *fang chang*):

Problem 3.3. *Three sheaves of a good harvest, 2 sheaves of a mediocre harvest, and 1 sheaf of a bad harvest yield a profit of 39 tous. Two sheaves of a good harvest, 3 sheaves of a mediocre harvest, and 1 sheaf of a bad harvest, yield a profit of 34 tous. One sheaf of a good harvest, 2 sheaves of a mediocre harvest, and 3 sheaves of a bad harvest yield a profit of 26 tous. How much is the profit from each sheaf of good, mediocre and bad harvest?*

The Chinese scribe set down the sheaves and profits (using the rod numerals and a counting board):

1	2	3	good
2	3	2	mediocre
3	1	1	bad
26	34	39	

Since Chinese is traditionally written from top to bottom and from right to left, it was only natural for the Chinese scribes to allow each column to correspond to an equation, just as we allow each row in a matrix to correspond to an equation. The first few steps in the reduction of this system of equations were to multiply the second and third columns (counting from the right) by 3, the number at the top of the first (rightmost) column:

1	2	3		3	6	3
2	3	2	$\rightarrow$	6	9	2
3	1	1		9	3	1
26	34	39		78	102	39

Then subtract the first column from the third column, and twice the first column from the second column:

$$
\begin{array}{ccc}
3 & 6 & 3 \\
6 & 9 & 2 \\
9 & 3 & 1 \\
78 & 102 & 39
\end{array}
\rightarrow
\begin{array}{ccc}
3 & & 3 \\
4 & 5 & 2 \\
8 & 1 & 1 \\
39 & 24 & 39
\end{array}
$$

Now multiply the leftmost row by 5, the number at the top of the middle row, and subtract 4 times the middle row from the leftmost row:

$$
\begin{array}{ccc}
& 3 & 3 \\
20 & 5 & 2 \\
40 & 1 & 1 \\
195 & 24 & 39
\end{array}
\rightarrow
\begin{array}{ccc}
& 3 & 3 \\
& 5 & 2 \\
36 & 1 & 1 \\
99 & 24 & 39
\end{array}
$$

which is then solved using back-substitution. Occasionally the procedure would result in negative coefficients; the Chinese solved this problem by using black rods to represent negative numbers and red rods to indicate positive numbers; rules for adding and subtracting signed numbers were also included.

The *Nine Chapters* includes 24 problems involving right triangles and the Pythagorean theorem. The two legs of a right triangle were referred to as the *gou* (usually the shorter) and *gu* (usually the longer), while the hypotenuse was the *xian*, so the rule relating the sides was referred to as the *Gougu* rule. Some of the problems involve being given quantities such as the difference between *gou* and *gu*, and require solving a quadratic equation. What is noteworthy is that the rule in the *Nine Chapters* describes how to set up the corresponding equation. According to *Nine Chapters* commentator ZHAO SHUANG (fl. 250):

Rule 3.3. *Subtract the square of the difference from the square of the hypotenuse. Halve the remainder [to form the constant term]; the difference is the* congfa *[linear coefficient]. Take the square root to find the* gou; *add the* gou *to the difference to find the* gu.

In short, given $b - a$ and c, then a is the root of $x^2 + (b - a)x = \frac{c^2 - (b-a)^2}{2}$, and $b = a + (b - a)$.

The direction to "take the square root" is key to the Chinese method of solving quadratics (and higher degree equations). In modern terms, the Chinese method for solving the quadratic (and similar equations) is equivalent to the following. Say we wish to solve the equation $x^2 + 5x = 1400$. By inspection we can see that the root must be a two-digit number; let $x = 10y$ to form the new equation $100y^2 + 50y = 1400$. By inspection we see that $3 < y < 4$, so let $y = 3 + z$ and substitute to obtain the equation $100z^2 + 650z = 350$. Since $0 < z < 1$, we may approximate the solution to $100z^2 + 650z = 350$ as $z \approx 350/650$. Let $z = v/10$ or $v = 10z$, which will give us the equation $v^2 + 65v = 350$, which we can solve exactly with $v = 5$. Hence $z = 5/10$, $y = 3\ 1/2$, and $x = 35$.

To solve the equation $x^2 + 5x = 1400$ on the counting board, Chinese mathematicians would set down the *shi* ("dividend"), 1400; the *fa* ("divisor") 5; and the *jiesuan* ("borrowed rod") 1; the last represents the coefficient of x:

Root, *Yi*	
Dividend, *Shi*	1400
Divisor, *Fa*	5
Borrowed Rod, *Jiesuan*	1

Reading from the bottom we have the equation $x^2 + 5x = 1400$. Chinese mathematicians were instructed to "reduce the root" until it was between 1 and 10: in this case, the root would have to be reduced tenfold. Then the *fa* would be increased by a factor of 10 and the *jiesuan* by a factor of $10^2 = 100$:

Root, *Yi*		
Dividend, *Shi*	1400	1400
Divisor, *Fa*	5	50
Borrowed Rod, *Jiesuan*	1	100

Reading from the bottom we have the equation $100y^2 + 50y = 1400$, where $x = 10y$. By inspection we see that the root is between 3 and 4 (and hence the root of the original equation is between 30 and 40). Thus we let our partial root (*yi*) be 3. To obtain the new *fa*, multiply the *yi* by 2 and the *jiesuan*, then add it to the original *fa*. In this case, the new *fa* would be $3 \cdot 2 \cdot 100 + 50 = 650$. To obtain the new *shi*, multiply the *yi* by the *jiesuan* and add it to the *fa*; multiply this by the *yi*; then subtract this from the *shi*. Thus the new *shi* is $1400 - (3 \cdot 100 + 50) \cdot 3 = 350$. This gives us the table:

Root, *Yi*		3	3
Dividend, *Shi*	1400	1400	350
Divisor, *Fa*	5	50	650
Borrowed Rod, *Jiesuan*	1	100	100

Reading from the bottom we have the equation $100z^2 + 650z = 350$. At this, the second stage in our process, the reason for using the term *fa* ("divisor") becomes evident: the root is $z \approx 350/650$, so it must be increased by a factor of 10 to be between 1 and 10. This means the *fa* is divided by a factor of 10 and the *jiesuan* by a factor of $10^2 = 100$:

Root, *Yi*		3	3	3
Dividend, *Shi*	1400	1400	350	350
Divisor, *Fa*	5	50	650	65
Borrowed Rod, *Jiesuan*	1	100	100	1

This gives us the equation $v^2 + 65v = 350$. The approximation for the root is $v \approx 350/65$, and we find that $v = 5$ works exactly. Hence the root of the original equation is 35.

There are several noteworthy features about this method, rediscovered a thousand years later in the West and known as Horner's Method because it was described in 1819 by WILLIAM GEORGE HORNER (1786–September 22, 1837).[3] First, it is evident that one can use this method to find a root as accurately as one desires; the only difference is whether or not the process terminates. In fact, since most rational numbers have non-terminating decimal expansion, the difference between rational and irrational never played an important role in Chinese mathematics. Second, by the thirteenth century, the method had been extended to solving equations of any degree.

---

[3] Horner was not even the first European to describe the method: Both Newton and Lagrange presented equivalent algorithms.

3.1.3   Partition and Unification

Around 190, a poor crop in Szechuan led to a peasant uprising, which turned into a general revolt against the Han. Tung Cho, appointed to suppress the revolt, instead deposed the Emperor and installed his own candidate. Tung Cho was assassinated in 192, which restored the Emperor Hsien. However, the real power was in the hands of the generals charged with suppressing the rebellion, none more notorious than Ts'ao Ts'ao. After his death in 220, his son Ts'ao P'i deposed the emperor and established the Kingdom of Wei in northern China. Other warlords established their own kingdoms, though Wei in the north, Wu in the south, and Shu in the west would be the dominant ones. This period is known as the Era of the Three Kingdoms and is a popular period for Chinese adventure novels, with Ts'ao Ts'ao usually playing the role of a villain.

Two of the main commentators on the *Nine Chapters* lived in this time period. Zhao Shuang lived in the Kingdom of Wu, on the shores of the China Sea. Wu traders, venturing southwards, encountered a remarkable drink: tea. Over the centuries, this would have a profound effect on Asian culture, because water boiled for tea was less likely to contain dangerous microorganisms. City-dwellers benefited the most: in a medieval European city like London, the river Thames served both as a supply of drinking water and as a waste disposal system.

LIU HUI (fl. 250) lived in the Kingdom of Wei. In his commentary on the *Nine Chapters*, Liu Hui found approximate areas for regular polygons with 6, 12, 24, 48, 96, and 192 sides inscribed in a circle with a diameter of 2 *chi* (which is 20 *cun*). For the last, he gave an area of $314 \frac{64}{625}$ square *cun*, and noted the area of a circle is greater than $314 \frac{64}{625}$ square *cun*. and less than $314 \frac{169}{625}$ square *cun*. This corresponds to $3.141 < \pi < 3.142$.

Liu Hui's most influential work was his *Sea Island Mathematical Manual*, consisting of nine surveying problems. The first problem is:

Problem 3.4. *There are two poles* 5 bu *high and* 1000 bu *apart. Viewed from ground level* 123 bu *behind the front pole, the top of a sea island coincides with the top of the pole. Viewed from ground level* 127 bu *behind the rear pole, the top of the island coincides with the top of the pole. Find the height of the island and the top of the pole.*

Today we might use similar triangles to solve these problems, but Liu's solution is as follows: Multiply the distance between the poles by the height of the poles, then divide this by the difference in the distances between each pole and the sighting point (i.e., $127 - 123$). Add this to the height of the pole to obtain the height of the island.

This is actually a clever application of geometric algebra as follows. In Figure 3.2, let PQ represent the height of the sea island and let the poles be situated at R, T, and let S be a point 123 *bu* behind the first pole, and V a point 127 *bu* behind the second; hence $RS = 123$ and $TV = 127$. It is easy to show (and known since at least the time of Euclid) that the rectangular regions A are equal in area; let $TU = RS$ to obtain rectangle B, which has the same area as A. Again, it is easy to show that the areas $A + C + D = B + F$; consequently $C + D = F$. But $C + D$ is a rectangle whose area is the distance between the poles, multiplied by the height of the poles; consequently the area of F is 5000 square *bu*. Since $TU = RS = 123$ and $TV = 127$, then $UV = 4$. Thus rectangle F must have a height of $5000/4 = 1250$ *bu*; thus $PW = 1250$ and $PQ = 1255$ bu. The distance can be found in a similar fashion, and we need not resort to the use of similar triangles.

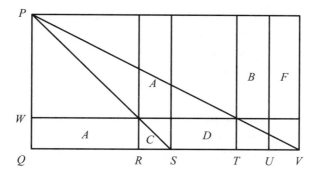

Figure 3.2. Out-In Method

The era of the Three Kingdoms ended with a de facto partition of China into numerous independent states. The next few centuries saw vigorous warlords founding new dynasties (often centered around the southern city of Nanking) and unifying large areas, while their less competent successors would invite rival claimants and civil war would shatter the unified state. After a period of anarchy the cycle would repeat. Because of the periods of chaos, much of the detailed history of the era has been lost; because of the periods of order, most of the remainder was rewritten, and many legends (possibly based on actual people) were created. For example, an anonymous poet of the fifth or sixth century wrote *Mu Lan Shi* ("The Ballad of Orchid") about a girl whose father is about to be conscripted into the imperial army; she disguises herself as a man, fights in his place, gains great honor and prestige, and returns home where she puts on women's clothes again to the amazement of her comrades (who had been fighting alongside her for twelve years).

SUN ZI (400–460?) is somewhat less legendary, and we know almost nothing about him. His main work is known as the *Sunzi Suanjing*; "Sunzi" simply means "Master Sun," so the book is literally "Master Sun's Mathematics Manual." There is another Master Sun in Chinese history: the author of *The Art of War*, written around 500 B.C. In the eighteenth century Chinese mathematicians believed that the two Master Suns were the same, but references to certain types of taxes in the *Sunzi Suanjing* suggest it was written between 280 and 474.

Sun Zi's most important contribution appeared in Problem 26 of the third chapter of the *Sunzi Suanjing*, where he solved the problem:

Problem 3.5. *Suppose we have an unknown number of objects. If counted by threes, two remain; if counted by fives, three remain; if counted by sevens, two remain. How many objects are there?*

This is the earliest example of the Chinese remainder theorem, and Sun Zi's solution is essentially our modern method, though without explanation. First, he finds the numbers 70 (presumably because $70 \equiv 1 \mod 3 \equiv 0 \mod 5 \equiv 0 \mod 7$), 21 (again presumably because $21 \equiv 1 \mod 5 \equiv 0 \mod 3 \equiv 0 \mod 7$) and 15 (where, again, $15 \equiv 1 \mod 7 \equiv 0 \mod 5 \equiv 0 \mod 3$). Since $223 = 2 \cdot 70 + 3 \cdot 21 + 2 \cdot 15$, then 223 is a solution. Because $105 = 3 \cdot 5 \cdot 7$, we can add multiples of 105 to obtain additional solutions.

By 624, most of China was unified under the rule of the Li family. Li Yüan designated his eldest son the heir apparent. The middle son, Li Shih-min, took offense, ambushed and killed his brothers, and forced his father to abdicate. Thus began the T'ang dynasty, which reunified China under the imperial banner, and attempted to expand east towards Korea and west towards the Turks. Neither venture succeeded very well, and in 751 at the Talas River (near Tashkent), the Turks defeated the Chinese. Some of the captured prisoners included conscripted papermakers, and soon after a flourishing papermaking industry appeared in Samarkand.

During the T'ang dynasty a type of civil service exam system became an integral part of Chinese civilization, institutionalizing a practice used haphazardly by the Han and later dynasties. By T'ang times, candidates could prepare for these exams at one of many schools or even the Imperial University. While the exams placed great importance on orthodox interpretation of Confucian classics, they tested practical knowledge as well, particularly mathematics.

For example, in 656 the Imperial University set a curriculum that included the *Sea Island Mathematical Manual*; the importance of this work can be judged by the fact that three years were to be devoted to the study of its nine problems. In theory, the system favored no one: candidates were anonymous, and to prevent identification of a candidate by their handwriting, an exam was recopied before being handed to two officials who graded it independently, while a third official reconciled the two grades and gave the final score. To make the system even more equitable, LI CHUNFENG (602–670), the director of the Imperial Astronomy Bureau, undertook the task of compiling, editing, and annotating twelve canonical texts (ten of which still exist), which gained the appellation *suanjing* (roughly, "mathematical classic").

One such text was the *Suan-ching* ("Arithmetical Classic") of CHANG CH'IU-CHIEN (fl. 6th cent.?), about whom nothing is known for certain. His *Arithmetical Classic* provides one of the earliest examples of a linear indeterminate equation:

Problem 3.6. *A rooster costs* 5 *coins, a hen* 3 *coins, and three chickens cost* 1 *coin. If we buy* 100 *birds with* 100 *coins, how many of each will we have?*

Chang gave one solution (4 roosters, 18 hens, and 78 chickens) and a general rule for finding other solutions (though not a general rule for solving linear indeterminate equations).

Another mathematician of the era, WANG HS'IAO-T'UNG (fl. 7th cent.), wrote a work that includes several problems that result in cubic equations. One problem is:

Problem 3.7. *The product of the lengths of two sides of a right triangle is* 706 $\frac{1}{50}$*, and its hypotenuse is greater than one side by* 30 $\frac{9}{60}$*. Find the three sides.*

Wang Hs'iao-t'ung gave no indication of how this problem was to be solved, but presumably it was by a variation of the method used to solve quadratic equations.

Another period of civil war destroyed the T'ang and produced several short-lived successor states. China was finally reunited under the Sung dynasty in 960, with its capital at K'ai-feng in the north (the so-called Northern Sung). The peace and prosperity in the early years of the dynasty had important side effects. Iron production expanded so rapidly that trees could not be turned into charcoal fast enough to meet the demand; consequently coal came to be used as a substitute. Even more significantly, commercial activity grew so rapidly that a coin shortage developed.

Fortunately one of the new industries was printing (possibly invented in Korea), and the Sung began the systematic use of paper money around 995. Printing also makes it possible to produce playing cards, which begin to appear around this time. The value of printing is easily seen in the fate of the twelve mathematical classics annotated by Li Chunfeng. In 1084, the ten surviving classics were printed for the first time, and though no copies of this original printing have survived, the ten classics themselves have survived thanks to numerous reprintings since the eleventh century, while the two lost classics have never been recovered.

Details of the manufacture of black powder also appeared during the Northern Sung dynasty; by 1100 the Sung arsenal included rockets, bombs, and primitive guns. Unfortunately this technological edge did not stop an invasion from the north by Jurchen tribesmen. In 1127 they took the capital K'ai-feng, captured the emperor and royal court, and established themselves as the Chin dynasty.

Li Yu was the secretary to a Jurchen officer who eventually settled in Ta-hsing (modern Peking), where his son LI ZHI (1192–1279) was born. Since "Zhi" (the boy's given name) is the same as the name of the third T'ang emperor, Li Zhi changed his name later in life to Li Ye, the name he is sometimes known by; we will use Li Zhi when necessary to distinguish him from other Lis. Li took the government exams and did well enough to be chosen to fill several official positions. Unfortunately he lost all his positions, not through any fault of his own, but because of a new group of invaders: the Mongols.

Temüjin was born into the royal Borjigin family of the Mongols. At the age of nine, his father was poisoned as a result of a longstanding clan feud, and he and his mother were abandoned. From this inauspicious beginning he collected followers, defeated and exterminated his enemies, and rose in power; by 1206, he had united all the previously warring tribes of Mongolia and was acclaimed *Genghiz Khan* ("Universal ruler"). Soon he would rule over the greatest land empire in world history.

To the south lay the ultimate prize: China. But the conquest of China would have to wait, for a grave insult offered by the governor of the Khwarezm led to an invasion and devastation of the Khwarezm, Transoxania and Khorāsān. Before he could turn back to China, Genghiz Khan died suddenly in 1227. His son and successor Ögödei conquered the Chin Empire by 1234, depriving Li Zhi of his job and driving him into seclusion for many years afterwards. He completed the *Ce Yuan Hai Jing* ("Sea Mirror of Circle Measurements") around 1248.

Most of the problems in the *Sea Mirror* are set around a circular walled town. For example:

Problem 3.8. *A person begins at the west gate of a circular walled town and walks south 480 paces. A second person leaves the east gate and walks east for 16 paces, at which point he just sees the first person. What is the diameter of the town?*

The problem gives rise to a fourth-degree equation with one positive real solution (240 paces). Li Zhi solved the problem without explaining his procedure; again, we can infer the existence of some variation of the method used for solving quadratic equations.

An interesting feature is Li Zhi's use of the "method of the celestial unknown," where the coefficients of a polynomial are recorded in descending order on a counting board; not only are negative coefficients allowed, so are terms of negative degree. For example, the expression $x^3 + 4x^2 + 15x + 45$ would be expressed on a counting board with the rod numerals equal to 45, 15, 4, and 1 placed vertically, and a special marker (called the *yuan*, "celestial unknown") set next to the 15, and another (called the *tai*, "excess") next to the 45. Recording the terms in this fashion emphasizes the similarity between operations on polynomials and operations on whole numbers, a key step in the arithmetization of algebra.

Meanwhile, the Southern Sung dynasty established itself in the south, with a new capital at Hang-chou. While the loss of the north to the Chin and the arrival of the Mongols was a serious blow, it was hardly a fatal one. Indeed, the remaining Sung lands included fertile agricultural lands, and the fact that Hang-chou was a port city helped turn the Sung into a maritime power. Sung trade routes expanded east towards Japan, south towards the Indonesian archipelago and the Philippine islands, and west to India, Syria, and Africa. The change from coastal trade routes, which required little navigational equipment, to transoceanic voyages caused the magnetic compass (known since Han times) to become standard shipboard equipment.

Around 1224 QIN JIUSHAO (1202–1261) and his father, a minor government official, moved to Hang-chou. Qin Jiushao gained proficiency in mathematics, poetry, fencing, archery, riding, and architecture, and did well enough on the state exams to be assigned many important positions. He built a fortune on bribes and kickbacks. For a time he served on the frontier as an army commander, facing the Mongols under Ögödei, but Ögödei's death in 1241 again postponed the Mongol conquest of the Middle Kingdom.

In between acquiring ill-gotten wealth, Qin wrote *Shushu Jiuzhang* (1247), "Mathematical Treatise in Nine Sections." In the first section he presented a remarkable generalization of the Chinese remainder theorem, which allowed for congruences to be solved even if the moduli were not relatively prime; Qin's method (which he claims he learned from calendar experts, though they did not understand the theoretical basis behind it) would be independently invented in the west nearly six hundred years later by Gauss and Legendre.

As we have seen, cubic and quartic equations appear in the works of Qin's predecessors. Qin himself was the first to explain the procedure; in his *Mathematical Treatise* he derives the equation $-x^4 + 763200x^2 - 40642560000 = 0$ and solves it, again using a procedure similar to Horner's method, but predating Horner's publication by several centuries. Qin gave the first clear explanation of how the method works.

The same method of solving higher degree equations is explained in the *T'ien Mou Pi Lei Ch'êng Ch'u Chieh Fa* (1275), "Practical Rules of Arithmetic for Surveying" by YANG HUI (fl. 1261–1275). What is noteworthy is that Yang Hui credits the method to Liu Yi, about whom nothing is known, and does not refer to Qin Jiushao or Li Zhi, nor does he use the "celestial element" terminology. This suggests that he learned his equation solving from a school of Chinese mathematics independent of the tradition represented by Li and Qin.

To understand the method, let us work with the simpler equation $x^4 = 100$. On the counting board, we would arrange the constant and coefficients vertically, with the coefficient of the highest degree term at the bottom, though for convenience we will write the terms horizontally (with the highest degree term on the left):

| 1 | 0 | 0 | 0 | 100 |

Trial and error tells us the root is between 3 and 4, so the first digit of the root is 3. Place this in the first column, then multiply 1 by 3 and add it to the next column; multiply the sum by 3 and add it to the next column, and so on, continuing up to the last column where the product is subtracted from the constant:

3	1	0	0	0	100
		+ 3	+ 9	+ 27	− 81
		3	9	27	19

Next, repeat this process, stopping at the next to last column:

3	1	0	0	0	100
		+ 3	+ 9	+ 27	− 81
		3	9	27	19
		+ 3	+ 18	+ 81	
		6	27	108	

The procedure is repeated several more times, each time stopping one column short of where the previous application ended:

3	1	0	0	0	100
		+ 3	+ 9	+ 27	− 81
		3	9	27	19
		+ 3	+ 18	+ 81	
		6	27	108	
		+ 3	+ 27		
		9	54		
		+ 3			
		12			

This sequence of operations has transformed our equation as follows: From $x^4 = 100$, we let $x = 3 + y$. Then $(3 + y)^4 = 100$ which, upon expansion and collection of the constant terms on the right side, gives us $y^4 + 12y^3 + 54y^2 + 108y = 19$. At this point, we can scale the root (as we did with quadratic equations) to obtain a new equation, which we would represent on the counting board as:

	1	120	5400	108000	190000

As before, an approximate solution to this new equation will be $190000/108000$, which we round down to 1. Applying the procedure again gives us:

1	1	120	5400	108000	190000
		+ 1	+ 121	+ 5521	− 113521
		121	5521	113521	76479
		+ 1	+ 122	+ 5643	
		122	5643	119164	
		+ 1	+ 123		
		123	5766		
		+ 1			
		124			

Thus the root is approximately 3.1. Clearly we can repeat this process as many times as necessary to find additional places of the root. Moreover, from the second step onwards, we are actually solving a polynomial equation; consequently, this procedure actually allows us to find numerical solutions to polynomial equations of any degree.

3.1.4 The Mongols

In 1264 Genghiz Khan's grandson Kublai overthrew the last of his brothers and established himself as the Great Khan. Kublai turned his attention towards China, and in 1276 the Mongols captured Hang-chou. By 1279 the last of the Sung dynasty had been eliminated, and Kublai fulfilled Genghiz's dream of conquering China, ruling the Middle Kingdom as the first Emperor of the Yüan dynasty.

Kublai brought China the first real stability it had had in a millennium. Indeed, though the fortunes of the Middle Kingdom have risen and fallen since the thirteenth century, the boundaries of modern China are essentially those created by Kublai Khan. The economic revival following the invasions led to a flowering of literature, art, and drama. Kublai himself tried to recruit Li Zhi for the Imperial University, though the aged scholar (by then in his seventies) declined.

The fabulous empire of Kublai Khan was described in 1298 by the merchant Marco Polo in his *Travels*. Polo's contemporaries disbelieved many of the things described in the book, such as paper money, coal, and cities that dwarfed any in Europe. There are some peculiarities: Polo nowhere mentions tea, printing, or chopsticks, nor does Polo show any familiarity with the Chinese language. No Chinese sources mention Polo, despite his claim to have risen to a high position within the government. These anomalies have led some

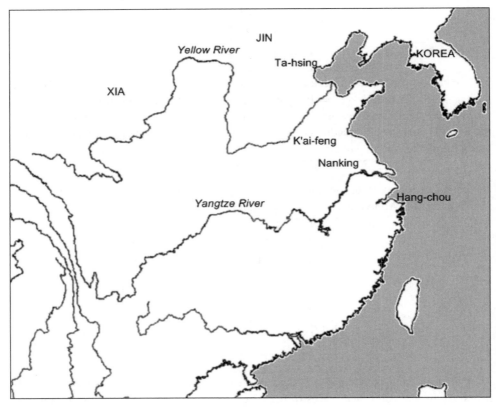

Figure 3.3. China on the Eve of the Mongol Invasions

scholars to suggest that Marco Polo only made it as far as the Black Sea, and pieced his *Travels* together from Islamic guidebooks.

Marco Polo also omitted any mention of the greatest mathematician of medieval China, though perhaps a merchant traveler would have remained unaware of ZHU SHIJIE (fl. 1300). Almost nothing is known about his life, though he apparently used the opportunity presented by a unified China to travel throughout the country, teaching everywhere he went for twenty years; it is interesting to note that the travels of Zhu Shijie and (if he made them) Marco Polo would have been roughly contemporaneous.

In *Suanxue Qimeng* ("Introduction to Mathematical Studies," 1299) and *Siyuan Yujian* ("Precious Mirror of the Four Elements," 1303) we see Chinese mathematics at its height. The first is largely an annotation of the *Nine Chapters*, though Zhu includes the method of the celestial unknown. The second is a remarkable work, and may have been the most sophisticated mathematics text in the world at the time. *Precious Mirror* opens with what he calls the "diagram for the eighth and lower powers," but which is obviously the first eight rows of the arithmetic triangle; he notes that the diagram is a very old one, not original with him.

Zhu extends the method of the celestial unknown so it can handle up to four variables (the "four elements" of the title, named heaven, earth, man, and thing). Remember that the method of the celestial element separated the terms of an equation by their degree; the four-element method of Zhu used the same idea, with the terms of higher degree indicated by moving outward from a central point. For example, one problem Zhu considered was:

Problem 3.9. *Consider a right triangle with an area of* 30 *and the sum of the two sides is* 17. *Find the sum of one side and the hypotenuse.*

If we let the sum of one side and the hypotenuse be represented by the (deliberately suggestive) variable N, and the side itself to be represented by E, then the hypotenuse is $N - E$ and can be represented on the counting board as

		1		
	×	−1		

where × indicates the center of the arrangement (where the constant term, if any, would be located). The counters are to be manipulated to obtain an equation in a single variable, though Zhu does not give detailed instructions for how this is to be effected.

Chinese mathematics reached a high level of sophistication in Zhu Shijie's *Precious Mirror*, but after Zhu Shijie, it stagnated for some centuries. There are no clear reasons, but a comparison with the history of Indian mathematics may prove illuminating. Hence we turn to the subcontinent.

3.2 India

India is separated from Asia by the Himalayas, the highest mountains in the world. Because of this imposing physical barrier, the flora and fauna of India are distinctly different from that of the rest of Asia; consequently India is usually classified as a subcontinent.

Agriculture in India began along the banks of the Indus River in northwest India around 3000 B.C. By 2500 B.C. a flourishing civilization existed, called either Indus Valley or Harappan, after its capital at Harappa. Unlike Mesopotamian or Egyptian civilizations, the existence of Harappan culture was unknown until 1922. At Mohenjo-Daro, excavations reveal a great city with a public bath, granary, and underground sewers. The streets are laid out in a grid pattern, and at its height the city may have had a population of 35,000. But by 1600 B.C., Mohenjo-Daro had been abandoned and Harappan civilization disappeared.

The invasion by the Aryans, a group of tribes speaking an Indo-European language, was once blamed for the fall of Harappan civilization, but more recent evidence suggests that the Aryans were, at worst, the final blow to a civilization already in decline. Most of what we know of Aryan culture comes from literary sources. Portions of the Mahābhārata (named after the Bharata, one of the Aryan tribes and now commemorated in the official Sanskrit name for India) date to 1000 B.C., and describe the struggle for succession among members of the royal family.[4] More important are the collections of hymns and stories known as the *Vedas* ("Books of Knowledge"), written in Sanskrit.

3.2.1 The Sulbasutras

The Vedas and their commentaries, known as the Brahmanas, Aranyakas, or Upanishads, depending on their authorship, include extensive descriptions of Aryan ritual. One of the key tasks is the precise construction of an altar, which is addressed by the *Sulbasutras* ("Cord rules"). All we know of mathematics of the Vedic age comes from the *Sulbasutras*. Hence we know nothing about BAUDHĀYANA (fl. 800 B.C.?), MĀNAVA (fl. 700 B.C.?), or ĀPASTAMBA (fl. 600 B.C.?), except their approximate dates and their authorship of a *Sulbasutra*. Their work suggests they were artisans who used mathematics, and not mathematicians who constructed altars. The main problems solved by the *Sulbasutras* are:

1. Constructing a square with area equal to the sum of the areas of two other squares,

2. Constructing a square equal in area to the difference in area between two squares,

3. Constructing a rectangle equal in area to a given square [given one side of the rectangle],

4. Constructing a square equal in area to a given rectangle,

5. Reducing a given square to a smaller square.

All these problems are solved by application of the Pythagorean Theorem, more than two hundred years before Pythagoras was born. Because Pythagoras spent some time in the Persian Empire, which bordered on India, it is plausible (but unproven) that Pythagorean mathematics had Vedic roots.

Baudhāyana was the first to make use of the converse of the Pythagorean Theorem, and gives the following directions for construction of a square altar. Begin with a cord twice as long as the side of the desired square, and mark out one side of the square in an east-west

---

[4]When transliterating Indian names, a bar over a vowel lengthens it, while a dot below a t, d, or n indicates a very minor change in the pronunciation.

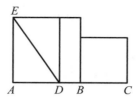

Figure 3.4. Baudhāyana's Sum of Two Squares

line.[5] Mark the midpoint of the cord; on half the cord, make a mark at the quarter-length
mark (from the midpoint). Tie the cord to the endpoints of the east-west line, take hold of
the mark, and stretch it towards the south to form one corner of the square. If the original
cord had length 8, then marking the cord as indicated would divide it into two parts of length
3 and 5; stretching the cord to the south will produce the hypotenuse and shorter leg of a
$3 - 4 - 5$ right triangle. This construction is especially elegant, since all of the markings
can be made by successive halvings; Baudhāyana gives an alternative construction that
requires dividing a segment into six parts that results in the construction of a $5 - 12 - 13$
right triangle.

To construct a square equal in area to the combined areas of two given squares, Baudhāyana does the following: If AB, BC are the bases of the two squares, mark off AD
equal to BC; the diagonal ED will be the side of the desired square (see Figure 3.4).
Baudhāyana notes that in general, the diagonal of a rectangle produces a square equal in
area to the combined areas of the squares on the two sides, and draws special attention to
rectangles with sides 3 and 4; 5 and 12; 8 and 15; 7 and 24; 12 and 35; and 15 and 36.

Using these methods we can construct, geometrically, the side of a square with twice
the area of a given square, but some *Sulbasutra* authors also discussed the *saviśeṣa*, the
calculated length of a diagonal such as ED in Figure 3.4. To obtain the *saviśeṣa* of a
square, both Baudhāyana and Āpastamba note that one should increase it by its third, a
fourth of its third, less a thirty-fourth of this last. In other words

$$d = a\left(1 + \frac{1}{3} + \frac{1}{3 \cdot 4} - \frac{1}{3 \cdot 4 \cdot 34}\right)$$

which can be interpreted as the approximation $\sqrt{2} \approx 1.4142157$. It is not known how the
Sulbasutra authors derived this approximation, though an intriguing possibility emerges
from the continued fraction expression for $\sqrt{2}$:

$$\sqrt{2} = 1 + \cfrac{1}{2 + \cfrac{1}{2 + \cfrac{1}{2 + \ddots}}}$$

---

[5]Because the altars had a specific orientation, the authors of the *Sulbasutras* often made use of the cardinal
compass directions.

For this expansion, the convergents (analogous to the partial sum of an infinite series) are:

$$a_1 = 1 + \frac{1}{2} \qquad a_2 = 1 + \cfrac{1}{2 + \cfrac{1}{2}} \qquad a_3 = 1 + \cfrac{1}{2 + \cfrac{1}{2 + \cfrac{1}{2}}} \qquad \cdots$$

The seventh convergent is equal to $1 + \frac{1}{3} + \frac{1}{3 \cdot 4} - \frac{1}{3 \cdot 4 \cdot 34}$.

To construct a square equal in area to a given circle, Baudhāyana used a method equivalent to finding the side of the square s from the diameter of the circle d via the relationship

$$s = d \left(\frac{7}{8} + \frac{1}{8 \cdot 29} - \frac{1}{8 \cdot 29 \cdot 6} + \frac{1}{8 \cdot 29 \cdot 6 \cdot 8} \right).$$

From this we may determine $\pi \approx 3.09$. As this method is rather complicated (requiring division of the diameter into eight parts, then one of those parts into twenty-nine parts, then one of those into six parts, and one of the last into eight parts), Baudhāyana and other authors also gave $\frac{13}{15}$ as an approximation to the actual square. It is unclear whether they realized that the first method also gave an approximation, though a better one.

3.2.2   Early India

The Vedas also guided the formation of the caste system. By the sixth century B.C. most members of society fell into one of four basic groups (properly called *varnas*): in descending rank they were the *brahmans* (priests); *kshatriya* (warriors); *vaishya* (merchants and landholding farmers); and *shudra* (peasants). Below the shudras were the outcastes or untouchables who were essentially in society but not part of it. Each of the main castes was subdivided, and by one estimate some 3000 castes and subcastes existed.

The caste system defies easy explanation. Very roughly speaking, the caste system provides a place for each individual; one's caste was hereditary, though the Vedas suggest that changes in caste occurred regularly. Each caste took responsibility for the welfare of its members, and in theory could exert considerable pressure through collective action: for example, if a member of one caste felt he was cheated by a merchant, all members of the caste might boycott that merchant.

The caste system may be responsible for making the history of India very different from that of China, Islam, Europe, or the ancient River Civilizations. In the histories of other lands, we see a tendency towards unification. Religions grow from insignificant philosophies to faiths that embrace an empire. Tribes that acknowledge a common ancestor unite under a single leader to form the nucleus of a great nation. Languages spread until large areas have a common tongue. Empires fall, but later generations seek to restore the unity they represented. Ultimately, coexistence occurs through assimilation. But in India, coexistence occurs through accretion: the parts are still identifiable, and never submerge into a common whole.

Nowhere is this more obvious than in the dominant religion of India, Hinduism, probably based on a primitive polytheism that predates the Aryan invasions. Christianity, Islam, and Judaism are all monotheistic religions: a single God for all people. Hindus acknowledge the existence of thousands or millions of gods, each appropriate for his or her followers, and each theoretically autonomous. Because of its implicit support of the caste system, Hinduism gained the support of the *brahmans*, and became the dominant religion of the subcontinent.

However, there were challenges to Hinduism. By default the secular rulers of India were *kshatriya*, inferior in principle to the *brahmans*. Thus it is perhaps no coincidence that the two greatest challenges to Hinduism came from religions founded by princes. Gautama came from the royal family of Sakya in northern India. Around 527 B.C. he came to the conclusion that suffering in this world was the result of desire. For example, most people desire happiness, but suffer when they do not achieve it. Consequently, Gautama argued, to eliminate suffering, one must eliminate desire. Gautama came to be known as Buddha ("enlightened one"), and Buddhism spread rapidly through India and beyond.

Around the same time Vardhamana, also born into a noble family, took up the life of a wandering ascetic. He became known as Mahāvīra ("Great Hero") and played a formative role in Jainism. Mahāvīra's key doctrine is that the ultimate goal of life should be to cleanse the soul (*jiva*) of impurities; the fastest way to contaminate the soul is by the killing of other living things. Jains practice nonviolence (*ahimsa*) in thought and deed and, of course, vegetarianism; in addition, since tilling fields might inadvertently kill living creatures in the soil, most Jains pursued commercial activities such as moneylending, becoming wealthy in the process.

An important component of Jain religion is the division of time, regarded as infinite, into *kalchakras* (also infinite); each *kalchakra* is divided into two equal halves, the *utsarpini* and the *avarsapini*; each of these is divided into six unequal (but finite) periods. Roughly speaking, during the *utsarpini* the condition of man gradually improves towards enlightenment, while during the *avarsapini*, man falls towards depravity. According to the Jains we are in the fifth period of the current *avarsapini*, expected to last about 21,000 years.

This viewpoint encouraged the Jains to consider enormously large numbers. They gave names to each of the decimal places (which suggests the use of positional notation), and around the first century B.C. the *Anuyogadwara Sutra* described the population of the world as being equal to a number requiring 29 places to describe, namely the product obtained by multiplying the "sixth square" by the "fifth square." Like the Greeks, the Jains considered 2 to be the first actual number; the sixth square is the result of squaring 2 six times (i.e., 2^{64}) and the fifth square is 2^{32}. Hence the Jains believe the population of the world to be $2^{64} \times 2^{32}$, which is indeed a 29 digit number.

Indian mathematicians also investigated combinatorics. In the sixth century B.C., the physician Susruta noted that there were 63 possible combinations of the six tastes (bitter, sour, salty, astringent, sweet, and hot); the number is small enough so that they could have been found by direct enumeration. Other writers during the the Vedic era considered the number of ways that poetic meter could be varied. By 300 B.C., the Jains identified the theory of permutations, which they called *vikalpa*, as a separate branch of mathematics.

Both Jains and Buddhists figure prominently in the rise of the first great Indian empire, the Mauryans. The founder of the empire, Chandragupta may have fought alongside Alexander the Great, and unified much of India before abdicating in 301 B.C. to become a Jain monk. Very little is known about the reign of his son and successor Bindusara, though he expanded the empire to the south and sent a request to Antiochus I (Seleucus's successor) for Greek wine, figs, and a Sophist. Antiochus sent the wine and figs, but patiently explained that there was no market for philosophers. Bindusura's son Ashoka was the greatest of the Mauryans. Ashoka completed the conquest of the south (leaving only the southernmost tip of India and Sri Lanka free of Mauryan rule); the bloody conquest of the last independent tribal group, the Kalinga, apparently caused him to rethink his policy of aggression, and he became Buddhist in the tenth year of his reign.

After Ashoka's death the Mauryan empire decayed rapidly, and vanished entirely by 184 B.C. No single power emerged to take its place for five centuries. Meanwhile Buddhism, Jainism, and Hinduism reached their mature forms. Stories about Krishna, the most popular of the Hindu gods, began to appear in the *Mahābhārata* around 200 B.C., where he heartens the hero Arjuna with the dialog subsequently known as the *Bhagavad Gita*, "Song of the Blessed One." The popularity of Buddhism led to its inclusion within Hinduism, for Buddha (along with Krishna) is credited with being the most recent *avatara* (roughly "earthly manifestation") of Vishnu, one of the chief Hindu gods. The *Manava Dharmashastra* ("Law Code of Manu") also appeared at this time, formalizing the caste system and making it an integral part of Indian culture.

Indian mathematics of this time period is represented by the Bakhshālī manuscript, discovered in 1881 near the village of the same name near Peshawar and probably written in the third or fourth century A.D., though some authors date it as late as the twelfth century (the issue is further confused over the likelihood that the existing manuscript is a copy of an older work). The author calls himself the SON OF CHAJAKA. The manuscript shows a highly developed mathematics with techniques for approximating square roots, summing arithmetic sequences, and solving systems of linear equations. For example:

Problem 3.10. *Three men have 7* asvas *(superior horses), 9* hayas *(inferior horses), and 10 camels, respectively. Each gives one of their animals to each of the two others, and then they are equally rich. How much is each animal worth?*

The Bakshālī manuscript also gave solutions to some linear and quadratic indeterminate equations, a type of problem that would fascinate Indian mathematicians for more than a millennium.

The dating of the *Yavanajātaka* of SPHUJIDHVAJA (fl. 3rd cent. A.D.) is more precisely known than that of the Bakshālī manuscript, because the author indicated it was written in a year corresponding to about 269 A.D. As in Jain mathematics, numbers are described using what seems to be a place-value system, with the smallest order of magnitude being given first. What is interesting is that each number is associated with an object or group of objects that normally appear in that quantity: for example, "Moon" or "Earth" represents one; "eye" or "twin" two, and so on. "Sky" or "void" or "dot" indicate the absence of an order of magnitude: thus "Moon-Dot-Eye" would represent the number one unit, no tens, and two hundreds. This is the verbal equivalent of base ten positional notation with a zero.

3.2.3 Āryabhaṭa

It was not until 320 that a new power emerged in northern India: the Guptas. The Gupta Empire stretched from the Arabian Sea to the Bay of Bengal, and reached its height under its third ruler Chandragupta II. Fa-Hsien, a Chinese monk on pilgrimage to India to find

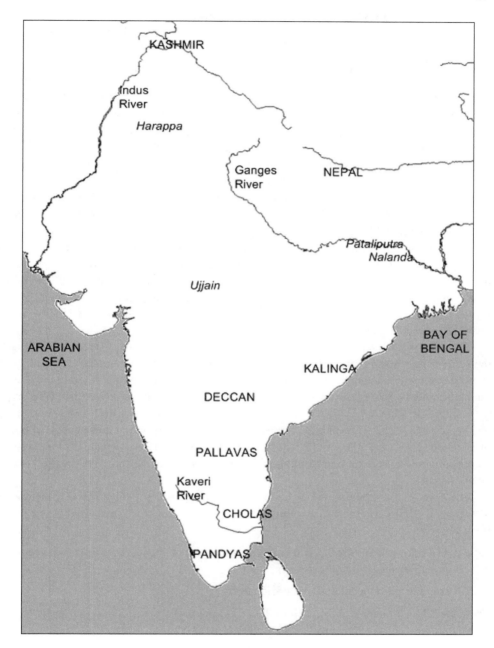

Figure 3.5. Medieval India

Buddhist texts, observed that the realm was peaceful and prosperous, with free hospitals at the capital Pataliputra, and that Chandragupta ruled without decapitations or floggings.

The Guptas established a great university at Nalanda in northeast India to study Buddhist texts. In time it would house more than 1500 teachers and several thousand students from as far away as Japan, Tibet, and Turkey. Profane literature also flourished:

the *Mahābhārata* came into its final form around this time: with 100,000 couplets, it is seven times longer than the *Iliad* and *Odyssey* combined. Even better known than the *Mahābhārata* is the *Kamasutra* of Vatsyayana.

Northwest of Nalanda is Pataliputra (modern Patna), which served both the Mauryans and the Guptas as capital, and was the birthplace of ĀRYABHAṬA (b. 476). By Āryabhaṭa's birth the empire had nearly collapsed under the assault of the Huns, and by the end of the fifth century the Guptas would be no more; Āryabhaṭa might have been a refugee, for he seems to have lived in Kerala, on the southwestern coast. His name is suggestive: one possible interpretation is "scholar warrior." The little that is known for certain about his life comes from the preface to his *Āryabhaṭīya*, where he notes that he wrote it at the age of twenty-three in the 3600th year of the Kaliyuga. Giving the date relative to the Kaliyuga ("Age of Kali," comparable to the Jaina *avarsapini*) suggests Āryabhaṭa was a Hindu who finished his work in 499.

The *Āryabhaṭīya* consists of 121 verse stanzas, thirty-three of which cover mathematical topics. One of the intriguing topics is the naming of each of the orders of magnitude, a practice used by the Jains: Āryabhaṭa himself expressed the number 57, 753, 336 as *cayagiyinusuchlr*, where the syllables represent the numbers 6, 30, 300, and so on. This is further evidence of the presence of position magnitude in Indian mathematics long before Sebokht's seventh century observation.

The other mathematical stanzas discuss topics such as the sums of series, including the sum of the whole numbers, their squares, and their cubes. In trigonometry, Āryabhaṭa showed that the Indian mathematicians had shifted focus from Ptolemy's chord to the *jiva*, the half-chord, by giving the differences between the half-chords for angles that differ by $3\frac{3}{4}°$: 225, 224, 222, 219, 215, 210, 205, 199, 191, 183, 174, 164, 154, 143, 131, 119, 106, 93, 79, 65, 51, 37, 22, 7. In modern terms, these values correspond to the rounded values of $3438 (\sin(3.75°(n + 1)) - \sin(3.75°n))$. In Book II, Stanza 12 Āryabhaṭa gives a purely algorithmic rule for calculating them, though the rule as stated does not give precisely the values listed.

The *Āryabhaṭīya* contains one of the earliest formulations of the rule of three, a method of solving the most commonly encountered type of linear equations (namely those of the form $ax = b$):

Rule 3.4. *In the rule of three, multiply the fruit by the desire and divide by the measure. The result will be the fruit of the desire.*

An example from the later work of Sridhara is:

Problem 3.11. *If one and one quarter* pala *of sandalwood cost ten and a half* pana, *how much does nine and a quarter* pala *of sandalwood cost?*

The measure is one and a quarter *pala*, which corresponds to the fruit of ten and a half *pana*; the desire is nine and a quarter *pala*, and the fruit of the desire is the cost. Essentially Āryabhaṭa is giving the solution to the equation $\frac{a}{b} = \frac{c}{x}$ as $x = \frac{cb}{a}$. Āryabhaṭa also solves quadratic equations arising from problems of compound interest. Perhaps the most interesting result in the *Āryabhaṭīya* is in stanzas 32 and 33. The verses are nearly incomprehensible, but can be interpreted as a method of solving linear indeterminate equations; this type was not examined by Diophantus, and represents a development original with the Indian mathematicians.

The *Āryabhaṭīya* lends weight to a statement by the eleventh century Persian historian al Biruni, who described Indian mathematics and astronomy as a mixture of "pearl shells and sour dates, or pearl and dung, or costly crystals and common pebbles." Nowhere is this assessment more valid than in Indian geometry. In the same sentence, Ārybbhaṭa tells us that the area of a circle could be found by multiplying half its diameter by half its circumference, and that this area multiplied by its square root gives the exact volume of a sphere. He goes on to tell us:

> Add 4 to 100, multiply by 8, and add 62, 000. The result is approximately the cir-
> cumference of a circle of which the diameter is 20, 000.[23, p. 28]

Thus $\pi \approx 3.1416$.

3.2.4 Varahamihira

Ancient Indian astronomers ran their prime meridian through the city of Ujjain, mentioned in the *Mahābhārata* as one of the holy cities of India. According to tradition Krishna, Ashoka, and other notables of Indian history and mythology received their education in Ujjain; certainly by the sixth century it was home to a thriving astronomical research community whose members included VARAHAMIHIRA (505–587). By then the Huns had been stopped and the Gupta empire had been broken into several independent kingdoms.

Varahamihira may have been the first person in history to consider the mathematics of zero. In his *Five Astronomical Canons*, he discusses the motions of the sun and planets. For example, as the Earth moves around the sun, the sun appears to move against the background stars (along the path known as the zodiac). Since the sun completes one circle of the sky in a year, the mean motion along the zodiac is easy to find. But because the Earth's orbit is elliptical, a correction factor (known as the mean anomaly) is needed, representing the difference between the easily computable constant rate and the observed rate. Varahamihira gave the mean anomaly as a quantity to be added or subtracted, depending on the month. Because the mean anomaly is sometimes zero (when the mean motion and actual motion of the sun coincide), Varahamihira sometimes directed that one should "add nothing" or "subtract nothing."

Varahamihira was also one of the first to treat combinatorial questions systematically. To find the number of perfumes that could be made by choosing four substances out of sixteen, Varahamihira used a variant of the arithmetic triangle: in his case, he formed a square with 1s along the left and bottom sides; all the remaining entries could be found by summing the entries immediately below and to the left. The desired number was the 4th entry of the 16th row (counting upward from the bottom).

3.2.5 Brahmagupta

In the seventh century, King Harsha of Thanesar reunited most of Northern India. Harsha was an enlightened ruler. He made numerous donations to the university at Nalanda, culminating in funds to build a wall around the university to protect it from attackers. During his reign the practice of Tantra entered India. Originally from Bengal in the northeast, it emphasized the primacy of a feminine power (*shakti*) that could be experienced through sexual intercourse. It is perhaps indicative of Indian culture that Tantra did not replace either Hinduism or Buddhism, but instead spawned Tantric forms of these two religions. Tantric practices, emphasizing very careful control over one's body, also led to the development of various forms of yoga.

BRAHMAGUPTA (598–670) worked at Ujjain around this time. His *Brahma Sphuta Siddhanta* ("The Opening of the Universe") appears to have been the original source of Islamic knowledge of Indian numeration. KANAKA (fl. 8th cent.), an Ujjain scholar credited with inventing amicable numbers (two numbers a and b where the sum of the proper divisors of one is equal to the other, such as 220 and 284), is said to have gone to Baghdad at the request of the Caliph al Saffāḥ, bringing a copy of the *Brahma Sphuta Siddhanta* to teach Indian astronomy and mathematics to the scholars there.

Brahmagupta treated linear and quadratic indeterminate equations using a method he referred to as *kuttaka* ("the pulverizer"). Using his method he solved equations of the form $ax^2 + b = y^2$: for example, Brahmagupta solved $61x^2 + 1 = y^2$, whose smallest whole number solution is $x = 226153980$ and $y = 1766319049$. In addition, the method of Brahmagupta allowed an infinite number of solutions to be created, rather than the single solutions of Diophantus.

Brahmagupta also gave a rule for the area of a quadrilateral with sides of length a, b, c, d: if

$$s = (a + b + c + d)/2,$$

then he claimed the area was equal to

$$\sqrt{(s - a)(s - b)(s - c)(s - d)}.$$

The formula is correct, provided one condition (unspecified by Brahmagupta) is met: the quadrilateral must be one inscribed in a circle (a so-called cyclic quadrilateral). It is not known if Brahmagupta realized that the formula did not generalize to all quadrilaterals.

Like Varahamihira, Brahmagupta explained the arithmetic of zero. He also included directions for operating with signed numbers, including the rule that "a debt subtracted from a zero is a fortune:" i.e., $0 - (-a) = a$. The discussion of the mathematics of zero in Varahamihira and Brahmagupta indicates the existence of the mathematical concept of zero and, possibly, a symbol for nothing, and some time after Brahmagupta the two concepts of a zero symbol and positional notation came together, and our modern numeration system was born.

Unfortunately Harsha's empire did not survive his death, and northern India fell apart into numerous warring kingdoms, obscuring the events of the period. Thus we know almost nothing about the history of our system of numeration between the time of Brahmagupta and its appearance, fully developed, a few centuries later. There are some intriguing possibilities, however.

After Buddhism entered China in the first century A.D., India became an important destination for monks seeking original texts: Fa-Hsien and the students at the university of Nalanda were not isolated examples. In the fifth and sixth centuries, Indian texts became so popular that one scholar complained they were being read ten times as often as Chinese classics. Between 629 and 645 the Buddhist monk Xuanzang visited India, stayed for a time in Harsha's domains, and returned to China with a collection of texts he would spend the rest of his life translating. Harsha himself initiated the first "official" diplomatic contact between T'ang China and northern India. In 718 the zero appeared in a T'ang astronomical text, translated into Chinese from an Indian original. The oldest surviving zero symbol in India dates to 876, on an inscription in Gwalior. Did Indian astronomers invent our modern numeration system and introduce it to China during the T'ang dynasty? Or did the Chinese introduce it to India during Harsha's reign? Or did Chinese and Indian scholars invent positional notation independently of one another? At this point, the historical record is too sketchy to answer this question definitively.

3.2.6 Southern India

Between the sixth and tenth centuries, southern India was dominated by three main kingdoms: Pallava on the southeastern coast north of the Kaveri River; Pandyas on the southeastern coast south of the Kaveri River; and the Rashtrakutas in the Deccan. Both Pallava and Pandyas spoke Tamil, a Dravidian language unrelated to Sanskrit, and even today southern India is known as Tamil Nadu, "Land of the Tamils."

The Rashtrakutas spoke another Dravidian language known as Kannada. The early Rashtrakutas patronized the arts and literature, and Muslim visitors ranked the splendor of the Rashtrakuta court on a level with that of the Caliphs and the Emperors of China and Byzantium. The earliest surviving Kannada literature, a treatise on poetic forms called the *Kavirajamarga*, was written in part by King Amoghavarsa.

MAHĀVĪRA (9th cent.), a Jain, lived in Mysore during this time, and worked at the court of Amoghavarsa. He is sometimes referred to as Mahāvīra Acharya ("Mahāvīra the Teacher"), and is the best-represented of all the Indian mathematicians, for his lengthy treatise *Ganita Sara Samgraha* has survived the centuries intact. In nine chapters Mahāvīra describes all of elementary arithmetic, as well as some advanced topics. Some topics are quite pedestrian: he lists a few palindromic products such as $139 \times 109 = 15151$. Some are interesting computationally: he gives four different ways of finding a^2 and for finding a^3. For example:

$$a^3 = a(a + b)(a - b) + b^2(a - b) + b^3$$

where b is some conveniently chosen number. He also described solutions to quadratic equations and some non-linear indeterminate equations: for example, constructing a rectangle whose area is (numerically) a linear combination of its sides and diagonal.

Mahāvīra made no claims to originality, crediting his predecessors (though he does not identify them). One of those predecessors might have been SRIDHARA (9th century), about whom almost nothing is known for certain except for the fact that he was a follower of Shiva. Sridhara's major surviving work is poem on arithmetic with exactly 300 verses; hence it is known as the *Trisitaka* ("Three Hundred"). There are too many similarities between the surviving work of Sridhara and that of Mahāvīra to believe there is no connection. Both devote more space to fractions and to the rule of three than any other writer, and many of their examples are similar. This alone would not be enough to conclude a relationship, but there are idiosyncratic choices made by Mahāvīra and Sridhara that are repeated by no other Indian mathematician. For example, both treat addition, subtraction, and series summation in the same section, while all other texts treat series summation in a separate section. Significantly, both give the same four methods of finding the square of a number, including

$$1 + 3 + 5 + \cdots + (2n - 1) = n^2$$

which again appears nowhere else in Indian mathematics.

Sridhara seems to have been highly regarded by his contemporaries and successors, and may have been the earliest Indian mathematician to solve quadratic equations algebraically (as opposed to geometrically) using a technique equivalent to completing the square. However, this method appeared in one of his lost works, and we know of it only because it was quoted by BHASKARA (1114–1185), who lived in the Chola Empire.

The Cholas became the dominant power in the south after the Rashtrakutas went into decline at the end of the tenth century. Chola influence extended south into Sri Lanka, which they conquered at the beginning of the eleventh century, and north into the Deccan. In Tanjore and other cities, Chola bronze casters produced some of the most recognizable artifacts of Hinduism: statues of Shiva Nataraja ("Shiva, Lord of the Dance"). These statues depict a four-armed Shiva, the Hindu god of creation and destruction. One hand holds a flame symbolizing destruction, another hand a drum symbolizing creation, a third makes a *mudra* (gesture) that indicates "forget fear," and the fourth points to the defeat of ignorance, shown as the demon dwarf Muyalaka being trod underfoot.

Bhaskara, like Mahāvīra, is also given the accolade Acharya. His *Lilavati* is a textbook summary of the works of others, including Brahmagupta and Sridhara, and represents Indian mathematics at its height. According to the Persian translator of the work (writing in 1587), Lilavati ("the beautiful") was the name of Bhaskara's daughter; he calculated the precise time she should be married to ensure a happy life, and constructed a water clock to ensure that the marriage would take place at the correct time. Unfortunately a pearl blocked the water clock, the marriage took place at the wrong time, and after a year her husband died. To console her, Bhaskara taught her mathematics and dedicated his arithmetic text to her. More likely, "Lilavati" referred to how Bhaskara felt about the content of the work, though it is possible that he chose the title to emulate that of the *Lilavati* of Nemichandra, a well-known Sanskrit grammar.

The *Lilavati* discusses the standard topics of arithmetic of zero, geometry (where correct formulas for the area and volume of a sphere are given), permutations and combinations, and linear and quadratic indeterminate equations. Bhaskara also included a proof of the Pythagorean Theorem based on the decomposition of areas.

In his *Vija Ganita*, Bhaskara discusses quadratic equations and some other Diophantine problems, such as:

Problem 3.12. *Find four unequal numbers whose sum is equal to the sum of their squares.*

Bhaskara solved the problem by noting $1 + 2 + 3 + 4 = 10$ and $1^2 + 2^2 + 3^2 + 4^2 = 30$, so the numbers are 1, 2, 3, 4 multiplied by $\frac{10}{30}$.

The work of Bhaskara showed that Indian mathematics had reached a very high state of sophistication. Unlike the Greeks, the Indians had no difficulties working with negative or even irrational quantities; even the comparison of a perimeter and an area (absolutely nonsensical in Greek mathematics) posed no difficulty, since the Indians viewed both as purely numerical quantities. The presence in Indian mathematics of a positional numeration system with a zero symbol, regardless of whether it was an indigenous development or an import, meant that the Indian mathematicians had a more effective way of representing number than any before them.

3.3   The Importance of Proof

Because the tools of algebra are much more flexible than those of geometry, the Indian and Chinese mathematicians could, in their own field, go farther than the Greeks ever did. But then their mathematics stagnated, and we might wonder why. No doubt the political fortunes of China and India played a role in preventing a strong mathematical tradition from arising, and geographical factors made it difficult to form a mathematical community. But Greek mathematicians faced equally turbulent times, and Islamic mathematicians faced even greater geographic separation, so other factors must be considered.

One key feature that stands out in Chinese and Indian mathematics is the minimal importance of proofs. While proofs appear in Indian and Chinese texts, they are nowhere as important as they are in Greek or Islamic works. This may be an inevitable consequence of focusing on algebraic problems: it is trivial to prove the validity of a solution to an equation, determinate or indeterminate, so it is easy to believe that proof is, in general, not important. Indeed, we even use a different word: a solution is checked, not proven.

This suggests that proof is important for the development of mathematics. But why? The fact that a "proven" result is one that is known with certainty seems insufficient to account for the explosive growth in mathematical knowledge during the Greek, Islamic and later European eras. An alternate possibility is that the act of proof invariably spawns new tools and new problems: those who work without proof answer their questions but do not generate new ones in the process. We might say that the Indians and Chinese created a powerful tool (algebra) that allowed them to solve problems, while the Greeks created a powerful tool (proof) that allowed them to create problems. Ultimately the Greek geometers created more problems than they could solve, while the Indian and Chinese mathematicians solved all the problems they had. Neither could progress without the other, and the two tools came together in the hands of Islamic mathematicians.

For Further Reading

For the history of China, see [2, 33, 36, 42, 82]. For the history of India, see [107, 134]. For Chinese mathematics, see [81, 64, 121, 137]. For Indian mathematics, see [26, 27, 64, 113].

4

The Islamic World

4.1   Early Islam

Before the seventh century, the Arabian peninsula was populated by nomadic Bedouin tribes who worshipped spirits they believed present in objects such as rocks, animals, and plants. In Bedouin mythology Allah was the supreme ruler of the universe, but there were others, such as Hubal the Moon God. But in 610 a merchant named Muhammad had a revelation: there is no God but Allah. In Arabic this is the sonorous refrain, "*la ilaha illa Allah.*" The denial of the very existence of other gods was made one of the five pillars of a new religion, soon known as *Islam* ("submission," i.e., submission to the will of God) and its adherents would be called Muslims.

Muhammad began preaching in his birthplace of Mecca. In one respect, it was a good place to begin, since many came to the city to worship at the cube-shaped shrine known as the *Ka'aba*, which housed relics honoring more than 300 gods. The most sacred relic was the "Black Stone" of Hubal which, according to legend, fell to Earth from the Moon; because of this story, most scholars believe the Black Stone is a meteorite. On the other hand, the merchant families that dominated the city feared that reducing the number of gods would jeopardize a valuable source of revenue, so the early Muslims faced considerable persecution. After ten years of steadily increasing opposition, Muhammad broke the bonds (*hijra* in Arabic) that tied him to Mecca; this event in 622 marks the beginning of the Muslim calendar. Muhammad settled in Yathrib, some 200 miles north of Mecca, but later renamed "The City of the Prophet of God:" *Madīnat Rasūl Allāh*, or Medina.

By Muhammad's death in 632, the entire Arabian peninsula was under Muslim rule. The fragile coalition hand-crafted by Muhammad looked as if it might fall apart, and it was only the efforts of Abu Bakr, Muhammad's father-in-law, that kept the Muslims united. Abu Bakr would be accorded the title of *Caliph* ("successor") and under him the Muslims began to expand beyond Arabia.

During the early years of the Caliphate, the Muslims were generally tolerant of Jews and Christians. For example, when the Muslim general Khalid ibn al-Walid took Damascus in 635, he personally guaranteed the lives, property, and churches of the inhabitants. Jerusalem surrendered under similar conditions, two years later. Indeed, the Muslims not only tolerated other religions, but discouraged conversion to Islam: it is estimated that as late as the eighth century, after a generation or more of Muslim rule, less than 10% of the inhabitants of Iran, Iraq, Syria, Egypt, and Spain were Muslim. The reasons were not entirely altruistic: non-Muslims had to pay a special tax, called the *jizya*, so the government discouraged proselytizing.

As a result, Christian and Jewish communities thrived in Muslim lands. Indeed, the Muslim lands were a safe haven for heterodox sects. The Bishop Cyril, who helped destroy the last remnants of Greek learning in Alexandria, also engineered the condemnation of the Nestorians in 431, driving them from the west. In 489 the last Nestorian school in the Byzantine world was closed by Imperial edict, and the remaining followers migrated east to Persia. There they would thrive for centuries, establishing a presence as far east as China and the shores of Lake Baikal in Siberia.

Nestorian scholars played a key role in bringing the learning of the Hellenistic world into the Islamic one. But contact with cultures farther east led the more sophisticated to realize that learning was not a Greek monopoly. Severus Sebokht, the Nestorian Bishop of Keneshra (on the Euphrates), felt that his contemporaries focused on the achievements of the ancient Greeks while completely ignoring the contribution of others, particularly the Indians. To point out that scientific thought was not a Greek monopoly, in 662 Sebokht wrote of the discoveries of the Indians in astronomy and mathematics and left an intriguing observation that their discoveries were made using their *nine* signs for numeration. It is not clear if Sebokht knew of the zero symbol but ignored it through misunderstanding its significance, or if the zero symbol had not yet appeared in Indian mathematics. The Indians called the symbol *sunya* ("void"), since it represented an empty space; the Arabs translated this as *aṣ-ṣifr*, which became *zephirum* as it entered medieval Europe, from which we get the word zero, as well as cipher.[1]

The Arabs wrote the numbers from smallest place value to largest: thus 1 thousand, 2 hundreds, 3 tens and 4 units would be written beginning with the '4'. However, because Arabic is written right to left, this meant that the rightmost digit represented the smallest place value. Thus the number one thousand two hundred thirty-four would be written 1234, while the Arabic authors, writing about the numbers, would naturally tend to list the numerals so they increased in magnitude from right to left: 9 8 7 6 5 4 3 2 1. We will see this footprint of the Arabic writers in the earliest European expositions of the new numbering system.

4.1.1 Umayyads and Abbasids

The murder of the third Caliph, Uthman of the Umayyad family, led to a civil war between two main groups. The *shi'at Ali* ("followers of Ali") supported Muhammad's son-in-law Ali: this group felt that only a relative of Muhammad could become Caliph, and came to be known as the Shi'ites. Others argued that anyone who emulated the habitual behavior, or *sunna*, of Muhammad had the right to lead the Muslims; adherents to this belief were consequently known as Sunnites. Ali became Caliph in 656, but was murdered in 661. Mu'awiya, the governor of Syria and Uthman's kinsman, thus became the first of fourteen Umayyad Caliphs, ruling from Damascus. Since the Shi'ite position questioned the very

---

[1] Arabic transliteration is a complex issue, but roughly speaking the following conventions apply. For vowels, *ā* is an elongated short "a" ("aaah"), *ī* is like the "i" in "machine," *ū* is like the "oo" in "school." For consonants, an underdot (*aṣ-ṣifr*) indicates an emphasized consonant, and should be pronounced separately: Isḥāq should be pronounced "Is-haaq" and not "Ish-aq."

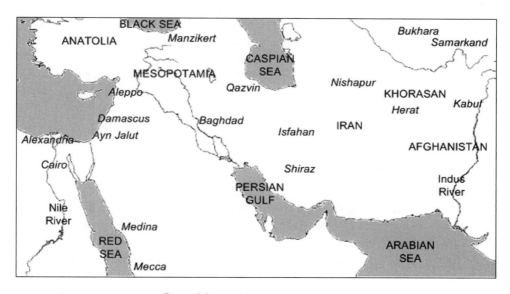

Figure 4.1. The Eastern Islamic World

legitimacy of Umayyad rule, the Umayyad Caliphs were merciless in their persecution of the Shi'ites.

The Shi'ites were (and still are) a minority faction within Islam. But very early on, the Shi'ites won the support of a key group of Muslims: the Persians. The Persians were the heirs to an ancient and sophisticated civilization, and despised the Arabs as barely literate barbarians. The Sunnite position was tantamount to declaring that the Arab conquerors of Persia were morally superior, an intolerable idea. Consequently the Persians adopted the more palatable Shi'ite position.

Resentment against the Umayyads exploded into open revolt, beginning in the Persian province of Khorāsān in June 747. By 750, the armies of rebellion overthrew the Umayyads and established the Abbasid Caliphate (named after Abbas, the military leader of the revolt). The second Abbasid Caliph, Abu Ja'far al Mansur, moved the capital to a small village on the Euphrates, which would be closer to the center of the empire than far-off Damascus. Construction to make the city the capital of an empire began in August 762, a date his court astrologer determined was particularly propitious. Al Mansur renamed the village "The City of Peace", *Madinat al-Salam*, which is how it is often referred to in the stories of the Arabian nights. But the inhabitants stubbornly continued to call the city by its old name: Baghdad.

Baghdad became a great city, the crossroads of the world, under the rule of the Caliph Harun ar-Rashid. Every day ships and caravans arrived carrying goods from far-off lands: silk from China; gems from India; furs from Scandinavia; leather from Spain; and slaves from Africa. With wealth and peace came time to reflect and think.

One of the key issues that must be confronted by any believer in an omnipotent God is the following: why are men allowed to perform acts of evil? Either men have no free will (in which case, it is not right for God to judge them for actions over which they have no control), or they choose to do evil and will be judged for it. But this implies the existence of a standard of justice, and such a standard can be applied to God's own actions. If God's actions are always just, then God is not omnipotent, for he cannot carry out an unjust action. The alternative is that God is free to carry out great injustices.

Because of arguments like these, most religions whose tenets include an omnipotent God deny that logic can be applied to theology. Islam was no exception, but in the eighth century the Mu'tazilites, a group of philosophers who believed that reason was a viable method for reaching truth (even in questions of theology), became prominent.

One of the two sons of Harun ar-Rashid, Abdallah al Ma'mun, subscribed to the Mu'tazilite position. Al Ma'mun was older, wiser, and more statesmanlike than his brother Muhammad, but since his mother was Persian, ar-Rashid chose Muhammad (whose mother was Arab) to be his successor. When Harun ar-Rashid died in 809, the long-standing conflict between Persians and Arabs was brought to the fore and shortly afterwards, al Ma'mun rebelled. In 813, his forces captured Baghdad and Muhammad was killed (probably against Abdallah's orders).

As Caliph, al Ma'mun would usher in an era of peace and prosperity. The Mu'tazilites and their belief in the power of reason received official sanction. Notable Mu'tazilite scholars of this era include Ya'qūb ibn Isḥāq aṣ-Ṣabāh al-Kindī, who wrote on subjects as diverse as astrology, Indian numeration, and sword making. Kindī argued, like Sebokht a century and a half before, that no culture or people had a monopoly on truth.

4.1.2 Al Khwārizmī

Al Ma'mun established a "House of Wisdom" in Baghdad, similar in intent to the great Museum of Alexandria. He was said to have been inspired by a dream in which Aristotle appeared to him; this spurred him to obtain and translate as many ancient Greek manuscripts as he could obtain, including the works of Euclid and Ptolemy.

Among the scholars at the House of Wisdom was ABU JA'FAR MUHAMMAD IBN MUSA AL KHWĀRIZMĪ (ca. 780–850). "Abu Ja'far" means "father of Ja'far"; "Ibn Musa" is "son of Musa" (the Arabic form of Moses); "al Khwārizmī" is "the man from Khwarezm" (a region near the Aral Sea, now in Uzbekistan); thus, his name can be translated as "Muhammad, father of Ja'far, son of Moses, the man from Khwarezm." He is usually referred to as al Khwārizmī.

Al Khwārizmī wrote on a variety of subjects, including a work on the numbering system of the Hindus, probably based on the work of Brahmagupta. A Latin translation of al Khwārizmī's *On the Indian Numbers* became one of the means by which Europe learned of the new system of numeration. The translations were not always of the highest quality. One fifteenth century work, *The Craft of Numbering*, turned al Khwārizmī into Algor, a king of India! More accurately, the author went on to say that the craft of computation was named after Algor, and the textbooks describing the Hindu-Arabic system of arithmetic came to be known as *augrims*, algorisms, or finally algorithms, a term we still use to describe any systematic computational procedure.

Around 825, al Khwārizmī wrote *The Condensed Book of Restoration and Reduction*. The "restoration and reduction" of the title refer to the processes of adding or subtracting terms to eliminate them from one side of the equation. In particular, from $3x - 5 = x + 8$ we obtain $3x = x + 13$ by restoration (adding 5 to both sides), and from this last we obtain $2x = 13$ by reduction (subtracting x from both sides). In Arabic, "restoration and reduction" is *al-jabr wa-l-muqābala*, and the first term ("restoration") gave rise to the word algebra.

The introduction to *The Condensed Book of Restoration and Reduction* gives high praise to Muhammad (the prophet) and Islam, and goes on to explain how the patronage of al Ma'mun encouraged al Khwārizmī to write his "short work." Al Khwārizmī makes it clear that he will be

> confining it [the book] to what is easiest and most useful in arithmetic, such as men constantly require in cases of inheritance, legacies, partition, law-suits, and trade, and in all their dealings with one another, or where the measuring of lands, the digging of canals, geometrical computation, and other objects of various sorts and kinds are concerned...[101, p. 3].

In other words, he intended to present the purely practical aspects of mathematics; nonetheless, this "short work" is about 200 pages in length.

The algebra of al Khwārizmī represents in many ways a step backward from the work of Diophantus. Notation is nonexistent, and even the numbers are spelled out:

> One square, and ten roots of the same, amount to thirty-nine dirhems ... The solution is this: you halve the number of roots, which in the present instance yields five. This you multiply by itself; the product is twenty-five. Add this to thirty-nine; the sum is sixty-four. Now take the root of which, which is eight, and subtract it from half the number of the roots, which is five; the remainder is three. This is the root of the square which you sought for; the square itself is nine [101, p. 8].

The problems are generally simpler than those posed and solved by Diophantus and the ancient Babylonians. However, what distinguished his work was one important feature: Diophantus, and before him the Babylonians, solved their problems using a variety of clever methods, each specific to a type of problem. Al Khwārizmī and his successors approached problem solving by *reducing* a problem to one of just a few fundamental types, the solutions of which are straightforward. Thus, the work of al Khwārizmī shifted the emphasis from solving equations, to transforming equations using what we would now call the rules of algebra.

Al Khwārizmī also took important steps towards the arithmetization of algebra: the treatment of algebraic quantities in the same way as numerical quantities. For example, to find the product of "ten and thing" by itself (i.e., $(10 + x)^2$), al Khwārizmī began by noting that "ten and one" multiplied by "ten and two" was ten times ten, plus ten times one, plus ten times two, plus one times two: in other words,

$$(10 + 1)(10 + 2) = 10 \cdot 10 + 10 \cdot 1 + 10 \cdot 2 + 1 \cdot 2$$

The product of $(10+x)(10+x)$ was found analogously. Al Khwārizmī went on to introduce the fact that the product of two negatives is a positive:

> Ten less one, to be multiplied by ten less one, then ten times ten is a hundred; the negative one by ten is ten negative; the other negative one by ten is likewise ten negative, so that it becomes eighty; but the negative one by the negative one is one positive, and this makes the result eighty-one [101, p. 22–3].

If we go by the space devoted to a topic, *Condensed Book of Restoration and Reduction* is really about inheritance law: slightly more than half the book concerns the problem of dividing an estate among heirs, a procedure known as *'ilm al-farā'iḍ*. For example, al Khwārizmī presents the problem:

Problem 4.1. *A man dies leaving one-ninth of the estate to a stranger, with the remainder to be divided among his mother, wife, and two brothers and two sisters. How should the estate be divided?*

The solution is:

Solution. Divide the shares [of the mother, wife, brothers, and sisters] into forty-eight parts. If you take one-ninth from any amount, eight-ninths remain. So add one-eighth of forty-eight, to obtain fifty-four [which is the entire estate]. Give the stranger a ninth, or six; leaving forty-eight to divide among the rest of the heirs proportional to their legal shares.

□

Why 48? In a later problem, it can be inferred that a widow was entitled to $\frac{1}{8}$ and the mother $\frac{1}{6}$ of what remained of an estate after a fraction had been disposed of. Hence by making the remaining shares (after $\frac{1}{9}$th is given to a stranger) equal to 48, the widow receives 6 and the mother receives 8 shares.

The space devoted to it in the work of al Khwārizmī and others makes it seem as if it was an important part of Islamic life. However, the fourteenth century Islamic scholar ibn Khaldūn noted that in most cases, the division of the estate was straightforward and criticized authors for inventing problems that required very complex mathematics but would have only occurred in very rare cases.

4.1.3   The Biblical Value of π

As al Khwārizmī noted in the introduction, his *Condensed Book of Restoration and Reduction* was meant to be more than a book on algebra, and it includes a section on geometry. The geometry is purely practical:

> In any circle, the product of its diameter, multiplied by three and one-seventh, will be equal to the periphery. This is the rule generally followed in practical life, though it is not quite exact. The geometricians have two other methods. One of them is, that you multiply the diameter by itself; then by ten, and hereafter take the root of the product; the root will be the periphery. The other method is used by the astronomers among them: it is this, that you multiply the diameter by sixty-two thousand eight hundred and thirty-two and then divide the product by twenty thousands [101, p. 71–2].

The first is from Heron; the second two are drawn practically verbatim from Hindu sources.

In contrast to the relative accuracy of the three values of π given by al Khwārizmī, the Bible, in the First Book of Kings, Chapter 7:23, describes a circular cauldron whose diameter is 10 cubits and whose circumference is 30 cubits; consequently the ratio of the circumference to the diameter is 3. However, some claim the Bible includes an even more accurate value of π, embedded in the words themselves.

The Hebrew word for circle is *qavah*, but in the Hebrew Bible it is spelled *qava*. The ancient Hebrews, like the ancient Greeks, used letters to write their numbers. If the numerical values of the letters are substituted, *qavah* represents the number 111, while *qava* represents 106. This gives us four numbers: the real value of π; the Biblical value (3); the "real value" of the word "circle" (111); and the Biblical value of the word "circle" (106). The mystically inclined can thus set up the proportionality π is to 3 as 111 is to 106; this gives a value for π of about 3.141509. This may be more than coincidence, but less than significant.

The current extant Hebrew version of the Bible, known as the Masoretic text, was produced by scholars in Iraq and Palestine between the sixth and tenth century A.D. The editors pointed out any peculiarities in spelling or grammar that were deemed part of the original canon (thus the discrepancy between the two spellings of the word "circle" is noted in any Masoretic text). To prevent future copyists from changing the text, the editors included statistical information (such as letter, word, and verse counts, as well as the number of times a particular word or spelling or letter was used) on the margin of each page. As a result of this early version of an error-detecting code, the first printed Masoretic texts are virtually identical to the oldest manuscripts.

The Masoretic texts themselves are based on copies of the Hebrew Bible that date back to the 3rd century B.C., which makes them contemporaneous with the works of Archimedes. Thus both the editors of the Masoretic texts and those who made the oldest known copies of the Hebrew Bible had access to an accurate approximation to π. The Bible may contain a good approximation to π, but no special insight would have been needed to obtain it.

4.2 The Rise and Fall of Dynasties

The successors of al Ma'mun, beginning with his brother al-Mutasim, began the practice of surrounding themselves with bodyguards with no local ties: the Turks, drawn from the steppes of Central Asia. In Baghdad, relations between the Turks and the local populace became so bad that al Mu'tasim relocated the capital to Samarra in 836. This further isolated the Caliph from his subjects, and made him even more subject to the whims of his Turkish bodyguards; as a result, the Turks became increasingly more important in the Islamic world.

One of al Khwārizmī's contemporaries may have been of Turkish descent: ABD AL-HAMID IBN WASI IBN TURK AL JILI (fl. 9th cent.). "Ibn Turk al Jili" means "son of the Turk from Gilan" (a region in northwestern Iran on the shores of the Caspian Sea); this is very nearly the total of our biographical knowledge of ibn Turk. His dates are uncertain, and his origin is variously ascribed to Iran, Afghanistan or Syria.

Ibn Turk wrote an algebra textbook, of which only the section entitled "Logical Necessities in Mixed Equations" survives. The style of the surviving fragment is very similar to al Khwārizmī's, suggesting that al Khwārizmī's place in the development of algebra is similar to that of Euclid's place, in other words, preparing a textbook for a science that had already reached a high level of development (and, unfortunately like Euclid's *Elements*, causing the other works to be neglected). Ibn Turk's work is superior to al Khwārizmī's from a theoretical standpoint, as he analyzed in greater detail the existence of solutions. For example, equations of the form $x^2 + p = qx$ are not always solvable. Ibn Turk gave

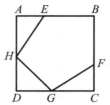

Figure 4.2. Abū Kamil's Pentagon

a geometric proof that identified the necessary conditions for equations of this type to be solvable (that is, to have a positive real solution).

Meanwhile, the Caliph's power continued to decline. In 868, the Turkish governor of Egypt, Ahmad ibn Tūlūn, organized an independent Egyptian army, effectively establishing himself as the ruler of an independent nation. Under the Tūlūnids, Egyptian power stretched north as far as the borders of the Byzantine Empire, and Egypt underwent a cultural renaissance.

ABŪ KĀMIL (850–930) came from Egypt around this time. We know almost nothing else about him, not even his name (Abū Kāmil means "father of Kāmil"). However, he had a great influence on the development of mathematics. His *Algebra* is primarily a commentary on al Khwārizmī—many of the problems are the same—but Abū Kāmil takes greater care in proving algebraic identities. More importantly, he appears to be the inspiration for the algebraic work included in Leonardo of Pisa's *Book of the Abacus*, so Abū Kāmil is of great importance in the development of western algebra.

Besides his work in algebra, Abū Kāmil also solved indeterminate equations in *Book of Rare Things in the Art of Calculation*. The work is noteworthy for having been written before Diophantus had been studied in depth by the Arabs, so it represents an independent development of methods of solving indeterminate equations.

One achievement of Abū Kāmil's is noteworthy for what it foreshadowed: inscribing an equilateral (but not equiangular) pentagon in a square. There are two ways this can be done; Abū Kāmil solved the case where one vertex of the pentagon $EBFGH$ coincides with one vertex of the square $ABCD$ (see Figure 4.2). Seven hundred years before Descartes, Abū Kāmil applied the tools of algebra to solving a geometric construction problem. In particular, he shows that if one side of the square has length 10, then the side of the pentagon will be $20 + \sqrt{200} - \sqrt{200 + \sqrt{320000}}$.

In 892, al-Mutaḍid became Caliph, and for a time, restored the power and prestige of the Abbasids. First, he returned the seat of government to Baghdad, which reduced the power of the Turkish guard. Next, he married the daughter of the Tūlūnid Caliph, which turned a mortal enemy into someone with a vested interest in preserving the Abbasid Caliphate. Finally, through a combination of shrewd diplomacy, vicious warfare, and clever intrigue, he played the factions within the Caliphate against each other. The result was a brief return to the glory days of al Ma'mun.

4.2.1   Translating and Transcending

One of the ongoing activities at the House of Wisdom was making the knowledge of classical antiquity available to the Islamic world. This involved much more than merely translat-

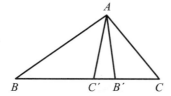

Figure 4.3. Ibn Qurra's Generalization of the Pythagorean Theorem

ing Greek texts into Arabic. In the days before the printing press, every copy of a manuscript was transcribed by hand. In some cases, the copyist was familiar with the subject matter and made a copy for their own personal use. But in most cases, the copyist had little to no understanding of the text they were copying. Consequently errors crept into the manuscripts.

In the early years of the House of Wisdom, MUHAMMAD IBN MUSA (ca. 800–870), AHMAD IBN MUSA (ca. 800–870), and AL HASAN IBN MUSA (ca. 800–870) played a crucial role in the development of Islamic mathematics. As their names suggest, they were brothers (Muhammad the oldest, and al Hasan the youngest), all sons of Mūsā ibn Shakir, a highway robber who became an astrologer for al Ma'mun. The brothers became known as the Banu Musa ("sons of Musa"), and their work with Apollonius's *Conics* shows the magnitude of the task of translating the works of the Greeks.

By the time of the Banu Musa, the surviving copies of Apollonius's *Conics* included so many errors that learning from them was impossible. To make sense of the available manuscripts, al Hasan had to rediscover many of the properties of the conic sections for himself; he would write a treatise on the topic before his death. Although such a treatise was invaluable, there was still interest in making a proper translation. To search for manuscripts, older brother Ahmad secured a government post in Syria. While there, he obtained a copy of a commentary by Eutocius on Books I through IV of Apollonius's *Conics*, as well as copies of Books V through VII from other sources.

Ahmad began writing his own commentary on this nearly complete manuscript. This required a more accurate translation of the Greek text, which he entrusted to two other members of the House of Wisdom. The first four books were to be translated by Hilāl ibn Abī Hilāl al-Himsī, and books five through seven by THĀBIT IBN QURRA AL-HARRANI ("Thābit [Tobit] from Harran, son of Qurra") (836–February 18, 901). Thus began ibn Qurra's illustrious career as a translator. Some works of the Greek geometers come down to us only in Arabic translations done by Thābit ibn Qurra: besides Books Five through Seven of Apollonius's *Conics*, ibn Qurra also translated Archimedes's *Book of Lemmas* and *[Inscribing a Regular] Heptagon in a Circle*, and greatly improved the translation of Ptolemy's *Almagest* done by ISHAQ IBN HUNAIN (d. ca. 910).

But ibn Qurra was no mere translator; like the Banu Musa, he mastered the work of the ancient Greek geometers, suggesting extensions and improvements. In *Quadrature of a Parabola*, Archimedes relied on inscribed triangles to show that the area of the parabolic segment was $\frac{4}{3}$ the triangle with the same base and vertex. Ibn Qurra proved the same result, though he relied on both inscribed and circumscribed figures. Moreover, ibn Qurra's approach is similar to our own: unlike Archimedes, who partitioned the region into triangles whose bases were the sides of other triangles, ibn Qurra did the equivalent of partitioning the domain at points in geometric proportion, then summing the areas of the corresponding

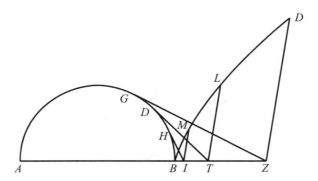

Figure 4.4. Hyperbola Construction

rectangles. Unfortunately ibn Qurra's work remained unknown in the West until long after the same method was reinvented by Fermat—eight hundred years later.

Ibn Qurra also discovered a generalization of the Pythagorean Theorem. Given any triangle ABC, draw AB', AC' where the angles $AB'B$ and $AC'C$ are both equal to A. Ibn Qurra stated without proof that the squares on AB and AC are together equal to the rectangle on $(BB' + CC')$ and BC. The result can be proved using the Law of Cosines (known as early as Euclid's time), and was not rediscovered until the 20th century.

Ibn Qurra's grandson, IBRAHIM IBN SINAN (908–946), proved to be a worthy geometer in his own right. Ibn Qurra's rather lengthy work on the quadrature of the parabola was reworked by ABU ABDALLAH MUHAMMAD IBN ISA AL-MĀHĀNĪ (820–880) into a shorter, more readable version. Ibn Sinan took the existence of this work as a challenge to his family's intellectual honor. In response, he wrote *On the Drawing of the Three Conic Sections*, which included procedures for pointwise construction of the parabola, hyperbola, and ellipse using compass and straightedge methods.

For a hyperbola, he gave three methods, the simplest of which is: given semicircle AGB with diameter AB, extend the diameter to Z and pick arbitrary points G, D, H on the semicircle; draw the tangents GZ, DT, HI. Draw IM at an arbitrary angle, and TL, ZD parallel to IM, with $IM = IH$, $TL = TD$, $ZD = ZG$ (see Figure 4.4). Then B, M, L lie on a hyperbola with AB as a diameter and whose parameter and transverse sides are equal to AB.

The drawing of a conic section had more than theoretical interest, for it solved a very practical problem: that of creating accurate sundials. Suppose a stick is placed in the ground; since the position of the shadow relates to the position of the sun, it can be used to determine the local solar time (which clock time approximates). Moreover, since the sun traces a very nearly circular path across the sky each day, the line between the position of the sun and the tip of the stick traces out, in an Apollonian fashion, a double cone, with the stick's tip as the cone's vertex. The plane of the ground cuts both sides of this cone, and thus the path traced out by the stick's shadow is a hyperbola.

4.2.2   The Geometers of the Buyid Court

Ibn Sinan lived only half as long as his grandfather, dying of a liver tumor at the age of 37. At the end of his life, he saw another upheaval in the Islamic world. In 945, the Buyids from

Gilan took Baghdad. Bowing to the inevitable, the Caliph conferred upon the Buyids the title of *ad-Dawlah*, which implied they were the political leaders of the Abbasid *dawlah* (state). Three Buyid brothers divided the remnants of the Caliphate, ruling from Shiraz, Isfahan, and Baghdad.

In 977 the last of the brothers died. Within a few years, 'Aḍud ad-Dawlah united the Buyid territories, built libraries, schools, and dams (the one near Shīraz, known as the Emir's Dam, remains to this day), and fostered a brief cultural revival. One of the oldest surviving descriptions of Hindu arithmetic came from around this time: the *Principles of Hindu Reckoning* by KŪSHYĀR IBN LABBĀN (fl. ca. 1000). Ibn Labbān might have come from Gilan, for he is sometimes referred to as "al-Jilī" (the Gilanite). His work illustrates a peculiarity in the adoption of positional notation: at the same time he explained the use of base-10 positional notation for numbers greater than 1, he continued to use sexagesimals to express numbers smaller than 1.

'Aḍud's court attracted a number of prominent mathematicians, including ABU SAID AHMAD IBN MUHAMMAD AL-SIJZI (945–1020) from Gilan. Al-Sijzi described a means of trisecting an angle using a circle and a hyperbola, and proved $(a + b)^3 = a^3 + 3a^2b + 3ab^2 + b^3$, using the decomposition of a cube into cubes and rectangular parallelepipeds (a method analogous to the one used by Euclid to prove that $(a + b)^2 = a^2 + 2ab + b^2$).

The brightest star of 'Aḍud's court was the Persian mathematician MOHAMMAD ABU'L-WAFĀ AL-BUZJANI (June 10, 940–July 997 or 998), who came to Baghdad in 959. One of his important works was a book on arithmetic for scribes and government officials. Abu'l-Wafā described a type of numeration known as "finger arithmetic," where the ten fingers are used to indicate numbers from 1 to 10,000 to store intermediate results from long computations. Abu'l-Wafā also wrote *On Obtaining Cube and Fourth Roots and Roots Composed of These Two*. Nothing whatsoever is known of this work beyond the title, which suggests it gave a procedure for finding roots of indices higher than 2 using a composition of roots. For example, we might find a sixth root by $\sqrt[6]{x} = \sqrt{\sqrt[3]{x}}$.

Abu'l-Wafā and his student (yet another Gilanite), ABŪ NAṢR MANṢŪR IBN 'ALĪ IBN 'IRĀQ (d. ca. 1036), invented trigonometry as we know it. Euclid and Ptolemy laid down some important computational principles, including the Law of Cosines and theorems equivalent to the half-angle, angle-sum, and angle-difference identities for sines. The Greeks, however, always spoke of the length of a chord in a circle. The Hindus were the first to speak of the "half-chord," the precursor to our sine function. But Abu'l Wafā was the first to use quantities equivalent to our tangent, secant, and cosecant functions (initially described in relationship to the lengths of the shadows cast by a stick, either placed perpendicular or parallel to the ground). In addition, Abu'l Wafā (or ibn 'Irāq, or both) discovered the Law of Sines, adding to the theoretical knowledge of the three trigonometric functions. Finally, Abu'l-Wafā created a table of sine and tangent values at $15'$ intervals, accurate to eight decimal places. Ptolemy's table of sines, by comparison, was only accurate to three decimal places.

The name sine itself comes from a misunderstanding of written Arabic. The Hindu word for "half-chord" is *jiva*, which became *jiba* in Arabic. Since Arabic is normally written without vowels, "half-chord" *jiba* was written in the Arabic letters equivalent to "J B." European translators frequently transliterated Arabic scientific terms: this gives us words like alcohol, algebra, azimuth, and zero. *Jiba* suffered a different fate, for an early translator,

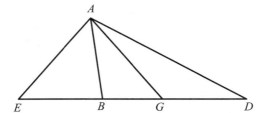

Figure 4.5. Abū Sahl's Construction of a Heptagon

Robert of Chester, knew just enough Arabic to misread the "J B" in one of al Khwārizmī's works as *jaib* ("bay"). Thus, rather than transliterate it (which would have given us the "jib" and "cojib" of an angle), he translated the word as *sinus* (Latin for bay). Thus the Hindu and Arabic "half-chord", which is perfectly descriptive, became sine, which is perfectly nonsensical.[2]

Aḍud's court also included ABŪ SAHL WAIJAN IBN RUSTAM AL-KŪHĪ (fl. 970–1000), a marketplace juggler turned geometer. Abū Sahl's geometric work included solving the problem of constructing a regular heptagon and inscribing an equilateral pentagon in a square. He accomplished both of these using conic sections. A remarkable feature about these solutions is that they are equivalent to solving fourth-degree equations by means of intersecting conic sections.

Archimedes's *[Inscribing a Regular] Heptagon in a Circle*, translated by ibn Qurra, did not actually produce a constructible figure; it only identified a key relationship between the side of a heptagon and another figure. Abū Sahl completed the construction, and described his work in a treatise dedicated to Aḍud.

First, Abū Sahl showed that the problem of constructing a regular heptagon can be reduced to constructing a triangle AGB with angles in ratio of $1 : 2 : 4$ (see Figure 4.5); if a circle is inscribed about triangle AGB, then BG will be the side of a regular heptagon inscribed in the same circle. The heptagon can then be scaled to inscribe it in any given circle.[3]

To construct this triangle, extend BG towards D and E so $BE = AB$ and $GD = AG$. It is easy to show that $\triangle DAB \sim \triangle AGB$, so $DB : AB = AB : GB$, or $AB^2 = GB \cdot DB$; since $AB = BE$, this is the same as $BE^2 = GB \cdot DB$. Likewise $\triangle GEA \sim \triangle AEB$, so $GE : AE = AE : BE$, or $AE^2 = GE \cdot BE$. However, triangle GEA is isosceles, with $AE = AG$, so this last becomes $AG^2 = GE \cdot BE$; finally, since $AG = GD$, we have $GD^2 = GE \cdot BE$. Thus the problem of constructing triangle ABG is reduced to the following: Given a line ED, find points B, G so $BE^2 = DB \cdot BG$ and $GD^2 = GE \cdot EB$.

To find the points B, G, Abū Sahl used conic sections. In modern terms, let $BG = 1$, $BE = x$, and $GD = y$. Then the first equation, $BE^2 = DB \cdot BG$, becomes $x^2 = 1 + y$, while the second equation, $GD^2 = GE \cdot EB$, becomes $y^2 = (1 + x)x$. The first equation corresponds to a parabola and the second to a hyperbola. These curves intersect in the first quadrant, and locating their point of intersection allows us to construct the heptagon.

---

[2] Secant and tangent both have sensible names: secant comes from the Latin *secare*, "to cut," (as the secant line cuts across the circle), while tangent comes from the Latin *tangere*, "to touch."

[3] Alternatively, $\angle AGB$ will be the central angle subtending one side of a regular heptagon.

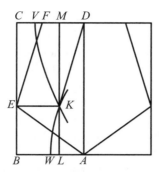

Figure 4.6. Abū Sahl's Equilateral Pentagon in a Square

This problem is very interesting from an algebraic perspective, because if we solve the first equation for y and substitute into the second equation, we obtain the fourth-degree equation $x^4 - 3x^2 - x + 1 = 0$. It is not clear if Abū Sahl realized this: on the one hand, the algebraic manipulations needed to derive this fourth-degree equation were well within the scope of the algebra of his time, and while a fourth power is a non-geometric quantity, Islamic geometers had already begun to view algebraic unknowns as pure numbers. On the other hand, there would have been no reason to *derive* the fourth-degree equation from the relationships needed to construct a regular heptagon.[4]

Another quartic equation appears in Abū Sahl's construction of an equilateral pentagon in a square. Abū Kāmil solved the case where a vertex of the pentagon and the square coincide. The second case inscribes a pentagon in the square as shown in Figure 4.6. In the figure, suppose BA is half the side of the square and FD half the side of the pentagon. Complete the parallelogram $FDKE$, with $FD = EK$ and $EF = DK$; draw ML through K parallel to BC. Let $AB = 2a$, $BL = x$, $KL = y$. By the Pythagorean Theorem, $BE^2 + AB^2 = AE^2$; hence $y^2 + 4a^2 = 4x^2$, which means that K is on a hyperbola with vertex W located at the midpoint of AB. In a similar manner, $DK^2 = MK^2 + MD^2$, or $4x^2 = (4a - y)^2 + (2a - x)^2$, which means that K is on a hyperbola with vertex V where $CV = \frac{1}{3}CD$. Hence K is specified as the intersection of two hyperbolas, and can be found geometrically. Once again, this geometric construction problem corresponds to a fourth-degree equation in x.

ABU BAKR AL KARAJI (fl. 1000) was probably also at ʿAḍud's court, though next to nothing is known about Abu Bakr's life. Even his first name is unknown: "Abu Bakr" simply means "father of Bakr," probably his first born son. "Al Karaji" indicates he came from Karaj, which would make him Persian—but he is also referred to as "al Karhki", which would place his origins in Karkh, a suburb of Baghdad.

Al Karaji wrote a work on algebra, *The Marvelous*, and dedicated it to the local governor, Fakhr al Mulk. Soon afterwards al Karaji left Baghdad for "the mountain countries," where history lost track of him. The two events were probably unrelated, though if the governor fell into disfavor, his supporters might have reason to absent themselves from Baghdad.

Al Karaji took an important step. Previous authors of algebra limited their attention to the square and cube of a variable, since these could be represented geometrically. Diophan-

---

[4]The equation has rational factors, but this fact plays no role in Abū Sahl's solution.

tus felt free to use higher powers, because to him the unknown was simply a numerical quantity. Al Karaji used Diophantine terms such as "square-cube" for x^5, but more importantly, he noted that the higher degree terms in a polynomial could be treated exactly like the digits of a multidigit number. Thus, to add two polynomial expressions, such as "three square-cubes plus four squares plus three units" and "two square-cubes plus three square-squares plus two-squares plus five things plus two units," al Karaji lined up the terms of the same power (just like one would line up the units in performing an addition), then added the coefficients. This seemingly simple but crucial step meant polynomial expressions could be added, subtracted, multiplied, and divided, using the same algorithms that had been developed for working with numbers in positional notation. Al Karaji failed to provide a complete system of arithmetic only because there was no generalized concept of a negative number; in particular, the procedure of subtracting a negative from a negative (i.e., $-a - (-b)$) had no analog among arithmetical algorithms.

4.2.3 Decimal Notation

In 935, the Caliph ar-Rādī gave Muhammad ibn Tughj, the governor of Egypt, the title of *Ikhshīd* ("prince" in Persian). It was the beginning of a short-lived Egpytian dynasty based out of Syria. The dynasty should have been called the Tughjids, after its founder, though history has strange twists, and instead the dynasty came to be known as the Ikhshīdids. The Ikhshīdids quickly became figureheads, and during the reign of Muhammad ibn Tughj, real power was held by Abū al-Misk Kāfūr, an Abyssinian slave who was emancipated and promoted to the position of vizier. Kāfūr managed the affairs of Egypt well, and the Ikhshīdid realm, which included Egypt and Syria, underwent a great cultural flowering.

The earliest surviving examples of decimal fractions come from the Ikhshīdid realm. Around 952 in Damascus, ABU'L-ḤASAN AL-UQLĪDISĪ (fl. 952) wrote the *Book of Chapters on Hindu Arithmetic*, known only from a single copy, made in 1157. "Al-Uqlīdisī" is Arabic for "The Euclidean scribe": he was someone who made his living by making copies of the *Elements* for students of geometry. This is all we know about the life of al-Uqlīdisī.

Before the ninth century, many computations were performed using an abacus, a word of uncertain derivation. The earliest abaci appeared in the Roman world, and the importance of this method of computation can be inferred from the word *calculus* ("pebble" in Latin).

In its simplest form, an abacus is a ruled table on which pebbles are placed in different positions to represent different numbers; addition is accomplished by adding pebbles in the appropriate positions (later Roman versions included the "beads on a wire" type found today). As the pebbles can simply be picked off the dusty ground, the abacus was also known as a "dust board" (and it is sometimes claimed that "abacus" comes from the Semitic word meaning "dust"). For example, suppose we wished to add three hundred eleven to two hundred sixty-four.

On an abacus, three hundred eleven would be represented by placing three pebbles in the hundreds, one in the tens, and one in the units:

We would add two hundred sixty-four by placing two more pebbles in the hundreds, six more in the tens, and four more in the units:

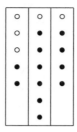

The sum is obviously five hundred, seven ten (seventy), and five. Subtraction would be accomplished by removing pebbles, "borrowing" where necessary. Multiplication could be accomplished through repeated addition, and division through repeated subtraction. In this way, all the ordinary computations could be performed.

After the fall of the Roman Empire, the abacus made its way east with such success that it is still used today throughout Asia; its use was, however, forgotten in the west and only reintroduced with the revival of commerce in the Middle Ages, when again it was a ruled table on which computations were performed. In shops, a customer's goods would be brought to the table and the sum would be totaled on the *counter*.

The abacus had several advantages. First, computation is instantaneous: addition can be accomplished as rapidly as one can set the pebbles down, and as late as World War II, skilled operators of the Japanese abacus (the *soroban*) could outperform American clerks using mechanical adding machines.[5]

But there were many disadvantages. First, the abacus must be on a reasonably level surface, lest the pebbles slide or roll out of position, introducing computational errors. Second, it took considerable manual dexterity to place the pebbles properly with any speed and, at the same time, *not* disturb any pebbles already placed. Third, and most importantly, it was impossible to check computations without redoing them: the abacus has no memory.

Obviously in this case a written record of the computation would be desirable. But one problem existed: in the west before the eighth century, *no* suitable writing material existed. Materials to write upon existed in plenty—but some, like stone, were difficult to work, and others, like vellum (made from sheep skins) were too expensive to use for so mundane a purpose as simple computation.

This began to change during the eighth century. At that time, the Islamic world had recovered from the civil war that brought down the Umayyads, and was rapidly expanding. The Byzantine Empire, fearing the new threat, sent emissaries to the leaders of the T'ang Dynasty in China, and encouraged them to attack the Muslim world from the east. But the T'ang dynasty had difficulty exerting itself so far to the west. At the Talas River in July 751, a T'ang military expedition was decisively defeated by the Muslim armies and the Chinese were expelled from Central Asia.

Some of the prisoners captured at the Battle of Talas were masters of an art unknown outside China: paper making. Made from rags that might otherwise be discarded, paper

---

[5]It was not until the 1970s, when electronic calculators became widely available, that the *soroban* began to be replaced in Japanese stores.

was the first cheap, disposable writing material, and within a century, flourishing paper mills existed from Samarkand (now in Uzbekistan) to Muslim Spain.

The existence of a new type of writing material encouraged the development of new methods of computation. By the tenth century, many of these methods existed, and al-Uqlīdisī collected the best of them in his *Book of Chapters*. The new writing material had many advantages, not the least of which was that the dust board (and dusty hands) were the hallmarks of astrologers (the "mathematicians" censured by Diocletian and Augustine). By using pen and paper, an honest calculator could distinguish himself from a charlatan.

One inconsistency in the way that positional notation was used in the tenth century was that for whole numbers, the places differed by powers of 10, while sexagesimals were used to express numbers smaller than the unit. An "obvious" solution is to use the same system for both: our modern form of decimal fractions is precisely such a solution. But this obvious solution took several hundred years to develop, and it was al-Uqlīdisī who was the first to describe purely decimal fractions by noting that half of 1 was 5 in the "place before." Unfortunately, Europeans would not learn decimal fractions from writers such as al-Uqlīdisī; instead, they would reinvent them—six hundred years later.

4.2.4 The Seljuqs

'Aḍud delayed but could not stop the decline of Buyid power. Upon his death in 983 the state fell apart into chaos; al-Sijzi even complained that in his location (which he did not specify), it was legal to kill mathematicians! By that he probably meant astrologers—which would hardly have been any comfort, for in his lifetime, al-Sijzi was well known as an astrologer.

The longstanding conflict between Persians and Arabs over who would rule the Islamic world would finally be ended in the eleventh century, though the resolution would prove unsatisfactory to both sides. The eastern part of Iran, including Khorāsān, was ruled by an Iranian dynasty called the Samanids. The Samanids appointed Sebüktigin, a freed Turkish slave, to be governor of the city of Ghazna (now in Afghanistan) and the guardian of their eastern border. Soon, Sebüktigin rejected Samanid authority and established himself as ruler in his own right, and by 1005, the Samanid lands were divided between the Ghaznavids in the west and the Qarakhanids (also Turks) in the east.

Under Sebüktigin's son and successor Maḥmūd, the Ghaznavids grew in power and importance, and contemporaries began to refer to him as *Sultan*, an Arabic word implying he had great authority. This would become the title of the ruler in many Islamic lands. Maḥmūd's main interest was India, and he led seventeen invasions to pillage the rich temple cities of the northwest. As a result this region, now the Punjab, is populated by Muslims and Sikhs, the latter adherents to a religion heavily influenced by Islam.

The wealth gained from these invasions allowed Maḥmūd to patronize intellectuals. Perhaps his most noteworthy courtier was ibn 'Irāq's student al-Bīrūnī, who arrived with his teacher ibn 'Irāq around 1016. They remained there for the rest of their lives. Al-Bīrūnī wrote extensively on many subjects, including mathematics and astronomy. The Ghaznavid invasions of India provided him with a first-hand look at the cultural achievements of the subcontinent, leading him to make his observation (quoted earlier) that Indian mathematics was a mixture of "pearl shells and sour dates, or pearl and dung, or costly crystals and common pebbles."

Maḥmūd is also known for his shabby treatment of the Persian poet Ferdowsi. Ferdowsi chronicled the history of the Persian kings in his great epic, the Shāh-Nameh. The nearly 60,000 couplets of the epic are written in very pure Persian, with only a tiny admixture of Arabic words, and Ferdowsi's place in Persian literature is similar to that of Shakespeare in English literature: a thousand years later, Ferdowsi's Persian is still readable by literate Iranians.

Ferdowsi finished the epic in 1010 and dedicated it to Maḥmūd, hoping to obtain the patronage of the Sultan. Unfortunately the Sunnite Maḥmūd was convinced that the Shi'ite poet deserved only a pittance. Disgusted, Ferdowsi gave the money to a bath attendant and a bartender, fled to Herāt, and eventually made his way to the court of Sepahbād Shahreyār at Mazanderan (on the southern shore of the Caspian Sea), a descendant of the kings commemorated in the Shāh-Nameh. Ferdowsi wrote a new introduction to the Shāh-Nameh, which included a 100-verse satire of Maḥmūd, and offered to rededicate the poem to Shahreyār. For his part, Shahreyār wished to maintain cordial relations with Maḥmūd, so he bought the satire (which has been preserved) on the condition that it be removed from the poem. Later Maḥmūd tried to atone for his poor treatment of the greatest poet of his age and sent a gift of 60,000 dinars worth of indigo, but Ferdowsi died before it arrived.

The Ghaznavids would be supplanted by another group of Turks, the Seljuqs (named after their leader). Maḥmūd allowed them to settle in Khorāsān itself, and in 1038, Seljuq's grandson Toghrïl Beg declared himself Sultan. With his brother Chaghri Beg, they drove the Ghaznavids out of northeastern Iran by 1040. Chaghri Beg stayed behind in Khorāsān to consolidate Seljuq rule, while Toghrïl Beg advanced westwards, conquering most of Iraq. In 1055, he entered Baghdad with the declared intention of ridding the Caliphate of the remaining vestiges of Buyid influence. The Caliph was all too happy to let him: the Seljuqs were Sunnites.

Toghrïl Beg realized the greatest danger to his own rule was the other Turkish tribes. Learning from Maḥmūd's mistake, he encouraged the remaining nomadic tribes to go west and attack the Byzantine empire. Subsequently, they conquered a large portion of territory in Asia Minor: modern Turkey.

Though Persians were no longer the rulers, they continued to be the bureaucrats. The most prominent Persian in the Seljuq Empire was Abū Ali Ḥasan ibn Ali, who is better known by his nickname: "The Order of the Kingdom", or Niẓām-al-Mulk. The son of a Ghaznavid tax official, Niẓām-al-Mulk and his father fled when the Seljuqs conquered Khorāsān. Later, Niẓām-al-Mulk returned and joined the service of the governor, Alp-Arslan. Eventually Alp-Arslan became Sultan, and Niẓām-al-Mulk became his vizier.

Niẓām-al-Mulk's policy was simple, and understandable: the Persians had governed an empire for centuries, while just a few generations before, the Turks were a group of nomads. Thus the best thing the Turks could do was to adopt the Persian forms of government and political practice. As this meant the establishment of a single, central, universal monarch, Niẓām-al-Mulk molded imperial administration along Persian lines. Under Niẓām-al-Mulk's guidance, the empire ran so efficiently that by 1092 he could boast that he was the ultimate source of the Sultan's power.

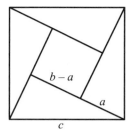

Figure 4.7. Abu'l-Wafā's Proof of the Pythagorean Theorem

4.2.5 Al-Khayyāmī

Alp-Arslan was succeeded in 1072 by his son, Malik-Shāh, who made his capital at Isfahan in Iran. He had a great interest in literature, art, and science, and had an observatory built in Isfahan. To staff it, he and Niẓām-al-Mulk extended an invitation to one of the rising stars of Islamic science: 'UMAR IBN IBRĀHĪM AL-NĪSĀBŪRI AL-KHAYYĀMĪ ("Omar from Nīshāpūr, son of Abraham the Tentmaker") (May 15, 1048–December 4, 1131).[6]

Al-Khayyāmī (better known in English as Omar Khayyam) is known for the Rubaiyat of Omar Khayyam, a series of short, four-line poems. Lines from the Rubaiyat include, "a loaf of bread, a jug of wine, and thou," as well as, "The moving finger writes, and having writ, moves on," though how much of the Rubaiyat was actually written by Al-Khayyāmī, and how by other authors (including FitzGerald, the English translator) is not known for certain.

While in Isfahan, al-Khayyāmī attended a meeting between geometers and artisans to discuss the decorative arts. These meetings were fairly common (Abu'l-Wafā described one he attended in Baghdad), and suggest that many of the design motifs in Islamic art originated after consultation with mathematicians. In particular, since a strict interpretation of the Shari'ah prohibits the depiction of living objects, intricate geometric patterns became a common design element in Islamic art.

Key to designing tessellations was finding polygons that fit together. Since the only regular polygons that can be used to tile the plane are equilateral triangles, squares, and regular hexagons, some method of generating other shapes was desirable. Abu'l-Wafā's work *The book on what the artisan requires of geometric constructions* discussed meeting a group of artisans in Baghdad. During the course of one discussion, the problem of constructing a square composed of three squares arose: in other words, given a square of side s, to construct another of side $\sqrt{3}s$. This led to a discussion of the Pythagorean Theorem, during the course of which Abu'l-Wafā gave a proof of the theorem tailored for the needs of the artisans: a dissection proof (see Figure 4.7), where a small square is surrounded by four identical right triangles. The five pieces together correspond to c^2, and they can be rearranged into a figure that consists of two squares, one of side a and the other of side b.

The figure held some intriguing possibilities for artisans. Suppose we take four more right triangles, of the same size and shape as the originals, and add them as shown in Figure 4.8. This is the beginning of a pattern that can be easily extended in all directions.

---

[6]Besides being where Toghïl Beg declared himself Sultan in 1038, Nīshāpūr was already well-known for a type of glazed pottery made there.

Figure 4.8. Beginning of a Tessellation

Moreover, the overall impression is easy to vary by changing the dimensions of the original squares and triangles.

Since all squares are similar, creating different tessellation patterns is a matter of varying the quadrilateral formed by joining the two right triangles along their hypotenuse. Joined as in Figure 4.9, the two form a figure Islamic artisans called an "almond." In general, let ABC, AEC be identical right triangles (with $\angle ABC$, $\angle AEC$ being right angles), and join them as in Figure 4.9. Take $AH = AB$, and draw FG through H and perpendicular to AC. It is easy to show that $BF = FH$, $EG = GH$, and thus the quadrilaterals $ABFH$, $AEGH$ form another pair of almonds. If the sides of the original right triangle satisfy certain conditions, then the resulting almonds can be used to tile the plane. Such a tiling may be seen on the west *iwan* (a semi-open vault) on the Great Mosque (actually a complex of several buildings) in Isfahan, among other places.

Although this dissection holds in general, a particularly interesting one emerges if we require that $FC = AB$. In order to determine the dimensions of the original triangle ABC, we might note that since $AH = AB$ and BC is perpendicular to AB, then BC will be the tangent to a circle with center at A and radius AH (see Figure 4.10). Al-Khayyāmī showed that the triangle can be constructed by finding the point B so that if BD is drawn perpendicular to the radius AH, then $AH : BD = AD : DH$.

An interesting feature of this triangle is that it appears to have been incorporated into the design of the so-called North Dome Chamber of the Friday Mosque of Isfahan, constructed while al-Khayyāmī lived in the city. The North Dome Chamber's construction is remarkable for its precision: the chamber is 9.9 meters across and the floor is almost perfectly level, with the center being only about 2 cm lower than the edges, while the walls deviate

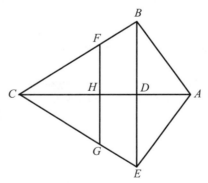

Figure 4.9. Artisan's Almond

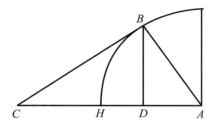

Figure 4.10. Al-Khayyāmī's Triangle

a maximum of about 3 cm from the vertical. In contrast, the precision in the construction of the South Dome Chamber is markedly inferior. The contrast between the two has led to speculation that al-Khayyāmī was involved in the construction of the North Dome Chamber (the construction of the South Dome Chamber, in contrast, was supervised by Abu'l Fath, the son of Muhammad the treasurer and, in all likelihood, a political appointee.

In an untitled treatise, al-Khayyāmī discussed this triangle. First, he proved that $AC = AB + BD$. Next, he noted the value of algebraic notation, which made multiplication and division of expressions much easier, and used it to show that the problem yielded a cubic equation. Specifically, if $AD = 10$ and $BD = x$, then we have $AB^2 = 100 + x^2$. By similar triangles, $\frac{AC}{AB} = \frac{AB}{AD}$, so $AB^2 = AC \cdot AD$, or $AC = \frac{100+x^2}{10} = 10 + \frac{1}{10}x^2$. Since $AC = AB + BD$, then $AB = 10 + \frac{1}{10}x^2 - x$. Squaring both sides gives $AB^2 = 100 + \frac{1}{100}x^4 + 3x^2 - 20x - \frac{1}{5}x^3$ or, since $AB^2 = 100 + x^2$, this equation can be reduced to $x^3 + 200x = 20x^2 + 2000$. Al-Khayyāmī solved this problem using conic sections. It is worth noting that in the course of deriving this cubic equation, al-Khayyāmī used the non-geometric fourth power.

In his untitled treatise, al-Khayyāmī expressed a hope of writing a treatise giving a complete geometric solution to the cubic, including a discussion of the existence of solutions. This was the focus of *Treatise on Demonstrations of Problems of Restoration and Reduction*. His work was probably the most rigorous of all the Arabic algebras, for he based his proofs solidly in geometry. Indeed, he noted that the work could not possibly be understood by those who did not understand Euclid's works thoroughly, both the *Elements* and the *Data*, as well as the first two books of Apollonius's *Conics*—a hefty prerequisite, considering that the *Data* and *Conics* were part of Pappus's domain of analysis!

In *Treatise*, al-Khayyāmī solves cubic equations geometrically. Recall that the problem of duplicating the cube had been reduced by Hippocrates to the problem of finding two mean proportionals, and that this problem was subsequently solved by Menaechmus using conic sections. Al-Khayyāmī presented a different method of solving the problem of finding two mean proportionals. In modern terms, Al-Khayyāmī's method is the following: Let a, b be the two given quantities, and x, y the desired mean proportionals, where $a : x = x : y = y : b$. From $a : x = x : y$, we have $ay = x^2$, which corresponds to a parabola. From $x : y = y : b$ we have $bx = y^2$, which corresponds to another parabola. Thus x, y will correspond to the intersection point between two parabolas.

Since the problem of inserting two mean proportionals is equivalent to the problem of solving a degenerate cubic (in this case, $x^3 = a^2b$), it seems plausible that other cubic equations could be solved in a similar fashion. According to the introduction, al-Māhānī came upon an equation involving cubes, squares, and numbers, which no one could solve. The problem originated from Eutocius's commentary on Archimedes's *On the Sphere and Cylinder, Part II*, and is equivalent to the cubic equation $x^2(a - x) = b^2c$ where a, b, and c are known positive quantities. ABU JAFAR AL KHAZIN (d. ca. 965) solved the problem using conic sections; other geometers found solutions to other types of cubic equations. Al-Khayyāmī's work is a summary of their results. It is worth noting that Al-Khayyāmī made no mention of Abū Sahl's work, and even denies the possibility of solving a fourth-degree equation, which suggests that the connection between Abū Sahl's geometric construction problems and the quartic equation had not been made.

For example, consider cubics of the form $x^3 + p^2x = p^2q$ (where p, q are positive quantities). This problem is equivalent to solving the proportionality $p : x = x : y = y : (q - x)$ as follows. From $p : x = x : y$ we have $p^2 : x^2 = p : y$. From $p : x = y : (q-x)$ we have $p : y = x : (q - x)$; hence $p^2 : x^2 = x : (q - x)$, or $p^2(q - x) = x^3$ or, rearranging, $x^3 + p^2x = p^2q$. Hence if we can solve the proportionality, we can solve the cubic. From $p : x = x : y$, we have $py = x^2$; hence x, y corresponds to a point on a parabola. From $x : y = y : (q - x)$ we have $y^2 = x(q - x)$; hence x, y corresponds to a point on a circle. Thus the desired point is at the intersection of a parabola with parameter p and a circle with diameter q. The other types of cubic equation can be solved in a similar fashion.

Al-Khayyāmī also attempted to prove Euclid's fifth postulate, using a quadrilateral with two sides AC and BD perpendicular to the base AB (what is now called a Saccheri quadrilateral). Without using the fifth postulate it is possible to prove the two remaining angles must be equal, and hence they are either both acute, both obtuse, or both right. Al-Khayyāmī proved that the two must be right angles by appealing to a principle of Aristotle that two converging lines necessarily intersect—a principle equivalent to the fifth postulate.

4.3 The Age of Invasion

In Clermont, southern France, Pope Urban II called for a war against the Muslims who held Jerusalem and Palestine. The war was called a crusade (from *crucis*, "cross" in Latin) to indicate its religious nature. In 1092, five waves of zealous crusaders swept across Europe. They fought anyone they came across, robbed anyone they could, and massacred entire communities of Jews—all before they left Europe. Two groups, consisting of perhaps 12, 000 people, made it to Asia Minor, where, for the first time, they encountered someone who could fight back. The Muslims annihilated them.

The nobles of France organized their own campaign. They, too, would have met with disaster but for a crucial event. The attempts by the Sunnite Turks to restore the political unity of the Caliphate met with failure, because by the eleventh century, the Islamic world consisted of many independent states and territories that fought unification. Once again, Gilan enters into the history of Islam, for out of Gilan came the most notorious opposition to the Seljuqs: the Nizari branch of the Isma'ili sect of Shi'ism, led by Ḥasan-e Ṣabbāh.[7]

---

[7] There is an oft-repeated story that al-Khayyāmī, Niẓām-al-Mulk, and Ḥasan-e Ṣabbāh were classmates. The fact that Niẓām-al-Mulk was born about 110 years before the deaths of al-Khayyāmī and Ḥasan-e Ṣabbāh makes this unlikely.

In 1090, the Nizari seized the fortress of Alamut and began to use it as a base of operations against the Turks. They expressed their dissatisfaction with the Seljuqs by periodically murdering key figures in the government and military. According to rumor, the agents of the Nizari would smoke hashish to induce visions of the Paradise they felt would be their reward for murder; on the basis of this rather improbable story, they became known as *assassins* ("hashish addicts").

In 1096, an assassin killed Nizām-al-Mulk on the road from Isfahan to Baghdad. Nizām-al-Mulk's boast that he was the ultimate source of the Sultan's power proved all too correct, and without his guidance, the Seljuq state fell apart into a dozen quarreling fragments. The chaos that followed allowed the Shi'ite Fatimids of Egypt to take Jerusalem from the Seljuqs in 1098. But before they could consolidate their hold on Palestine, the French arrived.[8] They overwhelmed the small garrisons and quickly captured Jerusalem. Six hundred years before, Jerusalem surrendered to the Muslims and the safety of the inhabitants was personally guaranteed by the Caliph. On July 15, 1099, the Crusaders entered the city and massacred the Jewish and Muslim inhabitants of the city.

The Seljuq world was too busy trying to reorganize itself after the death of Nizām-al-Mulk to expel the crusaders, so they were able to establish a number of states in Palestine by the beginning of the 12th century. But at least some of the Seljuq governors took the existence of the crusader states as an affront to their sovereignty. In 1144, Imad al-Din Zangi of Mosul conquered the northernmost crusader state, the County of Edessa (ruled by a Count). Zangi's son, Nur al-Din expelled the Franks from other portions of Palestine, but it would be one of Nur al-Din's generals who would become the greatest hero of the time: Yūsuf ibn Ayyūb (Joseph, son of Job), who later became known as *Saladin* ("Righteousness of the Faith"). Saladin reconquered Palestine and retook Jerusalem on October 2, 1187. A century before, the victorious Franks had massacred the Muslim and Jewish population of Jerusalem. Now the city was filled with Christians, and it would have been all too easy to allow a similar massacre, this time of the Christians. But Saladin allowed his troops no license—and they obeyed.

4.3.1 Al-Samaw'al

One of Saladin's contemporaries was AL SAMAW'AL BEN YAHYA BEN YAHUDA AL MAGHRIBI (d. 1180), a Jew born in Baghdad. Although Jews and Christians were generally tolerated in Muslim lands, they faced significant legal and financial discrimination. Consequently many converted for practical reasons. For al Samaw'al, conversion was a result of careful thought. After having examined the basis for the three main faiths, Samaw'al decided that Islam was the best supported, and he converted on November 8, 1163; he later wrote a book, *Decisive Refutation of the Christians and Jews*, supporting his conclusions.

By Samaw'al's time, the golden era of Islamic mathematics had passed. The state of mathematics education was so poor he could find no one to teach him more than the first few books of Euclid, so he studied the works of the ancient Greek geometers on his own. The arithmetization of algebra begun in the time of al Khwārizmī was completed when Samaw'al gave clear procedures for finding polynomial quotients, as well as an explanation of how to handle the subtraction of two negative quantities. The procedures for handling

---

[8]Since most of this first wave of crusaders were French, the Muslims soon began calling the crusaders, and westerners in general, "Franks."

these tasks appeared in his *Shining Book of Calculations*, a title that may seem odd, but even today we use "clear" to indicate both transparent and comprehensible, and we "shed light" on a subject that is hard to understand.

Notation had advanced little. Samaw'al used terms like "square-square" to mean x^4 and "part-square" to mean $\frac{1}{x^2}$; in the example below, we will use modern notation. Consider the quotient of $2x^4 + 3x^3 + 5x^2 + 8x + 5 + \frac{1}{x}$ divided by $x^2 + 2 + \frac{1}{x}$. First, write down the powers in their "natural order." Then, below them in the appropriate column, write the coefficients of the dividend, and below them, the coefficients of the divisor, starting in the leftmost column. If a term is missing, write '0'. Thus our division would begin by setting down:

x^4	x^3	x^2	x	1	x^{-1}
2	3	5	8	5	1
1	0	2	1		

Samaw'al's procedure was meant to be performed on a counting board, not paper; hence, it may seem more complicated when reduced to print. First, the leading term of the divisor, 1, is divided into the leading term of the dividend, 2; since this is really $1x^2$ into $2x^4$, the quotient is $2x^2$. This is written at the top of the x^2 column; then the divisor is multiplied by 2 and the product subtracted:

		2			
x^4	x^3	x^2	x	1	x^{-1}
2	3	5	8	5	1
2	0	4	2		
		3	1	6	

Now the divisor is written down starting with the leftmost column, and the whole procedure is repeated. Notice that the process of writing down the divisor again is exactly analogous to the procedure performed in the division of two whole numbers. The final result is

		2	3	1	
x^4	x^3	x^2	x	1	x^{-1}
2	3	5	8	5	1
2	0	4	2		
		3	1	6	
		3	0	6	3
			1	0	2
			1	0	2	1
			0	0	0

so the quotient is $2x^2 + 3 + \frac{1}{x}$.

Samaw'al had no difficulty handling negative expressions, using rules identical to our own. Most interesting of all is his example of dividing $20x^2 + 30x$ by $6x^2 + 12$. Just as with whole numbers, some divisions do not "come out even," and one ends with an infinite repeating decimal; Samaw'al performed the division of the two polynomials using the division algorithm and obtained

$$3\frac{1}{3} + 5\frac{1}{x} - 6\frac{2}{3}\frac{1}{x^2} - 10\frac{1}{x^3} + 13\frac{1}{3}\frac{1}{x^4} + 20\frac{1}{x^5} - 26\frac{2}{3}\frac{1}{x^6} - 40\frac{1}{x^7}$$

where he stopped, calling this the "approximate answer"; clearly, he recognized that the division would never terminate. Al Samaw'al was also responsible for introducing a form of decimal notation, centuries before its appearance in Europe.

4.3.2 The Mongol Era

In 1194 Tekesh, the Khwarezm Shah, helped the Abbasid Caliph An-Nāṣir defeat and kill the last Seljuq prince. Tekesh's son 'Alā al-Dīn Muḥammad expanded eastward into Afghanistan, south into Khorasan (now part of Iran), and northward into Transoxania (or Transoxiana, now Uzbekistan and Tajikistan). To offset the territorial ambitions of the Khwarezm, An-Nāṣir allied with the Nizari. By their murderous actions the Nizari had isolated themselves from the rest of Islam, so towards the end of the eleventh century the ruler of the Nizari around Alamut pledged to become a Sunnite as part of an effort at reconciliation. An-Nāṣir's role was to encourage the other Sunnites to accept this conversion as genuine, and not as part of some more elaborate plot.

It might seem remarkable that the followers of the most radical branch of Shi'ism would meekly accept their leader's decision to become Sunnite, and even more remarkable that he would survive such apostasy. However, there were two key factors at work. The first was that the Shi'ites, as a minority group whose beliefs invariably called into question the very legitimacy of the Islamic states they lived in, had long ago adopted the doctrine of *taqiyyah* ("concealment"). Most religions reserve terrible punishments for those who would deny their faith; under the doctrine of *taqiyyah*, denial was permissible in certain circumstances. Indeed, part of an-Nāṣir's task was overcoming Sunni belief that the conversion of the Isma'ili was nothing more than *taqiyyah*.

It took the conversion of a Shi'ite *imam* to indicate any real change. According to Shi'ite belief, only an *imam* ("leader") had the ability to correctly interpret the Koran, and their interpretations (and consequently their actions) were beyond question. Most Shi'ites were "Twelvers," who believed that twelve *imams* had existed since the time of Muhammad. The twelfth, who disappeared in 873, was expected to return some day as the *mahdi* ("righteous leader") and usher in a golden age. In 1164 the the Nizaris recognized their leader Ḥasan II as a thirteenth imam; his successors were also recognized as imams, and in 1210, the Nizari imam became a Sunni.[9]

NAṢĪR AL-DĪN AL-ṬŪSĪ (February 18, 1201–June 26, 1274) was the son of a prominent Twelver jurist. Born and educated in Tus (near Meshed, Iran), Naṣīr al-Dīn went to al-Khayyāmī's birthplace of Nīshāpūr to continue his study of philosophy, medicine, and mathematics.

Around the time Naṣīr al-Dīn went to Nīshāpūr, a caravan of 450 Muslim merchants arrived in Otrar, on the Syr Darya River (in Kazakhstan). The governor of the city, believing they were spies for a neighboring power, confiscated their property and ordered their execution. One escaped to tell the story to the local chieftain who had guaranteed their safety: Genghiz Khan. When he sent three ambassadors to Sultan Muḥammad to demand repara-

---

[9]Within Shi'ism sects would arise over the actual number of *imams*. The other important group, besides the Twelvers, are the Seveners, who were politically important because a later Persian dynasty would claim descent from the seventh *imam*.

tions and the punishment of the governor, Muḥammad executed one of the ambassadors and shaved the beards of the other two—a terrible insult—before sending them back.

Genghiz Khan exacted a terrible retribution for the mistreatment of his ambassadors. In 1219 the Mongols invaded the Khwarezm; Muḥammad fled and his army disintegrated.[10] The Mongols habitually spared the inhabitants of any city that surrendered without a fight—and mercilessly massacred the inhabitants of any city that resisted. According to a contemporary account Herāt (in Afghanistan) suffered 1,600,000 casualties and Nīshāpūr suffered 1,747,000. These numbers need not represent the actual populations of these cities—they may include refugees from neighboring regions—and were probably exaggerated, but it is clear that Genghiz Khan's invasion of the Khwarezm was a true catastrophe.

Transoxania and Khorāsān bore the brunt of the Mongol invasion. The catastrophe extended far beyond the civilian casualties. Much of the region is desert, and only irrigation makes agriculture possible. At the time most of the irrigation was accomplished through the *qanāt*, an underground canal. Whether or not the Mongols engaged in systematic destruction of the *qanāt* system, the deaths of so many people meant that the *qanāt*s soon fell into disrepair and the land reverted to desert relatively quickly. Moreover, unlike rainfall (which provides relatively pure water), canal water contains dissolved salts, which remain when the water evaporates and reduce the fertility of the soil. Hence the Mongol invasions not only depopulated the regions, but made it very difficult to *re*populate the regions after the Mongols left.

For example, Naṣīr al-Dīn's home town of Tus was destroyed during the Mongol invasions and never rebuilt. No living inhabitants remained in Samarkand, once the capital of the Khwarezm. The great cultural and educational center of Nīshāpūr took centuries to recover. The major remaining power in the region were the Nizari, relatively safe in their mountain hideouts. Naṣīr al-Dīn's scholarship and educational background, as well as descent from a prominent Twelver jurist, allowed him to become a prominent apologist for the Nizari, so by 1250 he had an honored place at the center of Nizari power: the fortress of Alamut.

Naṣīr al-Dīn wrote commentaries on all the major Greek geometers during this period of Nizari patronage. Like al-Khayyami, Naṣīr al-Dīn also attempted to prove the fifth postulate; again, like al-Khayyami, his proof relied on an assumption equivalent to the fifth postulate.

In addition to the theoretically important attempt to prove the fifth postulate, Naṣīr al-Dīn produced works on practical mathematics, particularly trigonometry. His commentary on the *Almagest* (1247) included further explanations of how chord lengths in a circle could be calculated. Moreover while the *Almagest* and other texts included trigonometry as an important component of the study of astronomy, Naṣīr al-Dīn treated trigonometry as an independent subject. His *Treatise on the Quadrilateral* has been called the first trigonometry text in history, and it includes one of the first proofs of the Law of Sines.

Naṣīr al-Dīn might have spent the rest of his life as a highly respected member of the Nizari court. But a dramatic change occurred in the 1250s. According to one story, which seems unlikely, the Nizaris sent 400 disguised assassins to kill the Great Khan, Möngke

---

[10]There is a tradition that an-Nāṣir invited the invasion of the Khwarezm as a means to eliminate his rivals. However there is no evidence to support this story, and Muḥammad's actions alone would have been sufficient to bring on the Mongol invasion.

(Genghiz's grandson). The assassination attempt failed, but brought the Islamic world back to the attention of the Mongols. Another story is that a *qadi* (judge) in Qazvin (near Alamut) complained to Möngke that he and other prominent citizens were obliged to wear armor underneath their clothes out of fear of the assassins.

In 1253 Möngke sent his brother Hulagu to Persia to deal with the Nizari. Hulagu scoured the mountains of Persia, eliminating one assassin stronghold after another and killing every member of the Nizari sect he could find. In 1256 Hulagu's troops stormed Alamut and wiped out the last of the Nizaris. Since Hulagu was intent on annihilating the Nizari, and since Mongol practice was to kill everyone in a city or fortress that refused to surrender immediately, it is odd that Naṣīr al-Dīn survived and even became one of Hulagu's most trusted (or at least, most listened-to) advisers.

Did Naṣīr al-Dīn help the Mongols conquer Alamut? It is a distinct possibility. Naṣīr al-Dīn's father was a Twelver. Under strict Twelver beliefs, the Isma'ili claim to having an *imam* had to be false, and the subsequent conversion of this false *imam* to the Sunni branch of Islam was nothing less than a betrayal of Shi'ism. In this light, Naṣīr al-Dīn's support of the Nizari was *taqiyyah*. Thus he would have no real allegiance to the Nizari, and when confronted with a choice between certain death at the hands of the Mongols, or betrayal of a group that never held his allegiance, he chose survival.

In 1258 Hulagu sacked Baghdad, massacred its inhabitants, destroyed its irrigation system, and burned most of the city. The Mongols continued west towards Syria and won quick victories against the Ayyubids (the dynasty established by Saladin). But then Möngke died in 1259, and Hulagu withdrew most of his army to await developments. The Mamluks, who had usurped Ayyubid authority in Egypt, defeated the the Mongol rearguard at 'Ayn Jalut (September 3, 1260). While his brothers Ariq-böke and Kublai fought a four-year war, Hulagu established his power in Persia and the surrounding regions. When Kublai emerged victorious in 1264, Hulagu acknowledged his supremacy. Kublai in turn allowed Hulagu to remain in control of Persia, establishing the Il-Khans ("subordinate Khan" in Persian).

Hulagu established his capital at Maragheh (now in Iran), and gave Naṣīr al-Dīn a prestigious position. Like the Turks before them, the Mongols were content to let the Persian bureaucracy run the country, and Naṣīr al-Dīn even wrote a treatise on finance for the edification of Hulagu's son and successor Abaqa, as well as tracts supporting Twelver Shi'ism, which helped cement his position as one of the great Shi'ite philosophers.

Naṣīr al-Dīn also persuaded Hulagu to fund a great research complex near Maragheh on a hill overlooking the city. The complex included a scientific library and school, and an observatory. Twelve years of planetary observations led to the publication of the *Il-Khanic Tables* and a new theory of planetary motion based completely on uniform circular motion.

Hulagu's successors were unable to weld the Il-Khanate into a unified state, in part because of the inability to create a workable system of taxation. Thus the Il-Khanate lurched from financial crisis to financial crisis and upon the death of the last Il-Khan in 1335 the state vanished and no centralized power emerged in Persia to succeed it.

Northeast of the Khwarezm lay Transoxania, still recovering from the initial incursions of the Mongols under Genghiz Khan and ruled by descendants of Genghiz's son Chagatai. But in 1364 a local chieftain named Timur began a revolt against the Chagatai Khans. Timur, who had been wounded in an early skirmish and walked with a limp, was known derisively as Timur Lenk ("the Lame"): Tamerlane. By 1370 he had overthrown the Chagatai Khans, and set himself to the task of reuniting the Mongol Khanates. For the rest of his life Timur dedicated himself to this goal, waging war on all his neighbors.

Timur took the tactics of terror and refined them to a horrific art. At Isfahan, he had 70,000 severed heads piled into a great tower; this became the characteristic feature of Timur's conquests. The region of Sīstān (now in western Iran) saw some of the worst excesses: the main city was completely eradicated, and after the conquest of one town, Timur built a tower out of bricks, mortar—and 2000 still-living men.

Frequently the only survivors of Timur's massacres were artisans. Many of these were brought back to his capital Samarkand (near his birthplace) or sent to rebuild and beautify the cities destroyed by Timur. The tomb of Ali in Mazār-e Sharif (Afghanistan) is a fine example of Timurid architecture, while Timur's mausoleum in Samarkand is another.

Upon Timur's death in 1405 his empire was divided among his sons. Shah Rokh received Khorāsān and established his capital at Herāt, but soon secured control over most of the rest of the Timurid Empire. Shah Rokh turned Herāt into a great artistic center. The Persians ignored the Shari'ah's prohibition against making images of living objects, and found a natural artistic ally in the Chinese who routinely incorporated natural motifs into their painting. The Mongols brought the Chinese sensibilities to Persia, and a new style of artistic expression emerged: miniature paintings.

According to one story, miniature painting was invented by Ahmad Musa at the court of the last Il-Khan, Abū Sa'īd. Under Shah Rokh and his son Baysunqur, a flourishing school of Persian miniature painting evolved and reached its high point during the period between 1420 and 1440. Some of the most beautiful examples of Persian miniatures appear in a copy of the Shāh-Nameh completed by Baysunqur's master calligrapher Ja'far.

Another of Shah Rokh's sons proved to be an even more remarkable patron of the arts and sciences. Given the city of Samarkand to rule, ULŪGH BEG (1394–October 27, 1449) turned it into a great cultural center, supporting poets, scholars, and astronomers.

In 1428 Ulūgh Beg began construction of an observatory at Samarkand, and collected a team of mathematicians and scientists to help prepare a new set of astronomical tables. He was no mere patron of the sciences: he collected his own observations and discovered errors in Ptolemy's tables of planetary position. He was said to possess the *second* greatest mind in Samarkand. This accolade came from the possessor of the greatest mind in Samarkand: GIYĀTH AL-DIN JAMSHĪD MASŪD AL-KĀSHI (d. June 22, 1429).

Al-Kāshi's least important major work was *The Calculator's Key*, completed in 1427 and dedicated to Ulūgh Beg. It included a method of finding the fifth root of a number that could be generalized to finding nth roots in general, and is yet another instance of Horner's method discovered centuries before Horner. To illustrate the basic principles behind al-Kāshi's method, suppose we wish to find $\sqrt[5]{12,798,253}$. First, we note

$$20^5 = 3,200,000 < 12,798,253 < 30^5 = 24,300,000.$$

Hence the root is $20 + b$, where b is some number between 0 and 10. To determine the root to the nearest whole number, find the whole number b where

$$(20+b)^5 - 20^5 < 12,798,253 - 20^5. \tag{1}$$

Expanding the left hand side and simplifying we see that:

$$5(20)^4 b + 10(20)^3 b^2 + 10(20)^2 b^3 + 5(20)b^4 + b^5 < 9,598,253. \tag{2}$$

Although the left hand side can be evaluated directly, al-Kāshi's evaluation essentially relied on noting it could also be written as:

$$b(5 \cdot (20)^4 + b(10(20)^3 + b(10(20)^2 + b(5(20) + b)))) < 9,598,253. \qquad (3)$$

By finding the highest b that satisfies this inequality, the next digit of the root may be found.

The most intriguing aspect of al-Kāshi's method is that it is virtually the same as that used by the Chinese centuries earlier. Did the Mongols bring the methods of Chinese mathematics to the Islamic world? The idea might seem far-fetched, as the Mongols were not particularly interested in mathematics or the other accouterments of civilization. However they frequently brought Chinese officials in to govern the areas they had conquered, and it is conceivable that these officials or their subordinates brought the Chinese computational methods to the Islamic world.

Far more interesting than al-Kāshi's computational procedures are what he did with them. Ptolemy's *Almagest* gave all the theoretical foundations for computing a table of sines of unlimited accuracy, including what we would now call the half-angle and angle-difference formulas. Since $\sin 30°$ and $\sin 18°$ represent geometrically constructible chords, then these two values may be found exactly; using the half-angle and angle-difference formulas, we can then find an exact expresson for $\sin 3°$. The problem is that no combination of half-angles and angle-sums will serve to determine $\sin 1°$ exactly, so it is impossible to construct a table of exact values of sines of angles differing by $1°$. Ptolemy solved this by approximating $\sin 1°$ by bounding it between $\sin \frac{3}{4}°$ and $\sin \frac{3}{2}°$. However, because the approximation is based on two fixed quantities, there is no systematic way to improve it; the use of these two values limits the approximation of $\sin 1°$ to about six decimal places.

Al-Kāshi found a different method which relied on an identity we would now write as $\sin 3\theta = 3 \sin \theta - 4 \sin^3 \theta$. If $\theta = 1°$, then the left-hand side is $\sin 3°$, which is known exactly; finding $\sin 1°$ then becomes a matter of solving a cubic equation. Although al-Kāshi knew no algebraic method of solution, he used a method of approximation to find $\sin 1°$ to sixteen decimal places. This allowed him to create a table of trigonometric chord values of unprecedented accuracy. Moreover, he could compute the perimeter of a regular $805,306,368$-sided polygon inscribed in a circle; if the circle has a radius of 1, al-Kāshi determined that the polygon would have perimeter 6.2831853071795865, giving an approximation of π accurate to sixteen decimal places—a world record at the time.

After al-Kāshi's death, Ulūgh Beg created a table of sine and tangent values accurate to eight decimal places; contemporary accounts indicate that he did much of the work himself. But the other Timurid princes did not appreciate Ulūgh Beg's work, and after Shah Rokh's death in 1447, they revolted, deposed him, and Ulūgh Beg's own son 'Abd al-Laṭīf had him executed in 1449. The death of the scholar-Khan Ulūgh Beg is symbolic, for Islamic learning was in decline. In 1453, the Turks conquered Constantinople, and refugees brought Greek primary sources in mathematics to the west. There they found a mathematical tradition that had been nurtured by Islamic discoveries from Spain, and in the years following, modern mathematics was created.

4.4 Andalusia

By the fifth century, Roman rule in Spain had effectively vanished and the peninsula was invaded by several groups of Germanic tribes. The Vandals crossed the narrow strait sepa-

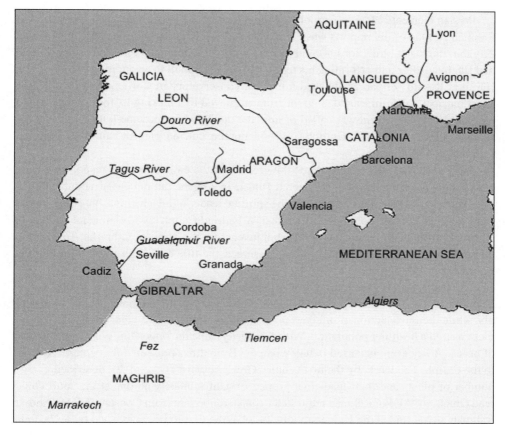

Figure 4.11. Andalusia

rating Europe from Africa and established a kingdom in what is now Libya and Tunisia. In 455, they sacked Rome itself, leaving such destruction in their wake that ever since reckless destruction of property has been known as Vandalism. Moreover, despite the fact that the Visigoths displaced the Vandals a few years later, the Iberian peninsula remained in name the homeland of the Vandals: Andalusia.

In 711 Mūsā ibn Nuṣayr, the governor of North Africa, sent an army under the command of Ṭāriq ibn Ziyād across the strait to the Iberian peninsula. The mostly Berber army made landfall at the foot of a massive rock thenceforth known as the Mountain of Ṭāriq: *Jabal Ṭāriq* in Arabic, which became Gibraltar in English. The Visigoths under their king Roderic were crushed on the shores of lake La Janda on July 19, 711; Roderic died soon after, and by 719 the conquest of the peninsula was virtually complete.

After the Abbasids overthrew the caliphate in Damascus, they murdered every Umayyad they could find. 'Abd-ar Raḥmān survived by fleeing to Spain in 755. The governor of Andalusia, Yūsuf al Fihrī, offered the former Umayyad prince his daughter in marriage and a prosperous estate, provided he refrained from meddling in political affairs. 'Abd-ar Raḥmān declined, and a brief civil war followed. On May 14, 756, he assumed the title of *amīr* ("prince" or "commander"), which signified a nominal allegiance to the Abbasid caliph.

Persian elements came to Cordoba by the efforts of the fourth Umayyad ruler, 'Abd-ar Raḥmān II. One of his imports was the musician Abū 'l-Ḥasan 'Alī ibn Nāfi', nicknamed Ziryāb ("the black bird", for his complexion). According to a rather dubious story, Ziryāb fled Baghdad after outperforming his teacher Isḥāq al Mawṣilī in an audition before Caliph Hārūn al Rashīd. Ziryāb established a musical conservatory in Cordoba and brought the latest musical fad from Baghdad: an instrument played by plucking its four strings with a plectrum. The instrument was called *al'ud* ("the oud"), which became lute. Ziryāb set the standard for high culture in Cordoba, introducing fine cuisine, elegant hairstyles, seasonal fashions, and toothpaste.

'Abd-ar Raḥmān III is considered the greatest of the emirs, and was responsible for a key change. After the Fatimids conquered Tunisia in 909, the Emirate faced a double threat. The Abbasid Caliph might claim to be the spiritual leader of all Muslims, but could hardly exert political control from far-off Baghdad. A Fatimid Caliph could claim and attempt to enforce spiritual *and* political control from just across the Strait of Gibraltar. To forestall this possibility, 'Abd-ar Raḥmān himself assumed the title of Caliph in 929, turning the Emirate into a Caliphate.

The tenth century was a time of great prosperity for the Caliphate of Cordoba. Crops newly introduced to Spain, such as rice and hard wheat, ensured a more reliable food supply, while the introduction of oranges, pomegranates, spinach, and other fruits and vegetables ensured a healthier population. With prosperity came the opportunity to pursue the arts of peace. A key event occurred in 949 when the Byzantines presented the *Materia Medica* to the Caliph. The work, by the first century Greek botanist Dioscorides, described a large number of plants and their medicinal properties. Unfortunately no one at the court could read Greek, so 'Abd-ar Raḥmān requested a translator be sent from Constantinople. 'Abd-ar Raḥmān also founded the University of Cordoba, the first to be established by an Islamic state.

His son al-Hakam II continued his legacy, and assembled a great library in Cordoba with some 400,000 volumes. Books were imported from as far away as Persia and duplicated by copyists, including at least one woman, a poetess named Lubna. The process was made easier because the paper industry arrived in Andalusia the century before. Cordoba and, by association, Muslim Spain became renowned amongst Europeans as a center for learning; according to some stories, Gerbert of Aurillac traveled as far as Cordoba in search of wisdom and learning.

If he did (and there is neither evidence to support nor reason to disbelieve), he might have met ABŪ'L-QĀSIM MASLAMA IBN AḤMAD AL MAJRĪTĪ (ca. 950–1007). "Al Majrītī" indicates that he was an Arab from Madrid. Judging by the works attributed to him, Maslama was probably al-Hakam's Court Astrologer. Maslama adjusted al Khwārizmī's astronomical tables so that they could be used in Cordoba and translated Ptolemy's *Planisphere* into Arabic. Maslama's original work included a text on the use of the astrolabe (later translated by Adelard of Bath) and a treatise (now lost) on commercial arithmetic. Maslama founded a school and some of his students wrote additional treatises (also lost) on commercial arithmetic; contemporary sources suggest that they had access to the works of Euclid, Archimedes, and Abū Kāmil. Given time, Cordoba would have become one of the great intellectual centers of the later middle ages. But time was the one thing the Caliphate of Cordoba did not possess.

When al-Hakam died in 976, his son and successor Hisham II was only eleven years old. The Caliphate was ruled by a triumvirate consisting of al-Hakam's prime minister, al Mushafi; his general, Ghalib al Nasiri, and the rising star in the bureaucracy, Muhammad Ibn Abi Aamir. Ghalib and Muhammad, both Arabs, deposed the Berber al Mushafi in 978. Three years later, Muhammad deposed his rival and became undisputed leader of the Caliphate. He assumed the nickname al-Mansur bi'llah: "Victorious by the grace of God," and is frequently referred to as Almanzor.

According to some estimates, al-Hakam left a treasury filled with forty million dinars, or about six years of state revenue. Al-Mansur spent much of it on construction projects, rewards to supporters, and payoffs to enemies. By cutting a tax on olive oil, he gained widespread support but greatly reduced revenue. To prevent financial crisis, he resorted to an age-old strategy: war. The arms factories of Cordoba could well support such campaigns: by mid-century yearly production reached 12,000 bows, 240,000 arrows, 1300 shields, and 3000 tents. A successful military campaign meant plunder and slaves. Remarkably al-Mansur, trained from birth to be an administrator, proved to be a successful general, leading at least fifty-seven victorious campaigns against the rest of Christian Spain.

Unfortunately, successful military campaigns require powerful armies. Al-Mansur's army drew heavily upon the Berber tribes of North Africa. Al-Mansur and his 'Abd al Malik maintained their control over the Berbers, but upon the latter's sudden and unexpected death in 1008, Hisham and the Berber generals saw an opportunity to retake control of the Caliphate of Cordoba. The result was civil war. In May 1013, Berber tribes sacked Cordoba, destroyed the palaces, and slaughtered thousands. The 400,000-volume caliphal library vanished.

4.4.1 The Mathematicians of the Party Kings

By 1031, the Caliphate of Cordoba had been replaced by more than sixty independent states whose rulers came to be known as the "Party kings," referring at first to the fact that each ruler presided over a political and religious party, or *tawā'if*. Some of the party kings were Berbers; others were Arab; most were professional soldiers. The predominate *tawā'if* states were Seville and Granada in the south; Badajoz, Toledo, and Valencia in central Spain; and Saragossa in the northeast.

The greatest mathematician of the era was ABU 'ABD ALLĀH MUHAMMAD IBN MU'ĀDH AL JAYYĀNĪ (989–1079). From his name we know he lived in the Emirate of Jayyan, a minor state north of Granada and east of Cordoba. Born in Cordoba in the last years of the Caliphate, al Jayyānī lived in Cairo between 1012 and 1017, and thus missed the sack of the city. Later he returned to Jayyān and became a *qāḍī* (judge), specializing in inheritance law.

Al Jayyānī's primary works were *Maqāla fi sharh al nisba* ("Commentary on the concept of ratio") and *Kitāb majhūlāt qisī al kura* ("Unknown arcs of the sphere"). The first was a commentary on Euclid's definition of ratios in Book V of the *Elements*, and might have been the basis for a rigorous formulation of the theory of the real numbers. The second was one of the first complete textbooks on spherical geometry, and is noteworthy for the fact that, for the first time, it treated the subject independent of its applications to astronomy. The work shows a thorough awareness and understanding of the great works in trigonometry carried out by the contemporaneous scholars in Baghdad, such as al Bīrunī, showing that communication between Baghdad and the rest of the Islamic world still existed.

In the Emirate of Valencia, ABŪ ZAYD ʿABD AL RAḤMAN IBN SAYYID (fl. 1063–1096) wrote a (now lost) work on arithmetic theory, apparently based on the *Arithmetic* of Nicomachus. Ibn Sayyid also investigated the conic sections and other, more complex curves, and wrote on the classical problems of trisecting an angle and finding two mean proportionals.

At least one party king achieved prominence as a mathematician: the Emir of Saragossa, ABŪ YŪSUF IBN AḤMAD AL MUʿTAMAN (d. 1085). The work of al Muʿtaman ("the trusted") suggests that some of the 400,000 volumes in the great library of Cordoba survived the sack of the city in 1013, since his personal library included works ranging from Euclid's *Elements* and *Data*, Menelaus's *Spherics*, Apollonius's *Conics*, Archimedes's *On the Sphere and Cylinder*, Thābit ibn Qurra's work on amicable numbers, al Haytham's *Optics*, and Ibrahim ibn Sinan's *Quadrature of the Parabola*. He used these works to write a comprehensive mathematical work entitled *The Perfection* (remember "perfect" also means "complete").

The introduction to *The Perfection* has been lost, so we do not know what al Muʿtaman intended. Some of al Muʿtaman's work verges on plagiarism: ibn Qurra's theorem regarding amicable numbers is reproduced almost verbatim, without attribution (in the surviving portions of *The Perfection*). But in other cases, al Muʿtaman introduces a new proof or a simplification of an existing solution, showing himself to be a highly competent mathematician. For example, he seems to have attempted the quadrature of the ellipse and hyperbola, based on the methods used by ibn Qurra and his grandson, ibn Sinān. Although he did not solve either problem, he discovered a number of theorems in the process.

Perhaps the best evidence of al Muʿtaman's ability (in contrast to his ambition) appears in connection with the work of two problems from ibn al-Haytham's *Optics*. The first is: Given a circle with diameter BG, a point A on the circle, and a line KE of fixed length. Find H on the circle so that if AH and BG are extended until they meet at D, then $DH = KE$. The second problem is to find the point H on the circle so that the chords AH and BG cross at D in such a way that $DH = KE$. Al-Haytham gave different solutions to the two problems. Analytically, however, the two problems are the same, and al Muʿtaman recognized this. Not only did he present a single solution to both problems, but he generalized the problem by allowing BG to be any chord in the circle.

The complex politics of Andalusia can be seen in the life of one of al Muʿtaman's retainers: Ruy Dias de Vivar, a member of the minor nobility in exile from the Christian kingdom of Leon for leading an unauthorized expedition against the Muslim Emirate of Toledo. The Arabic word for such minor nobility was *al Sīd* ("lord"), so in his own lifetime Vivar would become known as *El Cid*, the legendary hero of Spain.

Abraham bar Hiyya Ha-Nasi, known as SAVASORDA (d. 1136), provides another example of the complex politics of Andalusia. Both Christians and Muslims feared internal revolt (from their coreligionists) and external threats (from Muslim or Christian elements, respectively). Consequently both found Jews trustworthy choices for high positions within the government. Judging by his names alone, Savasorda rose to very high rank in the County of Barcelona: *Ha-Nasi* is Hebrew for "the prince," while Savasorda is a corruption of the

Arabic title *ṣāhib al shurṭa*, "chief of police."[11] This might not seem to be a glamorous position, but before the modern era, the police existed to protect the rulers from internal threats. Thus, the *ṣāhib al shurṭa* had to be someone absolutely trustworthy. About the only other thing that is known for certain is that on April 10, 1136, a scribe added "May the memory of the righteous be blessed" after bar Hiyya's name, a benediction ordinarily given to the recently deceased.

Savasorda was a crucial link between developments in Islamic mathematics and the Latin west. His *Foundation of Understanding and the Tower of Faith* was an encyclopedia of practical knowledge in mathematics. Its contents suggest it was written early in his career, since it omits any discussion of astronomy or astrology (subjects he would write several books on). In addition, his discussion of fractions omits any mention of sexagesimal (further evidence that he had not yet studied astronomy) and, though he describes decimal numeration in great detail, he omits any mention of the zero. Yet for all its omissions, Savasorda's *Foundation* was the first algorithm written in the West.

Savasorda also wrote the first algebra in the West: *Treatise on Measurement and Calculation*, which included solutions to quadratic equations. Both *Foundation* and *Treatise* were written in Hebrew, which limited the direct impact of the works, but in 1145 one of Savasorda's colleagues in Barcelona, PLATO OF TIVOLI (fl. 1140) translated the work into Latin. Plato translated other works from Arabic into Latin, and through these and other translations, the learning of the Islamic world came to the consciousness of western scholars.

4.4.2 The Mathematical Diaspora

The Berber tribes of North Africa were converted in the early years of Islam during the westward sweep of the Arab armies. But the Berbers were only given rudimentary religious instruction, and after a pilgrimage to Mecca, Yaḥyā ibn Ibrāhīm al Jidālī, the leader of a confederation of Berber tribes along the Atlantic coast of North Africa known as the Sanhaja, requested a teacher be sent to instruct them in orthodox Islam. Around 1039, 'Abd Allāh ibn Yāsīn al Jazūī, educated in the schools of Andalusia, arrived. Ibn Yāsīn founded a religious community or *ribat* and eventually unified the fractious Berber tribes under the banner of *al Murabitun*, ("The people of the *ribat*"): Almoravid. Since English usually prefaces nationality nouns with an article, this group is usually known as "the" Almoravids, though the usage of the term "Moravid" has become more popular in recent years. Ibn Yāsīn died in 1059, but his successors continued the process of unification until all of Morocco was under Almoravid control by 1083. By 1094 the Almoravids controlled most of southern and eastern Spain, and a new capital had been established at Seville. The twelfth century would see civilization in Muslim Andalusia reach its highest point.

The life of Abū Bakr Muhammad ibn al 'Arabī gives us a good picture of education at the dawn of the 12th century. Al 'Arabī's father was one of the advisers to al Mu'taḍid (his name indicates he was an Arab), and al 'Arabī himself received a good education: he reports being instructed not only in the reading of the Quran, but also in algebra, the geometry of

---

[11] The fact that the title is Arabic should not be interpreted to mean that it was obtained in an Arabic regime; many Spanish titles originate from Arabic terms. For example, the Arabic term for "commander of the fleet" is *amīr-al-mā*, which became *almirante* in Spanish and admiral in English.

Euclid, the use of the astrolabe, and Arabic poetry (both classical and contemporary), all by the age of 16. After spending some years with his father traveling through the Islamic world, al 'Arabī returned to Seville, became a qādī, organized the rebuilding of the city walls, wrote several legal treatises, and gained such a reputation for integrity and honesty that he attracted students from all around Andalusia, as well as the wrath of the unscrupulous, who arranged for his house to be ransacked and vandalized.

A contemporary resident of Seville was JABIR IBN AFLAH (1100–1150), who would play a major role in the development of trigonometry. Aflah's most influential work is *Correction of the Almagest*, a title that verged on scientific heresy. Jabir gave an alternative to Menelaus's theorem on spherical triangles, and his worked formed the basis of large portions of Regiomontanus's *On Triangles* (1484), the first trigonometry textbook written in the West. Regiomontanus neglected to give Jabir credit, a fact pointed out by Cardano.

Given time, the Almoravids might have unified Spain into a single, Islamic state. But just as the orthodox Almoravids swept away the party kings, the Almoravids were themselves swept aside by a new wave of religious orthodoxy from the Berber tribes of North Africa: the Almohads (from *al Muwahhidūn*, "those who believe in the unity of God"). War swept across the Iberian peninsula, and the last Almoravid Emir died during a siege on February 22, 1145. The Almohads consolidated their power in Morocco first, leaving Andalusia to fall into anarchy and chaos.

A refugee from the chaos in the Iberian peninsula was AHMAD AL AB'DARI IBN MUN'IM (d. 1228), originally from Denia in Spain. Mun'im moved to Marrakesh, the former Almoravid capital, and wrote one of the earliest works on combinatorics. The arithmetic triangle appeared when he considered the following problem: Given ten different colors of silk thread, how many bundles could be made consisting of one, two, three, or more strands of different colors. To determine this number Mun'im began by setting down a row of 1s to indicate the ten possibilities for a one-strand bundle:

1st	2nd	3rd	4th	5th	6th	7th	8th	9th	10th	Color
1	1	1	1	1	1	1	1	1	1	First Strand

To determine the number of color combinations in an orderly fashion, he noted that the second color had only one possible color it could be joined to (the first); the third color had two possible colors it could be joined to (the first or second); the fourth had three possible colors, etc. This gave a second row of the table (which he placed above the first):

1st	2nd	3rd	4th	5th	6th	7th	8th	9th	10th	Color
	1	2	3	4	5	6	7	8	9	Second Strand
1	1	1	1	1	1	1	1	1	1	First Strand

For the three-strand combinations, the third color could be combined with the one strand of two colors; the fourth color could be combined with the $2 + 1 = 3$ strands of two colors; the fifth color could be combined with the $1 + 2 + 3 = 6$ strands of two colors, and so on:

1st	2nd	3rd	4th	5th	6th	7th	8th	9th	10th	Color
		1	3	6	10	15	21	28	36	Third Strand
	1	2	3	4	5	6	7	8	9	Second Strand
1	1	1	1	1	1	1	1	1	1	First Strand

In this way the table could be filled out. Others had written about the table of combinations before, but ibn Mun'im was among the first to make a theoretical contribution by noting that the method of generating it meant that each entry was the sum of all the entries on the line below it and to its left.

The Almohads eventually consolidated their rule in Spain, and took a fateful step: They banned all religions but Islam in their territories. Jews were given the choice of exile, conversion, or death. Most chose exile to nearby Christian lands. The educated scholars among the exiles brought with them a knowledge of Arabic and a familiarity with Islamic science.

The refugees included ABRAHAM BEN MEIR IBN EZRA (ca. 1090–1164/7), born in Toledo in the Emirate of Saragossa. Ibn Ezra gained a reputation as a scholar and a poet.[12] Shortly after his only surviving son Isaac converted to Islam and moved to Baghdad, ibn Ezra left Spain with the intention of returning his son to Judaism. When this failed, ibn Ezra went to Rome and spent the rest of his life wandering through Europe. His written works usually listed the date and place they were finished. Based on this evidence, we know he was in Rome (1140), Lucca (1145), Mantua (1145–46), Verona (1146–1147), Beziers (1155), Narbonne (1158), London (1159), Rouen (1159), and Narbonne again (1160). But ibn Ezra always called himself a Spaniard (*Sephardi*), and longed to return home. In 1167 he died en route to Spain.

Ibn Ezra wrote three books introducing Hindu-Arabic numerals. He also translated al Birūnī's commentary on al Khwārizmī's astronomical tables into Hebrew. In the introduction, he gave an account of how Indian science made its way into the Islamic world. According to ibn Ezra, the first Abbasid caliph, al Saffāḥ had a Jew translate an Indian work, the *Lion and the Bull*, into Arabic. Al Saffāḥ was so impressed with the knowledge therein that he sent the Jew to India to try and bring back a Hindu scholar. The Jew recruited a Hindu whose name is given as "KNKH" (written Hebrew, like written Arabic, ordinarily omits vowels), who brought the Hindu system of numeration, as well as much of Hindu astronomy, to the Islamic world.

How reliable is ibn Ezra's account? In general ibn Ezra was a careful scholar, disinclined to accept tradition as historical fact. His position before leaving Spain would have allowed him access to the best Arabic resources available. There is a gratifying lack of hagiography in his account, and "KNKH" might be the Hebrew form of Kanaka, a Hindu mathematician living at the right time and credited with inventing amicable numbers. We may never know how Hindu astronomy and mathematics entered the Islamic world; ibn Ezra's story is at least plausible.

Savasorda and Ibn Ezra helped bring an appetite for Islamic science to Europe, and began a wave of translation that would revitalize mathematics and science in the west. Many libraries in Andalusia held Arabic copies of the great works of Greek science. Texts from these libraries seemed to have survived the chaos of the last years of the Almoravids, the first years of the Almohads, and the concurrent warfare between Christian and Muslim.

For example, Saragossa, which contained the library of al Mu'taman, was conquered by Alfonso I of Aragon in 1118; the remainder of the Emirate of Saragossa was taken, city by city, over the next few years. When Hermann of Carinthia arrived in the region around 1140, he found and translated Euclid's *Elements* and the astronomical tables of al

---

[12] Browning's "Rabbi Ben Ezra" has almost no connection with the historical ibn Ezra. Browning needed someone learned in religious and secular studies, and ibn Ezra fit the description.

Khwārizmī. Hermann and a collaborator, Robert of Ketton (later a canon of the church in Tudela), debated the merits of disseminating Islamic science: might the knowledge be cheapened by exposure to a larger audience? Hermann recounts a dream where Minerva, the Roman goddess of wisdom, appeared and assured him that knowledge should be made freely available (a story suspiciously similar to that told about al Ma'mun). Hermann and Robert thus began an ambitious program to translate as much Islamic mathematics and science into Latin as they could.

In Toledo, the Archbishop Raymond appointed his archdeacon Dominicus Gundissalinus to oversee an ambitious plan to systematically translate Islamic works into Latin. The most important member of the group was GERARD OF CREMONA (1114–1187), who produced Latin editions of Ptolemy's *Almagest*; ibn Aflah's *Correction of the Almagest*; al Jayyānī's astronomical tables; and the mathematical works of Thabit ibn Qurra, Abu Kamil, and others.

Despite their religious intolerance, the Almohads valued pagan Greek philosophy, particularly Aristotle. At the request of the Almohad caliph, Abu Ya'qub Yusuf, ibn Rushd, better known as Averroes, wrote new translations of most of Aristotle's works into Arabic. In most of his works he included a three-tiered system of commentaries: a summary, a "medium" commentary, and a "long" commentary. These were translated into Latin by Michael Scot, who arrived in Toledo around 1217, and hence Aristotle was reintroduced into the west.

The accumulated knowledge of the centuries thus made its way into western Europe, primarily through Andalusia and through the work of Islamic, Jewish, and Christian translators. Islamic scientists had written comprehensive commentaries on the more difficult works of the ancient Greeks, and added their own developments in algebra and trigonometry. A new system of numeration was beginning to make its influence felt in southern Europe. Mathematics in the west, which had been in decline for more than a thousand years, was ready to begin its revival, and around 1200, Scot began correspondence with the man who would become the first great European mathematician: Leonardo of Pisa.

For Further Reading

For the history of the Islamic world (including Andalusia), see [35, 54, 56, 57, 67, 82]. For the mathematics of Islam and Andalusia, see [7, 70, 71, 88, 105, 110].

5

The Middle Ages

5.1 Norman Sicily

In 911, Charles the Simple ceded a section of coastal France to the Viking chieftain Rollo in exchange for a promise to convert to Christianity and to withdraw his warriors from Paris. Within a century the land of the Northmen (*Nortmanni*) became the well-administered, highly centralized state of Normandy.

The Normans followed the rule of primogeniture, whereby the eldest son inherits all the lands of his father. Thus only one of the twelve sons of Tancred of Hauteville would inherit the family lands in western Normandy; the rest had to find their own fortunes. Luckily for them, southern Italy was a battleground between the Italians, Byzantines, and Muslims. Tancred's fourth son, Robert Guiscard ("cunning"), went from leading a gang of bandits to ruling southern Italy. In 1059, Pope Nicholas II recognized Robert's claim to southern Italy, and offered him the island of Sicily as well, provided he take it from the Arabs. Robert entrusted the task to his brother Roger, who conquered all of Sicily by 1091.

Roger's second son, also named Roger, ruled over a thriving Norman Kingdom in southern Italy and Sicily that combined the best of east and west. Salerno, on the mainland, hosted a medical school that attracted students from around the Mediterranean world. The students gained a much better understanding of medicine because they were allowed to perform dissection of human bodies, a practice that had been outlawed in the West since the early days of the Roman Empire. Palermo, where speakers of Greek, Latin, and Arabic mingled freely, became known as the "City of the Three Tongues." The Capella Palatina, finished around 1143, exemplified the union of cultures with its Norman doors, Islamic mosaics, Byzantine roof, and Egyptian murals. Even the coinage of the kingdom was cosmopolitan: Robert Guiscard issued gold *taris*, resembling coins of the same name struck in Egypt, while his successors added silver ducats in the Byzantine style and bronze coins with Latin inscriptions.

Roger's approach to scholarship was as cosmopolitan as his approach to politics. The works of Archimedes arrived from the Byzantines, while the Islamic scholar ash-Sharīf al-Idrīsī came to Sicily 1145, possibly attracted by reports from relatives that Muslims prospered under Norman rule, and possibly because, as a descendant of Mohammed, al-Idrīsī had a legal claim to the caliphate that made the unassuming scholar a threat to every Islamic ruler. Al-Idrīsī stayed in Sicily for the rest of his life, producing one of the great works of medieval geography, the *Kitab nuzhat al-mushtaq fi ikhtiraq al-afaq* ("The Pleasure Excursion of One Who Is Eager to Traverse the Regions of the World"), which is also tellingly known as the *al Kitab ar-Rujari*: Roger's Book.

5.1.1   Adelard of Bath

Rulers might not see the value of engineers, mathematicians, or scientists, but they invariably employ the best physicians they can find. In the middle ages, the best physicians were those educated in Sicilian, Byzantine, or Islamic lands. In 1110, a converted Jew, Pedro Alfonso, became the court physician to Henry I of England. Pedro stressed the importance of observation and conveyed some of the discoveries of Islamic scientists and mathematicians to England. It is probably no coincidence that around the same time, Henry sent ADELARD OF BATH (fl. 1116–1142) on a scientific pilgrimage. First, Adelard went to Tours and Laon, but before long he made his way to Salerno and Sicily, where he could study Islamic culture firsthand. Intrigued by the high achievements of the Muslims in mathematics and the sciences, Adelard traveled through Cilicia (in Asia Minor), Syria, and Palestine before returning to Bath around 1130 with a knowledge of Arabic and an Arabic copy of the *Elements*.

By this time, knowledge of deductive geometry had reached its lowest point in the West. One of the main texts was a long didactic poem, *The Marriage of Philology and Mercury*, written around 420 by Martianus Capella. The seven liberal arts are personified as bridesmaids, and give lessons as their wedding gifts. Geometry's lesson consists of a long discourse on the size and shape of the universe and the geography of the Mediterranean world, and a closing section on geometrical definitions. Arithmetic fares somewhat better: much of Nicomachus is reproduced, and the section ends with observations on divisibility (for example, Arithmetic observes that if a prime number divides any term of a geometric sequence beginning with 1, it must divide the second term).

Adelard would eventually produce three different versions of the *Elements* from his Arabic copy. He made the first complete translation of the *Elements* into Latin. Indeed, it was more than complete, for it contained two additional books and a number of propositions which, though rigorously proven, were added in the centuries since Euclid (one of these propositions, for example, was the proof that the side and diagonal of a square are incommensurable: in other words, that $\sqrt{2}$ is irrational).

Adelard's second version of the *Elements* was an abridged edition. Most of the propositions are stated without proof or, at best, a sketch of how the proof proceeded. But judging by the number of surviving copies it was by far the most popular, and it rapidly became the standard text used in medieval universities. Adelard's third version followed the outline of the second version, but included carefully reasoned proofs, adding an introduction and commentaries; we might call it a teacher's edition.

Adelard's student John (or Nicholas) O'Creat worked on Adelard's second edition of the *Elements*, and made one of the earliest translations of an Arabic algorism into Latin. This work may have been based on a (now lost) treatise of al-Khwārizmī. In O'Creat's and other algorisms of the era, the numerals are invariably referred to as Indian and never as Arabic, indicating that everyone knew their ultimate origin.

Adelard lived during the chaotic period called "The Anarchy" by English historians. In 1135, Henry I of England died, leaving the throne to his only surviving child, his daughter Matilda. But the barons threw their support behind Matilda's cousin Stephen, leading to a civil war. Despite some early successes, Matilda was eventually driven from England in 1141. War resumed when Matilda's son, Henry, invaded England in January 1153, and obtained recognition as Stephen's heir, becoming King Henry II in 1154. The civil war was over, but the country was in chaos.

Thus one of Henry's first tasks as king was re-establishing law and order in England. In the Assize of Clarendon (1166) Henry ordained that twelve men from every hundred (where a "hundred" was a measure of area) had to give the king's judges a list of men in the shire who were guilty of serious crimes, such as robbery or murder. In order to prevent abuses like false accusations, the twelve men had to swear (*iurare* in Latin) that their accusations were true; the practice evolved into the modern grand jury.

5.1.2 The Hohenstaufens

In 1189 William II of Sicily died without legitimate heirs. There were two claimants to the throne. The first was the German Emperor Henry VI, who married Roger II's daughter (and thus William's aunt) Constance in 1186. The other was Tancred, the illegitimate son of Constance's brother Roger (and thus Roger II's grandson). Henry invaded Sicily in 1191, but Tancred had the support of the Sicilians and the Papacy.

Henry, however, had the King of England. In 1192, Richard the Lionhearted. was on his way back from the Third Crusade when his ship was forced ashore near Venice by bad weather. Richard was almost immediately captured by Duke Leopold of Austria (the two had quarreled vehemently during the siege of Jerusalem). Leopold turned Richard over to Henry VI, who demanded Richard pay an enormous ransom or else Henry would turn Richard over to Philip II of France. This was a dire threat, for Richard and Philip were in the midst of a centuries-long struggle over how much of France would be English; if Richard fell into Philip's hands, Philip was sure to demand that Richard give up claims to land in France. The ransom was raised, Richard was released, and Henry launched another invasion of Sicily. This campaign might have failed like the first, but Tancred died just as Henry entered Sicily. The local barons threw their support behind Henry, expecting he would return to Germany and leave the day-to-day ruling of the country to them.

The barons got more than they could have hoped for. A few years after taking the crown, Henry died, and another civil war erupted, this time within the German empire. The two sides were Henry's family, the Hohenstaufens, and a rival family, the Welfs. In Italy, these factions became known as the Ghibellines, after Waiblingen, the name of a Hohenstaufen castle, and Guelphs, which is the Italianization of Welf. The barons took this opportunity to name Henry's son Frederick as the new king, crowning him in May 1198. The choice was deliberate: Frederick was four, and the barons intended to take advantage of Frederick's minority.

Frederick came of age in 1208, and by 1212 ruled in his own name. Within a few years, he curtailed the rampant abuses of the barons, restored order to Sicily, ended the civil war in Germany, and re-united the Hohenstaufen empire as Emperor Frederick II. Perhaps his greatest achievements were in Palestine. In 1099, the First Crusade captured the Holy

Land from a Muslim world distracted by civil war. The Muslim world soon rallied and expelled the crusaders. The Second and Third Crusades were epics of incompetence that accomplished nothing. The Fourth Crusade did not even make it to Palestine: the crusaders were unable to raise the money to pay the Venetians for transport and instead attacked (at Venice's request) the city of Zara, then went on to sack Constantinople in 1204; a quarter of the city was burned and countless treasures of antiquity were lost forever. During the Children's Crusade of 1212, two groups of children numbering perhaps 50, 000 altogether attempted to retake the Holy Land by love. Thirty-thousand or so made their way to Marseilles, where they boarded ships and were never seen again (they were probably sold into slavery). Several (including Leibniz) have suggested the episode as the origin for the story of the Pied Piper.

Frederick was about to embark on a crusade in 1227 when an epidemic broke out among his men. He canceled the expedition, preferring not to make war with sick and disabled soldiers. This earned him an excommunication from Pope Gregory IX. The next year, Frederick went to Palestine armed with a novel weapon: diplomacy. Through negotiations with the Sultan of Egypt, Frederick obtained a ten year truce, recognition as King of Jerusalem, and possession of the cities of Jerusalem, Bethlehem, and Nazareth, as well as a corridor through which pilgrims could visit the cities unmolested. Frederick returned to face an invading Papal army, which he defeated after some effort. When Frederick in turn threatened to invade the papal territories, the Pope rescinded the excommunication.

Like his maternal grandfather Roger, Frederick was fascinated by learning. Frederick was a scholar himself, and wrote a book on falconry, *The Art of Hunting with Birds*, based on his own observations, and considered one of the classic works on the subject. He attracted other scholars, and his court (which traveled with Frederick wherever he went) came to be known as the "Wonder of the World" (*stupor mundi*). One of these scholars was Michael Scot, the Court Astrologer, who might have been Frederick's tutor. Michael brought LEONARDO OF PISA (1170–1250) to Frederick's attention.

5.1.3   Leonardo of Pisa

Leonardo's best known work is his misnamed *Book of the Abacus*, a monumental tome of practical mathematics which dealt with every aspect of commercial arithmetic *except* the use of the abacus (and a better translation of its Latin title might be *Book of the Calculator*). Though born in Pisa, Leonardo grew up traveling around the Mediterranean with his father Guglielmo, who was a factor: the agent and representative of a group of merchants. Guglielmo's primary posting was in the North African port town of Bugia (modern Bejaïa, Algeria), where Leonardo learned of the Islamic achievements in mathematics and in particular, the Hindu-Arabic system of numeration.

When Leonardo returned to Pisa in 1200, he began a correspondence with Michael Scot. Around 1202, Leonardo summarized what he had learned of practical arithmetic in his *Book of the Abacus*, and dedicated the work to Scotus; a second edition appeared in 1227. Although not the first to introduce the Indian numerals to the west, Leonardo's treatise was by far the largest work to incorporate them: a printed copy runs to more than 400 very large pages, filled with endless examples of how to use the Indian numerals to perform commercial computations, particularly those that might be encountered by the medieval merchant.

The goods available to the merchant were diverse: Leonardo's problems include references to linen from Syria and Alexandria; oil from Constantinople; furs (probably from Scandinavia via Constantinople), cheese from Pisa, figs, mastic (a type of resin), saffron, cinnamon, and black pepper. Perhaps the most striking feature of medieval trade was its incredible complexity, mainly due to the multitude of currencies in common use: a Mediterranean merchant might expect to be paid in Byzantine bezants, French deniers, Islamic taris, Imperial soldis, Pisan denaris, Genoan januses, or Venetian pounds. Not surprisingly, Leonardo included many examples of currency exchange:

Problem 5.1. *An Imperial soldi is worth* $31\frac{1}{2}$ *Pisan denarii, and a hundred peppers cost* $11\frac{11}{20}$ *Pisan pounds [where one pound is 12 Pisan denarii]. How much pepper can you buy with* $57\frac{1}{4}$ *Imperial soldi?*

A more subtle problem was that of weights and measures. During the Roman Empire, the standard unit of length was a foot, which was divided into 4 handspans, each of which was divided into 3 *uncia* ("twelfth part", named because one *uncia* was one-twelfth of a foot).[1] Unfortunately this uniform system of measurement passed away with the Roman Empire, and by Leonardo's time, every region had its own local units. Thus cloth (one of the most important commodities in the Middle Ages) was measured in canes and handspans. But:

> A cane in Pisan is ten handspans or 4 cubits, but a Genoan cane is defined as 9 handspans. There are also provincial canes, Sicilian, Syrian, and Constantinopolitan which measure 8 handspans...[2]

The complexity of the system of weights and measures is most obvious in what seems to be a nonsensical question: which weighs more, a pound of gold or a pound of feathers? Gold and other precious commodities were measured in Troy units, named after the semiannual trade fairs at Troyes in Champagne, France, where goods from throughout Europe could be exchanged. The Troy pound is divided into twelve troy ounces, and each ounce into twenty pennyweights, and each pennyweight into 24 grains: thus, a Troy pound is equal to $12 \times 20 \times 24 = 5760$ grains. An avoirdupois pound (from the French "having weight") was defined as having a weight of 7000 grains: thus a pound of gold (5760 grains) weighed less than a pound of feathers (7000 grains). Even more confusingly, the avoirdupois pound was divided into 16 inappropriately named ounces, so an ounce of gold ($20 \times 24 = 480$ grains) was heavier than an ounce of feathers ($7000 \div 16 = 437.5$ grains).

The irregularity of the subdivisions of a unit of measurement were the probable inspiration for one of Leonardo's least successful innovations: a form of fractional notation, where $\frac{1}{5}\frac{2}{10}\frac{3}{20}$ meant $\frac{1}{5 \cdot 10 \cdot 20} + \frac{2}{10 \cdot 20} + \frac{3}{20}$. The purpose of this notation is clear: a quantity expressed using differently sized units can be written compactly, thus 3 (Troy) pounds, two ounces, five pennyweight, seven grains can be written as $\frac{7}{24}\frac{5}{20}\frac{2}{12}3$ (where Leonardo placed the fractional part of the number before the number itself, echoing the Arabic practice of writing the symbols representing the least values first). Although little came of Leonardo's invention, it foreshadowed the development of decimal fractions: in one problem, Leonardo obtained an answer of $\frac{7}{10}\frac{1}{10}8$, and it is a small step from this to the decimal equivalent 8.17.

---

[1] The word *uncia* comes down as both "ounce" and "inch" in English.

[2] Leonardo, p. 111

Leonardo's work brought him to the attention of Frederick's court, and an invitation was soon extended. One of Leonardo's colleagues in Frederick's court was Master John of Palermo, who posed the following problem to Leonardo:

Problem 5.2. *To find a square number that, if five is added or subtracted, gives a square number.*

Leonardo's basic approach is the following: it had been known since antiquity that the sum of the first n odd numbers was equal to n^2. Thus, Leonardo needed to find m, n, and p so:

$$\text{Sum of first } m \text{ odd numbers} + c = \text{Sum of first } n \text{ odd numbers,}$$
$$\text{Sum of first } n \text{ odd numbers} + c = \text{Sum of first } p \text{ odd numbers,}$$

which would make $n^2 + c = p^2$, and $n^2 - c = m^2$ (and, obviously, $p > n > m$). Rearranging gives us:

$$c = \text{Sum of } m + 1 \text{ to } n\text{th odd number,}$$
$$c = \text{Sum of } n + 1 \text{ to } p\text{th odd number.}$$

Thus c can be written as the sum of consecutive odd numbers in two different ways. Leonardo showed that c must be a multiple of 24. Moreover, if c has a square factor, the square factor can be removed. Thus, since:

$$31^2 + 720 = 41^2,$$
$$41^2 + 720 = 49^2.$$

Then, since $720 = 5 \cdot 12^2$, we can divide by 12^2 to obtain the solution to Master John's problem:

$$\left(\frac{31}{12}\right)^2 + 5 = \left(\frac{41}{12}\right)^2,$$
$$\left(\frac{41}{12}\right)^2 + 5 = \left(\frac{49}{12}\right)^2,$$

and thus the solution is $\frac{41}{12}$, which if 5 is added or subtracted, will yield a square. Leonardo summarized his results in his *Book of Squares*, dedicated to his patron, Frederick II.

5.1.4   The Translators

Frederick's son Manfred eventually inherited both the Kingdom of Sicily and the Guelph-Ghibelline conflict. In addition, the presence of a strong king in southern Italy limited the Pope's power in central Italy. Consequently the Pope preferred either a weak ruler, or a distant one; Manfred was neither. Urban IV excommunicated Manfred and offered the Kingdom of Sicily to Charles of Anjou, the brother of King Louis IX of France. At Benevento in 1266, Charles and Manfred met in the field of battle. By the end of the day, the Sicilians were defeated, Manfred was dead, and Charles of Anjou was the king of Sicily.

With Sicily came the manuscripts of Archimedes, carefully compiled during the reign of the Byzantine Emperor Leo VI, given to Normans, and passed on to Hohenstaufens. Charles had no use for the manuscripts, so he sent them as a gift to the Pope, where they were placed in the papal library in Viterbo.

Urban surrounded himself with notable scholars, most of whom stayed on during the reigns of his successors. Perhaps the best known scholar at his court was Thomas Aquinas, who had just earned a degree in theology from the University of Paris. Aquinas helped reconcile the doctrines of the pagan philosophers, particularly Aristotle, with those of the Catholic Church. Aquinas's work might be taken as the culmination of the process of using logic and rational thought in support of faith and belief.

There are two problems with reconciling Aristotle and Catholic doctrine. First, there was the problem of reconciling reason and faith: can matters of faith be subject to logical analysis? If they are, then it follows that God must be limited by the rules of logic, which contradicts the doctrine that God is omnipotent. Thus rational theologians must be very careful lest they wander into heresy. John Duns Scotus proposed that this problem could be resolved by distinguishing between matters of faith amenable to reason, and matters of faith that are not; however, the average person had no use for the logic chopping and hairsplitting of the dunces. Ultimately, the applicability of reason to faith is a matter of doctrine.

The second problem was the poor translations of Aristotle available to the west. At Viterbo, the Belgian-born WILLIAM OF MOERBEKE (ca. 1220–December 26, 1286) prepared a new translation of Aristotle from Greek originals, and after the fall of the Hohenstaufens, William prepared the first translations of Archimedes into Latin.

Moerbeke's translation of Archimedes would be of little use until the general level of mathematical knowledge in the west was raised. This may have prompted Urban IV's chaplain, CAMPANUS OF NOVARA (b. ca. 1220–1296), to write an edition of Euclid that would prove popular for centuries. Campanus's *Elements* was more a textbook on Euclidean geometry than a translation of Euclid, and showed that Europeans had begun moving from merely studying the works of the ancients to creating original works of their own.

Around the time Campanus finished his version of the *Elements*, an open revolt broke out against the French in Palermo. In the hour of vespers (early morning) of Easter Monday, March 30, 1282, the Sicilians rose up and massacred nearly two thousand French men, women, and children. The Sicilian Vespers continued to war against the French with the help of mercenary soldiers called *mafie*, and eventually expelled the French with the help of Pedro III of Aragon. But this only exchanged one master for another, and set the stage for future disaster, for the French would never forget they once held the crown of Sicily.

5.2   Early Medieval France

Two of the Germanic tribes that dominated Europe after the fall of the Roman Empire were named after their weapon of choice: the Saxons, who preferred the shortsword (*sax*), and the Franks who relied on the javelin (*franca*). The Franks originated along the Rhine in modern Germany, but their settlement in Roman Gaul would eventually make that region known as France. The greatest of the Frankish kings, Charles the Great or Charlemagne, would rule over a vast empire that included most of modern France, Germany, and northern Italy. The Roman title "Emperor and Augustus" was revived and conferred upon Charles

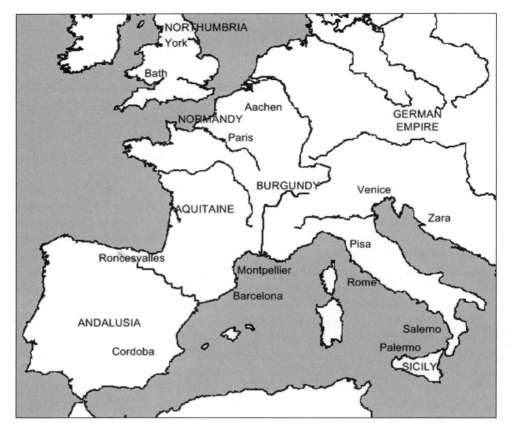

Figure 5.1. The Carolingian Empire

on Christmas Day, 800 by Pope Leo III. Since the Latinized form of Charles is *Carolus*, the lands ruled by Charlemagne became known as the Carolingian Empire.

One of the important mathematical problems of the Middle Ages was calculating the date of Easter, defined by the Council of Nicaea as falling on the first Sunday after the first full Moon on or after the vernal equinox. Since predicting the actual motion of the Moon is an extraordinarily complicated problem, the Computists (those charged with performing this determination) used an "ecclesiastical Moon," based on a mathematical model of the Moon's motion. This included defining the "full Moon" as occurring on the fourteenth day of the month, and the vernal equinox as occurring on March 21.

In Northumbria, BEDE (672/3–May 25, 735) noted some of the confusion that resulted from this definition of Easter and the requirement that its date be calculated in his *History of the English Church and People*. This chronology is noteworthy for several reasons, including the fact that it is one of the first to use the A.D. system of dating. Bede also noted that the vernal equinox actually occurred around March 18, which reflects the fact that the Julian calendar's mean year length of 365.25 days is slightly too long, causing the beginning of the year to occur later and later in the season.

Around 725, Bede wrote *The Reckoning of Time*, where he described timekeeping in general. Among other things, Bede explained that the spherical shape of the Earth caused

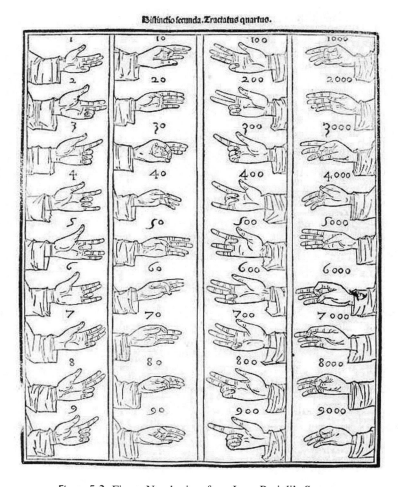

Figure 5.2. Finger Numbering, from Luca Pacioli's *Summa*.

the variation in the lengths of days over the course of the year, and gave a detailed explanation of how the date of Easter is to be computed. In the opening section he described a system whereby intermediate computational results can be represented by various finger positions (one through ninety-nine on the left hand, and numbers up to nine thousand nine hundred on the right); because of this, the Hindu-Arabic number symbols came to be known as digits (from *digitus*, "finger" in Latin). Bede's is the oldest complete description of this system of finger numeration, which remained substantially unchanged from the time of the ancient Romans (see Figure 5.2 for a sixteenth century illustration of the method). For the numbers one through ten on the left hand:

1. Little finger bent at joint.

2. Little and ring fingers bent at joint.

3. Little, ring, and middle fingers bent at joint.

4. Ring and middle fingers bent at joint.

5. Middle finger bent at joint.

6. Ring finger bent at joint.

7. Little finger bent to palm.

8. Little and ring finger bent to palm.

9. Little, ring, and middle fingers bent to palm.

10. Tip of forefinger to middle joint of thumb.

Bede taught Egbert of York, who in turn taught ALCUIN of York (735–May 19, 804). In 781, Alcuin received an invitation from Charlemagne to go to his capital at Aachen and found a school for the education of the future rulers and administrators of the empire. Alcuin probably brought Bede's work into the Empire, for a licensing examination dating to 809 indicates that Computists were expected to know and apply the material from *The Reckoning of Time*.

Alcuin established a regimen of study based on the seven liberal arts of Boethius, focusing on grammar, rhetoric, logic, and astronomy. Where textbooks did not exist, he wrote them. His *Problems to Sharpen the Young* includes a number of problems, such as:

Problem 5.3. *A son and a father marry a widow and her daughter, respectively. What is the relationship of their sons?*

which, of course, is not a mathematical question (Alcuin's solution is that the sons are uncle and nephew of one another; he misses the possibility that the son is his own grandfather). Many are mathematical, including the first European appearance of the Diophantine problem:

Problem 5.4. *A man buys* 100 *pigs with* 100 *pence. A boar costs* 10 *pence, a sow* 5 *pence, and two piglets cost a pence. How many of each animal does he buy?*

Alcuin also posed the problem:

Problem 5.5. *A staircase has* 100 *steps. On the first step there is* 1 *pigeon; on the second,* 2 *pigeons, on the third,* 3 *pigeons, and so on. How many pigeons are there altogether?*

Alcuin's solution is to note that the first and 99th step together have 100 pigeons, as do the second and 98th, and so on. The 50th and 100th steps remain unpaired. Thus there are 49 pairs of steps with 100 pigeons, 50 more pigeons from the 50th step, and 100 more from the 100th step, for 5050 pigeons altogether.

Alcuin also presented a number of area problems, from which we can derive the rule that the area of a quadrilateral with sides of length a, b, c, and d is given by $\left(\frac{a+c}{2}\right)\left(\frac{b+d}{2}\right)$ and that the area of a circle with circumference C is given by $(C/4)^2$. These are, of course, only approximations, though it is not clear if Alcuin realized this.

These advanced topics required prepared students, so in 787 Charlemagne ordained that every diocese was to have a cathedral school to teach reading and writing; the cathedral schools would be the foundation of European education for a thousand years. Charlemagne himself studied Latin, Greek, and astronomy, but, though he kept a writing pad beneath his pillow so he could practice during sleepless nights, never acquired the manual dexterity needed to write legibly.

It was in the field of writing that Alcuin made one of his most important contributions. The printing press was centuries in the future, so the readability of any manuscript depended on the ability of the copyist to produce readable letters. Roman *majuscule* ("large letters" in Latin), essentially the same as our modern capital letters, was very easy to read, despite the practice of omitting interword spaces, though if absolutely necessary a · might separate words: THREE·SEPARATE·WORDS. By Carolingian times, interword spaces were common, but letters were written in a cursive script that joined letters with ligatures that followed no regular pattern, producing a very elegant script that is painfully difficult to read and makes learning needlessly complex. Alcuin and others invented a new type of script, called Carolingian minuscule ("small letters" in Latin), consisting of essentially the same letters used in this sentence.

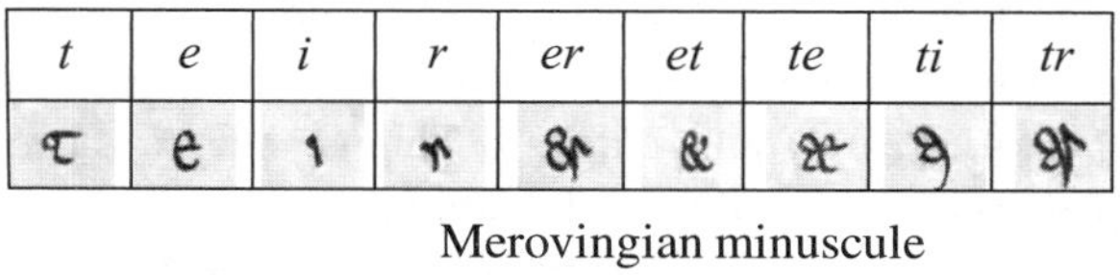

t	*e*	*i*	*r*	*er*	*et*	*te*	*ti*	*tr*
τ	ℯ	⟍	⟋	ℯ⟋	ℯℇ	ℯτ	ℯ	ℯ⟋

Merovingian minuscule

Figure 5.3. Merovingian Minuscule, from the Luxeuil manuscript (8th century). Note radical differences between individual letters (e, r) and the appearance of the script when these letters appear together (er).

The practice of numbering kings did not emerge until the Renaissance; instead, kings were differentiated by being given a nickname. The nicknames give us a sense of what the kings were best known for, and some insight into the subsequent history of France. Charlemagne's successors included his son Louis the Pious; Charles the Bald; Louis the Stammerer; Charles the Fat; Charles the Stupid; and finally, Louis the Do-Nothing. In 987, Louis died in a hunting accident, and his uncle and Charlemagne's only surviving descendant, Charles of Lorraine, was poised to succeed him. But Charles's claim to the throne of France was rejected, thanks to the work of two churchmen: Adalbero, the Bishop of Reims, and his assistant, GERBERT OF AURILLAC (945–May 12, 1003).

5.2.1 Gerbert of Aurillac

At the end of the ninth century Aurillac, in Auvergne, found itself trapped between the growing power of the Duchy of Aquitaine to the north and the County of Toulouse to the south.[3] Count Gerald of Aurillac carefully played Aquitaine and Toulouse against each other and maintained the independence of Auvergne, but there was no one who could take his place after his death. Thus he conceived a clever stratagem: he would give his lands to the Pope, then receive them back and rule over them as an abbot, ensuring that after his death Auvergne would remain independent of both Toulouse and Aquitaine. While the Pope would in theory become the overlord of Auvergne, the papacy was too weak and too far away to demand more than a nominal allegiance. Gerald's strategy worked so well that his adversary William of Aquitaine used a similar tactic, placing an indefensible patch of his inheritance in the hands of an abbey at Cluny.

---

[3] A county is ruled over by a count, as a duchy by a duke. In terms of prestige, a Count, from the Latin *comes*, "companion," is lower than a Duke, from the Latin *ducem*, "leader."

At Gerald's abbey, Gerbert received the rudiments of reading and writing, but the nearest place a first-class education could be found was in the County of Barcelona. In 967, Count Borrell of Barcelona visited Aurillac and was persuaded to take Gerbert back to Spain with him. For a while, Gerbert studied astronomy and Hindu-Arabic numerals at the abbey of Santa Maria de Ripoll. His knowledge of such arcane arts led to accusations by his enemies that he had entered into a pact with the devil, and a series of medieval myths grew up around Gerbert. According to the myth, Gerbert apprenticed to a Muslim magician. To steal his secrets, he seduced the magician's daughter, persuaded her to drug her father's wine, stole his book of spells, and fled, leaving the daughter behind. The story is clearly a retelling of the Greek myth of Jason and Medea with Gerbert playing the role of Jason.

Gerbert continued to learn at the cathedral school in Vic (or Vich), coming to the attention of Bishop Atto (or Hatto). In 969, they left on pilgrimage for Rome, where Atto was assassinated, leaving Gerbert without a sponsor. Fortunately he impressed Pope John XIII, who introduced him to Otto I, the reigning Emperor. Though heir to the title held by Charlemagne, Otto presided over a much more modest empire consisting primarily of Germany and northern Italy, the core territories of what would become known as the Holy Roman Empire. This posed a legal problem: the Roman Empire still existed, in the form of the steadily shrinking Byzantine Empire, ruled from Constantinople. When Charlemagne claimed the imperial title, the weak-willed nobodies reigning in Constantinople protested but did nothing. By Otto's time, that situation had changed.

When the Carolingian Empire was divided amongst Charlemagne's three grandsons, Michael III was Byzantine Emperor, but Michael left the administration of the empire in the hands of his uncle Bardas. Bardas believed in fiscal responsibility, rebuilding the treasury, and strengthening the defenses of the Empire. Michael believed in debauchery, promiscuity, and frivolity; Byzantine historians refer to him as Michael the Drunkard.

There were some noteworthy events that occurred during Michael's reign. In the ninth century, the Vikings made their way east and west from their Scandinavian homeland, interested in two things: trading and raiding. On June 8, 793 the Vikings raided the Monastery of St. Cuthbert on Lindisfarne island, just off the coast of Northumbria in England. Over the next two centuries the Vikings would raid the coastal communities of Ireland and Spain, as well as cities as far inland as Hamburg and Paris. The Vikings who went east tended to concentrate on trading. Some of them settled in the region around Kiev, which became known as the land of the Rus, a word that might be derived from *rothr* ("to row" in Norse). By extension the whole region came to be known as Russia.

The Viking Rus made their way as far south as Constantinople. Because some of them took an oath (*varar* in Norse) to serve in the imperial guard, the Vikings in general came to be known as Varangians. The Byzantines traded the pen for the sword: during Michael's reign two missionaries, Cyril and Methodius, created the Cyrillic alphabet, a written language for the Slavic languages based on letters of the Greek alphabet.

Meanwhile, Bardas's fiscal economies proved too much for Michael, so in 865 he had Basil, a Macedonian wrestler and court favorite, murder Bardas. To bring his mistress Eudocia into the palace without arousing suspicion (or at least, to preserve the appearance of propriety), Michael forced Basil to divorce his wife and marry Eudocia. Michael foolishly made Basil co-emperor in 866, and less than a year later, Michael was murdered in the midst of a drinking spree. Basil became sole emperor and founder of the Macedonian dynasty.

Basil I proved to be just what the Empire needed: a capable, conscientious (albeit ruthless) ruler. He spent most of his reign replenishing the treasury, rebuilding the army and navy, and codifying the complex system of Byzantine law. This project was completed during the reign of Leo VI, ostensibly Basil's son but (since his mother was Eudocia) just as likely Michael's.

There is some confusion about Leo, whose continuation of Basil's work earned him the eponym Leo the Wise. Earlier in the century another Leo attained fame for his learning: LEO THE PHILOSOPHER (790–869). One of Leo's students was captured by the Arabs around 831; the student's mathematical knowledge impressed his captors, and the Caliph al Ma'mun attempted to recruit the teacher. To keep his internationally renown subject, the Byzantine Emperor Theophilus arranged a position for Leo at the Church of Forty Martyrs in Constantinople. Leo stayed, and eventually compiled the extant works of Archimedes. Leo's compilation eventually formed the basis of William of Moerbeke's translation into Latin.

Faced with a resurgent Byzantine Empire, the German Emperor Otto I sought to appease them. In 967 he arranged for his son (also named Otto) to marry Theophano, a Byzantine princess, described as the niece (or granddaughter) of John Tzimiskes, though her exact lineage is uncertain. The cultural differences were immense: the Byzantines educated women, wore silk, ate with forks, and bathed daily, and Theophano came under heavy criticism for these and other decadent habits.[4]

At the wedding of Otto II and Theophano in 972, Gerbert met the logician Archdeacon Gerann of Reims. Gerbert, who felt himself to be weak in logic, obtained a leave of absence from imperial service. Back in Reims, Gerbert studied with Gerann and Bishop Adalbero of Reims; soon Gerbert began teaching alongside them at the cathedral school. There Gerbert prepared texts on music, arithmetic, and astronomy, based on Boethius. When Gerbert returned to imperial service, Otto II made him the Abbot of Bobbio in 980. Gerbert proved an unpopular administrator and when Otto II died in 983, Gerbert returned to Reims.

While in Spain, Gerbert learned the use of the Hindu-Arabic numerals, and tried to popularize them. Gerbert's introduction, one of the earliest European treatises, was incomplete, for he omitted any mention of the zero symbol. To make positional notation more comprehensible, Gerbert may have invented a type of abacus, consisting of twenty-seven columns grouped into sets of three columns labeled S, *singularis* or units; D, *decem* or tens; and C, *centum* or hundreds; this was probably the first time that the orders were grouped like this (as they are today when writing large numbers like $1,000,000$). Gerbert had over a thousand counters inscribed with the nine Hindu-Arabic numerals, and a number like three hundred forty-six could be represented by placing a 3 in the C column, a 4 in the D column, and a 6 in the S column.

Unfortunately Hindu-Arabic numerals are ill-suited for computation on the abacus, where adding two and three is as simple as adding •• to • • • to get • • • • •. Since all counters are identical, it is merely a matter of reaching into the bag to pick up the appropriate number of counters. On Gerbert's abacus, however, one not only had to remember

---

[4]It is not true that medieval Europeans never bathed. The frequency of bathing varied from place to place, from century to century, season to season, and, of course, person to person, but very roughly speaking, it was not uncommon to bathe at least once a week.

that '2' and '3' made '5', but also one had to search through the bag to find the '5' counter to indicate the sum. Thus Gerbert's abacus was a failure.

Gerbert was more successful with a form of division he introduced around 980: division by differences. Gerbert's method, applied to the division of 96 by 8, works as follows. First, find the difference between the divisor 8 and 10 (the nearest "easy" divisor), which is $10 - 8 = 2$. Set up the division as follows:

C	D	S	
		2	Difference
		8	Divisor
	9	6	Dividend

Divide 10 into 90, obtaining 9. Set this down as the partial quotient. Multiply the partial quotient 9 by 10 to obtain 90, and subtract this from the dividend (we will simply cross out the 9 in the tens column).

C	D	S	
		2	Difference
		8	Divisor
	9̸	6	Dividend minus 90
		9	Partial Quotient

Then multiply the partial quotient 9 by the difference 2 to get 18, and *add* 18 to the remainder to obtain the new dividend.

C	D	S	
		2	Difference
		8	Divisor
	9̸	6	
	1	8	Product of 9 and difference
	2	4	Sum and new dividend
		9	Partial Quotient

Now repeat the process by dividing 10 into the new dividend. This gives 2, so the product of 2 and 10 is subtracted (again, we will simply cross out the 2 in the tens column) and the product of the partial quotient 2 and the difference is added:

C	D	S	
		2	Difference
		8	Divisor
	9̸	6	
	1	8	
	2̸	4	Dividend minus 20
		4	Product of 2 and difference
		8	Sum and new dividend
		9	Partial Quotient
		2	Partial Quotient

Finally the quotient of the new dividend 8 and the divisor is 1, which is the third partial quotient; the quotient is the sum of the partial quotients.

C	D	S	
		2	Difference
		8	Divisor
	~~9~~	6	
	1	8	
	~~2~~	4	
		4	
		~~8~~	Dividend minus 8
		9	Partial Quotient
		2	Partial Quotient
		1	Partial Quotient
	1	2	Quotient

In 987, the French King Louis the Do-Nothing died and Gerbert's superior Adalbero declared the French monarchy to be an elective one. In other words, the king had to be chosen by a group of electors, usually the most powerful nobles in the land. The nobles rejected the claim of Charles of Lorraine, and instead elected Hugh Capet (named after the cape that was his characteristic article of clothing). Hugh would be the first of fourteen Capetian kings of France.

Gerbert played an important role in Hugh's election, for his connections to the Ottonians were used to guarantee support (or at least, neutrality). Gerbert became one of Hugh's advisors and in 991, Hugh made Gerbert the Archbishop of Reims. However, the appointment was never approved by the Pope, and a few years later, it was declared invalid. Gerbert made the mistake of pointing out that Hugh's son and successor (and, incidentally, Gerbert's student at Reims), Robert II, married his second cousin Bertha, making the legality of their marriage questionable.[5] Hence, shortly after Robert's succession, Gerbert left France for the German Empire, and became the tutor and advisor to Otto III.

The Empire was in the midst of a political struggle with the papacy. The key issue was lay investiture, which centers around the question of who is entitled to choose the high church officials, particularly bishops. In theory the Pope was the head of the Church, so the Pope alone could decide who could become a bishop. But since bishops generally ruled over territory that was part of a kingdom, the kings and emperors held that they, too, had the right to select bishops (as Hugh did by appointing Gerbert the Archbishop of Reims). The problem was aggravated because the Popes involved themselves in regional struggles, and could appoint bishops who served as a focal point for opposition to the king. Thus, those who controlled the Papacy controlled Catholic Europe. By the tenth century, the Papacy was little more than an extension of the powerful families of Rome, who chose and deposed Popes at will. Since the most notorious figures of these families were the mothers, mistresses, and daughters of the Popes, this period is known as the pornocracy ("rule by mistresses") and is considered the nadir of papal prestige and power. Centuries later these times would enter popular culture as the myth of the female Pope Joan.

Otto I compelled the Romans to accept his candidate for Pope, and Otto III continued the tradition. Backed by the imperial army, Otto installed his cousin Bruno as Pope Gregory V in 996. But as soon as the imperial army withdrew, the Crescentine family revolted and

---

[5]Their common great-grandparents were the German Emperor Henry I and Matilda, the parents of Otto I.

deposed Gregory. Crescentius II Nomentanus offered the Papacy to Giovanni Filagato (Otto III's godfather); after some thought, he accepted and became Pope John XVI. Otto marched back into Italy, beheaded Crescentius, mutilated John XVI, and restored his cousin. When Gregory V died a few years later in 999, Otto put Gerbert forward as a papal candidate. With Otto's army on their doorsteps, the Romans grudgingly allowed Gerbert to ascend as Pope Sylvester II. Clearly a better system for choosing the pope was needed: as long as the pope was chosen by force, the pope could be deposed by force. Thus within a generation, a College of Cardinals was established and in theory it alone had the power to choose or depose a pope.

Gerbert intended to initiate a reform of the papacy, but died before he could make any significant changes. His main achievement as Pope was granting the royal title to Stephen of Hungary, marking the founding of the Kingdom of Hungary. Significantly, though Otto III dreamed of restoring a universal Roman Empire, Gerbert made Hungary an independent kingdom that owed no allegiance to the Emperor.

Gerbert reawakened interest in the study of logic by preparing copies of Boethius's treatises. As in the Muslim world, logic soon became an important element of faith. Perhaps the best-known application of logic to theology was the ontological argument of Anselm of Canterbury. Anselm argued as follows: A concept either exists in reality as well as in our minds, or the concept exists only in our minds. "God" can be used to denote something greater than anything that can be conceived. If "God" exists only in our minds, then we may conceive of a new concept, a God that is existent in reality. But this new concept is greater than "God" (because "God" exists only in our minds, which is less than a God that exists in reality). This contradicts our definition of "God." Hence "God" must have a physical reality.

5.2.2 The University of Paris

Anselm's argument attracted little notice, and is better known today than it was during his lifetime. Much more influential was the work of Peter Abelard. Abelard had an affair with one of his students, Heloise, who gave birth to a son. They named the boy Astrolabe, after the latest Islamic import. Abelard and Heloise married in secret, but Heloise's uncle, the Canon (Church lawyer) Fulbert, was so incensed that he had Abelard castrated. Abelard retired to the royal abbey of Saint-Denis and Heloise was forced into a convent; they would never see each other again, and would communicate only by letters, which are considered by some to be the finest examples of medieval love letters.

In Abelard's youth, higher education consisted of a Master (from the Latin *magister*, teacher) lecturing from an authoritative text, such as the Bible or the writings of the early Church fathers, to a student or group of students. The Master provided commentary and explanations of difficult passages, and the students memorized the work and commentary. The well-educated man was one who could recite the authoritative text, commentary, and explanation. Nowadays we do not equate learning with memorization, and it is easy for us to deride medieval scholars for their reliance on authority. But printing was still centuries in the future: books were rare and expensive, and a scholar's reference library consisted of what he carried in his head. Thus the emphasis on memorization served a very important purpose. Moreover, until the basic facts can be established, progress is impossible. The situation was summarized by Bernard of Chartres at the dawn of the twelfth century: he compared himself and his colleagues to "dwarves standing on the shoulders of giants."

Memorization of facts and authoritative opinions continued to be important, but Abelard also stressed the value of debate. His *Sic et Non* (Yes and No) is a compilation of apparent contradictions between the Bible and the teachings of the Christian church, and how the contradictions could be analyzed and resolved. It was the beginning of what is now called rational theology; Aquinas's work was the culmination of this process. More importantly, *Sic et Non* established that logical arguments were essential for arriving at the truth. After Abelard, the disputation became an academic rite of passage: the student or master would expound a thesis, and defend it against all objections using sound, logical arguments as well as citations from authoritative texts.

Students from all over Europe flocked to Paris to study with Abelard. In turn, the presence of so many students drew teachers to the city, and soon, they organized a university ("guild" in Latin). Exactly when this occurred is unknown, but the University of Paris first appeared on a police report of 1200, when the Provost of Paris (the equivalent of the Chief of Police) led an attack on a group of German students, killing one of them. The guild complained to King Philip II, who reprimanded his Provost and granted the university a Royal Charter. The charter gave the teachers and students the rights and privileges of clerics. In particular, this meant that the students and faculty could only be tried by ecclesiastical courts, and in general, could not be prosecuted by the civil authorities. But these privileges came at a price: students, as *de facto* members of the clergy, were compelled to remain unmarried. Hence, the degree they earned would in time be called a bachelor's degree.

5.2.3   Paladins, Troubadours, and Translators

In the north, the popular entertainment of the time included tales of the great battles of Charlemagne and his paladins ("palace companions" in French). The most famous of these *chansons de geste* ("songs of history") was the *Song of Roland*, written around 1100. The *Song of Roland* commemorates the Battle of Roncesvalles (August 15, 778), when Basques attacked Charlemagne's army, then in the process of withdrawing from Spain. The unknown author (probably Norman, possibly named Turold, whose name appears in the last line of the poem) turned Charlemagne's inconclusive Spanish campaign into the conquest of the entire Iberian peninsula, changed the attackers to Muslims, added a dose of treachery and betrayal, and included a happy ending when Roland and his companions were avenged by Charlemagne.

The *Song of Roland* marks one of the earliest appearances of a code of behavior eventually called chivalry. Chivalry has been much romanticized and defies easy definition, though its main components are strict adherence to Christian moral sensibilities, loyalty to one's liege and a rigorous code of honor that, above all, considers one's word to be an unbreakable pledge, regardless of the circumstances under which it was given. In later centuries chivalry was synonymous with knighthood, though in *Roland* the identification was not yet complete: the warriors were chivalric, *and* they were knights.

While northerners told tales of war and battle, southerners sang songs of love and courtship. Provence, on the Mediterranean coast of France, was home to a flourishing culture whose best known products were the odes about courtly love sung by lyric poets known as troubadours (from the words meaning "to find" or "to create"). William IX, Duke of Aquitaine was one of the first troubadours, and wrote playful bawdy verses that are some of the earliest examples of Provençal literature. William's granddaughter Eleanor of Aquitaine (Philip's ex-wife) was entertained by the ballads of Bernard de Ventadour, and Eleanor's own daughter, Marie of Champagne, encouraged Chrétien de Troyes to write down the tale that epitomizes courtly love: the story of Lancelot and Guinevere. Most of the troubadour ballads were written in a southern French dialect known as Languedoc, after its word for yes (*oc*, in contrast to the northern French *oïl*). The language is still spoken today in Provence and Languedoc by about $1, 500, 000$ speakers.[6]

One of the characteristic poetic forms of the troubadour era is known as a *sestina*, invented by the Occitan poet Arnaut Danièl. The sestina is a thirty-nine line poem that consists of six stanzas of six lines (or sextets), plus a final three line stanza (tercet) called an "envoi." Unlike a poetic form like a sonnet or a quatrain, the terminal words of the sestina's lines do not rhyme with one another. Instead, the terminal words of each of the six lines of the first stanza recur as the terminal words of the other stanzas.

In particular, suppose the terminal words of the lines of the first stanza are A, B, C, D, E, and F (thus we can describe the first stanza as having rhyme scheme $ABCDEF$). Then the second stanza's rhyme scheme is $FAEBDC$; the third stanza has $CFDABE$; the fourth $ECBFAD$; the fifth $DEACFB$; and the sixth $BDFECA$. Note that in every case, the succeeding stanza's rhyme scheme can be obtained from the preceding stanzas's rhyme scheme through the application of the permutation we would designate as (615243). In fact, since the next application of the permutation would return the letters to the arrangement $ABCDEF$, then the sestina's rhyme scheme corresponds to a cyclic subgroup of the permutation group on six elements.

A careful study of contemporaneous troubadour and Arabic works shows significant influences on one another. This is not surprising: because of its location, southern France maintained close ties with the Islamic world. Barcelona, Valencia, and Sicily could be reached by a short ocean voyage (Leonardo of Pisa mentioned spending time in Provence), and the merchants of Montpellier and Marseilles actively traded with the Levant. Benjamin of Tudela, a Jew, traveled from Castile to Palestine and back in the 1160s, and contrasted the oppressed Jews of Thessalonica and Constantinople with the thriving Jewish communities in Provence; he also noted the cosmopolitan nature of Montpellier (in Languedoc), where Jews, Christians, and Muslims intermingled peacefully.

This made it a perfect environment for translators, and around 1150, Judah ben Saul ibn Tibbon settled in Lunel, about 20 kilometers east of Montpellier; the Tibbonides would become an important family of translators over the next century. Judah's son Samuel translated many Arabic treatises into Hebrew, including Maimonides's *Guide to the Perplexed*, an attempt to reconcile religion, philosophy, and science; to ensure the most accurate translation possible, Samuel corresponded with Maimonides himself.

Islamic science flowed into southern France, especially after the rise of the Almohads forced many Jews to leave Andalusia. As early as 1137, Montpellier had a reputation as a prestigious center to learn *phisica* (medicine and natural science). In 1140, Raimon of Marseilles adjusted the planetary tables of Toledo so they could be used locally; Raimon

---

[6]The Languedoc language is also known as Provençal or Occitan, the latter derived from Aquitaine, another region that once spoke it.

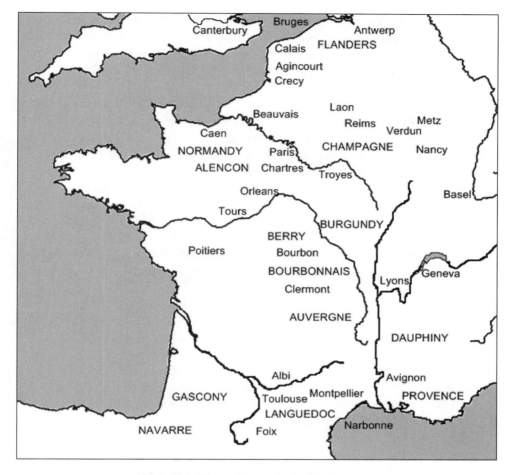

Figure 5.4. Western Europe in the Middle Ages

boasted (incorrectly) that he was the first westerner to learn the science of the Arabs. In 1143, Hermann of Carinthia, working in Toulouse, completed a translation of Ptolemy's *Planisphere* from the Arabic.

Around the time William IX of Aquitaine wrote the first troubadour songs, Provence produced another, more fateful product: Catharism. The Cathars believed in the inherent sinfulness of the material world, and sought salvation by purging themselves (*catharsis* in Greek) of all material goods and desires, a doctrine reminiscent of Buddhism (though there is no evidence of a direct connection). An inevitable consequence of the Cathar belief was that the vast territorial holdings and wealth of the Catholic Church stood in the way of true salvation. This put them into a collision course with the papacy.

The Cathars were particularly notable in the city of Albi, in Toulouse, so they are frequently called the Albigensians. In 1208 Pope Innocent III called for a crusade—not to the far-off Holy Land, but to the south of France. The prospect of taking a short trip over familiar territory to conquer land in Europe drew thousands of adventurers. For his part, Philip was content to let nobles loyal to him take on the difficult task of conquering the south. The result was the long and bloody Albigensian Crusade, the most notorious incident of which occurred at the city of Beziers. In 1209 the city was taken, and the Papal legate, Arnald-Amalric, happily reported the massacre of 20,000 men, women, and children; this inspired a later story that he told the crusaders, "Kill them all, and God will know his own."

Attempts were made to convert the Albigensians back to Catholicism by peaceful means as well. The most successful were the ministrations of the Spaniard, Dominic of Guzman, who used a combination of logical argument and disputation. This required thorough knowledge of Christian doctrine as well as considerable skill in the art of debate, so as early as 1218 we find seven Dominicans studying at the University of Paris. The next year the Dominicans opened a school in Toulouse in the heart of southern France, and in 1229 the school was reorganized as the University of Toulouse by Pope Gregory IX. Since a degree in theology was a virtual guarantee of a prestigious position after graduation, the study of mathematics and the natural sciences, which might have undergone a resurgence with the new translations available, stagnated. On the other hand, the need for maintaining orthodoxy meant a close study of the newly translated works, paving the way for the future.

With the conquest of the south, modern France began to take shape. Paris grew rapidly. The city grew from a population of about 30,000 in 1200 to about 150,000 a century later. This rapid expansion caused problems for students attending the University of Paris. Students wrote plaintive letters home asking for more money, citing the cost of books, lodging, and other expenses, though Alvarus Pelagius, a Grand Penitentiary at Avignon, noted wryly that they spend their money "in taverns, conviviality, games, and other superfluities ... They contract debts and sometimes withdraw from the university without paying them" [135, p. 174]. Consequently the University sponsored the creation of colleges ("assembly" in Latin): boarding houses where the students could live. Generally speaking the residents of a college were either studying the same subject or originated from the same region. Students were also organized into nations ("tribes" in Latin): originally, there were the French, Norman, Picard, and English nations, though additional nations were organized as more students from other lands arrived in Paris.

The years following the end of the Albigensian crusade also saw the appearance in France and northern Europe of many algorithms, works dedicated to explaining the use of the Hindu-Arabic system. One of the earliest algorithms was by ALEXANDER OF VILLEDIEU (d. 1240), better known for a work on Latin grammar, *Doctrines* (1199). *Doctrines* became the standard intermediate text on Latin grammar and underwent numerous editions (nearly 300 in the fifteenth and sixteenth centuries alone!). Around 1209, while teaching at the University of Paris, Alexander composed *Ode to Algorithm*, a poem explaining the use of the Hindu-Arabic numerals. It begins:

> Here follows the art of algorithm, which is
> Born from the twice five figures of India
> 0 9 8 7 6 5 4 3 2 1
> The first signifies one, the second indeed two
> The third signifies three; thus leftwards proceed
> Until the end, the one which "cipher" is called... [115, p. 72]

Note that Alexander refers to the right-most figure as the "first" figure—precisely the way they would be introduced in Arabic. The poem continues for many stanzas giving careful descriptions of how to perform the fundamental arithmetic operations using the Indian numerals. Today we might find it quaint that arithmetic operations are described in verse form; in the Middle Ages, this was a very practical solution to the lack of printing, for the meter and rhyme of a verse make it easier to remember the content.

One of Alexander's contemporaries was John of Holy Wood, known as SACROBOSCO (1195–1256) after the Latinization of "Holy Wood." Sacrobosco studied at Oxford University in England, which also made its first appearance in a police report. In 1209, two students had been lynched after the rape and murder of a local girl. The king (John of England) sided with the university and ordered the townspeople to establish two scholarships for poor students.

On June 5, 1221, Sacrobosco became a teacher at the University of Paris. His fame rests primarily on two works. *Treatise on the Sphere* soon became the standard text by which students learned Ptolemaic astronomy, and was required at the University of Paris; while *Algorithms* provided students with yet another introduction to computation using Hindu-Arabic numerals. Like his contemporaries, he listed the symbols in decreasing order:

> Here they are: 0.9.8.7.6.5.4.3.2.1. The tenth is called the circle or zero or the nothing figure for it signifies nothing; instead it keeps the place for the other figures...[104, p. 1]

Theoretical mathematics was still represented by Boethius's *Arithmetic*, listed around 1235 as one of the key texts students must know before taking their degree.

5.2.4 Architecture

Paris continued to grow, with new buildings springing up everywhere. One of the main problems facing master masons (who are now called architects) is how to span a roof. The Romans solved the problem using arches or vaults (half-cylinders), groined vaults (two half-cylinders intersecting at right angles), or hemispherical domes. The Pantheon, built in Rome during the second century A.D., remains the largest unreinforced concrete dome in the world, and in size it would not be surpassed until the completion of Brunelleschi's *Duomo* in Florence in 1436. In the first century B.C., the Roman architect Vitruvius wrote *Ten Books on Architecture* to give later generations vital information about building massive structures. But Vitruvius discussed more than the technical aspects of how to construct a building, and went into detail about the purpose of architecture, how buildings have to be designed to fit their inhabitants, and why architects should be more famous than wrestlers. Knowledge of Roman construction techniques was rediscovered in the tenth century, allowing Romanesque structures to arise throughout Europe, incorporating arch, vault, and dome.

A dome or vault is very, very, heavy so the walls of a Romeanesque structure must be extremely thick. Getting light into such a structure is problematic. The Roman solution was to leave an opening, called an *oculus* ("eye" in Latin), in the center of the dome. This left the interior subject to weather, but in the warm, southern European climate, this posed no serious problems; indeed, most Roman houses had an open central area called the *atrium*. But in northern Europe, December snow might not melt until March, so it was essential to close off the oculus. This left the inside in perpetual gloom—hardly a suitable environment for a house of worship.

Figure 5.5. Flying Buttress, from Villard de Honnecourt's sketchbook (13th century).

One solution is to leave openings in the walls for windows. But this posed a new problem. The roof applies two forces on the walls. The first is compression, from the actual weight of the roof itself. Stone is very good at resisting compression, and even with material removed to make room for windows, the stone walls would be strong enough to avoid collapse. But the roof also applies a sideways force (called shear) that pushes the walls outward. When material was removed to make windows, the walls were less able to resist shear and the danger of collapse increased. Master masons solved this problem by adding supports, called flying buttresses, to push inward on the walls. The result was a new style of architecture later called Gothic by Renaissance builders, who believed erroneously the structures were built by the barbarian Goths.

With glass becoming a common building material, new problems arose. Glassmaking requires tremendous amounts of heat supplied by burning wood, either as firewood or charcoal. Thus glass was generally manufactured in the country, where trees were plentiful, and transported to the city, where glass was in demand.

Figure 5.6. The flying buttresses of Notre Dame Cathedral. Picture by Beverly Ruedi

However, transporting large panes of glass over the rough medieval roads invariably resulted in losses through breakage. One solution was to make the glass near a river or ocean and transport it by ship. In Venice, the glass industry moved to the island of Murano in 1291 (fear of fires was the main reason); to the present day glassmaking is one of Venice's main industries. Alternatively, smaller panes might be manufactured and transported, then pieced together to make a large window. Thus the windows for Gothic structures were intricate glass mosaics.

Today we think of glass as a transparent substance, but during the Middle Ages, only the Venetians knew how to make clear glass, and they kept their methods secret. Most glass had a green tint, and by adding other colorants, glass of almost any color could be made. It was not long before someone thought to deliberately piece together glass of different colors to create a picture: the first stained glass windows appeared in the German city of Augsburg in the twelfth century.

In 1129, the Abbot of Saint-Denis, Suger, began a project to reconstruct the abbey. He had to proceed carefully: according to popular belief, the abbey had been consecrated by Christ, so demolishing it was sacrilege. Suger's elegant solution: rebuild one section at a time. Beautiful stained glass windows allowed plenty of light inside the abbey; indeed, Suger went further and, in an age where most people were illiterate, he contrived to have the pictures tell a story. These were the first narrative stained glass windows. With its graceful flying buttresses and stunning stained glass windows, the Abbey of Saint-Denis, consecrated in June 1144, became the best-known building in Paris.

But success breeds imitators, and the Abbey's fame would be quickly eclipsed by the Cathedral of Notre-Dame, whose construction began in 1163. Other Gothic churches sprang up throughout France, each city striving to outdo its rivals by building a structure larger or more decorated or higher than all the others. Height quickly became the measure of success. Notre-Dame's vault set a record of 32.8 meters above the ground in 1163. The masons of the cathedral at Chartres built theirs 36.55 meters high (1194), to be outdone by Rheims at 37.95 meters (1212), Amiens at 42.3 meters (1221), and finally Beauvais at 48 meters (1225).

5.2.5 Jordanus Nemorarius

The 66-page sketchbook of the thirteenth century architect Villard de Honnecourt includes many things of interest to the master builder: elevations, floor plans, useful mechanical devices, and fantastic beasts to inspire sculptors. But it lacks a scientific theory of *how* to construct a building. The need for such a theory was demonstrated in a spectacular fashion in 1284, when the vault at Beauvais collapsed. Clearly architects had to have a better understanding of the forces acting on a stable structure.

This is the domain of statics, a branch of mathematical physics first investigated by Archimedes with works on the principles of the lever, and extended by Greek, Roman, and Islamic authors. The first significant medieval European contribution to the subject was the *Book on the Computation of Weights*, written by JORDANUS NEMORARIUS (fl. 1225) between 1246 and 1260. About all we know about Jordanus himself is that he lived during the heyday of cathedral construction, and might have studied in Paris and taught at the newly founded University of Toulouse. Even the year and circumstances of his death are uncertain.

Jordanus was the first to correctly state the law of the inclined plane, essential if the shear force on a supporting wall is to be computed correctly. Imagine an object on a plane inclined at an angle of θ to the horizontal. According to Jordanus, the "positional weight" of the mass was inversely proportional to the obliqueness (the ratio of the oblique distance to the vertical distance); this is equivalent to saying $F \sim \sin \theta$, and is one of the key formulas of modern statics. In addition, Jordanus stated and proved several propositions relating to the law of the lever and was the first to give a correct mathematical statement of the law of the bent lever (like a crowbar): both of these were critical tools of the construction industry. The actual impact of Jordanus's work is hard to determine; he seemed to have no successors in the field of statics, and may have simply been describing a field already developed, rather than establishing the basis for a new one. However his work was copied many times over the next few centuries.

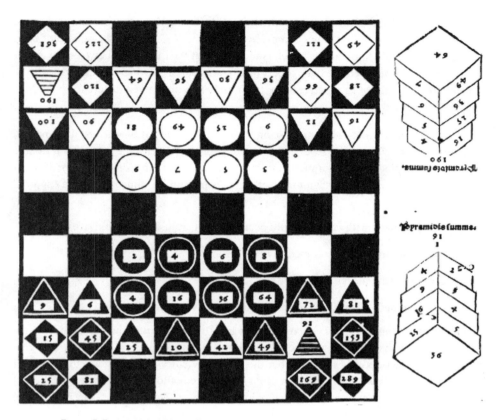

Figure 5.7. A 1496 edition of a Rithmomachy Board, from Jordanus's Treatise.

Besides his work on mathematical physics, Jordanus wrote *On the Given Numbers*, which is essentially a treatise on algebra. Jordanus's format is to give a proposition, such as:

Proposition 5.1. *If a given number is separated into two parts with a given difference, the two parts can be found.*

The proof is by construction: Jordanus solved the system $x + y = N$ and $x - y = M$ by addition, with $x = \frac{N+M}{2}$ and $y = N - x$. This particular proposition is important for most of Jordanus's later propositions, such as:

Proposition 5.2. *If a given number is separated into two parts with a given product, the two parts can be found.*

which gives rise to a quadratic equation.

What made Jordanus's work significant is not the type of problems that he solved, but his approach. First, his propositions are all existence theorems: they assert the existence of a solution to a given type of problem. Second, he took the first steps towards algebraic notation. The steps were very tentative and every time a quantity was changed, Jordanus gave it a new name. Thus Jordanus's proof of the previous proposition was as follows (note the geometric inspiration of his notation):

Proof. Let abc be the given number, which is divided into the parts ab and c. Let the given product be d. Let four times the given product be f. Let e be the square of abc, and subtract f from e to obtain g, which is the square of the difference between ab and c; thus the difference between the two parts is the square root of g, which we call b. Thus the given number abc has been separated into two numbers with a given difference b, and the two parts can be found.   □

Nemorarius also wrote a short treatise on Rithmomachy, one of the earliest documented educational games. There are several extant rules, but most play the game on a double chessboard. The pieces have numbers on them (the two sides have different sets of numbers), and capture is effected in a number of ways. A piece with value n could be captured if:

1. An opposing piece with value n could occupy its square using a legal move,

2. Adjacent opposing pieces summed to n,

3. Adjacent opposing pieces multiplied to n,

4. An opposing piece with value m was k squares away, where $km = n$.

Captured pieces changed sides. Each side had a set of pieces that moved as a unit (the "pyramid") but whose pieces could be captured or could attack individually. The first goal was to capture the opposing pyramid, but the greatest victory was accorded to the player who could arrange four pieces in a row from which one could (using three of the pieces) form an arithmetic, geometric, and harmonic progression. For example, the player who could line up the pieces $4, 6, 9, 12$ would be reckoned great indeed, for these pieces include the geometric sequence $4, 6, 9$, the arithmetic sequence $6, 9, 12$, and the harmonic sequence $4, 6, 12$.

We have very little information about the last years of Jordanus's life. There is some evidence he died at sea on the way back from a pilgrimage to the Holy Land. If so, he might have been caught up in the crusading fervor inspired by the French king Louis IX. In 1244, Jerusalem fell to the Muslims. Thus Louis organized the Seventh Crusade and invaded Egypt. The invasion was a disaster, and on April 7, 1250, the Egyptians captured Louis and his army. Undeterred, Louis organized the Eighth Crusade, this time invading Tunisia. Unfortunately, a plague swept through the army and Louis died in August 1270. Twenty-seven years later the Catholic Church made him Saint Louis, the only king of France to be canonized.

5.2.6   The Encyclopedists

Louis was a better patron of the arts and sciences than crusader, and collected a library one contemporary chronicler compared to that of an oriental (i.e., Byzantine) monarch. In all likelihood, the library contained less than a thousand books.

Louis's chaplain Vincent of Beauvais completed the *Great Mirror* (ca. 1244), an encyclopedia that covered the whole of European culture in eighty books. Vincent discussed history, science, law, and other subjects, and drew upon a wealth of sources that included Hebrew, Arabic, and Greek authors, making it clear that the accumulated knowledge of the

Mediterranean world was making its way to the west. The great French epic of love, the monumental *Roman de la Rose* ("Romance of the Rose"), was finished by Jean de Meun around 1280, and though its main topic was love, there were frequent digressions into classical mythology, economics, astronomy, and many other topics; de Meun also translated some of Abelard and Heloise's love letters from Latin into French.

Bartholomew the Englishman, a Lecturer in Divinity at the University of Paris, wrote a 19 volume encyclopedia mostly on theological subjects, though he, too, described the discoveries of the Arabs and Jews in medicine and science. The encyclopedia proved very popular among the scholars at the University of Paris. Even earlier was the remarkable encyclopedia written by Herrad of Landsberg, the Abbess at the convent of Hohenburg in Alsace. The *Garden of Delights* (1195), written for her nuns, was a compilation of biblical, theological, and horticultural knowledge, with an extensive collection of poetry and music at the end.

Even the illiterate could partake in the educational movement that inspired the encyclopedists. The stone that made up the walls of the cathedral had to be very rugged to bear the weight of the building, but along the lintels and interior surfaces a softer, more pliable mineral called "freestone" could be used and carved into elaborate and imaginative sculptures that provided a veritable encyclopedia in stone, giving lessons in history, science, morality, and the arts to those who could not read.[7]

5.3   High Medieval France

Names can be misleading. Saint Louis was a dismal failure as a crusader, and his grandson, Philip IV was known as Philip the Fair. But "Fair" referred to his appearance and not his equity, for he would be among the most unscrupulous of the French kings when it came to finding new sources of revenue.

First, Philip turned to the County of Flanders (roughly modern Belgium) whose towns, notably Bruges and Ghent, profited immensely from the lucrative wool trade. Evidence of Bruges's wealth can be found in its civil legislation. Fire was the greatest threat to a medieval city, with its wooden buildings and thatched roofs. In Bruges, the city required buildings to have tile roofs, with a city subsidy paying for one out of every three tiles. Sanitation was non-existent in most medieval cities, where animals of all sorts trod on city streets and refuse (including human wastes) was simply thrown out the window after shouting a warning to passersby. In Bruges, the streets were cleaned weekly at city expense. Congestion was a problem in most medieval cities, whose expansion was frequently limited by the city walls. In Bruges, the city widened the roads and reimbursed the owners for the demolished property (one of the earliest examples of eminent domain). The city of Bruges also financed the construction of a large warehouse over the river, so goods could be unloaded more easily.

The Counts of Flanders had a nominal allegiance to the French king, but Philip, intent on raising money, increased taxes until the townspeople revolted. They raised a makeshift army, consisting mainly of weavers and other artisans, untrained in the skills of war and

---

[7]It is possible but by no means certain that the freemasons were originally the freestone masons who, since they had to have some understanding of the subjects they were carving, might be perceived as having access to certain arcane secrets.

poorly equipped besides. The rebels, mostly on foot, armed themselves with a cheap, improvised weapon called a *goedendag* ("good day" in Flemish, though the name only appears in later sources): little more than a five-foot wooden club with a short iron spike at one end. At a time when military strategists considered one mounted knight to be worth at least 10 footsoldiers, Philip sent 2500 mounted knights, 1000 crossbowmen, 1000 pikemen and about 2000 other infantry to meet the 9000 poorly armed, untrained, unmounted footsoldiers of the Flemish militia.

On July 11, 1302, the two armies met outside the city of Courtrai. Convinced of their overwhelming superiority, the French knights charged the militia across a swollen river and a muddy field. The Flemish routed the French and killed the flower of French chivalry: sixty barons, hundreds of knights, and thousands of squires. The battle is frequently called the Battle of the Golden Spurs, from the 500 pairs of spurs recovered from the fallen French knights. Flemish independence was assured, and July 11 has been celebrated as a holiday in Flanders ever since. Philip had to give up his designs on Flanders and turn to other sources of revenue. Thus he turned to a target both wealthy and unpopular: the Jews.

5.3.1   Exiles and Captives

A number of discriminatory laws defined the position of the Jews in medieval Christian Europe. Many places prohibited Jews from owning land or entering a wide variety of skilled professions. Since church law forbade Christians to lend money at interest, many Jews turned to money lending. Philip's great-great grandfather, Philip II, tried to expel the Jews but found the financial life of France depended too much on Jewish capital, so he was forced to allow them to return. By the thirteenth century, the Lombard bankers of northern Italy obtained Papal dispensation to lend money at interest, breaking the Jewish monopoly. In 1290, Edward I of England earned the dubious distinction of being the first European ruler to expel the Jews from his territory, after confiscating their property, including the right to collect on loans they had made. In 1306 Philip IV followed Edward's example, confiscated the property of French Jews, and gave them a choice: they could convert to Christianity, or they could leave. Many chose exile.

The family of LEVI BEN GERSON (1288–April 20, 1344), who was born in Bagnols on the Rhone, chose exile. "Levi" is the Hebrew equivalent of Leo, while "ben" means "son of"—we might call him Leo, son of Gershwin, though he was known to his contemporaries as Leo the Jew. When the expulsion edict came, he and his family moved downstream to the city of Avignon, part of a papal enclave acquired as a result of the Albigensian crusade.

In 1321, Levi ben Gerson finished *The Art of the Calculator*. The first half of the treatise is on practical arithmetic. The second half, however, is more theoretical, and includes some of the first proofs using mathematical induction, which he called *hadragah* ("stepped-cliffs"). One example was his proof that the number of permutations of n objects was $n!$:

Proposition 5.3. *If the number of permutations of n things is K, then the number of permutations of $n + 1$ things is $(n + 1)$ times K.*

Proof. Suppose there are K permutations of n things. Let an additional thing be added. Take any permutation of the n things, and place the new thing first. There are K permutations with the new thing first. Take any other element, and place it first. There are K

permutations of the remaining n elements. Proceeding this way with all $n + 1$ elements placed first gives $(n + 1)K$ permutations of the $n + 1$ elements. $\qquad\square$

Between 1336 and 1344, Levi ben Gerson wrote a treatise on astronomy, part a philosophical work called *Wars of the Lord*. In it, he used a procedure he called heuristic reasoning as a means of approximating solutions to very complicated equations; it is a variant of the method of double false position.

Ben Gerson's greatest use of his method was to find a value of $\sin \frac{1}{4}^{\circ}$ accurate to eight decimal places. He proceeded as follows. Using the half-angle formula and known values like $\sin 30^{\circ} = \frac{1}{2}$ and $\sin 18^{\circ} = \frac{-1+\sqrt{5}}{4}$, we can obtain $\sin 15^{\circ}$ and $\sin 1 \frac{1}{2}^{\circ}$. Hence we can find $\sin\left(16 \frac{1}{2}^{\circ}\right)$, and apply the half-angle formula repeatedly to obtain $\sin\left(\frac{1}{4}^{\circ} + \frac{1}{128}^{\circ}\right)$. Likewise, a repeated application of the half-angle formula to $\sin 15^{\circ}$ will give us $\sin\left(\frac{1}{4}^{\circ} - \frac{1}{64}^{\circ}\right)$. For convenience, designate $a = \sin\left(\frac{1}{4}^{\circ} + \frac{1}{128}^{\circ}\right)$ and $b = \sin\left(\frac{1}{4}^{\circ} - \frac{1}{64}^{\circ}\right)$. Then $\sin \frac{1}{4}^{\circ} \approx b + \frac{2}{3}(b - a)$.

With an accurate value of $\sin \frac{1}{4}^{\circ}$, ben Gerson could create a table of chord lengths accurate to five decimal places, which appeared in *On Sines, Chords, and Arcs*. This work and another, *Treatise on Astronomical Instruments*, was dedicated to a fellow resident of Avignon: Pope Clement VI.

The Papacy was another victim of the rapacity of Philip IV, who tried to tax the church. Pope Boniface VIII refused to allow this. The University of Paris became the focus of demonstrations for and against the king, so Philip sent commissioners there to investigate the loyalties of its faculty. As a result, eighty Franciscans were expelled from France (including Duns Scotus). In retaliation, Boniface revoked the University's right to issue degrees in law and theology, its two most lucrative programs.

Philip in turn prohibited the export of precious metals from France, which threatened Boniface's ability to pay for his many construction projects (he employed the artist Giotto for a time). Despite this, Boniface continued to forbid Philip to tax the church, so on September 7, 1303, French knights under William of Nogaret, the king's magistrate, and Sciarra Colonna, a member of the rising Colonna family, burst into the papal palace at Anagni, accused Boniface of all sorts of crimes, and demanded his resignation as a prelude to a trial. The people of Anagni freed the Pope a few days later, but Boniface died shortly thereafter. His successor, Pope Benedict XI, excommunicated Nogaret and Colonna and demanded *their* trial. A few days later, Benedict died suddenly and unexpectedly: suspicious timing, considering the circumstances.

After Benedict's death, Clement V ascended to the papal throne. Clement was born in France (though in the region controlled by the English), and was a personal friend of Philip the Fair. Clement moved the Papal court from Rome to France, ultimately settling in Avignon in 1309. Though not part of France, Avignon was close enough to make Clement reluctant to challenge Philip or his policies. Thus when Philip turned on the Knights Templar, a group that had grown rich through international banking, Clement cooperated, turning all the force of the Inquisition on the knights and allowing Philip to seize their assets.

The papacy remained in Avignon until 1377. The period period when the Popes lived in Avignon is known as the Babylonian Captivity of the Papacy. The fourth of the Avignon Popes, Clement VI, was the first to actually rule Avignon itself, having bought it from Queen Joan of Provence in 1348. The pope welcomed Jews into his territory. The need was great, for in the 1350s, many of the German city-states followed the examples of Edward and Philip, and expelled the Jews after seizing their wealth. With the papal enclave one of the few islands of tolerance in western Europe, the region soon boasted a thriving Jewish population.

Just downstream from Avignon itself is Tarascon, the home of IMMANUEL BEN JACOB BONFILS (fl. 1340–1377). Among the earliest discussions of decimal fractions in the west was his *Method of Division by Rabbi Immanuel and Other Topics*. The title emerges from a misunderstanding of the copyist, who read the first line—"Unity is divided into ten parts"—and thought the treatise would be about division. In fact, the treatise gives a systematic treatment of decimal arithmetic, anticipating the work of Simon Stevin by nearly two centuries.

Bonfils also produced an important set of astronomical tables, *The Wings of Eagles*, used for several centuries to help determine important dates in the Jewish calendar. He included the correction factors necessary to make use of the calendar in places as far away as Constantinople or the Crimea, nearly two thousand kilometers away. It might seem audacious of Bonfils to write a work expecting it would be used over such a wide area, but he had good reason: just south of Tarascon was the Republic of Genoa whose tendrils of trade extended to the Black Sea. It would be a deadly connection.

5.3.2 Plague and War

In 1347, one of Genoa's Black Sea trading posts was under siege by the Khanate of the Golden Horde. During the siege, the Mongol attackers contracted a disease, and many died. In the hopes that the disease could be transmitted to the defenders, they catapulted the corpses into the fortress. Their hopes were soon realized and the defenders began to die from a disease they had not encountered before: plague.

Because the most characteristic feature of the plague is pustules, or buboes, that form on the groin and armpits, the disease is usually referred to as the bubonic plague. People of the time referred to it as the Great Death, though it is better known by its 19th century name: the Black Death. Plague would break out several times in Europe over the next few centuries, but the magnitude of the first outbreak is hard to comprehend. As many as 25 million people—one-third of the population of Europe—died during the first few years of the plague. Thousands of villages were completely depopulated. In Avignon, half the population died, including a quarter of the papal staff and a third of the cardinals. Dead bodies accumulated faster than they could be buried, posing a monstrous health hazard; Clement VI was forced to consecrate the Rhone so the bodies could simply be thrown into the river. Plague reached Paris in the spring of 1348. At its peak, 800 people a day died from the plague, and by 1349 around half the population—fifty thousand people—had died.

But even these stark figures do not convey the real impact of the plague. It struck rapidly and lethally: of those infected, 90% were dead within a week. The only way to avoid plague was to avoid the sick. One eyewitness, the historian Giovanni Villani, described the plague's effects in Florence, and wrote, "The plague lasted until   ", leaving a space for him to fill in the date the plague abated. Villani never completed the sentence, for he died from the plague in 1348. Giovanni Boccaccio survived, and later told a tale of how three men and seven women fled to a country villa to take refuge from the plague raging in Florence. To pass the time, they would tell each other stories: ten tales a day for ten days. The result was the *Decameron* ("ten parts" in Greek), one of the great works of western literature.

The literary device of having a group of unrelated people, thrown together by happenstance, and entertaining each other with tales, attracted many imitators, including Geoffrey Chaucer. Many of the stories in the *Canterbury Tales*, such as those of the Clerk, Franklin, Merchant, and Reeve, are clearly related to those in the *Decameron*

Chaucer drew from other sources as well. The Nun's Priest's Tale is a retelling of an Aesopic story by Marie of France (fl. 12th century), the earliest known French female poet (though it is believed she wrote in England). In the story a rooster, Chanticleer, has a dream he is seized by a fox; one of his wives convinces him that dreams cannot foretell the future. When later a fox seizes Chanticleer in his jaws and carries him off, the household in pursuit, the rooster is saved when he persuades the fox to shout insults back at his pursuers. The story ends with a number of morals: when the fox tries to lure Chanticleer back to his grasp, the rooster replies that no amount of flattery will lead him back, and the fox bemoans the fact that he foolishly opened his mouth when he ought be silent.

During the course of the story the issue of fatalism arises, and the Nun's priest, after noting that the dispute has occupied "a hundred thousand" men, says he is not wise enough to go into the philosophical details, for he is neither Augustine, nor Boethius, nor Bishop Bradwardine.

THOMAS BRADWARDINE (1295–August 26, 1349) became Archbishop of Canterbury on June 4, 1349, was consecrated in Avignon on July 10, 1349, and died in England on August 26 from the plague. He studied at Merton College in Oxford, and until about 1335, occupied himself with questions on logic, mathematics, and philosophy. One of the key areas of study at Merton is now called kinematics, the physics of moving objects. Aristotle claimed that motion was only possible when the force acting on an object exceeded the resistance. Based on statements in his *Physics*, the scientists of the Middle Ages further believed the velocity of an object was proportional to the force acting on it divided by the resistance. Bradwardine showed that the two are logically incompatible: Imagine some initial force, velocity, and resistance. Doubling the resistance while keeping the force the same will halve the velocity, but no matter how many times the velocity is halved, it will never become zero. But if the resistance is doubled enough times, it will eventually exceed the force, and the velocity (by Aristotle's actual statement) would have to drop to zero.

To resolve the contradiction, Bradwardine assumed that an arithmetic increase in velocity corresponded to a geometric increase in the ratio of the force to the resistance: thus to double the velocity required squaring the ratio, tripling the velocity required cubing the ratio, and so on, which is equivalent to saying (in modern terms) $v \sim \log \frac{F}{R}$.

Bradwardine was the first to study star polygons (the result of joining every nth vertex in a regular polygon), and might have made important contributions to mathematics, but decided instead to pursue a career as a churchman. He became the canon of Lincoln in 1333, the Chancellor of St. Paul's Cathedral in London in 1337, and shortly thereafter became chaplain to King Edward III of England. This gave him a first-hand view of the start of the longest war in history.

Edward's mother Isabella was the daughter of Philip the Fair. In 1328, the last of her brothers died and Isabella claimed the French throne. But since Edward was Isabella's heir,

French nobles used a dubious interpretation of the law of the Salian Franks to deny her claim, instead offering the crown to her cousin, Philip of Valois.

In 1339, Edward invaded France to make good on his mother's claim to the throne, starting the Hundred Years' War. Unfortunately he could not win a decisive victory, so he returned to England to raise money. Edward's normal income, from which he had to pay all the costs of the war, was only about 30,000 pounds a year, whereas the campaign of 1339 cost over 100,000 pounds. To raise money he used the courts to levy fines on the nobles and confiscate their property. This highhanded behavior nearly led to a revolution in 1341, and Edward made concessions to parliament, the most important of which was that the House of Commons, the branch of the English parliament consisting of representatives from the poorer knights and richer towns, was given the exclusive right to levy new taxes.

With a new army, Edward invaded Normandy in 1346 (Bradwardine, as royal chaplain and confessor, accompanied Edward). The invasion began ignominiously when Edward, disembarking, fell face forward into the sand. After this inauspicious beginning, the army then ravaged much of northern France before withdrawing towards Flanders. Philip, hoping to inflict a decisive defeat upon the English, caught up with Edward's army just north of Paris at Crécy. Edward was outnumbered, 36,000 to 12,000, but his forces included 7000 English longbowmen. On August 26 the French attacked, making over a dozen charges against the English position, every one of which failed disastrously. By day's end, the English longbowmen had fired an estimated half million arrows and killed 1500 French knights and a thousand other French soldiers. Contrary to popular myth, an armored knight cannot easily be killed by an arrow fired from a longbow; most of the French casualties at Crécy occurred when wounded knights fell from their horses and were trampled by their own side, or had their wounded horses fall on them, leaving them easy prey for the daggers of English men-at-arms. The deaths of so many French knights infuriated Edward: a captured knight could be forced to pay a ransom for his release, which would have allowed Edward to prosecute the war without having to go to parliament for more money.

5.3.3 Oresme

Caen, the birthplace of NICHOLAS ORESME (1323–July 11, 1382), was the first town sacked during Edward's invasion. At the time, Oresme was probably at the University of Paris, for by 1348, he was one of the Masters of the Norman Nation.

In 1350, Philip died and was succeeded by his son John II. At a time when the kingdom was facing military disaster, John threw lavish tournaments and extravagant parties, making him known as "John the Good"—as in good company. In 1356, an English army, under the command of Edward, the Prince of Wales, ravaged Aquitaine in western France.[8] John fielded an army to crush the English raiders and met them at Poitiers on September 19, 1356. As at Crécy, the English were outnumbered three to one. As at Crécy, French knights once again charged English longbowmen. As at Crécy, French casualties were staggering. Over 3000 French knights were killed, while English casualties were less than 100. John was captured, and forced to sign a treaty that relinquished so much French territory that the French refused to accept it, resulting in the more modest Treaty of Calais (Chaucer played

---

[8]Edward the Prince of Wales is also known as the Black Prince, though no one knows why; the nickname did not emerge until the sixteenth century.

some role in its drafting): Edward relinquished his claim to the French throne in exchange for mastery of Aquitaine.

In addition to Aquitaine, John had to pay an enormous ransom, equivalent to about 2,000,000 pounds (compare this to Edward's ordinary annual income of 30,000 pounds!). The task of raising the ransom fell to John's son, the future king Charles V. One way to raise this enormous sum was to increase taxes, particularly the taille. All taxes are odious to those who pay them, but what made the taille particularly hated was the fact that it was in theory a monetary payment in lieu of military service. Thus the nobles, whose feudal obligations included military service, and the clergy, whose position exempted them from military service, did not have to pay the taille. The burden fell on the middle and lower classes, who had already born the brunt of the war's devastation. On May 21, 1358, discontent exploded into revolt. Because the nobles contemptuously referred to all peasants as "Jacques Bonhomme"—James Goodfellow—the revolt came to be known as the jacquerie. The revolt, the first of many, was mercilessly crushed.

Another way to raise money was by loans. Oresme, who befriended Charles around 1356, was one of his chief negotiators and by 1360 the ransom was raised and John was released. Since most of the currency in France had just been exported to England, John issued new coins, made of gold.

Ideally, a coin represented a quantity of precious metal whose purity and weight were guaranteed by some trusted authority, usually the government. Unfortunately it was not long before princes and potentates realized they could declare their coins to have a higher weight or purity than they actually had. In doing so they could effectively increase their wealth. Since lesser metals are known as "base" metals, the process of mixing in cheaper metals with silver or gold and producing coins from the alloy is referred to as debasement. The French kings were notorious debasers, and "him who falsifies his coins" (Philip IV) is listed by Dante as one of the condemned. By the time of John II, French currency was worth perhaps a third of what it had been before the war.

His first-hand experience of debased currency gave Oresme the subject material for *On the Origin, Nature, Uses, and Alteration of Money* (1360), the first scientific treatise on the theory of money. The first half is purely descriptive. The second half, beginning with the section entitled "On Alterations in General", discusses the debasement of currency and other alterations of coinage, outlining their effects in clear detail.

The problem with debased currency is that it fools no one. When the debased currency appears, prices rise, trade is interrupted, and the value of finer (more pure) coins increases: it is as if a quarter can be exchanged for twenty-five cents one day, and forty cents the next. When that occurs, consumers hoard the finer coins, which disappear from circulation: "Bad money drives out good." This observation is usually called Gresham's law, after Thomas Gresham, a sixteenth century economist, but Oresme noted the same thing two centuries earlier.

Because of its grave effects on the economy and the people, Oresme declared that debasement was worse than usury, itself considered one of the gravest crimes in the Middle Ages (Dante placed usurers in the Sixth Circle of Hell). This was a perilous position for Oresme to take, for John II had already debased the currency several times during his reign, and Oresme's condemnation of debasement might be taken as a criticism of the monarchy. Thus in the final chapter, Oresme admits that there are conditions under which debasement is permissible—even desirable. The debasement of currency is presented as a particularly equitable way to raise money: it needs neither assessors nor tax collectors; it exempts no one; fraud is impossible to perpetrate; and those with the most money bear the greatest burden. With all the negative effects of debasement, however, Oresme recommends that it be used only under unusual circumstances: in war, or to ransom a prince captured in war, *especially* if the money is to be transported to distant lands.

Oresme's final chapter notwithstanding, it is clear that in general he opposed debasement of currency, and it is interesting to note that Charles V was the only French king of Oresme's lifetime who scrupulously avoided further debasement of the currency and instead raised money by sound financial policy.

About the time he completed his treatise on money, Oresme also finished his *Treatise on the Configuration of Qualities and Motions*, based on work he had begun in the 1340s with *Questions on the Geometry of Euclid*. In his earlier work, he introduced the idea that the magnitude of some changing physical quantity, such as the velocity of an object under acceleration, could be represented by a two-dimensional figure (the "configuration of quality"). We may refer to the position of any point on the "summit" of the figure by its longitude and latitude (or altitude): in modern terms, its x- and y-coordinates. In effect, Oresme described the two-dimensional graph of a physical quantity varying over time. Oresme was anticipated in this idea by GIOVANNI DI CASALI (fl. 1346), an Italian Franciscan, who wrote down similar ideas in 1346, though Oresme's influence was greater due to his position at the University of Paris.

But Oresme did more than invent (or reinvent) the graph: in a sense, he invented the definite integral. Consider an object that accelerates uniformly from rest during some interval. The configuration of the quality (of velocity) will be a triangle, whose altitude is nothing at the beginning of the interval (the object is at rest) to some definite altitude at the end of the interval (when the object reaches its greatest velocity). Oresme identified the quantity of quality with the distance traveled by the object, and its quantity could easily be found by calculating the area of the triangle. Scholars at Merton College in Oxford (including Bradwardine) had already enunciated this result as the Merton Rule: the distance traveled by an object accelerating uniformly from rest is the same as if it traveled for the entire interval at the velocity it had at its midpoint.

Oresme went far beyond the Merton Rule and considered other cases, including some remarkable ones that correspond to evaluating improper integrals. For example, Oresme considered the configuration of an object which moved for some interval of time. If its velocity was constant for half the interval, then twice the original velocity for a quarter of the interval, then three times the original velocity for an eighth of the interval, and so on, then by the end of the interval of time the object would be moving with an infinite velocity. Oresme showed, however, that even in this case the object would only travel a finite distance.

He did so by referring to the configuration of quality (see Figure 5.8). In this case, let the extension be AB, and let the rectangle on AB correspond to configuration of an object that travels with the original velocity for the entire length of time; the quantity of this quality corresponds to the distance traveled. Let there be a second rectangle CD identical to the first, and let CD be divided into proportional parts (so ED is half of CD, FD is half of ED, GD is half of FD, and so on). Take the rectangle on base CE and stack it

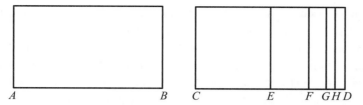

Figure 5.8. Oresme's Infinitely Fast Object

onto the rectangle on AB; take the rectangle on base EF and stack it onto the new figure, and so on; the resulting figure will correspond to the configuration of quality of the object whose velocity becomes infinite. It is clear that the quantity of quality will equal twice the rectangle on AB, and thus the object, though its velocity becomes infinite, will only travel twice as far as an object with a constant velocity.

Oresme recognized that the configuration solved a related problem: An object might travel for an *infinite* amount of time yet move only a finite distance. This would be the case if the object traveled for some velocity for some amount of time; half the velocity for an equal amount of time; a quarter of the velocity for yet another amount, and so on; the configuration would be identical, but rotated a quarter-turn.

Key to Oresme's work is the ability to compute areas. Oresme dealt with quantities that were either uniform (constant) or uniformly difform (linear), which gave rise to rectilineal figures; he noted the existence of difformly difform (non-linear) quantities, and divided these into sixty-six categories, most of which could not be rectified until the invention of calculus, three hundred years later.

Meanwhile, the Hundred Years War began to turn against the English. First, Charles found a competent commander, Bertrand du Guesclin, and made him the Constable of France in 1370. This gave du Guesclin supreme command of the French armies. Du Guesclin realized that charging English archers with French knights was ineffective, so he changed tactics: he would allow the English free rein in France, harrying them wherever possible, wearing them down bit by bit, but always denying them a great victory. Du Guesclin's tactics worked, and by Charles's death in 1380, the English controlled only a few coastal towns in Normandy and Bordeaux. It is probably no coincidence that the stories of Robin Hood became popular around the same time real English longbowmen could no longer achieve great victories against the French.

Secure in his realm, Charles V could afford to be a great patron of the arts and sciences, earning him the name Charles the Wise. He commissioned translations of many works into French, and had Oresme translate Aristotle's *Politics* and *Ethics*. As a reward for this and other services, Charles made him Bishop of Lisieux (in Normandy) in 1377, where Oresme spent the rest of his life. Jean Corbechon translated the great encyclopedia of Bartholomew the Englishman, and Guillaume Tirel (Charles's head cook, better known as Taillevent) wrote the earliest French cookbook. Charles soon acquired one of the largest libraries in western Europe: nine hundred volumes. Books still had to be copied by hand, and the greatest book of the era was known more for its illustrations than for its text: the *Book of Hours* of the Duke of Berry (Charles's brother). The *Book of Hours* was a combination of prayer book and calendar, lavishly illuminated and illustrated by many artists.

Charles left a stable, secure realm to his successors. Unfortunately when he died his son, Charles VI, was a minor, so the kingdom was ruled by his three uncles: the Dukes of Anjou, Berry, and Burgundy. They ruled for the benefit of Anjou, Berry, and Burgundy, to the detriment of France as a whole; the gorgeous *Book of Hours* was one result, but the Battle of Agincourt (1415) was another. At Agincourt, French knights would once again charge English archers across a muddy field, and once again, French knights died by the thousands. Charles was forced to recognize Henry V as his heir, though to maintain some continuity in the royal bloodlines, Henry married Charles's daughter Isabella; their son would be Henry VI of England. Agincourt was the greatest English victory of the Hundred Years' War.

It would also be the last one. The French were saved by two factors: a peasant girl, Jeanne Darc (later given a patent of nobility, changing *Darc*, a name of no particular meaning, into *d'arc*, "of Arc")—and gunpowder. Joan rallied the French, lifted the siege of Orléans, and helped arrange the coronation of the new king, Charles VII at Reims (1429). Charles in turn allowed her to be captured by his Burgundian enemies, turned over to the English, tried for witchcraft, and burned at the stake. It helped the English not at all, and by 1453, the English had been driven from all of France except Calais.

No formal treaty marked the end of the Hundred Years' War: the combatants simply found they had more important battles to fight. In England, Henry V's death in 1422 led to the long minority of his infant son, Henry VI, and in March 1454, Richard, the Duke of York, became Regent during one of Henry's periodic bouts with insanity. When Henry recovered later that year, Richard's refusal to step down began the War of the Roses, so called because the emblem of the House of York was a white rose, while the emblem of their rivals, the House of Lancaster, was a red rose.

In France, the great trade fairs of Champagne never recovered, and merchants sought safer locations to conduct their business. In 1462, the French King Louis XI granted the city of Lyon in southern France the right to run four annual trade fairs. These fairs proved so successful that when Louis tried to encourage the development of the silk industry in Lyon, local merchants objected because domestically produced silk would reduce the profits they earned from imported Italian silk.

In 1470, three German printers were invited to the University of Paris to set up a press at the Sorbonne. The scholars selected the books and oversaw all aspects of production. They objected to the German Gothic type fonts and insisted that the printers use Roman type fonts, greatly contributing to the success of the Roman over the black letter fonts. Printing arrived in Lyon shortly thereafter, and the first book published in French, *The Golden Legend*, appeared in 1473. Thanks to its trade fairs and relative freedom from religious and political censorship, Lyon became one of the most active publishing centers in Europe.

5.3.4 Chuquet

Ironically one of Lyon's residents did not take advantage of the printing press, and as a result was forgotten for centuries: NICHOLAS CHUQUET (1445–1488). Chuquet describes himself as a Parisian, though by 1480, the tax registers for the city of Lyon record him among its inhabitants. Very little is known about him, but Chuquet wrote a very sophisticated work, the *The Science of Arithmetic in Three Parts*, usually referred to as the *Triparty*.

The first part discussed the procedures for computing with Hindu-Arabic numerals. Merchants of the fifteenth century had to contend with a greater volume of trade and more wealth than ever before in history, and Italian merchants spoke of the "large thousand", the million. By analogy, Chuquet named even larger numbers: the *bimillion* or billion, the *trimillion* or trillion, and so on. Chuquet's definition of these higher numbers corresponds to the definitions used in England and Germany though, ironically, not in France: Chuquet's billion is $(10^6)^2$ or 10^{12}; his trillion is 10^{18}; and so on.

The *Triparty*, however, is largely a work on algebra. His notation for the powers of the unknown was particularly interesting: $5x^3$ was written as 5^3, and negative exponents were written using $\tilde{m}$: thus for $5x^{-3}$, Chuquet would have written $5^{\tilde{m}3}$. Chuquet also explained the rules of exponents, and wrote pure numbers like 5 as 5^0, a clear anticipation of the modern idea that $5x^0 = 5$. Perhaps his most interesting contribution is a method of interpolation he called the rule of mean numbers: in modern terms, if

$$f\left(\frac{a}{b}\right) < m < f\left(\frac{c}{d}\right),$$

then

$$f\left(\frac{a+c}{b+d}\right) \approx m.$$

Chuquet never printed his work, and as a result he was forgotten for centuries. It was up to his student and fellow citizen of Lyon, ETIENNE DE LA ROCHE (1470–1530), to disseminate Chuquet's work (without attribution) in his *On Arithmetic* (1520)—which was *printed*. Until the discovery of Chuquet's *Triparty* in manuscript form in the nineteenth century, de la Roche received credit as the first French algebraist.[9]

For Further Reading

For medieval history, see [3, 28, 35, 55, 73, 90, 93, 97, 108, 123, 125, 129, 131]. For the mathematics and science of the era, see [21, 37].

---

[9]After the discovery of Chuquet's manuscript, de la Roche was branded a plagiarist. A more charitable assessment is that as Chuquet's student, de la Roche was also heir to his intellectual property.

6

Renaissance and Reformation

6.1 The Italian Peninsula

Europe slowly recovered from the plague. In part this was due to the fact that most of the susceptible population died in the first few years of the plague, but in part it was due to drastic measures taken by cities to prevent the plague from entering. In Italy, divided into a number of independent states, the port cities instituted the practice of interning a vessel's cargo and crew for a period of thirty days before they were allowed to mingle with the general population: the original quarantine (from *quarantina*, "forty" in Italian).[1]

The plague had two important effects. Since it spread faster among the densely packed urban population, the cities were particularly hard hit, and the surviving artisans could ask for premium wages. Drawn by higher wages, many agricultural workers left the farm and migrated to the city.

The higher wages could be paid easily, for often the wealth of an entire family fell into the hands of a single survivor. In the past, the wealthy had given money to the church and its charities. But piety provided no protection from the plague. Thus, many took to living as if there would be no tomorrow—as well there might not be, if the plague came again. Some spent their inheritance on lavish, drunken orgies. Others spent their money in the manner of the great medieval kings, patronizing artists and scholars. Both of these were common features of the Renaissance, the "rebirth" of patronage and scholarship.

The exact causes of the Renaissance have long been debated and no definitive answer has yet emerged. However, the effects of the Renaissance are much more clear-cut. Key to the era was the attempt to revive the glories and practices of classical antiquity: Greece and Rome, particularly Republican Rome. An important concept was *virtue*: this had nothing to do with morals, and everything to do with competence. The great divide between the Renaissance concept of virtue and our own is best represented by the definition given by Machiavelli: a virtuous ruler is one who maintains his state's existence and prosperity by all necessary means.[2]

The interest in classical antiquity caused a resurgence in the study of Latin and, more importantly, Greek. Latin posed no special problem, for it was the language of scholars throughout Europe; recall that the grammar and rhetoric of the trivium were the study of

---

[1]The first use of the word in English is in Pepys's diary, where he mentions the thirty day quarantine.

[2]It is not quite correct to say that Machiavelli advocated "The ends justify the means." It is more correct to say that Machiavelli advocated, "Certain ends justify any means."

Figure 6.1. The Italian City-States

Latin grammar and rhetoric. Greek, on the other hand, is a complex language that is very difficult to learn, and few western scholars could read it: "It was Greek to me" (Shakespeare, *Julius Caesar*, Act I, Scene 2) echoed the sentiment of most Europeans.

However, unlike Latin, Greek was (and is) a living language, and thus a scholar interested in learning Greek could rely on native speakers of Greek. Alternatively, a native speaker of Greek could learn Latin, and translate the works of the Greek masters for others. The revival of Greek was crucial for the revival of higher mathematics.

6.1.1   Mathematics and the Liberal Arts

Unfortunately for the study of mathematics, the Renaissance saw a dramatic change in higher education. Originally, mathematics was not *a* liberal art: it was *all* of the liberal arts.

In particular, the four "upper division" liberal arts were either mathematics (arithmetic and geometry) or applied mathematics (music and astronomy).

In 1428 a committee in the Republic of Florence suggested that the studium (the nascent University of Florence) be expanded to include two new chairs: moral philosophy; and rhetoric and poetry. Cosimo de'Medici, a Florentine banker living in Rome, convinced his client Pope Martin V to levy a tax on the clerical institutions of Florence to endow the new chairs. The first holder of the chair of rhetoric and poetry (and also that of moral philosophy, until a suitable candidate could be found) was Francesco Filelfo, who launched an ambitious new curriculum that would become the basis of what Florentines called the study of humanity. Florentines began to apply the term liberal arts to the humanities only, beginning a trend of minimizing the mathematical content of a liberal arts education.

Perhaps the most important effect of the Renaissance and the revival of Greek and Latin learning had to do with a change in viewpoint. For a thousand years, the west focused on theological matters, and one's life depended on Providence or Divine Will. But classical antiquity predated Christianity, and emphasized the importance of man, not the gods, not the afterlife: in mythology, for example, heroes like Achilles and Hercules were admired, while the gods existed primarily to help or hinder the hero. As a result, the individual became more important: "Man is the measure of all things" (originally said by Protagoras, an ancient Greek philosopher) became the slogan of the Renaissance.

The changing viewpoint was reflected in changing artistic techniques. If man was indeed the measure of all things, then the universe ought to be shown in its proper relationship to man: this required a more realistic portrayal of objects, which required the use of perspective. Some Greek and Roman artwork appears to use perspective, though the application tends to be haphazard at best. Some time before 1415, the Florentine Filippo Brunelleschi invented an algorithmic method of drawing objects in perspective. Brunelleschi drew two paintings (now lost) of Florentine street scenes.

The first treatise on perspective was written by LEON BATTISTA ALBERTI (February 18, 1404–April 1472), who might have been Brunelleschi's student. Alberti was primarily an architect, though his formal training was in law and he was renown for his ability to hold a heavy weight at arms' length for several minutes. Alberti's *On Painting* (1435), dedicated to Brunelleschi, is considered the first book on the theory of painting, describing not only *how* to paint, but *why*. The first section describes how to create pictures that looked as if they were views through a window, using a technique now known as one-point linear perspective. In the third section, Alberti discussed the education of an artist, and emphasized the importance of a good, liberal education. Mathematics in particular was key to an artist's success.

Alberti described how artists *should* be, and not how they *were*: he hoped to raise the status of the artist from craftsman to intellectual. He saw mathematics as the key to turning painting into a liberal art, and originally explained perspective to his friends using geometry to validate its techniques. However, for reasons of brevity, Alberti omitted such demonstrations in his *On Painting*.

One of Alberti's mathematically inclined friends was PIERO DELLA FRANCESCA (1415–October 12, 1492), who moved to Florence from his birthplace in Sansepolcro shortly after the completion of *On Painting*. Francesca wrote a treatise on the use of the abacus and another on the five regular solids. His treatise on perspective includes a proof

Figure 6.2. Boscoreale mural. Roman mural done "in perspective." Note the lack of a vanishing point.

of the validity of Alberti's methods. For example, Francesca proved that a horizontal plane rendered in perspective has the shape of a trapezoid. In 1460 Francesca put theory into practice in his painting, *The Flagellation of The Christ*. Curiously, the flagellation of the title takes place in the background, while three men stand in the foreground unaware of the action behind them.

LUCA PACIOLI (1445–1517) was another of Alberti's mathematically inclined friends. Like Francesca, Pacioli was born in Sansepolcro. Pacioli entered the family business and by 1470 he was working for Antonio Rompiansi, a Jewish merchant of Venice. While there he wrote his first arithmetic text, probably for the education of Rompiansi's sons. Rompiansi died soon after and his sons, who had no interest in following the profession of their father, dismissed Pacioli. Thus he went to Rome, where he stayed as a guest of Alberti, took Franciscan vows, and soon appeared at the University of Perugia, where he wrote a second arithmetic text. In 1481, he wrote yet a third, this time while in Zara. None of these exist today, for they were hand-written and few, if any, copies were made.

In 1478, the first *printed* arithmetic text appeared, by an anonymous author. Called the *Treviso Arithmetic*, after its publication place in Treviso, part of the Venetian empire, it shows the highly developed state of commercial arithmetic: the arithmetic operations using the Hindu-Arabic numerals are clearly explained, and several methods of multiplication and division are given. Pacioli's first printed mathematical work appeared in 1494: the

Figure 6.3. Francesca's *Flagellation of the Christ*.

Summary of Arithmetic, Geometry, Proportion, and Proportionality which, as its name suggests, surveyed virtually all of the mathematics known at the time. Pacioli made no claim to originality; instead, he presented the work as a compendium of the mathematics known at the time. He dedicated the work to his former student, Duke Guidobaldo of Urbino, whose court was the setting for an influential work on Renaissance manners, Baldassare Castiglione's *Book of the Courtier*.

One section of the *Summary* dealt with a way of recording business transactions that Pacioli claims had been used in Venice for two hundred years: double entry bookkeeping. In single entry bookkeeping, a single ledger is kept, making notes of credits and debits. In double entry bookkeeping, two sets of ledgers are kept, one recording all debits, and the other all credits. The important difference between the two is that single entry bookkeeping only records transactions as they occur: when a bill is paid, or a payment is received. Double entry bookkeeping tracks all outstanding debits and credits. Because of his description of this practice in the *Summary*, Pacioli is often regarded as the father of accounting. The difference between the two is mathematically significant. In single entry bookkeeping, negative numbers do not exist: the ledger total is simply the sum and difference of a set of positive numbers. Moreover, since one cannot pay out money one does not have, the total of all the entries must be non-negative. In contrast, in double entry bookkeeping, the debit ledger corresponds to a sum of negative numbers, and one's net worth can correspond to a negative amount.

Printing played another important role in Pacioli's career. Around 1460, based on his extensive analysis of surviving inscriptions, Felice Feliciano described what is now known

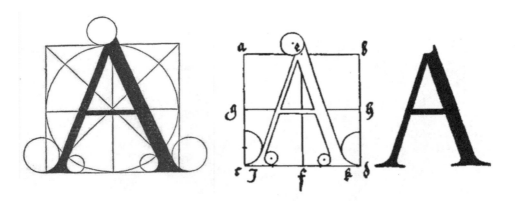

Figure 6.4. Geometrically designed A by Pacioli (left) and Dürer (right).

as Roman font; ironically, his description remained in manuscript form. The printing house of Aldo Manuzio used a font largely based on Feliciano's work, and the *Hypnerotomachia* (1499) of "Francesco Colonna" (possibly Alberti's pseudonym), helped establish the Aldine Roman font as a standard.

These fonts might be classified as the products of craftsmen, based on the work of scholars. However, a different approach to type fonts appeared in the work of Damiano Moille's *On Antique Letters* (1480), which described a geometrical basis for creating letters by inscribing them within an appropriate square or circle. This reflected a Renaissance belief that aesthetics could be described by a set of precise rules. Pacioli took the next step in his *On the Divine Proportion* (1509), and attempted to identify some of these rules. In the first section, Pacioli discussed the golden ratio, and gave it a central role in what might be called the laws of aesthetics. Pacioli identified the golden ratio in a number of different contexts, and used it as the basis of type font he created. However, Aldine fonts proved popular and became the basis for many later fonts, while Pacioli's mathematically precise letters are generally only found in isolation, such as the "M" logo of New York's Metropolitan Museum of Art.

In 1496, Pacioli received an invitation to teach mathematics at the court of Ludovico Sforza, the Duke of Milan. The Sforzas had a tenuous legal claim to the Duchy: In 1450, the *condottieri* (mercenary captain) Francesco Sforza captured the city and declared himself the new Duke of Milan. Sforza's claim was based on his 1441 marriage to Bianca Maria, the only daughter of the Duke of Milan, Filippo Maria Visconti.

However, Bianca Maria was illegitimate. In 1498, Louis XII became King of France. Louis could trace his ancestry to Gian Galeazzo Visconti, a former Duke of Milan, and thus he had a somewhat stronger claim to the Duchy than the reigning Sforza, Ludovico (known as *Il Moro*, the Moor, because of his dark hair and complexion). By then, Milan was a particularly tempting target, because *Il Moro* preferred to leave military affairs in the hands of his 10,000 Swiss mercenaries, and the fanciful war machines of Leonardo da Vinci.

Leonardo's notebooks include designs for parachutes, helicopters, tanks, and other mainstays of the modern battlefield, and on paper provided *Il Moro* with an advanced war machine. Unfortunately there was no practical means of *building* these devices using existing technology. The French invaded and occupied Milan in February 1499, driving

Ludovico out. Leonardo and Pacioli left soon afterwards, eventually arriving in Florence where they lived together and collaborated for several years. Leonardo produced the illustrations of the Platonic solids used by Pacioli in the third part of his *On the Divine Proportion*; the text itself was largely based on Francesca's treatise on regular solids.

6.1.2 The Secret of Solving the Cubic

Around 1501 Pacioli lectured on mathematics at the University of Bologna. The university was known mainly for its law school, one of the most highly regarded in all of Europe. At the end of the fourteenth century mathematics had been added to the curriculum, and by 1501 it boasted a flourishing mathematical community that included not only Pacioli, but SCIPIONE DEL FERRO (February 6, 1465–November 5, 1526), who had been teaching there since 1496.

Very little is known about del Ferro except for his presence in Bologna. In the *Summary*, Pacioli claimed that the cubic was algebraically unsolvable. Del Ferro and Pacioli may have discussed the solution to the cubic, and shortly after 1501, del Ferro discovered a method of solving at least one type of cubic, the form we would today write as $x^3 + px = q$, where p and q are positive constants. Italian algebras of the time referred to the unknown as *cosa* ("thing" in Italian), so outside Italy algebra was known as the cossick art, its practitioners were cossists, and an equation of the form $x^3 + px = q$ was classified as "cube and cosa equal to number." Though none of Ferro's writings have survived to the present day, it is believed that he was able to solve cubics of the form $x^3 = px + q$ as well.

Del Ferro kept his method secret, for in the sixteenth century, a mathematician's reputation depended on his ability to solve problems. With his knowledge of the solution to the "cosa and cube" equation, del Ferro could pose questions that no others could answer. Before he died he passed on his secret to his best and most talented pupils, one of whom was ANTONIO MARIA FIORE (fl. 1535) of Venice.

With Milan in French hands, Venice was the most powerful city-state in Italy. However, it faced nation states like France and Spain, and the Papacy, which could draw on much of Europe for assistance. Pope Julius II sought to make the Papacy the dominant power in Italy, and was responsible for several of the best-known features of the modern Papacy. First, he deemed it proper that the Pope have a grand palace. One of his early actions was hiring a headstrong sculptor to paint the ceiling of the Sistine Chapel: Michelangelo. Though the ceiling was over 6300 square feet in area, Michelangelo completed the work in just four years with only two assistants (who mainly mixed plaster). Julius also began the construction of St. Peter's Basilica, whose cost would be so great that in time all European Catholics would be called upon to pay for it. Finally he recruited men from the Swiss cantons loyal to him alone: the Swiss Guard.[3]

Julius led several military campaigns during his reign; he is known as the "Warrior Pope" in consequence. First, Julius joined forces with France and the Holy Roman Empire to crush Venice. But if Venice were destroyed, France would dominate northern Italy, so before Venice was completely smashed, Julius switched sides and organized a new Holy

---

[3]Despite claims to the contrary, Michelangelo did not design the uniform of the Swiss guards; these were designed in 1914 by Commandant Jules Repond. Repond did take inspiration from one Renaissance artist, Raphael, who painted a picture of Julius II being borne on a litter by the guard.

League, with Spain and Venice, to attack France. The Venetians expelled the French from the city of Brescia, but in 1512 the governor of Milan and leader of the French armies, Gaston de Foix, defeated the Venetians outside of Brescia and turned on the city itself.

Among the first soldiers over the wall was Pierre Terrail, the Lord of Bayard. Bayard was a national hero who had distinguished himself in every Italian campaign since 1494. But at Brescia, Bayard was seriously wounded when he entered the city. French troops exacted a terrible revenge on the city that dared to fight back, and massacred 8000 of Brescia's citizens. The town, once a thriving center of trade, would not recover for centuries.

One of Brescia's inhabitants was the thirteen-year-old Niccolo Fontana. A French soldier slashed him across his face, leaving him with a speech impediment that caused him to be known as "the Stammerer:" TARTAGLIA (1499–December 13, 1557). Tartaglia learned to write as far as the letter "K" when money for tuition ran out; thereafter, Tartaglia was self-taught. As a result, there were large gaps in his knowledge: his Latin was poor, and thus he would write scholarly works in the Italian dialect of Venice, where he moved in 1534.

A century before, Venice was the mistress of the Mediterranean, sitting astride the trade routes linking India and Europe. Spices, like black pepper, cinnamon, nutmeg, and cloves, were the most important commodities traded between east and west. It is not true that spices were used to "cover up" the taste of spoiled food: salt works quite well as a preservative, and is far cheaper. Spices were used for a simple reason: flavor.

However, spices were expensive. They were produced in the Far East, but by the time they arrived in western Europe, they had passed through the hands of many middlemen: the Arabs, Byzantines, and Venetians, each of whom marked up the cost before passing it on to the next. This increased the final cost to the point where peppercorns were bought one at a time, both because they were expensive and because a sensible buyer wanted to make sure that they were actually getting peppercorns. For example, in fifteenth century England, a pound of black pepper cost 2 shillings—a week's pay for an agricultural laborer.

But in the last years of the fifteenth century, Portuguese navigators discovered a route to India around Africa and broke the Venetian monopoly on trade with the East. Very quickly the Portuguese realized that they could undercut Venetian prices *slightly* and still make a fortune: the net result was that spice consumption in Europe doubled in the sixteenth century, while the price remained more or less constant. The spice centers of Europe moved to the Atlantic nations, particularly Portugal and the Netherlands.

Venice, a great power during the Middle Ages, entered a long period of decline from which it would never recover. Still, Venice had far to fall and her decline would take centuries. She still controlled many important trade routes with the Byzantines and the Ottomans, and her wealth attracted those seeking to make their fortunes. Finally, Venice was helped by geography: the sixteenth century invaders of Italy came from the north, west, and south. Venice, in the east, was an island of calm and became a refuge for those fleeing the chaos in the rest of Italy. Others who lived in Venice at the same time as Tartaglia include the artist Titian, the satirist Pietro Aretino, the sculptor Jacopo Sansovino, and the musician Adriaan Willaert.

Tartaglia came to a Venice of constant spectacle. Visitors told tales of festivals and celebrations, of bear-baiting and bull-running, and of street performers and musicians. While the rest of Italy collapsed into ruin because of constant warfare, the Venetians continued to throw festive galas and wild parties. It was in this town of continuous merrymaking he would make his reputation.

One way to achieve notoriety and fame during the sixteenth century was to win a scholarly debate. In the festive atmosphere of Venice, these public debates were a cross between erudite discussions and mudslinging political arguments, and were as eagerly attended by leading citizens as sporting events are today. In 1535, Fiore challenged Tartaglia to a mathematical duel.

Why Fiore, an established mathematician, would challenge Tartaglia, a relative nobody, is unknown. Tartaglia may have made a nuisance of himself: he was never above self-promotion, and mocked those he considered to be his intellectual inferiors. Fiore, the sophisticated Venetian, might have sought to put the poor stutterer from Brescia in his place. Fiore challenged Tartaglia to a problem-solving duel: each would pose thirty questions to the other, the loser to pay for thirty banquets.

We have no record of the questions, but Tartaglia apparently posed a variety of questions, while Fiore posed those of a single type: the cosa and cube. This put Tartaglia in a dreadful position, for he could solve none of them. But on the night between February 12 and 13, 1535, shortly before the time limit expired, inspiration struck, and Tartaglia invented a method for solving all problems of the cosa and cube type. Fiore lost, but Tartaglia declined the banquets, saying that winning was glory enough. At a stroke, Tartaglia established his reputation as the greatest mathematician in Italy.

We might know nothing more about Tartaglia but for a contentious chain of events. One of Leonardo's mathematically inclined acquaintances was Fazio Cardano, a Milanese lawyer. In 1501, plague broke out in Milan, and those who could left the city. Fazio Cardano convinced his pregnant mistress, Chiara Micheria, to take refuge in the nearby city of Pavia, where their son GIRALAMO CARDANO (September 24, 1501–September 21, 1576) was born. Sadly Chiara's other three sons, who stayed in Milan, died in the course of the plague.

Milan was now in French hands, but French fortunes in Italy were turning. After the siege of Brescia, Gaston de Foix turned on Ravenna in April 1512. The battle was hard-fought, and eventually the French prevailed—but at a terrible cost: Gaston de Foix was killed during the combat, and no competent leader emerged. Thus the French withdrew from Italy, and in 1513, Ludovico's son Massimiliano Sforza became Duke of Milan. Francis I launched a new offensive, and at Marignano on September 13–14, 1515, the army of Charles, the Duke of Bourbon, crushed Massimiliano's forces and conquered Milan. Francis made Charles the governor of Milan, and pensioned off Massimiliano, who spent the rest of his life in Paris.

In 1522, the French faced their newest and toughest adversary for the first time: Charles V of the Holy Roman Empire, the grandson of Maximilian I. Charles inherited the Holy League's mandate to drive the French out of Italy, so he invaded Italy, captured Milan, and handed it over to Massimiliano's brother Francesco in 1522. It was the opening phase of the first of Valois-Habsburg wars that would devastate Italy for a generation.

By then, Francis had lost a key supporter: the very Duke that had given him Milan. Charles of Bourbon had been named Constable of France in 1515, a position that made him supreme commander of the French armies, but Francis had neglected to pay him. Perhaps Charles complained too loudly about his back pay, because Francis began proceedings to seize the property of the Bourbons. In response, the Duke conspired with Charles V to partition France between the Empire, England, and himself. Francis learned of the plot and attempted to capture the Duke, who fled just ahead of the royal forces. The Duke would soon return at the head of an army of Charles's mercenaries. Despite early victories, the Duke's forces were soon driven out of France and towards Pavia.

By this time Cardano had enrolled in the University of Pavia and, against his father's wishes, began the study of medicine. Renaissance medical practice relied heavily on astrological notions; consequently all medical doctor had to be trained as *mathematicians* in the ancient sense. Vestiges of these astrological notions remain: we still speak of the disease *influenza*, caused by the supposedly malignant influence of the stars, and psychology has its *mercurial* and *jovial* temperaments.

With the armies of the Empire and France bearing down on Pavia, the University closed and the students were sent away. Cardano went to Padua to continue his studies, and missed one of the critical battles of the sixteenth century. On February 24, 1525, the armies of Francis I and Charles V met just outside the city. The French were defeated, and Francis himself was captured by the rebellious Duke of Bourbon and taken to Madrid as a prisoner. On January 14, 1526, Francis signed the Treaty of Madrid, renouncing all French claims in Italy, and incidentally restoring to the Duke of Bourbon all his former titles and granting him immunity from retaliation. But as soon as Francis was back on French soil, he denounced the treaty, arguing that it was obtained under duress and therefore not binding. This infuriated Charles: by the standards of the day, there were no mitigating circumstances for breaking one's word.

Meanwhile Cardano received his medical degree from the University of Padua in 1525, and applied for membership in the College of Physicians. Despite his excellent record, he was denied: Cardano was illegitimate, and thus ineligible for admission.

The Duke of Bourbon had his own problems. Charles made him Governor of Milan, but neglected to give him money to pay his troops. To keep them from mutinying, the Duke led them on a pillaging expedition through central Italy, eventually laying siege to Rome in 1527. The defenders fought valiantly, and the Duke of Bourbon was killed by a shot from a sniper: Benvenuto Cellini, a goldsmith turned soldier. Cellini would become famous for a gold salt-shaker he would make for Francis I a few years later, and even more famous for his *Autobiography*, where he extolled the virtues of the man he considered to be the leading figure of the day.

The mercenaries sacked the city of Rome, leaving Pope Clement VII under the control of Charles V. The same year, Henry VIII of England petitioned the Pope to annul Henry's marriage to Catherine of Aragon. Since Catherine of Aragon was Charles's aunt, the Pope refused to grant Henry the annulment. Consequently Henry broke with the Church and established a separate English Church.

Around 1539, Zuanne de Tonini da Coi came to Milan. Da Coi, like Tartaglia, was from Brescia, and when he met Cardano, told the tale of Tartaglia's triumph. Unfortunately da Coi knew nothing about Tartaglia's method. Thus Cardano visited Tartaglia. At first, Tartaglia refused to talk, hoping to preserve his secret and retain an unbeatable advantage over any opponent. But Cardano flattered Tartaglia, and after making Cardano swear an oath not to publish the method until after Tartaglia did so, he explained his solution.

There the matter may have rested, but for LUDOVICO FERRARI (February 2, 1522–October 1565). Ferrari entered Cardano's household as a valet (a gentleman-in-training) but Cardano recognized Ferrari's potential and proceeded to teach him Latin, Greek, and mathematics. Ferrari also obtained a lectureship at the University of Milan in 1540. Soon, Ferrari became Cardano's secretary—one entrusted with his master's "secrets." Tartaglia may have only known the solution to the cosa and cube equation. Cardano and Ferrari extended Tartaglia's method so that all cubics could be solved.

Today we need only consider a single type of cubic, namely $x^3 + Bx^2 + Cx + D = 0$. In the sixteenth century, however, mathematics was still based on geometry, so a term like x^3 represented an actual cube with a side of x. It is self-evident that if B, C, and D are all positive, the cubic will have no geometric solution (corresponding to a positive real solution). Cardano and Ferrari knew that the x^2 term could be eliminated (via the substitution $x = y - \frac{1}{3}a$); hence a general solution need only deal with three distinct types of cubic equations. If p and q are positive coefficients, the three types are:

1. The cosa and cube equal the number. $[x^3 + px = q]$

2. The cube equal to cosa and number. $[x^3 = px + q]$

3. The cube and number equal to cosa. $[x^3 + q = px]$

To solve the different types of cubic equations, Cardano first gave a conceptual outline, then gave an algorithm. Thus, to solve $x^3 + px = q$, Cardano gave a procedure equivalent to:

Rule 6.1. *Let* $u^3 - v^3 = q$, *and* $uv = \frac{1}{3}p$. *Then* $x = u - v$.

For example, consider the equation $x^3 + 6x = 20$. Using Cardano's procedure, we need to solve the system of equations

$$u^3 - v^3 = 20, \qquad\qquad uv = 2.$$

To solve this, Cardano made use of the relationship we would express as

$$\left(\frac{u^3 - v^3}{2}\right)^2 + u^3 v^3 = \left(\frac{u^3 + v^3}{2}\right)^2.$$

Hence an expression for $\frac{u^3 - v^3}{2}$ and $\frac{u^3 + v^3}{2}$ could be found; their sum would be u^3, while their difference would be v^3.

Applying this to the above, we have:

$$\frac{u^3 - v^3}{2} = 10, \qquad\qquad u^3 v^3 = 8.$$

Squaring $\frac{u^3 - v^3}{2}$ and adding 8 gives us 108, which is the square of $\frac{u^3 + v^3}{2}$. Thus $\frac{u^3 + v^3}{2} = \sqrt{108}$, and thus:

$$u^3 = 10 + \sqrt{108}, \qquad\qquad v^3 = \sqrt{108} - 10.$$

Thus $x = \sqrt[3]{10 + \sqrt{108}} - \sqrt[3]{\sqrt{108} - 10}$. A problem that arises (which Cardano did not address) is that the cubic $x^3 + 6x = 20$ has a unique real solution, namely $x = 2$. Hence it is necessary that $2 = \sqrt[3]{10 + \sqrt{108}} - \sqrt[3]{\sqrt{108} - 10}$, and there must be some way of simplifying the radical expression to a whole number.

For $x^3 = px + q$, Cardano gave the rule:

Rule 6.2. *Let $u^3 + v^3 = q$ and $uv = \frac{1}{3}p$. Then $x = u + v$.*

As in the preceding case, Cardano's method found an expression for $\frac{u^3 - v^3}{2}$ and $\frac{u^3 + v^3}{2}$; their sum would be u^3, while their difference would be v^3. It is worth noting that since the value of $\left(\frac{u^3 - v^3}{2}\right)^2$ is equal to a difference, we have the potential of running into the square root of a negative number.

For $x^3 + q = px$, Cardano's method was:

Rule 6.3. *Solve $y^3 = py + q$. Then $x = \frac{y}{2} \pm \sqrt{p - 3(\frac{y}{2})^2}$.*

Shortly after Cardano learned Tartaglia's method, da Coi posed the following problem to Cardano: to divide 10 into three parts in continued proportion so the first, times the second, is 6. Cardano replied that he could solve the problem, though this was (at the time) an empty boast. In the terminology of the times, the equation that resulted was

Problem 6.1 (Da Coi's Problem). *Sixty things equal to one square-square plus 6 squares plus 36.*

Cardano gave the problem to Ferrari, who eventually solved it. This was a landmark discovery in mathematics, for the equation corresponds to $60x = x^4 + 6x^2 + 36$, a fourth degree (biquadratic or quartic) equation.

The conceptual basis of Ferrari's method is to add a quantity that will make both sides of the equation perfect squares, and then reduce the degree of the equation by taking the square root of both sides. In modern notation, we are trying to solve the equation

$$x^4 + 6x^2 + 36 = 60x.$$

We can make the left-hand side a perfect square by adding $6x^2$ to both sides:

$$x^4 + 12x^2 + 36 = 6x^2 + 60x.$$

If the right-hand side was also a perfect square, the roots of both sides could be found and the degree of the equation reduced. In order to make both sides perfect squares, add some quantity y inside the parentheses on the left-hand side. Since $(x^2 + 6)^2 = x^4 + 12x^2 + 36$, then $(x^2 + 6 + y)^2 = x^4 + 12x^2 + 36 + 2yx^2 + 12y + y^2$. Moreover, since $x^4 + 12x^2 + 36 = 6x^2 + 60x$, we can reduce the right-hand side, obtaining:

$$\left(x^2 + 6 + y\right)^2 = (6 + 2y)x^2 + 60x + (12y + y^2).$$

We want the right-hand side to be a perfect square. Note that $(ax + b)^2 = a^2x^2 + 2abx + b^2$: hence the square of half the coefficient of x is the product of the first and last coefficients. Thus y satisfies the equation $(6 + 2y)(12y + y^2) = \left(\frac{60}{2}\right)^2$ which, if we expand it, is the cubic equation

$$2y^3 + 30y^2 + 72y = 900.$$

Since Cardano and Ferrari had, at their disposal, a method of solving any type of cubic, they could solve this problem. Cardano wanted to include Ferrari's solution in an upcoming

work on algebra, *The Great Art*, but found himself in a quandary. If Cardano published
Ferrari's solution, he would have to explain how to solve a cubic. This meant he would also
have to publish Tartaglia's work, yet he had sworn that he would not publish until Tartaglia
did so, and Tartaglia had not yet done so. What could be done?

Cardano recalled the debate between Fiore and Tartaglia. Fiore must have known how
to solve the cubic, for if he did not, he would not risk posing to Tartaglia questions that he
himself could not solve. Fiore was the student of Scipione del Ferro. Del Ferro was dead,
but his nephew, Annibale dalla Nave, had his papers and his position at the University
of Bologna, and might be able to shed some light on the solution to the cubic. Around
1542, Cardano and Ferrari went to Bologna, where dalla Nave showed them his uncle's
unpublished manuscripts.

Based on his perusal of the manuscripts, Cardano realized that del Ferro had to be able
to solve cubics before Tartaglia. Thus, Tartaglia's oath to secrecy had been exacted under
false pretenses. Hence Cardano felt that the oath to Tartaglia was no longer binding, so he
published the solution to the cubic in *The Great Art* (1545). Cardano gave Tartaglia credit
for having independently discovered the solution to some types of cubic equations, though
gave priority of discovery to Ferro.

Tartaglia was furious. By the standards of the sixteenth century, no mitigating circumstances allowed for the violation of an oath. Moreover, by giving credit to Ferro, Cardano's
work denied Tartaglia the honor of being the *first* to solve cubics algebraically. Tartaglia
challenged Cardano to a public debate. Cardano declined, but Ferrari took up the challenge.
The two met on August 10, 1548 in the Church of Santa Maria del Giardino dei Minori Os-
servanti in Milan, with town citizens and various dignitaries, including the Mayor of Milan, attending. We do not know exactly what happened, but afterwards, Tartaglia returned
to Brescia, and Ferrari was offered a number of positions, including the job of tutoring
Charles V's son Philip in mathematics. Ferrari chose to stay in Italy, and took a position
with Ercole Gonzaga, the Cardinal of Milan. Hence, we may infer that Tartaglia lost the
debate.

Besides the solution to the cubic and quartic, *The Great Art* brought complex numbers
to the attention of mathematicians. Cardano probably ran into complex numbers when applying his method of solving cubics of the form $x^3 = px + q$, but he presented them to the
reader in a problem which asked to find two numbers that add to 10 and multiply to 40. As
before, we have $a + b = 10$ and $ab = 40$. Hence $((a + b)/2)^2 - ab = ((a - b)/2)^2$, so
$((a - b)/2)^2 = -15$ (written by Cardano as $m : 15$), or $(a - b)/2 = \sqrt{-15}$. Thus the two
numbers will be $5 + \sqrt{-15}$ and $5 - \sqrt{-15}$. "Putting aside" the "mental tortures" involved,
we may verify through direct addition and multiplication that these two numbers solve the
problem.

Interestingly, bad luck seemed to plague all involved. Tartaglia never gained a distinguished position. Ferrari was poisoned a few years later, possibly by his sister over an
inheritance dispute, and Cardano was arrested by the Inquisition and placed under house
arrest on charges of heresy.

6.2 Central Europe

Charles V was a Habsburg (also spelled Hapsburg), whose family name comes from the
Habichtsburg, or "Hawk's castle," overlooking the Aar River in Switzerland. Originally

they were rulers of Switzerland, but the taxation policies of Rudolf IV of Habsburg led to a revolt in 1274. According to legend, the cantons were spurred to action when Gessler, a Habsburg governor, forced William Tell, a native of the Canton of Uri, to shoot an apple off his son's head as punishment for Tell's insolence. In 1291 three "forest" cantons, Uri, Unterwalden, and Schwyz, formed the Everlasting League to preserve their independence.

The revolt eventually drove the Habsburgs from Switzerland, but not before they acquired Austria. By the time of Rudolf IV of Austria (not to be confused with Rudolf IV of Habsburg, who is reckoned as Rudolf I of Austria), the Habsburgs were a powerful family. But in 1356, Rudolf's father-in-law, Holy Roman Emperor Charles IV, dealt the Habsburgs a humiliating blow.

Since 1273, the Holy Roman Emperor had been chosen by a group of electors, but the exact number and identity of the electors had never been formalized, leading to disputed elections and civil war. To settle this problem, Charles IV issued the Golden Bull in 1356, which definitively established the electors as the Archbishops of Mainz, Trier, and Köln; the Margrave of Brandenburg; the Count Palatine of the Rhine; the Duke of Saxony, and the King of Bohemia.[4] Conspicuously absent was the Duke of Austria.

The denial of electoral rank caused Rudolf IV to look for other means to enhance the power and prestige of his family. Others waged war, but Rudolf used a more potent weapon: lawyers. He discovered a document bearing the seal of Frederick I Barbarossa that gave great privileges (*Privilegium Maius* in Latin) to the Dukes of Austria. According to the document, Rudolf was not merely a duke but an Archduke, a title never before conferred on a ruler.[5] As Archduke, Rudolf would be of an equal rank with the electors of the Empire. Charles IV was suspicious of Rudolf's sudden promotion—and rightly so, for in 1856 the *Privilegium Maius* was exposed as a forgery. However, Charles took the pragmatic position that the Habsburgs could give themselves a meaningless title.

Rudolf sought other means to enhance the glory of the Habsburgs, and above all, sought to outshine Charles. Charles began building a cathedral in Prague in 1344, so Rudolf began building one in Vienna. Charles founded the University of Prague in 1348, so Rudolf founded the University of Vienna in 1365. It was the first university in the German speaking world.

Unfortunately, Rudolf's ambitions were beyond his abilities. The University of Vienna began with inadequate funding and a limited curriculum, focusing mainly on the quadrivium. By 1380, the school was on the verge of bankruptcy, and Archduke Albert III offered generous incentives for any scholars who would come to Vienna to revitalize the school. He would benefit from dramatic upheavals in the Papacy.

In 1378, the Avignonese Pope Gregory XI died during a visit to Rome. The Romans seized the chance to return the papacy to Italy and pressured the College of Cardinals to choose Bartolomeo Prignano, a Neapolitan, as the new pope. Prignano took the name Urban VI and almost immediately alienated the cardinals by railing against their vast accumulations of wealth and their ostentatious displays of luxury. In retaliation, the cardinals

---

[4]The term "bull" has nothing to do with livestock but instead comes from the Latin *bulla*, a golden seal, indicating the authenticity of the declaration. Thus the Golden Bull is really the Golden Gold Seal.

[5]The title of "Duke" comes from the Latin *ducere*, "to lead:" a duke was a leader of troops. The prefix "Arch" is from the Greek for "leader", so an Archduke is a Leader of leaders.

Figure 6.5. Central Europe

elected a second Pope, who took office in Avignon as Clement VII. It was the beginning
of the Great Schism, and the decline of the French domination of the Papacy.[6] Henry of
Hesse, a student of Oresme's at the University of Paris, took the unpopular position that the
papacy should return to Rome; as a result, Hesse found himself in Vienna.

---

[6]"Schism" is properly pronounced "SIZ-um;" it is etymologically related to the word "scissors." Because
Clement's Papacy was disregarded by later Popes, the next Clement to become Pope was also known as
Clement VII.

An effort to heal the Schism brought more scholars to Vienna. In 1409, the Council of Pisa deposed both popes and replaced them with a new pontiff, Alexander V. Predictably, neither pope stepped down, and for a while there were three Popes.

At Prague, the faculty was divided. Most of the German faculty denied the authority of the Council; they were supported by the Archbishop of Prague (who, by the university charter, was also the Chancellor). In contrast, most of the Czech faculty supported the Council's decision. Leading the Czechs was Jan Hus (or John Huss), the Dean of the College of Philosophy.

However, the voice of the Czech faculty was muffled because the university charter gave German faculty members three votes to one vote for the Czech faculty; thus, Germans dominated university affairs and held every important office. But the conflict over the schism was so great that the king of Bohemia, Wenceslas IV, intervened and rewrote the university charter, declaring that henceforth it would be the Germans who would receive one vote to the Czech's three. With the electoral balance shifted in their favor, the Czechs elected Hus as Rector of the University, and many German scholars left.

Hus took up the cause of religious reform. He argued that the liturgy should be preached in the vernacular and that the Pope could be removed for reasons of moral turpitude. In 1415, Hus was promised safe passage to Constance in southwestern Germany to discuss his views. Once there he was arrested, tried, and burned at the stake. Hus's betrayal and execution triggered the Hussite Wars, a nationalistic revolt in Bohemia that pitted Germans against Czechs. The Hussite Wars continued until 1452, when a shaky truce emerged between the parties.

6.2.1   Peuerbach and Regiomontanus

Chaos in Bohemia, new scholars from Prague, a reorganized curriculum, and above all a School of Theology helped to make the reformed University of Vienna a success. Henry of Hesse brought Oresme's configuration doctrine to the university, and Henry's student Nicolaus of Dinkelsbühl made it an integral part of the curriculum. Nicolaus's student, JOHN OF GMUNDEN (1380–February 23, 1442) would turn Vienna into a center of astronomical research. This was a critical change, for astronomy required trigonometry, so Vienna had to improve its mathematics curriculum. Four years after John of Gmunden's death, GEORGE PEUERBACH (May 30, 1423–April 8, 1461) came to Vienna.

Very little is known about Peuerbach's life prior to his arrival in Vienna in 1446. Over the next few years he gained great fame as an astronomer, and spent the 1450s traveling through France and Italy, giving well-attended lectures and receiving many offers from schools in Italy. But Peuerbach declined them all, returning to Vienna in 1453 to lecture on the classics: Virgil, Juvenal, and Horace.

Like many of his time (though, notably, not John of Gmunden), Peuerbach believed in the value of astrology. In 1454, Peuerbach became the Court Astrologer to Ladislas, Duke of Austria and King of Bohemia. Upon Ladislas's death a few years later, Peuerbach accepted a position with Ladislas's uncle, the Holy Roman Emperor Frederick III. But service to the crown was not for him, and by 1457, Peuerbach returned to Vienna to begin a fruitful collaboration with Johann Müller, known as REGIOMONTANUS (June 6, 1436–July 8, 1476).

Regiomotanus was born in the Archbishopric of Mainz in the city of Königsberg (Regiomontanus is the Latinized form of Königsberg, "King's mountain"). He enrolled at the University of Vienna on April 14, 1450, and earned his Bachelor's degree on January 16, 1452. He might have earned his Master's degree shortly thereafter, but the fifteen-year-old Regiomontanus had to wait: University statutes prohibited the granting of masters degree to anyone under 21. On November 11, 1457 he was simultaneously awarded his Master's and given a position at the University. By then, he and Peuerbach were already collaborating. Three years later their lives would be dramatically changed by the arrival of Cardinal Bessarion in Vienna.

In 1437, John Bessarion and the Byzantine Emperor John VIII Paleologus arrived in Europe to seek allies against the Ottoman Turks. The French, who began the century with a disastrous cavalry charge across a muddy field at Courtrai (1302), continued with a similar charge at Crécy (1346) and Poitiers (1356), showed their heroic disregard of elementary tactics by charging uphill against the Turks at Nicopolis (September 25, 1396). It was the last great effort of the west to slow the tide of Turkish expansion.

Bessarion failed to win allies, but the Pope made him a cardinal in 1439. Thereafter, Bessarion stayed in Italy, continuing his calls for action against the Turks. Few accepted the challenge, and only one proved even marginally successful: János Hunyadi, whose troops included battle-hardened veterans of the Hussite wars. In 1443, Hunyadi launched an offensive against the Ottomans. After some initial successes, he faced disastrous defeat at Varna in 1444 and at Kosovo in 1448, and could not prevent the Turkish conquest of the Balkans. The Turks installed a number of puppet rulers in the Balkans, including a seventeen-year-old whose father was a member of the Order of the Dragon, a group of knights sworn to uphold Catholicism and defeat the Turks. The boy's name was Vlad, but his contemporaries called him *Dracula* ("son of the Dragon" in Romanian). Vlad would eventually turn against his Turkish patrons and become a hero of Romania, but the Ottomans were unstoppable and in 1453, Constantinople itself fell to the Turks. The city took on a new name, owing to the Greek habit of referring to the wonders that could be found *eis ten polis* ("in the city"); hence, the city was known to the Turks as Istanbul.

After the fall of Constantinople, Bessarion tried to rally support for a crusade to retake the city and its environs. He arrived in Vienna on May 5, 1460. While there he spoke to Regiomontanus and Peuerbach about another crusade: this one against the faulty translation of Ptolemy's *Almagest* made by George Trebizond in 1450. Bessarion suggested that Peuerbach and Regiomontanus make a new translation from the Greek. They agreed, but Peuerbach died before he could do much more than get started. On his deathbed, he made Regiomontanus pledge to complete the work, and the *Epitome of Ptolemy's Almagest* appeared in 1462. The new edition destroyed Trebizond's reputation as a translator.

Upon Peuerbach's death, Regiomontanus went to Rome with the Cardinal, where he could consult not only Bessarion's vast collection of classical manuscripts, but also those of Pope Nicholas V, who had acquired many from Constantinople before its fall to the Turks: these manuscripts form the core of the Vatican Library. With access to original sources, Regimontanus hoped to complete translations of other works, including Ptolemy's *Geography*, as well as works of Heron, Apollonius, and Archimedes. One tantalizing project was a translation of Diophantus: Regiomontanus claimed to have seen a Greek manuscript of the *Arithmetic* containing *fourteen* books. As only six books are known in Greek (and four more from another source), such a manuscript would be of immense value, but it has never been found.[7]

Regiomontanus's most important work by far was *On Triangles* (1464), which established trigonometry as a mathematical science. Most of *On Triangles* was taken without attribution from Jābir ibn Aflah's *Correction of the Almagest*, leading Cardano, a century later, to accuse Regiomontanus of plagiarism; since Regimontanus had copied the work for his own reference twice before 1460, there is no doubt that he made extensive use of Aflah's work.

In 1468, Regiomontanus became the Court Astrologer to Matthias Corvinus ("raven", because his heraldic symbol was a raven), the King of Hungary. The younger son of Janos Hunyadi, Matthias was an active patron of the arts, and played an instrumental role in bringing the Renaissance of Florence to Hungary: he was called "friend of the Muses" by contemporaries. One of his greatest achievements was personally amassing the *Bibliotheca Corviniana*. The manuscripts in the Corviniana were stunning works of great artistry, lavishly illuminated by master craftsmen in Florence and other cities. The books in the Corviniana were beautiful—and few in number, perhaps 2500 altogether. Each had to be hand copied, line by line and page by page. But the era of the hand-copied book was coming to an end, and Regiomontanus arrived in Vienna just as the first printing presses were being set up in Mainz and other German towns.

For centuries, the Chinese, Koreans, Japanese, and even Europeans used woodblock printing, where a design was laboriously carved into a block of wood that would then be used as a very large stamp. Printers could use a variety of materials (textile printing was the most common), but the arrival of paper allowed for new possibilities. In particular, playing cards, which wear quickly and therefore require frequent replacement, spread rapidly, and by 1382 an ordinance in Lille (in Flanders) banned card games because they distracted from archery practice.

In the 1440s, a Mainz artisan, Johannes Gutenberg introduced several crucial inventions, including an oil-based ink that would not run when wet, and a new type of press modified from those used in winemaking. But Gutenberg's greatest insight was realizing that a page could be assembled, line by line and letter by letter, and thus Gutenberg is considered the inventor of movable type. In 1454, he began printing the 42-line Gutenberg Bible, traditionally designated the first printed book. Gutenberg probably made a good deal of money off the print run, but apparently neglected to pay back his investors, so he was sued by one of them, Johann Fust, in 1455. Gutenberg lost and had to pay back (with interest) the money Fust had lent him. Fust also gained control of the printing blanks for the Bible as well as the equipment necessary to print it. According to tradition, the legal settlement bankrupted Gutenberg, though there is no evidence for this.

Movable type made text easy to reproduce, and printing presses sprang up throughout Germany. By the end of the century printing presses existed in over two hundred cities. The output was prodigious: in just fifty years, some thirty thousand titles were published,

---

[7]Diophantus's *Arithmetic*, according to the introduction, contains only thirteen books, but this should not be taken as evidence that Regiomontanus was mistaken: it was a common practice to include other works by the same author (or by someone the copyist thought was the same author) and to count the extra books as part of the whole work. Moreover, others, including Luca Pacioli, also claim to have seen a complete Diophantus in the Vatican Library.

with an estimated six million copies—more than were produced during the entire Middle Ages. Many of these books were on religious topics, but many dealt with secular subjects: fantastic tales, reports of the discoveries in the New World, and etiquette guides like *How to Write to German Princes and Lords.*

William Caxton learned the art of printing in Köln between 1471 and 1472; set up a press in Bruges; then returned to London to publish *Dictes and Sayenges of the Phylosophers* on November 18, 1477. It was the first book printed in English, and Caxton is reckoned as the first English printer. Unfortunately English was in the midst of a great linguistic shift. As a result, many words are spelled today the way they were pronounced in the fifteenth century. For example, the original pronunciation of 8 was similar to the German *acht*, and in 1425, it was usually spelled *aucht*; Caxton spelled it *eyght*, which became *eight* by 1549. Words like *plate* and *knife* had all their letters pronounced, and it was only later that some of the letters became silent.

Printers also played a role in developing mathematical notation. The $+$ and $-$ first appeared in the 1489 arithmetic of JOHANNES WIDMAN (1462–1498), who used the signs to indicate addition and subtraction. He also used the signs to indicate deviation from a quantity: 5 pounds more than 4 hundredweights was 4 ct $+$ 5 lb, while 12 pounds less than 3 hundredweight was 3 ct $-$ 12 lb: in other words, it was used to represent a number, and not to indicate an arithmetic operation.

The most successful presses were established in the city of Nürnberg. The city achieved great prominence after the Golden Bull of 1356 required the Holy Roman Emperor to hold his inaugural Diet (parliamentary meeting) there, making the city a legal and mercantile center; moreover, the city housed the oldest paper mill in Germany, established in 1390 by Ulman Stromer. In 1470, Anton Koberger established the first printing press in Nürnberg, and a year later, in 1471, Regiomontanus set up the first press to specialize in scientific and mathematical works. In 1474 Regiomontanus printed an ephemerides where, among other things, he listed the lunar eclipse of February 29, 1504. On his fourth voyage to the New World, Christopher Columbus claimed to have used his foreknowledge of the eclipse to terrorize the natives of Jamaica into supplying his men with food, water, and labor.

As a maker of astronomical tables, Regiomontanus was in an ideal position to help the Catholic Church solve a problem: the date of Easter. The Council of Nicaea established the date of Easter as the first Sunday after the first full Moon on or after the vernal equinox (the first day of spring). There were two problems with this definition. First, predicting the actual motion of the Moon is an extraordinarily complicated problem. To solve this problem, the Church used an "ecclesiastical Moon," based on a mathematical model of the Moon's motion. The second was the date of the vernal equinox. The Council of Nicaea defined this date as March 21.[8] However, the Julian calendar's mean year length of 365.25 days is slightly longer than the actual solar year. Consequently seasons appeared to move backwards, and by Regiomontanus's time, the vernal equinox was occurring around March 12.

Thus in 1475 Pope Sixtus IV called Regiomontanus to Rome to help reform the calendar. Unfortunately Regiomontanus died before he could do anything. According to one rumor, he was poisoned by George Trebizond's sons for having destroyed their father's

---

[8] Hence, Easter can occur as early as March 22, and as late as April 25.

Figure 6.6. Dürer's *Melancholia*

reputation, though it was more likely that Regiomontanus died from the plague (probably typhoid or malaria) then raging in Rome.

6.2.2 Albrecht Dürer

Movable type made text easy to print, but illustrations had to be done the old-fashioned way: by laboriously carving a design onto a wooden block, then inking the block and stamping the paper. Thus printing centers like Nürnberg became home to wood engravers as well. At Nürnberg, the most important workshop was that of Michael Wolgemut. On December 29, 1491 Wolgemut and his partner Wilhelm Pleydenwurff accepted a contract for the more than 2000 woodcuts in Hartmann Schedel's *Weltchronik* (1493); significantly, they accepted full responsibility for any flaws in the final printed illustrations. The *Weltchronik* and other products of the Wolgemut workshop proved influential in developing a new woodcut style, best represented by Anton Koberger's godson ALBRECHT DÜRER (May 21, 1471–April 6, 1528), taken as an apprentice by Wolgemut on November 30, 1486.

Figure 6.7. Magic Square in *Melancholia*

Dürer soon surpassed his master, and Wolgemut advised him to travel to broaden his artistic horizons. After four years of travel and study at the leading workshops in northern Europe, Dürer returned to Nürnberg, married Agnes Frey on July 7, 1494, and left her six months later to travel through Italy. The Italian artists convinced Dürer of the importance of mathematics in art, and he began a study of Euclid's *Geometry*, Vitruvius's *On Architecture*, and the works of Pacioli and Alberti. He returned to Nürnberg (and his wife) in 1495.

Dürer made a second visit to Italy from 1505 to 1507 to study mathematics. He met Pacioli in Bologna and when he returned he began collecting material for a monumental study of the application of mathematics to art. The work was never completed, but portions of it appeared in his *Introduction to the Intersections of a Circle and a Straight Line* (1525), the first advanced mathematical work printed in German. Dürer explained how to construct many curves, such as the spiral, conchoid, and epicycloid, and discussed approximate and exact methods of constructing several types of regular polygons.

Dürer's most recognizably mathematical work is his woodcut *Melancholia*. Obvious mathematical elements include the compass held by the winged figure and a polyhedron in the background. The polyhedron has spawned a remarkable amount of discussion over its exact nature. Consider the 4-by-4 grid of numbers just above the figure's shoulder, in which the numbers 1 through 16 are arranged (Figure 6.6 and 6.7).

This is an example of a magic square, an arrangement of the numbers from 1 to n^2 in an $n \times n$ array where the sum of the numbers in any row or column is $\frac{n(n^2+1)}{2}$. Magic squares had been known for millennia in China, but Dürer's square is the first printed European example. Dürer's is remarkable for another reason: not only does every row or column add to 34, but so do the two main diagonals; the four corners; the four central squares; and several other groups of four squares. Finally, the two center squares of the bottom row give the date: *Melancholia* was printed in 1514.

Nürnberg was the crossroads of the Holy Roman Empire, and many of the key figures of the sixteenth century would pass through the city and meet Dürer. The Duke of Saxony, Frederick the Wise, visited Nürnberg in 1496 and commissioned Dürer to paint his portrait; Frederick was so pleased with the result that he offered Dürer the position of Court Painter. Dürer declined, preferring to stay in Nürnberg. The Holy Roman Emperor Maximilian I was more successful in securing Dürer's services. But instead of paying Dürer directly, the cash-strapped Maximilian persuaded the town council of Nürnberg to exempt Dürer from taxes and to pay Dürer a pension. The Council did so—for a few years, then tried to renege and, after Maximilian's death in 1519, the Council stopped paying the pension entirely, insisting it would have to be renewed by the next Emperor—whoever it was.

Fortune seemed to favor Maximilian's grandson Charles. Maximilian's wife Mary was heir to the lands of Burgundy (roughly northeastern France and the Netherlands). Their son Philip married Joan, the daughter of Isabella of Castile and Ferdinand of Aragon, and in 1500, Charles was born, heir (on his father's side) to Austria and the Netherlands and (on his mother's side) Spain—and the Spanish New World. It was an enormous empire acquired not by conquest, but by marriage, and the Habsburg motto became, "Bella gerant alii, tu felix Austria nube:" Let others wage war, you, oh happy Austria, marry.

Sadly, Philip died in 1506 and shortly thereafter Joan had a mental breakdown (she is known as Joan the Mad), so Charles was raised in the Netherlands by his aunt, Margaret of Austria. There he was tutored by Adrian of Utrecht, who instilled in him a desire to form a universal European monarchy. Four regions remained outside Habsburg control: England, the Holy Roman Empire, France, and Italy. In 1509, Charles's aunt Catherine of Aragon married Henry VIII of England; after the birth of their daughter Mary, negotiations began to marry Tudor and Habsburg. This would unite England and Spain, leaving France, Italy, and the Empire the sole remaining non-Habsburg lands in Europe. Indeed if, as seemed likely, Charles became the Emperor, France would be surrounded by the Habsburgs on all sides.

Thus the king of France, Francis I, stepped forward as a candidate for the imperial throne, and the Pope encouraged the candidacy of Henry VIII and Frederick of Saxony. Frederick actually won an early ballot, but declined the dubious honor of a powerless title. Francis had the Pope's support, but Charles had New Spain, whose silver mines had just gone into production. Charles paid the electors 850,000 florins, 500,000 of which was borrowed from the Fuggers, an Augsburg banking house. On June 28, 1519, Charles became Holy Roman Emperor Charles V. Shortly after his election, Charles began to look for ways to unite Europe, which meant bringing France into the Habsburg sphere, by marriage if possible, by force if necessary. But Charles's efforts would come to nothing, thanks to a monk—and the printing press.

6.2.3   The Lutheran Reformation

Sixtus IV was a lavish patron of the arts who is considered the second founder of the Vatican Library due to the extensive renovations he began on the library in December of 1471. But his greatest project was constructing and decorating a chapel in the Vatican that he named after himself: the Sistine Chapel. A beautiful fresco over the altar was painted by renowned artists like Botticelli and Piero di Cosimo. Unfortunately, the fresco is gone forever, destroyed not by war or vandalism or accident—but to make way for Michelangelo's Last Judgment.

To raise money for the beautification projects, Sixtus and his successors turned to the age-old practice of simony (selling church offices to the highest bidder). But Sixtus invented a new and fateful way of raising money. According to Catholic doctrine, sins must be confessed, and penance done, but even then, the soul is condemned to spend time in purgatory before entry into heaven. For centuries, the church sold indulgences to reduce the time a soul spent in purgatory, but in 1476, Sixtus claimed that indulgences could reduce time spent in purgatory even for those who had died (and thus could neither confess nor do penance). The repercussions of this doctrinal change would shake the very foundation of the Church and plunge Europe into a century of nearly continuous warfare.

In 1514, Albert, the Margrave of Brandenburg, became the Archbishop of Mainz. However, Albert was already the Archbishop of Magdeburg, and church law forbade any one person from holding more than one diocese. Thus Albert appealed to Pope Leo X for a special dispensation. The Pope agreed, provided Albert made a very large donation to help rebuild St. Peter's Basilica in Rome. In order to raise money, Albert was granted the right to sell indulgences in Saxony and Brandenburg. In theory, the money was supposed to go to the Church, but through a secret arrangement, Albert could keep half the proceeds for himself.

In 1517, Albert brought in an expert salesman: a Dominican, John Tetzel. Tetzel was a practiced hand at selling indulgences, giving graphic descriptions of the suffering that would be inflicted upon the deceased in purgatory—unless, of course, their relatives bought indulgences: "As soon as the coin in the coffer rings, the soul from purgatory springs" was one of the jingles used by Tetzel to promote the sale of indulgences.

Frederick banned the sale of indulgences in Saxony, mainly because every indulgence sold drew money out of Saxony and into the coffers of his rival Albert. However the lure of indulgences was too great, and many flocked over the border to Brandenburg. The mania for buying indulgences, combined with the outrageous claims of the salesmen, prompted Martin Luther to circulate 95 theses for discussion towards the end of 1517.[9] Number 5 argued that the Pope could remit only those penalties imposed by the Church, and not those imposed by God. Number 26 denied the validity of the idea that the soul would spring from purgatory as soon as payment was made. Number 45 argued that those who gave money to indulgence salesmen incurred God's wrath, not redemption.

Luther was the first to call into question church practices after the invention of printing. The printing of the 95 theses (some say against Luther's wishes) began a continent-wide demand for a Reformation of the Church.

The Lutherans posed a dire threat to Charles V's hopes of a united Europe so in 1521, Luther was summoned by the Emperor to the city of Worms. Frederick negotiated safe conduct for Luther, and on April 17, 1521, at the Diet of Worms, Luther defended himself against the Emperor. The Imperial delegation wanted to arrest and try Luther, but Charles felt bound by his word, and allowed Luther to return to Saxony unharmed. But the subsequent Concordat of Worms declared Luther an outlaw, banned his writings, and required that he be turned over to the Emperor for trial and punishment. Fortunately for Luther, so long as he was under the protection of Frederick the Wise, the emperor was powerless. Frederick could not afford to oppose the emperor too openly, so he arranged for Luther

---

[9] According to legend, Luther nailed the 95 theses to the door of the castle church in Wittenberg. The legend is dubious for many reasons: it would have been overtly antagonistic, and at the time, Luther had no interest in an open conflict; moreover, the earliest reference to the event was written by Philip Melanchthon a few years afterwards—and he did not arrive in Wittenberg until 1518.

to be "kidnapped" and taken to his castle at Wartburg, where he stayed until 1522. While there, Luther began translating the Bible into German: the first time the Bible had been translated into a living European language.

Since Luther would have been arrested and tried if he left Saxony, it was the humanist Philip Schwartzerd who represented him at the Diet of Speyer (1526). Schwartzerd, who became the University of Wittenberg's first professor of Greek in 1518, is better known as Melanchthon, the Greek equivalent of his name (which means "black earth" in German). Melanchthon became the public face of the Reformation, and at Speyer, the Lutherans won a key victory when the Diet issued a declaration that each prince of the empire should enforce the provisions of the Concordat of Worms according to his own conscience. In effect, Luther would be safe so long as he remained in Lutheran territory.

To establish Lutheran churches throughout Germany, Melanchthon helped draft a set of instructions which included, almost as an afterthought, detailed directions for establishing schools. These became the model for the school system in Saxony and other German states, and Melanchthon suddenly found his services in great demand. At least fifty-six cities asked for his help to found schools, and he proved so influential that Melanchthon became known as the "First Teacher of Germany" (*Praeceptor Germaniae*); the modern German institution of the *gymnasium* (a college preparatory school) can be traced to Melanchthon's reforms. He also helped found the University of Königsberg (in Prussia, 1544), the first Protestant university; and at Jena in 1548, Melanchthon founded an academy (which became a university in 1577).

Both Luther and Melanchthon gave mathematics education a high priority. As early as 1524, Luther voiced an opinion on the importance of mathematics, listing it (along with languages, history, singing, and instrumental music) among those things he would have his (then hypothetical) children learn, while Melanchthon understood enough mathematics to deliver an address in 1536 on the history of mathematics, as well as to edit Sacrobosco's *De Sphaera* and Peuerbach's arithmetic.

6.2.4 German Arithmeticians

Charles acquiesced to the terms of Speyer because of a more pressing threat. After Hunyadi's defeat at Kosovo, the Turks swept across the Balkans and threatened Habsburg territory. The Ottomans captured Belgrade in 1521, defeated the Hungarian army at Mohàcs on August 29, 1526, and occupied Budapest itself shortly after. On September 1529, the Turks arrived at the gates of Vienna. By then they were overextended, so they withdrew in mid-October. It is probably no coincidence that in the same year, a new Diet at Speyer demanded the return of Lutherans to the Catholic Church. When they objected, they were labeled Protestants.

Heinrich Schreiber, known as GRAMMATEUS (b. second half 15th cent.) taught at the University of Vienna between 1517 and 1525. His *New Skill Book* (finished around 1518 but not published until 1521) introduced double-entry bookkeeping to Germany. He was also the first to limit the use of + and − to indicating addition and subtraction only.

One of Grammateus's students was CHRISTOFF RUDOLFF (fl. 1525). Rudolff stayed in Vienna to give private lessons (he was not otherwise affiliated with the university, though he was allowed to use the library) and by 1525 he published his own work, *Coss*; the title referred to the Italian practice of calling the unknown *cosa*, or "thing." It was the first German algebra, and Rudolff was the first to use $\sqrt{}$ to indicate square roots. *Coss* ended with a discussion of the cubic equation. Rudolff gave no solution, saying he wanted to stimulate research (though he didn't have an algebraic solution in any case). It would be twenty years before the first algebraic solution to the cubic was published.

Grammateus also taught (albeit indirectly) ADAM RIESE (also spelled Ries) (1492–March 30, 1559). Riese came from Erfurt, where he spent years fraternizing with the group of humanists (which at one point included the young Luther) that met at the house of the physician Georg Sturtz. Sturtz suggested Riese study the works of Widmann and Grammateus, and soon Riese began to teach algebra.

In 1523, Riese became Inspector of the Mines at the newly established town of Annaberg, founded in 1496 to take advantage of the silver mines discovered in Bohemia at the dawn of the sixteenth century.[10] Riese established a school there and wrote textbooks on computation so successful that ever since his time, Germans have used "In the manner of Adam Riese" to praise proficiency in arithmetic, while Annaberg touts itself as the city of Adam Riese.

The town might also call itself the city of MICHAEL STIFEL (1487–April 19, 1567). Stifel received his Masters from the University of Wittenberg, and became a monk at the monastery in Esslingen. After an attack of conscience that made him question the virtue of accepting donations from the poor, he was expelled in 1522 and sought refuge with the Lutherans, eventually returning to Wittenberg and staying with Luther himself. In 1528, Luther arranged to have him appointed pastor in Annaberg.

Stifel was originally a mathematician in the classical sense. After studying the books of Daniel and Revelation, he became fascinated with numerology and—no surprise—found that the "number of the beast", 666, was embedded in the name of Luther's great enemy, Leo X. Stifel's argument was as follows. In Latin, Leo X is Leo DeCIMVs, and if the capitalized letters are read as numbers, one has the number MDCLVI (1656). Since Leo was the tenth pope by that name, Stifel added ten to make 1666. Finally, since "M" is the first letter in the Latin word for mystery, Stifel subtracted one thousand to obtain the dreaded 666. Stifel went on to predict the world would end at 8 P.M. on October 18, 1533. Since the Bible specifically denies any possibility of foreknowledge of the date of the end of the world, such predictions are sacrilege, and Luther warned Stifel not to announce his results. But Stifel went ahead, and by October 19, his reputation was destroyed and he was dismissed from his post. Luther and Melanchthon managed to procure for him another pastorship at Holzdorft in 1535.

This last failure, it would seem, turned Stifel away from mathematics in the classical sense and towards mathematics in the modern sense. Rudolff's *Coss* had been out of print for many years, leaving Germans without an algebra book in their own language, so Stifel convinced Christoff Ottendorffer to sponsor a reprinting. Stifel's great original work was *Arithmetic of the Integers* (1544) which included, among other things, a version of Pascal's triangle (one of the first European versions, though it had been known for centuries in the Islamic lands and in China).

---

[10]The nearby town of Joachimsthal (Valley of [St.] Joachim), now Jachymov, Czech Republic, turned the silver into coins called *Joachimsthalers*, which was shortened to *thalers* and, eventually, dollars.

Stifel's most interesting contribution to mathematics concerned the relationship between the terms in an arithmetic and a geometric sequence, such as:

1	2	3	4	5	6	...
2	4	8	16	32	64	...

Stifel extended the arithmetic sequence backwards to include 0 and negative numbers, identifying 0 with 1; -1 with $\frac{1}{2}$, -2 with $\frac{1}{4}$, and so on: a clear precursor to the use of zero and negative exponents, as well as foreshadowing the development of logarithms.

6.2.5   Trigonometry

In 1536, Melanchthon helped RHETICUS, born Georg Joachim von Lauchen (February 16, 1514–December 4, 1574), find a position at Wittenberg teaching arithmetic and geometry. Von Lauchen, like Regiomontanus, took his academic name from his birthplace: Feldkirch, Austria was once part of the Roman province of Raetica. Melanchthon's support was a show of tremendous faith: Rheticus had not yet obtained his Master's degree. In 1539, Rheticus went to Frauenberg (now Frombork, Poland) to study with the son of a copper trader: NICHOLAS COPERNICUS (February 19, 1473–May 24, 1543).

Around 1514 Copernicus began to consider a heliocentric universe, which posited that the sun, not the Earth, was the center of the planetary motions. Rheticus had heard of Copernicus's work, but aside from a handwritten account, *Little Commentary*, Copernicus had not yet put his ideas into print. During his visit, Rheticus convinced Copernicus to allow him to publish an introduction to the heliocentric universe, the *First Report*, which appeared in 1541.

Religious opposition to the Copernican theory was immediate, for a literal reading of the Bible requires one to assume the sun moves around the Earth. Contrary to popular belief, there was very little initial opposition to the heliocentric theory from the Catholic church (which, for the next generation, considered it to be an unimportant question of philosophy); instead, the most vociferous objections came from the Protestants. Luther himself said, "This fool wishes to reverse the entire science of astronomy." As a result of Lutheran opposition, Rheticus was forced from his position at Wittenberg. Fortunately, Melanchthon, despite his own opposition to the heliocentric theory, obtained a position for Rheticus at the University of Leipzig (one of those Melanchthon had helped to reform).

In the meantime, Rheticus convinced Copernicus to publish his masterwork, *On the Revolution of the Heavenly Bodies*, where the full details of the heliocentric theory were laid out. Rheticus took the completed manuscript to Nürnberg to have it published. He left the final editing in the hands of Andreas Osiander. Osiander, a Lutheran minister who was instrumental in making Nürnberg a Protestant stronghold, added an anonymous preface to the work. The preface declared the heliocentric model was by no means intended to describe the actual universe, but instead was little more than a mathematical fiction to simplify the computationally difficult problem of calculating the future positions of the planets.

Rheticus kept his post at Leipzig for only a few years. In 1551, he was accused of having a homosexual affair with one of his students. Rheticus fled, was tried in absentia, and sentenced to 101 years in exile. After spending a few years studying medicine at the University of Prague, Rheticus went to Krakow where he would spend the next twenty years. Just as Copernicus overthrew Ptolemy's astronomy, Rheticus overthrew Ptolemy's trigonometry. Rheticus would be the first to define the trigonometric functions as ratios between the lengths of the sides of a right triangle, instead of as chords in a circle. He also realized that the sine of an angle was equal to the cosine of its complementary angles, which allowed him to halve the size of his trigonometric tables. Thus he began constructing a table of trigonometric function values for angles from $0°$ to $45°$ in intervals of $10''$ using a right triangle with a hypotenuse of $10,000,000$; this would give him seven decimal places of accuracy. For the tangent and secant values, Rheticus used a right triangle with a base of $10,000,000$, again giving him the values to 7 decimal places.

In the meantime, Charles V finally abandoned all hope of eradicating the Lutherans, and on September 25, 1555, he agreed to the Peace of Augsburg, which put an end to the religious warfare in Germany by allowing the religion of the prince to be the religion of the territory. A remarkably liberal provision allowed those who refused to convert to emigrate with their possessions.

Ironically, this forced Ulrich Fugger, whose loans helped Charles become Emperor, into exile. Fugger, who was an avid collector of manuscripts, eventually settled in the Palatinate. The Counts Palatine had their own book collection, noteworthy because of the relatively easy access they granted to the texts; Ulrich's will bequeathed his own collection as well as funds to maintain and expand the library. By the early 1600s, the library housed more than 3500 manuscripts, including one of the most complete copies of the Greek Anthology, a collection of short poems and pithy sayings, including the epigram that supposedly forms a brief biography of Diophantus's life.

Heidelberg, the capital of the Palatinate, boasted the second oldest German-speaking university, and in 1563 the University became officially Calvinist. Thus the Palatinate soon became the intellectual center of Protestantism in general and Calvinism in particular. BARTHOLOMEO PITISCUS (August 24, 1561–July 2, 1613) studied theology at Heidelberg, and in 1584, Pitiscus became the teacher of the eleven-year-old Count Palatine Frederick IV. When Frederick reached his majority, he made Pitiscus Court Preacher, and Court Chaplain at Breslau.

In addition to his theological work, Pitiscus continued the work begun by Rheticus. Because the work of Rheticus focused on the measure of triangles and not the chords in a circle, Pitiscus invented the term trigonometry to describe the new science in a new book published in Heidelberg in 1596. By 1597, trigonometry "in the manner of Pitiscus" was being taught at the University of Wittenberg and other places in Germany, and translations of Pitiscus's work into English (1614) and French (1619) appeared.

Another member of Frederick's court was L. VALENTINE OTHO (1550–1605). The lives of Rheticus and Otho bear a curious similarity. Otho paid a visit to Rheticus in his old age, just as Rheticus visited Copernicus. Rheticus entrusted Otho with his masterwork, just as Copernicus entrusted Rheticus with his. And like Rheticus, Otho had been forced from his university position. In Otho's case, the forced resignation occurred as a result of the Peace of Augsburg.

The provisions of the Peace of Augsburg made Catholicism and Lutheranism the only permissible choices of religion within the Empire. From 1574 onward, the Elector of Saxony, Duke Augustus, attempted to enforce orthodox Lutheranism. Professors at the University of Wittenberg were required to swear to a declaration of faith. Otho's refusal led to his dismissal in early 1581 and his exile from Saxony shortly thereafter.

On August 24, 1587 Otho became Frederick IV's Court Mathematician (in the astrological sense). It was there Otho finished Rheticus's monumental work: *Palatine Work on Triangles* (1596), named in honor of the Count Palatine. In the meantime, Otho had begun to create an even more accurate table of sines and cosines, though he, like Rheticus, would not live to see it published and, like Rheticus, entrusted its publication to a confidante: Pitiscus. In 1613, the new trigonometric tables appeared, accurate to 15 decimal places and including the sines and cosines of acute angles at intervals of $10''$ (with the angles from $0°$ to $1°$ and from $89°$ to $90°$ at intervals of $1''$).

Pitiscus died shortly after the publication of Otho's manuscript. He was fortunate: the Peace of Augsburg merely covered the problems facing the empire, and they would soon explode into the most devastating war yet seen in Europe. Within a few years, Heidelberg would be besieged and sacked by the Catholics, the Palatinate library removed to the Vatican, and the University shut for a generation.

For Further Reading

For the Renaissance and Reformation, see [25, 44, 58, 79, 135]. For the mathematics and science, see [31].

7

Early Modern Europe

7.1 France

In 1533, Francis I's son Henry married the fourteen-year-old Catherine de'Medici, the granddaughter of Lorenzo the Magnificent. Catherine introduced gloves from Italian fashion to the French court, as well as Italianesque palaces (she herself designed the Tuileries, in Paris), comedy troups known as *comédie Italienne* and later as the commedia dell'arte; dance troups; and haute cuisine. Catherine also popularized a new habit at the French court. She suffered from migraines, so in 1550, the French ambassador to Portugal sent her a local remedy: powdered tobacco. The ambassador's name was Jean Nicot.

The most famous person associated with Catherine's court was the astrologer Nostradamus, a provincial physician before he turned to fortune telling. One might expect that as a physician, Nostradamus would use astrology to make his predictions; instead, he apparently made them by gazing into bowls of water (hydromancy) or flames (pyromancy). He foretold that Henry II would have a long and prosperous reign. Later he turned his prophecies into cryptic verses. One of the more famous quatrains is:

> The young lion will defeat the old one
> In the field of battle by single combat
> He will pierce his eyes in a cage of gold
> Two wounds in one, and then a cruel death.

In 1559, Henry II of France died through a freak accident at a jousting competition: the lance of a competitor, Gabriel de Montgomery, shattered and flew through the eyeslit of his helm; it took Henry over a week to die a painful death. If we call Henry at 40 "old" and Montgomery at 30 "young," ignore the fact that the French king is represented by the fleur-de-lis and not the lion, interpret wounds to the forehead and throat as "pierced eyes", and turn Henry's iron helm into a "cage of gold," then Nostradamus gave an amazingly accurate prophecy. Presumably Nostradamus's earlier, non-cryptic prophecy that Henry would have a long and prosperous reign must have a hidden meaning.

By 1572, incessant warfare between French Protestants, known as Huguenots (a name of uncertain origin, though it might originate from an early Protestant leader named Hugues) and the Catholic Party, dominated by the powerful Guise family, had devastated the kingdom. In that year a wedding was arranged between Margaret, the sister of Charles IX, and

Henry of Navarre. Navarre was a Protestant stronghold in southern France, but Henry of Navarre was also a descendant of Louis IX (Saint Louis), and thus a potential claimant to the throne.

Protestants from all over France came to Paris to celebrate the wedding. On August 23, 1572, the night before the wedding, radical anti-Huguenots struck, murdering nearly three thousand Protestants in what would become known as the St. Bartholomew's Day massacre. Henry of Navarre was spared only by an immediate conversion to Catholicism. Though her exact role in the planning of the massacre is uncertain, Catherine de Medici shed no tears over the deaths of thousands. But violence breeds violence, and for the next generation, France would be torn apart by religious warfare.

7.1.1 Viète

Charles IX, never in good health, died in 1574, and his brother became King Henry III. Henry summoned FRANÇOIS VIÈTE (1540–February 23, 1603) to Paris. Viète, a lawyer, had made a name for himself by dispensing good advice to local legislatures around France, and Henry may have felt the need of someone who could find a way to tread the dangerous path between the Catholic Party and the Huguenots. Soon, Viète became one of the king's most trusted advisers, being sent on important and confidential missions.

Unfortunately, all Viète's good advice could not save Henry. Because the Estates General would not pay for a war of extermination against the Huguenots, Henry was unable to eliminate French Protestantism. The Catholic Party took this as a sign that Henry was *unwilling* to do so. Hence the Pope and Philip II of Spain organized a new Holy League, which aimed to place Henry of Guise on the throne. In 1584, to appease the radicals, Henry revoked all the concessions he had made to the Huguenots, leading to the War of the Three Henrys: Henry III, the king; Henry of Guise; and Henry of Navarre, who had recently renounced his conversion to Catholicism.

Guise entered Paris, forcing Henry III to flee; in 1588, Henry III had Guise murdered. Naturally, this placed Henry's own life in great danger, and, oddly enough, the one place he could find support and safety was with the Protestant Henry of Navarre. Unfortunately, nothing could protect the king from fanatics, and in 1589 the monk Jacques Clément murdered Henry III. Ironically, this made the Protestant Henry of Navarre the sole surviving legitimate claimant to the throne of France. The Holy League denied the validity of Henry's accession, and the war continued.

Henry of Navarre recognized Viète's value and kept him as an adviser. Some of Viète's earliest work was in cryptanalysis, the making and breaking of "secret messages," which has since become one of the more practical pursuits of mathematics. While still in the employ of Henry III, Viète decoded messages to Alessandro Farnese, then commander of the forces of the Holy League.

In 1589 some messages from Philip of Spain fell into Henry of Navarre's hands; he turned them over to Viète to see what he could do. Viète took advantage of a peculiarity of the Spanish codes: they did not encrypt numbers. A very large number, such as $100,000$, probably represented an amount of money, so one might guess the next word to be "ducat" or some other currency. In a message with military content, a large number, such as 4000 probably refers to non-elite troops like infantry, while a more modest number, such as 500, might refer to cavalry. One message Viète decrypted contained a plea for 6000 infantry to be spent from the Spanish Netherlands (modern Belgium), to help the Holy League's cause in France, lest it suffer a disastrous defeat. This information would have been of vital strategic importance—except by the time Viète gave the decoded message to Henry, he had already dealt the League a serious defeat.

Viète continued to read the dispatches of the Spanish, and made no secret of it, being confident that however they changed their codes, he could always decipher the messages. Philip, having discovered Viète could read messages he thought inviolable, complained to the Pope, claiming that the only way Viète could do so was to use black magic. The Pope, whose own cryptanalysts had been reading Spanish communications for years, ignored the complaint.

The war dragged on, and Henry of Navarre was shrewd enough to realize that the radicals would cause him no end of trouble. Thus, Henry converted back to Catholicism, saying (according to his enemies and detractors), "Paris is worth a mass." Pope Clement VIII forgave his apostasy and recognized him as the rightful King of France on September 17, 1595. Since France was still divided between Catholic and Huguenot, Henry, now Henry IV of France, announced the Edict of Nantes on April 15, 1598, which granted the Huguenots the right to practice their religion without persecution.

Shortly afterwards, Viète wrote to Clement VIII to complain about the newly adopted calendar, promulgated by Pope Gregory XIII but designed by a committee headed by Christopher Clavius, a Jesuit astronomer from Germany. Viète accused Clavius of having corrupted the Roman calendar in a "profane and unfortunate" attempt to carry out the desire of Gregory XIII to reform the calendar.

There were two problems. The first was making the mean length of the calendar year better approximate the solar year. Under the Julian calendar, years divisible by 4 are leap years. The Gregorian calendar makes century years (ordinarily divisible by 4) non-leap years unless they are also divisible by 400. Thus 2000 was a leap year, while 2100 will not be one. This gives a mean calendar year of 365.2425 days, which closely approximates the actual solar year of 365.2422 days. The second problem was that the Council of Nicaea fixed the date of the vernal equinox as March 21, and by 1582, the vernal equinox was actually occurring on March 11. One solution would be to redefine the date of the vernal equinox. The solution actually employed was to drop ten days from the calendar: October 4, 1582 would be immediately followed by October 15, 1582. Protests followed in some areas: while some may have really believed they had lost ten days of their lives, most objected for the purely practical reason that lenders and landlords treated October as a full month for purposes of calculating interest owed and rent due.

A rather more dramatic effect occurred because of the Protestant Reformation: in 1475, when Sixtus IV brought Regiomontanus to reform the calendar, all Europe was Catholic, and a calendar instituted by the Pope would have been adopted without difficulty. In 1582, Protestant countries refused to adopt a Papal calendar, so for a time, different areas of Europe had different dates. For this reason, many dates between 1582 and 1917 are indicated using "O.S." (Old Style, i.e., based on the Julian calendar) or "N.S." (New Style, based on the Gregorian calendar).[1]

---

[1] Russia, never part of Catholic Europe to begin with, only changed after the 1917 Revolution.

The situation in the British Isles was particularly confusing. England did not make the change until 1752. Thus Miguel Cervantes died in Spain on April 23, 1616, while William Shakespeare died in England ten days later—on April 23, 1616.[2] To further complicate matters, the English began their year on March 25 (known as Lady Day, and corresponding to the Feast of the Annunciation). Thus an English date between January 1 and March 25 of a year between 1582 and 1752 corresponds to a Gregorian date ten (or eleven, after 1700) days later in the next year. As a final complication, in 1600 the Scots (who were a separate kingdom until 1603) switched so the year began on January 1: thus January 1, 1600 in England was January 1, 1601 in Scotland and January 11, 1601 in France.

Viète's main claim to fame in mathematics stems from his *Introduction to the Analytic Art*, where he introduced a new form of algebraic notation, but its significance is often misunderstood. Algebraic notation existed before Viète: for x^3, Diophantus wrote K^Y, Cardano wrote *cubus*, and Viète wrote A cub., all of which are abbreviations of "cube." For example, the expression we might write as $5x^2 + 4x + 8$ would be written by Cardano as 5 quad. and 4 cos and 8, while Viète would write it as $5A$ quad. and $4AB$ and $8Z^{\text{pl.}}$ (where the B and $Z^{\text{pl.}}$ are factors necessary to make all the terms of the same degree). Clearly there is very little difference between the two; in fact, by insisting on homogeneity of terms, Viète is actually taking a step backwards.

Viète's real contribution was recognizing that the constants in an equation could *also* be represented by letters: essentially, Viète invented parametrization. Thus, whereas Cardano had to resort to a cumbersome verbal description of the procedure for solving an algebraic equation, Viète could simply give the quadratic formula:

Rule 7.1. *If A square $+ B$ times $2A$ is equal to Z plane, then , let $A + B$ be E. Therefore E square is equal to Z plane $+ B$ square. Therefore $\sqrt{Z + B}$ square $- B$ is equal to A.*

In other words, the solution to $x^2 + 2bx = z$ is $x = \sqrt{z + b^2} - b$. There is some speculation that Viéte's codebreaking work informed his development of algebraic notation: in both cases, one is trying to determine the meaning of an arbitrary symbol. For example, in the code phrase $100,000A$, A might mean "ducats." More significantly, there is no reason why the $100,000$ itself could not be encoded: hence if we knew A meant "ducats," then in the code phrase $B\ A$, B would represent some unspecified but definite number.

Viète's innovation opened up a new possibility: investigating the relationship between the solutions of an equation and its coefficients. For example, he considered the equation $BA - A^2$ equal to Z. If E is another solution to this equation, then $BE - E^2$ is equal to Z. Equating the two gives us $BA - A^2 = BE - E^2$. Hence:

$$BA - BE = A^2 - E^2.$$

Removing the common factor of $A - E$ gives us:

$$B = A + E.$$

Hence B (the coefficient of the unknown) is simply the sum of the two roots. This began the transformation of algebra from a tool for solving equations, to a tool for investigating the relationship between quantities.

---

[2]The actual date of Cervantes's death is open to some interpretation. April 23 is the date listed on his tombstone, but Spanish custom of the time was to list the date of burial and not necessarily the date of death.

Viète's association with Henry IV led to an unusual challenge. In 1593, the Belgian mathematician ADRIAAN VAN ROOMEN (September 29, 1561–May 4, 1615) posed a problem equivalent to:

Problem 7.1. *Solve:*

$$45x - 3795x^3 + 95634x^5 - 1138500x^7 + 7811375x^9 - 34512075x^{11} + 1105306075x^{13}$$
$$- 232676280x^{15} + 384942375x^{17} - 488494125x^{19} + 483841800x^{21} - 378658800x^{23} +$$
$$236030652x^{25} - 117679100x^{27} + 46955700x^{29} - 14945040x^{31} + 3764565x^{33} - 740459$$
$$x^{35} + 111150x^{37} - 12300x^{39} + 945x^{41} - 45x^{43} + 1x^{45} = C$$

where C is a given constant.

This daunting problem had been solved by a Dutch mathematician, Ludolph van Ceulen, but, the Dutch ambassador to Henry's court boasted, no one in France could solve it. Henry passed the problem onto Viète, who immediately gave one solution and then, the next day, twenty-two others.

Viète solved the problem by recognizing the expression was the result of expanding $\sin 45\theta$ in terms of powers of $\sin \theta$; Viète's discovery amounts to an early form of De Moivre's theorem. The basic idea behind Viète's method of solution can be illustrated as follows. Suppose we wish to pose a third degree challenge problem. We begin by letting $C = \sin 3\theta$, where C is some constant. This equation can be transformed using the identity $\sin 3\theta = 3 \sin \theta - 4 \sin^3 \theta$, which gives:

$$3 \sin \theta - 4 \sin^3 \theta = C.$$

If we let $x = \sin \theta$, we obtain the cubic $3x - 4x^3 = C$. For example, in the equation $3x - 4x^3 = \frac{\sqrt{2}}{2}$, we could make the substitution $x = \sin \theta$, and thus we would have $\sin 3\theta = \frac{\sqrt{3}}{2}$. Hence $3\theta = 45°$, and thus $\theta = 15°$ with $x = \sin 15°$. In a similar way, if we make the substitution $x = \sin \theta$ in Romanus's equation, the left-hand side can be reduced to $\sin 45\theta = C$, which can be solved by consulting a table of sines. Viète explained his solution in *Response to a Problem Posed by Adrianus Romanus* (1595). This led to a correspondence and close friendship between the two mathematicians that lasted until Viète's death.

7.1.2 The Age of Fermat and Pascal

On May 14, 1610, Henry IV became the second king of France in a generation to be murdered. Henry's son and heir, Louis XIII, was just eight, so France was effectively ruled by his mother, Marie de Medici and her court favorites, including Cardinal Armand Jean du Plessis, the Duke of Richelieu. The *Three Musketeers* notwithstanding, Richelieu's constant goal throughout his tenure as prime minister (from 1624, mainly through the machinations of Marie) was to increase the power and prestige of the French monarchy. Marie soon realized this, and on November 10, 1630, she made the mistake of forcing the king to choose between her and Richelieu. Louis sided with his Prime Minister, and sent Marie into exile from Paris for the rest of her life.

Soon afterwards, Marie's favorite son, Louis's brother Gaston, raised an army to over-throw Richelieu. One of Gaston's main supporters was Henry de Montmorency, the gov-ernor of Languedoc and a member of a distinguished family. Montmorency tried to rally Languedoc against Richelieu, but at the battle of Castelnaudary (September 1, 1632) his forces were soundly defeated and he was taken prisoner. Gaston surrendered shortly there-after and in exchange for a pardon (he was, after all, the king's brother), he named his co-conspirators. Louis ordered Montmorency tried by the *Parlement* of Toulouse. Mont-morency was found guilty and despite pleas from the nobility to pardon the member of such a distinguished family, Richelieu persuaded the king to stand his ground, and Mont-morency was executed on October 30, 1632.

The French *parlements* served a role similar to that of the U.S. Supreme Court: while they could judge criminal cases, their main function was reviewing the law. In theory, they could refuse to register a royal edict they viewed contrary to the principles of law or jus-tice. In practice, the position had more prestige than power, so cash-strapped members of *parlement* often sold their seats to the highest bidder. PIERRE DE FERMAT (August 20, 1601–January 12, 1665) purchased his seat in the Toulouse *parlement* for $43,500$ livres, and began serving on May 14, 1631.

Though Fermat, a graduate of the University of Toulouse, served in the *parlement* dur-ing Montmorency's trial, he was not directly involved. At the time, he was in the Chamber of Petitions, responsible for deciding whether a case was worthy of being sent to a higher court. But Fermat rose rapidly through *parlement*, becoming a member of the Chamber of Inquiry (roughly comparable to a Congressional Investigation Committee) in 1638, and in 1652 he became a member of the Grand Chamber, which actually sat in judgment over criminal cases. By then, that was all that the *parlements* were empowered to do, since even their slight ability to interfere with the king's power was a thorn in Richelieu's side. Louis, at Richelieu's insistence, called on the *parlements* to restrict their activities to judging cases, not the law.

In his spare time Fermat studied mathematics. In 1636, while reading Apollonius's *Plane Loci*, Fermat came to the fundamental principle of analytic geometry: the relation-ship between two unknown quantities translates into the solution of a locus problem. In particular, Fermat began with an algebraic relationship and identified it with a curve. For example, the hyperbola we would express as $xy = a^2$ is the locus of points P where the rectangle with opposite vertices at the origin and the point $P(x, y)$ has area a^2. Fermat's ideas on coordinate geometry are actually closer to our own ideas than those of Descartes, since he used perpendicular distances and also considered the curves corresponding to a given equation.

Fermat also solved the three main problems of calculus around 1636. If $n > 1$, Fermat showed (in modern terms) that $\int_1^\infty \frac{1}{x^n}\, dx = \frac{1}{n-1}$ and $\int_0^1 \sqrt[n]{x}\, dx = \frac{n}{n+1}$. He did so by partitioning the domain at points in geometric proportion; the inscribed rectangles had areas which were also in geometric proportion, and thus their total area could be found easily. Fermat's method of adequation could be used to solve optimization problems, and a variation of it could be used to find the tangent to a wide variety of curves.

His best known contributions are in number theory, though his fame in this field is overrated: he made interesting conjectures, but produced few proofs. His most important conjecture is sometimes known as Fermat's Lesser Theorem, which he presented in a letter dated October 18, 1640 to Frenicle de Bessy:

Theorem 7.1. *Given any prime p and a not a multiple of p, then p divides $a^n - 1$ for some n; moreover, n divides $p - 1$; finally, p will divide $a^{kn} - 1$ for all p.*

Fermat is also known for the (true) conjecture that all primes one more than a multiple of 4 can be written as the sum of two squares, and the (false) conjecture that all numbers of the form $2^{2^n} + 1$ are prime. Fermat's best-known conjecture, that $x^n + y^n = z^n$ has no whole number solutions x, y, z for $n \geq 3$, was only discovered and published after his death, though while alive he claimed the result for $n = 3$ and $n = 4$ in letters to Wallis and others.

A second mathematician of note served in the Toulouse *parlement* with Fermat: PIERRE DE CARCAVI (1600–April 1684). In 1636, Carcavi arrived in Paris, after having purchased an office in the Great Council that advised the king. Carcavi mentioned Fermat's work on falling bodies to MARIN MERSENNE (September 8, 1588–September 1, 1648), a Minimite friar.[3] From his cell, Mersenne wrote letters and served as a central clearinghouse for seventeenth century French mathematics and physics.

Mersenne brought the works of others to a broader audience. For example, Galileo's *Dialog on the Two World Systems*, arguing strongly in favor of the heliocentric theory, was written in Italian at a time when most scientific works were in Latin. Mersenne translated parts of the *Dialog* into French and brought Galileo's ideas to a broader audience.

One of Mersenne's more celebrated achievements concerns Euclidean perfect numbers, numbers that are the sum of their proper divisors. In the *Elements*, Book IX, Proposition 36 Euclid proved that if $2^n - 1$ is prime, then $2^{n-1}(2^n - 1)$ is perfect. Elementary algebra suffices to show that $2^n - 1$ is composite if n is composite, so finding Euclidean perfect numbers comes down to determining the prime numbers n for which $2^n - 1$ is also prime. In 1644 Mersenne conjectured that $2^n - 1$ was prime if $n = 2, 3, 5, 7, 13, 17, 19, 31, 67, 127$, and 257, and for no other values less than 257. Mersenne was only partially correct: $n = 61$, 89, and 107 correspond to prime numbers, while $n = 67$ and 257 give rise to composite numbers. Because of his work, primes of the form $2^n - 1$ are now called Mersenne Primes.

Fermat's first letter asked Mersenne to pose anonymously two problems in optimization:

1. Given a sphere, inscribe the right circular cone of greatest surface area.

2. Given a sphere, inscribe the right circular cylinder of greatest surface area.

Mersenne passed the problems along to GILLES PERSONNE DE ROBERVAL (August 10, 1602–October 27, 1675), who had just been appointed to the Ramus Chair of Mathematics at the Collège Royale. Roberval found the problems extremely interesting, because they could not be solved using known techniques, and had Mersenne write back to Fermat, asking him to divulge his methods. Fermat obliged in later letters.

Roberval occupies a singular position in the history of mathematics, for his *Treatise on Indivisibles* shows that he might have invented calculus a half-century before Leibniz and

---

[3] The Minimites consider themselves to be the "least" of the religious orders, and in addition to the traditional vows of poverty, chastity, and obedience, they take another vow to abstain from meat, eggs, fish, cheese, and milk.

Newton. But in 1634 he became the Ramus chair of mathematics at the Royal College in Paris. Every three years the position had to be renewed by competitive examination, with questions posed by the current holder of the chair. Consequently Roberval kept most of his methods secret, since this would guarantee his reappointment. Thus *Treatise on Indivisibles* did not appear in print until 1693, nearly twenty years after his death, though it was almost certainly completed before Roberval became Ramus chair.

In *Treatise*, Roberval considered the cycloid, generated by the motion of a point on a circle as the circle rolled along a line without slippage, and noted that the motion of the point consisted of two components: first, its motion along the circle; and second, the motion of the center of the circle parallel to the line. The second motion can be treated as motion at a constant rate, while the first motion gives rise to a curve that Roberval called the "companion of the cycloid:" the sine curve.

Meanwhile, the policies of Richelieu were carried on by his handpicked successor, the Cardinal Giulio Raimondo Mazzarino, better known as Mazarin. Unfortunately all that Richelieu had done to curb the power of the nobility looked as though it would be lost when Louis XIII died in 1643, leaving the throne to his son Louis XIV. Louis XIV was not yet five at the time, so his mother, Anne of Austria, governed France in his name, with Mazarin's help. Disaffected elements in French society took the opportunity of this period of "foreign rule" to launch a series of civil wars. Since much of the original rhetoric involved aspersions cast on the character of Mazarin and Anne (including unfounded claims they were lovers), the civil wars became known as the Fronde ("slingshot," used by Parisian children to hurl mud and other debris at passing coaches).

The first phase, the Fronde of the *parlement*, began over a sequence of highhanded royal acts to raise money. Between 1648 and 1649, the *Parlement* of Paris sought to place limits on the king's authority, and eventually won many concessions, but at a cost they would later regret: Louis gained a lasting distaste for the *parlements* and Paris. Moreover, the nobles took the opportunity to raise a revolt of their own to recover the special privileges they had lost under Richelieu. By skillfully playing the nobles and *parlements* against each other, Anne and Mazarin managed to defeat both by 1653, establishing the principle of central authority. Louis was crowned on June 7, 1654.

To further reduce the power of the troublesome nobles, Louis excluded them from the government, drawing his ministers from the middle class. As a result, the nobles were left with very few responsibilities, and they could devote their time to leisure activities, which generally included hunting, drinking, and gambling. Around the time of Louis's coronation, the Chevalier Méré ran into the following problem. In many games, one player must "make his point" (by rolling a specified number on a die) within a certain number of throws. Suppose a game is interrupted after the player has made a few unsuccessful throws; how should the stakes be divided? This is known as the problem of points.

Méré was no mathematician, but he knew one who might be able to answer the problem: BLAISE PASCAL (June 19, 1623–August 19, 1662). Pascal's father Etienne had moved to Paris from Clermont (now Clermont-Ferrand) in 1631, and by 1639 he had become a local administrator in Rouen. Among other duties, he had to collect taxes, which meant that he also had to perform a vast number of computations. Thus between 1642 and 1644, Blaise Pascal designed and built the first mechanical adding machine to help his father perform tax computations. Pascal designed over 50 models and on May 22, 1649, the eve of the

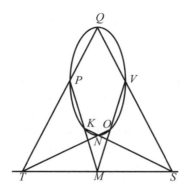

Figure 7.1. Pascal's Hexagon Theorem

Fronde of the *parlement*, Pascal received a monopoly on the manufacture and sale of the adding machines in France.

Pascal was already well known in scientific circles. At the age of 16, he wrote *Essay on Conics* (published 1640), a short treatment where he announced a remarkable theorem. As Pascal stated it, let there be a conic section through points K and V, and let there be two other points M and S. Let the lines MK, MV, SK, SV intersect the conic section at points P, O, N, and Q. Then the lines MS, NO, and PQ all meet at the same point. One possible interpretation is known as Pascal's hexagon theorem (see Figure 7.1):

Theorem 7.2. *Given a hexagon inscribed in a conic section. If the opposite sides are extended so they meet at points A, B, and C, the three points lie on the same line.*

After Méré wrote to Pascal about the Problem of Points, Pascal in turn wrote to Fermat. The original letter has been lost, but Fermat's reply, dated July 1654, has been preserved. The particular problem considered by Fermat is to make a specified point (for example, to roll a "6") in eight throws of a die. Suppose the stakes have already been entered. Then, since the player has one chance in six of making the point (and winning all the stakes), then it is worth one-sixth of the stakes to that player not to make the first throw.

Now if the player actually makes the throw and fails to make the point, he has effectively given up one-sixth of the stakes, leaving $\frac{5}{6}$. By the same argument, not making the next throw ought to be worth one-sixth of the stakes: hence if the game is interrupted after one (unsuccessful) throw, the player who is attempting to make the point ought to receive $\frac{5}{36}$ of the stakes, and the other player should receive the rest. Once more, if the players makes the throw and fails to make the point, he has effectively given up $\frac{5}{36}$ of the stakes: altogether, he has given up $\frac{1}{6} + \frac{5}{36} = \frac{11}{36}$ of the stakes, leaving $\frac{25}{36}$. Not making the next throw is worth one-sixth of this amount, and so on.

Pascal's response, dated July 29, 1654, commends Fermat's work on the subject, but expresses concern over the difficulty of actually performing the computations. Thus he presents a different method. In this case, Pascal considers the problem where each player must win three games to win a stake of 64 pistoles. First, suppose one player is ahead, two games to one. This player may make the following argument:

> I am sure to have 32 pistoles, since even a loss will give them to me; as for the other 32, perhaps I will have them and perhaps you will, the chances are equal. So divide

these 32 in half and give them to me, in addition to those 32 of which I am certain to have.[4]

Thus the stakes should be divided 48 to 16 in favor of the player who is ahead.

The other cases could then be reduced to this case. For example, if the score is 2 games to nothing, the player who is ahead is assured of 48 pistoles (since losing the next game would put the score at 2 to 1, where the two would split the stakes 48 to 16), while a win would give him the whole 64, or 16 more. Thus the player who is ahead should take 48 plus half of 16, or 56, leaving the other player 8. If only one game has been played, then the player who is ahead is guaranteed to win 32 and might win 56 or 24 more. Hence the stakes should be divided $32 + 12 = 44$ to 20.

Even this method (a form of recursion) is tedious to apply, so Pascal continued with a discussion on methods of determining numbers of combinations so the probabilities could be calculated directly. An interesting sidelight of the times is that when Pascal begins to describe a solution to this problem of determining the number of combinations, he switches to Latin, "because French is good for nothing" when it comes to scientific discourse.

In this letter Pascal refers to a table that he used to calculate the number of possible combinations, which he promised to send to Fermat in a later letter. This table is almost certainly what is usually referred to as Pascal's triangle, though Pascal himself called it the arithmetic triangle in a work of the same name. While the arithmetic triangle and its uses appeared in Chinese, Indian, and Islamic sources centuries before Pascal, and even in the work of Michael Stifel in the sixteenth century, Pascal was the first to elucidate and prove many of its properties. Interestingly from a modern standpoint, Pascal's "triangle" actually begins as a rectangular grid of cells. In each cell of the first row, place a "1." Then in general the kth cell of the nth row will contain the sum of the first k cells of the $n - 1$st row. After describing the generation of the triangle, Pascal then stated and proved eighteen propositions about it.

Overall Pascal's work shows he was a mathematician of high caliber, and might have become one of the greatest mathematicians in seventeenth century France. Unfortunately on the night of November 23, 1654 he experienced a "night of fire," and overcome by religious passion, entered the convent of Port Royale where he spent the next four years in meditative contemplation. One of the main results of this period was Pascal's Wager, an attempt to convince some of his friends that belief in God was consistent with rationalism. It was one of the earliest instances of a game theoretic argument: God either exists or does not exist, and one can believe in God or not believe. If one believes in God, then one has the potential for infinite reward (eternal life) if God exists and no reward (non-existence) if God does not exist; if one does not believe in God, then one can lose an infinite amount (eternal damnation) if God exists and no reward (non-existence) if God does not exist. Of course the flaw in Pascal's argument is that many religions require exclusivity, so one could not (for example) be Catholic *and* Muslim, both of which promise great rewards to believers and great penalties to unbelievers.

In 1658, a persistent toothache kept Pascal from sleeping, and to distract his mind from the pain, he began working on several problems relating to the cycloid. Miraculously, the toothache subsided, and Pascal took this as a sign from God that he should continue work-

---

[4]Fermat, *Œuevres*, Vol. II, p. 291

ing on mathematical problems. He eventually wrote a treatise on the properties of the cycloid that caused a great uproar because its preface, which described the history of the cycloid, completely neglected the work of Torricelli. But Pascal would do little more mathematics, for he died in 1662 at the age of 39.

7.2 The Low Countries

In 1556, worn out by years of campaigning to unite Christian Europe in a crusade against the Muslim Turks, Charles V retired. Spain and the Netherlands (which included Belgium and Luxembourg) went to his son, Philip II of Spain. It was a grave mistake. The Dutch tolerated Charles V, who grew up in Flanders and even announced his abdication in Brussels. However, the Protestant Dutch detested the Catholic Philip, a situation that was worsened by several of the governors Philip appointed to rule the Netherlands in his name. The most notorious, the Duke of Alba, once suggested to Philip II that he behead all who opposed his rule in the Netherlands. This highhanded behavior affronted Catholics and Protestants alike, and in 1568, the Netherlands revolted, beginning the Eighty Years' War.

In 1578, Alessandro Farnese became governor of the Netherlands. Farnese was the grandson of Charles V (through Charles's illegitimate daughter Margaret, the former governor of the Netherlands), and was a dangerous combination of brilliant military strategist and shrewd politician. Of the thirteen provinces in revolt, the southern six were inhabited mostly by Catholic, French-speakers known as Walloons, who might be reconciled to a Catholic King. Farnese reconquered many of the key southern provinces, and his offer of amnesty to Protestants who returned to Catholicism kept them from rebelling again. By 1585, the southern provinces were firmly back under Spanish control, and form the core of modern Belgium.

Protestants who refused to convert were forced into exile. Many went to Holland. In 1579, the seven northern provinces still in revolt concluded the Union of Utrecht, and under the leadership of William of Orange, announced their independence from Spain. William survived three assassination attempts, only to fall victim to the fourth when a Burgundian Catholic shot him on July 10, 1584, and leadership of the revolt fell on the shoulders of his son, Maurice of Nassau.

At the time, Maurice was in school at the University of Leiden (founded by his father in 1575 as a reward for the city's heroic defense against the Spanish). A less dedicated individual might have taken this opportunity to interrupt his studies, but Maurice had no desire to do so, and continued his studies with the help of a fellow student at Leiden: SIMON STEVIN (1548–March 1620).

7.2.1 Simon Stevin

There are very few facts known about Stevin: he was born around 1548, and entered the University of Leiden in 1583. Though he spent most of his life in the Dutch Netherlands, he always signed himself "Simon Stevin of Bruges," a city in the Spanish Netherlands (now Belgium). This suggests he was born in Bruges and chose exile over conversion. Stevin accompanied Maurice on his campaigns, and composed several texts while in the field. Perhaps worried that the manuscripts would be lost or destroyed, Stevin had them published; one unverified story is that Maurice paid for the publication of the works.

The Netherlands poses a unique tactical challenge for an invader. Most of the country lies below sea level—the region is called the "Low Countries" for this reason. Enormous dikes were built to hold back the sea, and windmills were used to pump water, forming polder land. But the sea cannot be long denied, and since the Middle Ages, the Dutch thwarted invaders by breaking the dikes and flooding the polders. Stevin designed a system that allowed the polder to be flooded quickly and easily.

Most of Stevin's works were published in the 1580s. The writing of whole numbers in Hindu-Arabic numerals was by then well established, but the Arabic use of decimal fractions was not yet known in Europe. In *The Tenths* (1585), Stevin introduced a form of decimal fractions, indicating place value using circled numbers. Thus, 3.141 would be written as 3 ⓪ 1 ① 4 ② 1 ③. Arithmetical operations were handled just as we handle them today, with the added feature of indicating the explicit place value using the ◯s.

In 1586, Stevin published *The Elements of the Art of Weighing*, written in Dutch at a time most scholarly works were written in Latin. Based on this work, Stevin should be considered one of the founders of calculus: not only was he one of the first to use the infinitesimal rectangle, but Stevin reintroduced the limit concept from the ancient Greek geometers, using an argument that is essentially the following:

1. Between any two different quantities there may be found a quantity less than their difference,

2. If two quantities differ by an amount less than any given difference, the two quantities are equal.

A less important contribution of Stevin, though one that showed how far ahead of his time he was, appeared in his *Arithmetic*, in which he defined number to be that which measures the quantity of any thing, which meant that there was no distinction, in Stevin's mind, between whole, rational, and irrational numbers. Moreover, since one measured the quantity of one thing, then, Stevin declared, "ONE IS A NUMBER" (capitalization is Stevin's). A generation after Stevin, mathematicians such as Wallis and Fermat would still debate over whether or not one was a number.

Stevin's view that it was meaningless to distinguish between rational and irrational played a role in his theory of tuning. In 1558 Gioseffo Zarlino published *Institutioni Harmoniche*, which presented one of the first significant advances in music theory since Pythagoras. Zarlino noted that the consonant ratios were all formed from the first six whole numbers: the octave (2 : 1), fifth (3 : 2), fourth (4 : 3), major and minor thirds (5 : 4 and 6 : 5), and major sixth (5 : 3). To deal with the minor sixth (8 : 5), Zarlino made a somewhat pedantic argument that comes down to $8 = 2 \cdot 4$, hence the minor sixth is made of 4, 2, and 5. Since 6 is a perfect number, Zarlino argued in effect that pleasant music is a product of pure mathematics. Zarlino's musical theories were savagely criticized by one of his own students: Vincenzo Galilei, the father of Galileo.

Rather than try and justify the thirds based on number mysticism, Stevin took a bold step: he argued that the adherence to the Pythagorean ideal of small whole number ratios was flawed. Stevin argued that the octave (2 : 1) was the fundamental ratio, and that it was divided into twelve equal semitones (instead of five tones and two semitones). Consequently the real ratio between semitones had to be $\sqrt[12]{2} : 1$. This meant that the Pythagorean fifth (3 : 2) was simply a whole number approximation to real ratio, $\sqrt[12]{2^7}$: 1. This is the basis for the modern theory of equal temperament, but Stevin's ideas remained largely unnoticed until after equal temperament became standard tuning practice.[5]

7.2.2 The Changing Battlefield

Some time during the tenth century the Chinese invented black powder, a mixture of charcoal, saltpeter and sulfur. By filling a tube and closing one end, they made rockets; by closing both ends, bombs. But it was a third use of black powder that proved most popular: guns (hence gunpowder). The Arabs and Europeans began using gunpowder within a few centuries, though whether by independent invention or learning from the Chinese is unknown. The proliferation of firearms (literally, weapons that made use of fire) changed the face of warfare.

Cannon appear in Florentine records as early as 1326 (the Turks used them somewhat earlier), and led to a dramatic change in castle architecture. Imagine a cannonball striking a wall. If the angle of impact is perpendicular or nearly so, most of the cannonball's momentum and force can be directed into the wall, breaking it down. On the other hand, a shallower angle of impact allows less momentum and force to be delivered, and if the angle of impact is shallow enough, the cannonball will actually ricochet off the wall. Alberti pointed out in *On Fortifications* (1485) that the geometry of a fortress should be star-shaped, so that most cannon shots would strike a wall at a very shallow angle.

Gunpowder did more than change the ideal shape of a castle. No wearable armor could withstand a shot from a cannon, so the use of heavy armor declined. This in turn made soldiers vulnerable to hand-held cannon: originally called a *donderbus* ("thunder tube" in Dutch), then later a musket (after a type of sparrowhawk).

In order to shoot a round, the main charge had to be ignited, usually by setting fire to a smaller charge in a pan adjacent to the touchhole (a small hole bored into the barrel that allowed access to the main charge). In early guns, this was accomplished by holding a burning brand (usually a rope) and touching it to the charge by hand, but inventors quickly devised a mechanism that would hold the rope until it was needed: a firelock (after its resemblance to the lock mechanism of a door). The lock mechanism was a simple spring and its operation was certain (this is one possible origin of the word "cocksure").[6] However, the spread of the fire from the pan charge to the main charge was uncertain, and it was not uncommon for the pan charge to ignite without touching off the main charge: this was a "flash in the pan," and necessitated repriming the pan and trying again.

Even if the weapon fired, it could not be reloaded quickly: a typical 16th century firearm took half a minute to reload, and a running man, especially one unburdened by heavy armor, can cover over a hundred yards in that time—greater than the effective range of the weapon. Thus most musketeers carried a secondary weapon, usually a sword.

---

[5] It should be pointed out that the Pythagorean adherence to small whole numbers has a physical basis. The length of a string is a rough measure of the fundamental frequency it will produce when plucked; however, the string will also generate other frequencies as overtones. Because of destructive interference, the frequencies of the fundamental tone and the overtones always form a whole number ratio. Consequently any other string whose fundamental frequency is a small whole number multiple of the first will reinforce a component of the natural sound of the original, and sound pleasant.

[6] The entire gun consisted of the lock, which ignited the pan charge; the stock, which helped brace or hold the gun; and the barrel, which contained the bullet. Hence the origin of the phrase "lock, stock, and barrel" to signify the whole of anything.

The sword, too, underwent great changes due to the invention of gunpowder. The medieval broadsword is designed to be swung so that the blow could penetrate heavy armor. Consequently, most broadswords had a blunt tip (and in most cases, not much of an edge, either). But the absence of heavy armor meant a lighter, more versatile sword could be used; moreover, the thrust—which could penetrate deep within the body and strike the vital organs—became more important. Thus the broadsword evolved into the rapier, requiring the invention of a totally new style of fighting: fencing.

The first fencing schools opened in Italy during the sixteenth century. At first the idea of a "scientific" fighting technique was slow to catch on: in Shakespeare's *Romeo and Juliet*, Mercutio ridiculed Tybalt's insistence on keeping "time, distance, and proportion" while striking with the "immortal passado, the punto reverso, the hay."[7] But elsewhere scientific schools of combat became well-attended. In Spain, Don Luis Pacheco de Narvaez applied geometry (including proofs!) to show that a particular stance or maneuver is in fact the best possible under a given circumstance. German schools, such as that of the Marxbrüder (the Brothers of St. Marcus of Löwenberg) became famous.

LUDOLPH VAN CEULEN (January 28, 1540–December 31, 1610) was a student of the new arts of war. He studied fencing, probably in Germany, and began teaching in Delft in the 1580s; in 1600 he accepted a position at the Engineering School at the University of Leiden, teaching fortification methods and mathematics.

Ordinarily, a name like "Van Ceulen" would indicate an origin from a town or region called Ceulen—except there is no such place. Instead, van Ceulen (the man) was German, and his original name was probably Ackerman, "man of the field." In Roman times, such men were *colonus* (the word "colonist" has the same root), and thus Ackerman Latinized his name to Colonus, thence van Ceulen.

He was probably a self-taught mathematician. His reputation in mathematics stems from his work calculating the value of π. In the 1590s, van Ceulen received a copy of Archimedes's *Measurement of the Circle*, translated by Jan Cornets de Groot, the curator at the University of Leiden. Van Ceulen made it his life work to continue in the tradition of Archimedes. With the help of Adrianus Romanus, who gave van Ceulen equations for finding the perimeter of a regular inscribed n-gon, van Ceulen published in 1596 a value for π accurate to 20 decimal places, using a polygon with 15×2^{31} sides. By van Ceulen's death, he correctly determined the first 35 decimal places of π using a polygon of 2^{62} sides. The value was first published on his tombstone, though it subsequently appeared in a 1621 work by one of his students, the physicist Willebrord Snell.

Meanwhile, tacticians continued to adapt to the new realities of gunpowder warfare. The key limitation of the new weapons was their slow rate of fire. But if it takes thirty seconds for one man to load and fire a musket, then five men can load and fire in six seconds... provided they were working on different muskets.

On December 8, 1594, William Louis of Nassau gave his brother Maurice a solution to the musket's slow rate of fire. Imagine a line of men (five in William's original diagram) where the first man in line has a loaded musket. He fires and moves to the end of the line to reload. The second man in line now becomes the first, and because he has been reloading, he is ready to fire as soon as the first man gets out of his way. The result is a

---

[7]*Romeo and Juliet*, Act II, Scene iv.

nearly continuous hail of fire; since the line will slowly move backwards, away from the enemy, the formation is known as a countermarch.

The countermarch changed the geometry of warfare. A pike square consisting of a hundred men might present a front only ten men across, and a single officer could shout orders that could be heard over the din of battle. But against gunfire, a dense formation simply gave a bullet more potential targets. A countermarch formation five men deep presented a much wider front to the enemy, and fewer potential targets if the shot missed the first man. But its greater width made it more difficult to command from a single position. As a result, the ratio of officers to non-commissioned soldiers doubled during the first years of the 17th century as the front lines widened.

The volley fire of the countermarch was only effective when soldiers moved, loaded, and fired together, as a unit. Thus constant drill and practice became a major component of the soldier's life. Maurice's cousin John wrote up the first illustrated training manual showing a thirty-two step procedure for reloading a musket (later increased to forty-two). The new tactics, Spain's commitments elsewhere, and above all the inability to inflict a decisive defeat led Spain to offer the Twelve Years' Truce to the rebels in 1609. Maurice won a respite, but soon afterwards found himself at war again, this time by choice.

Maurice's new war came about as a result of the Reformation, which gave Bohemia (modern Czech Republic) a pivotal role in Central European politics. Of the seven electors who chose the Holy Roman Emperor, three (the Archbishops of Mainz, Trier, and Köln) invariably supported the Catholic candidates, while three (the Count Palatine of the Rhine, Margrave of Brandenburg, and Duke of Saxony) invariably supported the Protestant candidate. Thus the seventh elector, the King of Bohemia, held the deciding vote.

Like the Emperor himself, the king of Bohemia was decided by a group of electors. In 1617, one candidate for the Bohemian throne was Ferdinand of Styria, who was both a Habsburg, alienating the Bohemian nationalists, and an archconservative Catholic, guaranteeing the opposition of the Bohemian Protestants (the Utraquists). Unfortunately, the nationalists and Utraquists were more suspicious of each other than their common enemy, and their inability to cooperate led to Ferdinand's election on June 17, 1617.

Ferdinand ruled Bohemia through a network of Catholic governors. The closure of two Protestant chapels under construction led to protests, and matters exploded on May 23, 1618, when the Utraquists met in Prague and threw out their Catholic governors—literally, tossing the two deputy governors and their secretary out of a fifty-foot high window of Hradčany Castle. Miraculously, the three survived the defenestration of Prague, either because they were borne on the wings of angels (the Catholic viewpoint) or because they had landed in a dung heap (the Protestant viewpoint).[8] In any case, Bohemian nationalists and Protestants joined forces, declared Ferdinand deposed, and on August 26, 1619 elected Frederick V, Count Palatine of the Rhine, as the new King of Bohemia.

Meanwhile, the seven electors were meeting in Frankfurt to choose the next Holy Roman Emperor. Ferdinand was a candidate, and since (as far as anyone knew) he was still King of Bohemia, then his vote for himself, plus the votes of the three Archbishops meant that he had a majority. Bowing to the inevitable, the three Protestant electors cast their vote in Ferdinand's favor as well; ironically, this meant that Frederick's vote helped put Ferdi-

---

[8] It was actually the second defenstration of Prague; the first occurred during the Hussite Wars and, unlike the second, ended fatally for seven council members thrown out of a window by Hussite radicals.

nand into power, but if the real situation was known, Frederick's vote (doubled, because he would have been both the King of Bohemia and the Count Palatine of the Rhine) would have blocked Ferdinand's accession.

When Ferdinand learned of his deposition, he called upon the Catholic powers to support his restoration, and in turn, Frederick called for help from the Protestant states to keep Ferdinand deposed. Frederick was immediately underwhelmed with support. In all of Protestant Europe, only Bethlen Gabor of Transylvania and Maurice of Nassau provided troops or money.

7.2.3 Descartes

RENÉ DU PERRON DESCARTES (March 31, 1596–February 11, 1650) had been a sickly child, though by the time he reached adulthood, his health had improved considerably, and after graduating with a degree in law from the University of Poitiers, Descartes joined the Dutch army towards the beginning of 1618. Descartes met Isaac Beeckman, who interested him in a variety of scientific and mathematical topics. On March 26, 1619, Descartes wrote to Beeckman, announcing that he had gotten the "first glimpse of a new science": what would eventually become analytic geometry.

A month later, Descartes was confronted with the most fundamental reality of a soldier's life: boredom. The constant drill necessary for soldiers to perform an effective countermarch proved too dull for the contemplative Descartes, so he left the Dutch army to travel through Europe over the next ten years. His travels are not well documented, but by his own account he observed the coronation of Ferdinand in Frankfurt in August 1619. Shortly thereafter, he joined the army of Maximilian, the Duke of Bavaria, fighting on the side of the Catholics to depose Frederick. On November 8, 1620, Descartes got his first taste of war when the Bavarians, under the command of the Count von Tilly, dealt the Protestants a decisive defeat at the Battle of White Mountain, just outside Prague. Frederick fled for the Netherlands the next day, his short reign as king of Bohemia earning him the epithet "The Winter King."

Descartes's fighting days were over after the Battle of Prague. Afterwards he started building his philosophy, which began by rejecting everything that could not be derived logically. The starting point is the acceptance of your own existence, since if you did not exist, you could not even ask the question of whether or not you existed: "I think, therefore, I am." This is an early version of the anthropic principle, an important (if controversial) element of modern cosmology. At the end of 1628, Descartes moved to Holland, and would stay there for twenty years.

The Netherlands was in the midst of an economic boom. In 1610, the Dutch East India company brought the first cargo of tea from China to Europe. Almost immediately a tea craze developed, and the Dutch East Indiamen—enormous, three-masted vessels weighing up to 1600 tons—dominated the seas and built the Dutch commercial empire. Soon, tea would replace alcoholic drinks as the national beverage of many European countries.

Since tea will be damaged or destroyed by damp, it must be stored in the upper levels of a ship's hold, making the ship top-heavy and liable to capsize. To balance ("trim") the ship, ballast had to be added to the lower decks. Ideally, the ballast itself could be sold at the destination port. Since it would be stored in the lower decks, where water and damp would accumulate, the ballast could not be tea or silk (the two major Chinese exports).

Traders quickly found a solution. For centuries, the Chinese had drunk tea from vessels made out of a translucent substance that reminded Marco Polo of a type of cowrie shell called a *porcellana* (possibly derived from "little pig", which the shell resembles). Thus, along with tea, the Dutch began importing porcelain, much of it produced in the great factories of the Kiangsi province, particularly the town of Ching-te-chen.

The influx of wealth from the tea and porcelain trade fueled the first speculative bubble in modern history: the tulip mania. Tulips had been described as early as 1559 by one of Rheticus's classmates, the Swiss botanist Konrad Gesner. Demand for the colorful new flowers grew, and by 1634 single bulbs were selling for thousands of florins (in comparison, a thousand pounds of cheese—part of a consignment traded for a single bulb—was valued at 120 florins). The era records some amusing anecdotes, such as the sailor who thought the bulb an onion and ate it, only to find the bulb cost over 3000 florins. One burgomaster earned 60,000 guilders in four months buying and selling tulips, whereas his annual salary from his position was only 500 guilders. Tulips were sold on the stock markets of Amsterdam, Rotterdam, Haarlem, Leiden, and other Dutch towns.

The promise of quick riches encouraged many middle-class and poor families to speculate in the tulip market, mortgaging their possessions to buy bulbs. Speculators bought and sold bulbs sight unseen, and purchased futures (the promise to buy bulbs at a certain price at a future date). The tulip mania reached its height in November 1636, when suddenly some investors realized they were paying incredibly high prices for objects they had never seen and had no use for, and sold their stocks. Other investors followed suit, and within days, the market collapsed. Thousands who had spent all their wealth to buy a few bulbs were wiped out; those who bought futures found themselves committed to buying bulbs at wildly inflated prices and reneged on their contracts. The Dutch economy took years to recover.

In the aftermath of the financial collapse caused by the tulip craze, Descartes published his most important work, *Discourse on the Method* (1637), presenting his philosophy. The real importance of the *Discourse* was its three appendices. One explained the shape (but not the colors) of the rainbow; one discussed the laws of optics; and the third was the *Geometry*, which began with a claim that every problem in geometry could be reduced to the problem of finding the lengths of certain straight lines. In effect, Descartes identified the problem of geometric construction with the problem of solving an algebraic equation, which marked the beginning of coordinate geometry.

For example, consider the following construction problem, which gives rise to the golden ratio: Given a line segment AB, find a point C on the line segment so that the rectangle with sides equal to AC, AB is equal in area to the square with side BC. We can convert this geometric construction problem into an algebraic equation in the following manner: if we let $AB = a$ and $BC = x$; then the rectangle with sides $AC = a - x$ and $AB = a$ is equal in area to the square with side BC: hence $a(a - x) = x^2$. Solving this equation gives us $x = a\left(\frac{-1+\sqrt{5}}{2}\right)$ (where, since x is a length, we discard the negative solution). Thus we know the relationship between AC and BC, which can (because the basic operations of arithmetic and the extraction of square roots correspond to specific Euclidean construction techniques) be turned into a solution to the geometric construction problem.

One of the important problems examined by Descartes was finding the tangent circle to a given curve. Descartes's method is interesting, for it is a purely algebraic one. In modern terms, if we wish the curves $f(x, y) = 0$ and $g(x, y) = 0$ to be tangent at a point (x_0, y_0), then we want the system of equations $f(x, y) = 0$, $g(x, y) = 0$ to have a double root at (x_0, y_0). For example, suppose we wish to find the circle tangent to the parabola $y = x^2$ at $x = a$. For simplicity, assume the center of the tangent circle is at $(0, k)$ on the y-axis; then the system of equations that describes the intersection of the circle and the parabola is:

$$y = x^2,$$
$$x^2 + (y - k)^2 = r^2.$$

Since $y = x^2$, we can substitute this into the second equation to eliminate one variable. Note that if we substitute x^2 for y, we will end with a fourth degree equation in x, while if we substitute y for x^2, we will end with a second degree equation in y. Hence we choose to do the latter, giving us:

$$y + (y - k)^2 = r^2,$$
$$y^2 + (1 - 2k)y + (k^2 - r^2) = 0.$$

We seek to find a value of k that will make this equation have a double root at $x = a$ (and consequently $y = a^2$). If $y = a^2$ is a double root, then $(y - a^2)^2$ is a factor, and since the equation is quadratic, we must have

$$y^2 + (1 - 2k)y + (k^2 - r^2) = y^2 - 2a^2 y + a^4.$$

Comparing coefficients we have $1 - 2k = -2a^2$; hence $k = a^2 + \frac{1}{2}$.

Descartes's method of tangents works quite well with the conic sections, but becomes cumbersome for curves as simple as $y = x^3$. In this case we would need $x = a$, $y = a^3$ to be a double root of the system

$$y = x^3,$$
$$x^2 + (y - k)^2 = r^2.$$

Substituting $y = x^3$ into the second equation and expanding gives us:

$$x^2 + (x^3 - k)^2 = r^2,$$
$$x^6 - 2kx^3 + x^2 + (k^2 - r^2) = 0.$$

As before we wish $x - a$ to be a double root; hence the left-hand side must factor as

$$x^6 - 2kx^3 + x^2 + (k^2 - r^2) = (x^2 - 2ax + a^2)(Ax^4 + Bx^3 + Cx^2 + Dx + E) \quad (1)$$

where A, B, C, D, and E are constants to be determined. This gives rise to a system of seven equations in seven unknowns (A, B, C, D, E, k, and r). While this system can be solved in a straightforward fashion, it requires tedious algebraic manipulations. Thus Descartes's method of tangents, while theoretically rigorous and conceptually simple, is impractical.

Figure 7.2. Rembrandt's *The Night Watch*.

7.2.4   The Dutch Masters

In 1649, the Dutch artist Frans Hals painted the most commonly reproduced portrait of Descartes. The portrait was indicative of an important change in artistic patronage. Historically, artists had been patronized by the nobility or by the church. The nobility demanded paintings that glorified their great deeds (usually in warfare). Thus, one of the most famous paintings of the era is a group portrait showing Captain Frans Banning Cocq ordering his lieutenant, Willem van Ruytenburch, to march out his troops. This piece, by Rembrandt van Rijn, is usually called "The Night Watch", owing to a 19th century misunderstanding (it actually depicts a daytime scene). The church, on the other hand, demanded artwork dealing with religious themes: the sufferings of Christ, portraits of the saints, and depictions of Biblical stories.

The success of the merchants of the Netherlands played an important role in the history of Dutch art, for instead of a relatively small group of nobles and churchmen, Dutch artists painted for a much larger group of burghers and guildmasters, who wanted art to decorate their homes and guildhalls. Flourishing art markets grew up in many cities throughout the Netherlands. At Delft, it has been estimated that as many as two-thirds of the households owned paintings. Less than a fifth of the paintings in the city had religious motifs. Nearly half of the paintings sold in Delft were landscapes like those of Jan van Goyen, which

Figure 7.3. Vermeer's *Astronomer*.

showed cities like Dordrecht and Den Haag at mid-century, or still-life paintings like those
of Pieter Claesz, that showed Dutch tables set with lemons, oysters, turkeys, and cups made
from a tropical oddity: the nautilus. Others artists like Jan Vermeer painted everyday people
carrying out everyday activities: women writing letters, artists painting pictures, and—
perhaps in recognition that Dutch prosperity depended on their work—the astronomer and
geographer at their trades. These paintings contain a fantastic amount of detail: the book in
the *Astronomer* has been identified as Adriaen Metius's introductory text on astronomy and
geography; the opened page shows the beginning of Chapter III, "On the Investigation or
Observation of the Stars," and is sufficient to identify the book as the second edition, while
the globes in the paintings are identified as a paired set made by Jodocus Hondius.

Figure 7.4. Vermeer's *Geographer*.

A sizable number were the so-called genre paintings that show scenes of daily life in the Netherlands. Egbert van der Pol recorded one of the grimmer events of the era, the aftermath of the explosion of a powder magazine in Delft on October 12, 1654; as van der Pol buried a daughter two days later, he might have been directly affected by the catastrophe. Happier memories of the era are recorded by Jan Steen: tavern scenes, a village school, and gatherings of rhetoricians meeting to discuss their latest works of poetry. A cheerful chaos pervades Steen's works, leading to the Dutch expression a "Jan Steen household."

Both Rembrandt and Steen studied at the University of Leiden before dropping out to become artists (in 1620 and 1646, respectively). A second group of "Dutch Masters" emerged at the University in the 1650s, led by professor of engineering FRANS VAN

SCHOOTEN (1615–May 29, 1660). Van Schooten succeeded his father (who died in 1645).[9] Van Schooten's importance in the history of mathematics stems primarily from his association with the printing house of the Elzevir (or Elsevier) family.

When Farnese offered Protestants in the Spanish Netherlands a choice between reconversion or exile, many chose exile. The printer Louis Elzevir moved from Louvain to Leiden in 1581. Louis's son Bonaventure, and grandson Abraham became the printers for the University of Leiden, publishing Stevin's mathematical works in 1634: it was one of the first "collected works" ever printed for a mathematician. In the 1640s, the Elzevirs sent van Schooten to Paris with the task of collecting the manuscripts of Viète for publication. The result was Viète's *Opera Mathematica* (1646), which included all of Viète's works save *Rules of Mathematics*, which van Schooten omitted because he thought it was unfinished.

In 1637, van Schooten met Descartes when he visited Leiden to prepare *Discourse on the Method* for publication (through the rival printing press of Johan Maire). Van Schooten eagerly adopted Descartes's version of analytic geometry, and he began teaching it to his students, who included CHRISTIAN HUYGENS (April 1, 1629–July 8, 1695), son of the diplomat and poet Constantijn Huygens. There was one problem with the *Geometry*: it was written in French, whereas Latin had the advantage of being understood by every scholar in Europe. Thus, van Schooten began translating it into Latin. The first Latin edition appeared in 1649 (again, under the Johan Maire imprint).

At Leiden, van Schooten gathered a group of geometers who used Cartesian methods, and in 1659, he added the discoveries of the Leiden group to the second Latin edition of Descartes—this one published by the Elzevirs. The two-volume edition (the second volume was published in 1661) included not only the *Geometry*, but appendices written by van Schooten, Jan de Witt, Hudde, and Hendrik van Heuraet, all of Leiden, as well as FLORIMOND DE BEAUNE (October 7, 1601–August 18, 1652), author of the introduction to the 1649 edition and discoverer of a number of results relating to the limits of the solutions of equations.

JAN DE WITT (September 24, 1625–August 20, 1672) probably learned Descartes's analytic geometry from van Schooten, with whom he lived for a few years while at Leiden. In the years preceding 1650, de Witt wrote *Elements of Curves and Lines*, which would appear in van Schooten's 1659 edition of Descartes. De Witt developed conic sections analytically, and was the first to use the term directrix when describing a conic section.

Meanwhile, de Witt turned to politics. One problem facing the Dutch Republic centered around the office of the stadholder, the commander of the provincial armies. In theory, the stadholder was subordinate to the provincial government. In practice, the stadholders wielded enormous power, and during the war of Dutch independence, the stadholderships were increasingly monopolized by the House of Orange: William of Orange held four of the seven stadholderships, while after his death, William's son Maurice held five and William Louis of Nassau held the remaining two.

The concentration of the office of the stadholder in fewer and fewer persons made sense during the Eighty Years' War, but after Dutch independence was recognized by the Treaty of Münster in 1648, many, particularly in Holland, had no desire to maintain costly armies

---

[9]The elder Frans van Schooten had in turn succeeded van Ceulen at the university.

at the disposal of the stadholder. Others, particularly William II of Orange (Maurice's cousin), wanted the war to continue until *all* of the Netherlands was freed from Spanish control. The decision was based on purely military considerations: the southern provinces had a different religion and spoke a different language, but the very presence of Spanish forces in the southern provinces meant that Spain could attempt at any time to retake the Dutch Netherlands. Civil war loomed, and Descartes, after living in Holland for twenty years, accepted an invitation from Queen Christina to join her in Stockholm in 1649. It was a fatal mistake for Descartes, for Christina insisted on taking philosophy lessons at five in the morning, and the rigors of working in the early hours of a cold, Scandinavian winter proved too much for Descartes.

The fundamental question was whether the stadholder could force the states to maintain their armies, or did the states have the right to raise and disband armies as they saw fit? In 1649, Holland tried to disband its costly and unnecessary army. William intervened, at first politically, then militarily; he arrested many of the states rightists (including de Witt's father) and laid siege to Amsterdam. William's sudden death from smallpox in 1650 ended the struggle for supremacy without resolving it, though de Witt and his party scored a temporary victory.

Unfortunately, while the danger of Spanish invasion receded, a new threat appeared on the horizon: the English. Because the Dutch could offer extremely competitive rates for shipping, much trade from the English colonies went through Amsterdam and other Dutch ports. Since this meant that Dutch merchants, not English, profited from English colonies, this prompted the English parliament to pass the Navigation Act in 1651, which mandated that English products be carried on English ships to English ports. Trade war drifted into real war, and during the First Anglo-Dutch War (1652–1654), the underfunded Dutch navy did poorly. De Witt took a prominent role in arranging the Treaty of Westminster. Since Cromwell and de Witt found a common enemy in William III of Orange—the posthumous son of William II of Orange and the grandson of Charles I of England—they included a secret provision that Holland would not elect a stadholder; later, de Witt helped pass the Act of Seclusion (1654) prohibiting any member of the House of Orange from holding office.

In 1660 the situation in England changed with the restoration of the monarchy. The trade issue was still unresolved, so in 1664 an English fleet seized the Dutch settlement on the island of Manhattan, renaming the region in honor of the king's brother: James, the Duke of York. The Second Anglo-Dutch War began the next year (1665–1667). De Witt himself took to sea with a fleet, and through a combination of de Witt's diplomacy and organization of the Dutch navy, and the brilliant victories of Michiel Adriaanszoon de Ruyter, the Netherlands were able to secure the favorable Treaty of Breda in which they renounced claims to New York (which had never been very lucrative anyway) in return for several crucial changes to the Navigation Act. The treaty was signed hastily, for the Netherlands were about to be invaded by the French.

In 1661, Louis XIV of France declared that he alone would govern, without the assistance of the nobles or the *parlements*: "I am the state." Louis would be the first of the absolute monarchs, subject to no limitations at all, and came to be known as the Sun King, about whom all of France—and by the end of his reign, all of Europe—would revolve.

It would be some time before France became the dominant power in Europe, however. French finances were a disaster. Richelieu and Mazarin established deficit spending and borrowing (at interest rates as high as 33% annually) as a means of financing the government. When Louis assumed absolute power in 1661, the royal income for 1661, 1662, and part of 1663 had *already* been spent. The national debt stood at 11 million *livres* (in comparison, a skilled laborer might earn half a *livre* a day), and taxes, such as the hated *taille*, were so inefficiently collected that the difference between expected and actual revenue was more than 8 million *livres* a year. Within a few years, French finances were restored by Jean-Baptiste Colbert, the son of a merchant. One of Colbert's first actions was to reform the *taille*; he was so successful that he was able to simultaneously lower the rate while increasing the actual revenue. Though Colbert hoped to use as much of the money as possible to build French commerce and industry, he knew that most of the money was earmarked for military expansion.

In 1670, Louis negotiated the Treaty of Dover with Charles II of England, guaranteeing English support for a campaign in the Netherlands, and shortly thereafter Louis negotiated an alliance with Sweden, also against the Dutch. This paved the way for the Dutch War (usually reckoned as the Third Anglo-Dutch War). In May 1672, the French invaded the Netherlands with the help of the British navy. In three weeks, the French occupied three provinces. The people demanded that William III of Orange be allowed to take the position of stadholder, and their anger over the ineptitude of the Dutch armies found a target in de Witt and his brother Cornelius. When they were accused (unjustly) of conspiring against William, the two were killed by an angry mob on August 20, 1672.

More practically, the Dutch fell back to a time-honored tactic to slow the invaders: flooding the polders. Around Amsterdam, the flooding was directed by the burgomaster, another van Schooten mathematician turned politician: JOHANN HUDDE (May 1628–April 15, 1704). Hudde's major mathematical discovery provided a general method for solving the tangent and optimization problems.

In general, applying Descartes's method of tangents to an nth degree polynomial curve gave rise to a multilinear system of $2n$ equations in $2n$ unknowns. Solving the resulting system is tedious (though not particularly difficult). By investigating the properties of polynomial equations with double roots, Hudde made a remarkable discovery, now called Hudde's First Rule, which first appeared in the 1659 Latin edition of Descartes. In modern terms:

Proposition 7.1. *Let* $f(x) = a_0 + a_1 x + a_2 x^2 + \ldots + a_n x^n$ *be an nth degree polynomial with a double root* $x = r$, *and* $b_0, b_1, b_2, b_3, \ldots$ *be any arithmetic sequence. Then* $g(x) = b_0 a_0 + b_1 a_1 x + b_2 a_2 x^2 + \ldots + b_n a_n x^n$ *has a root of* $x = r$.

For example, $x^3 - 5x^2 + 8x - 4$ has a double root $x = 2$, and $x = 2$ is a root of $1 \cdot x^3 - 2 \cdot 5x^2 + 3 \cdot 8x - 4 \cdot 4$. We will call this second polynomial the Hudde polynomial (and note that it is not unique).

Hudde's rule greatly simplifies Descartes's method of tangents. Finding the tangent to $y = x^3$ at $x = a$ required the equation

$$x^6 - 2kx^3 + x^2 + (k^2 - r^2) = 0$$

to have a double root at $x = a$ (see Equation 1). By Hudde's method if this equation has a double root at $x = a$, then any corresponding Hudde polynomial will have a root of

$x = a$. For simplicity, let us use the arithmetic sequence $\{0, 1, 2, 3, \ldots\}$, which generates the Hudde polynomial $6 \cdot x^6 - 3 \cdot 2kx^3 + 2 \cdot x^2 + 0 \cdot (k^2 - r^2) = 6x^6 - 6kx^3 + 2x^2$. If this has a root at $x = a$, then it is necessary that $6a^6 - 6ka^3 + 2a^2 = 0$ or $k = a^3 + \frac{1}{3a}$.

Hudde was probably well aware this discovery could be used to find tangents to curves, though the first person to do so was another member of the van Schooten group, HENDRIK VAN HEURAET (1633–1660); van Heuraet's only mathematical publication, which led to the first rectification of a curve, included the use of Hudde's procedure to find tangents, also appeared in the 1659 edition of Descartes.

Hudde himself applied his discovery to locating the maxima and minima of functions. If $x = a$ corresponds to a maximum or minimum value of $f(x)$, then $f(x) - f(a)$ has a double root at $x = a$. In modern terms, Hudde's Second Rule is:

Proposition 7.2. *Let $f(x) = a_0 + a_1x + a_2x^2 + \ldots + a_nx^n$ be an nth degree polynomial. If $x = m$ corresponds to a maximum or minimum value of $f(x)$, then $x = m$ is a root of $g(x) = a_1x + 2 \cdot a_2x^2 + \ldots + n \cdot a_nx^n$.*

Hudde even worked out a variation whereby the extreme values could be found if $f(x)$ was a rational function or even the case where f was a function of two variables. Superficially this is equivalent to the modern method of setting the derivative $f'(x) = 0$; however, Hudde's rules were purely algebraic results.

7.2.5 Leibniz

The expansion of French power concerned those who feared they might be Louis's next targets: the Rhineland states. An unsigned satire, *Most Christian Mars* (referring to the Roman god of war), explained that the "most Christian majesty" (Louis XIV) must necessarily subdue the Christian Germans and the Flemish before he could even think to attack the infidel Turks. In 1672, the Archbishop of Mainz, Johann Philip von Schönborn, sent a delegation to Paris to encourage Louis to attack Egypt, rather than the Rhineland, and included the author of the pamphlet among the delegates: GOTTFRIED WILHELM VON LEIBNIZ (July 1, 1646–November 14, 1716).

Leibniz came to the attention of the Archbishop through his chief minister, the Baron Johann Christian von Boineburg (or Boyneburg). Leibniz himself gave the following account of their meeting. He had become interested in alchemy, and attempted to join an alchemical society in 1660s. However, the society was unwilling to accept an unknown twenty-year-old like Leibniz. Thus Leibniz obtained some alchemical texts, selected the passages that made the least sense, wrote a letter based on those passages that was completely incomprehensible to himself, and asked for admission on the basis of the knowledge he demonstrated in the letter. He was not only offered a membership, but the position of Secretary and a pension. One of the alchemists introduced Leibniz to von Boineburg, who persuaded Leibniz to turn his enormous talents towards public service. While waiting for an appointment with Louis, Leibniz met Christian Huygens and other mathematicians and scientists living in Paris.

Huygens was a founding member of the French Royal Society (one of Colbert's pet projects) and had lived in Paris since 1666. A few years earlier, he had learned of the Fermat-Pascal correspondence on probability, and wrote *On Reckoning in Games of Chance*

(1657), the first probability textbook. *Games of Chance* concluded with five problems, the last of which is:

Problem 7.2. *Let there be two gamesters A and B with 12 coins apiece, and let three dice be rolled repeatedly. If the dice total 11, then A gives B one coin, while if the dice total 14, B gives A one coin. The game is played until one gamester has all the coins. What are the respective chances of A and B to win all the coins?*

Without explanation Huygens gave the answer as $244, 140, 625$ to $282, 429, 536, 481$. He probably hoped that these five problems would inspire other mathematicians to investigate probability.

Huygens suggested that if Leibniz was truly interested in mathematics, he would do well to read the works of Pascal (which Leibniz did); Huygens then gave Leibniz a mathematical problem. One wonders what might have happened if Huygens suggested a probability problem; instead, he gave Leibniz the problem of finding the sum of the reciprocals of the triangular numbers:

$$\frac{1}{1} + \frac{1}{3} + \frac{1}{6} + \frac{1}{10} + \frac{1}{15} + \cdots.$$

To sum the series, Leibniz noted:

$$\frac{1}{1} + \frac{1}{3} + \frac{1}{6} + \frac{1}{10} + \cdots = \left(\frac{2}{1} - \frac{2}{2}\right) + \left(\frac{2}{2} - \frac{2}{3}\right) + \left(\frac{2}{3} - \frac{2}{4}\right) + \cdots,$$
$$= 2.$$

Based on this and other early successes, Leibniz believed that *every* series could be summed. Within a few years, he would develop this into his version of calculus, and his notation would reflect this early preoccupation with sums ($\int$) and differences (d).

By late 1672, Charles II of England began to question the wisdom of allying himself with Louis, especially since the Dutch were proving a tougher adversary than anticipated, and hinted that England was interested in a separate peace. Thus, Leibniz accompanied a German delegation (which included Boineburg's son) to London in January 1673, where they negotiated the Treaty of Westminster (1674). While in London, Leibniz made his first visit to the Royal Society to present his plans for an adding machine—one of the earliest mechanical calculators. Leibniz also made the acquaintances of Pell, Oldenburg, Collins, and others. He tried to induce Wallis to teach him cryptanalysis, but Wallis demurred, saying that he could not in good conscience teach the art to a foreigner, perhaps the earliest example of a fundamentally mathematical discovery being designated a national secret.

While in London he obtained a copy of Barrow's recently published *Geometrical Lectures*, and developed his own methods for solving the tangent and quadrature problems. There is very little to support the notion that Leibniz plagiarized Newton; for the most part, Leibniz seemed unaware of Newton's work, and what similarities there are between the calculus of Newton and the calculus of Leibniz can be attributed to the fact that both solved similar problems.

Today there is an ironclad basis for determining priority of discovery: date of publication. In 1704, Newton published "Treatise on the Quadrature of Curves" as one of two appendices to his work *Optics*: it was the first widespread appearance of his method of fluxions and fluents. Even earlier, Newton had revealed elements of his calculus in his *Mathematical Principles of Natural Philosophy* (1687), though not in any systematic fashion:

the mathematics was presented as needed, since Newton's primary focus was on the development of physics as a mathematical science. But Leibniz published "A New Method for Maxima and Minima, and also for Tangents, which is not Obstructed by Irrational Quantities" in 1684. Leibniz's paper introduced the dx notation of modern calculus, and gave explicit rules for differentiation, including the quotient and product rules. Thus, by modern standards, Leibniz deserves credit for being the first inventor of calculus.

Unfortunately, the issue is rather more complex. Although Newton did not publish his work in a scholarly journal, his position as Lucasian lecturer required him to deposit copies of his lectures in Cambridge University's libraries, where they would have been accessible to anyone who cared to look at them. Newton's written work on calculus dated back to at least 1669. The priority dispute would become a major point of contention between British and continental mathematicians, and would lead to a wasteful split between Britain and the rest of Europe.

Leibniz's publication of his method of differential calculus was prompted by EHRENFRIED WALTER VON TSCHIRNHAUS (April 10, 1651–October 11, 1708). Leibniz communicated some of his methods to Tschirnhaus, and in 1683 Tschirnhaus revealed some results he had obtained on the rectification of curves. Tschirnhaus did not reveal his methods, but presumably used Leibniz's calculus. Concerned that Tschirnhaus might publish some of Leibniz's results as his own, Leibniz submitted "A New Method" to establish his priority.

Tschirnhaus, like most of the other continental mathematicians of the era, was not a professional mathematician: he was the Count of Krelingswalde and Stolzenberg, member of one of the oldest and most prestigious noble families in Europe. When the Dutch War broke out in 1672, Tschirnhaus—then in Leiden—volunteered for the Dutch army, and served for 18 months. In 1674, Tschirnhaus began a European tour, going to London, where he met Wallis and Collins, and Paris, where he met Leibniz and Huygens. Tschirnhaus would later go on to discover what are now called Tschirnhaus transformations that can be used to eliminate the $n - 1$ and $n - 2$ degree terms from an nth degree polynomial; most usefully, this reduces a cubic into an equation of the form $y^3 + p = 0$ and allows a solution to be found easily. However, Tschirnhaus's most celebrated discovery was the secret of manufacturing of porcelain.

Between 1604 (when a Portuguese ship, the *Catherina*, was captured by the Dutch and its cargo of 100,000 pieces of porcelain impounded) and 1657, Europeans imported over three million pieces of Chinese porcelain. But in 1644, the Manchus, a group of tribes along China's northern borders, conquered Beijing, and within a few years, had most of China under their control. Kiangsi province was invaded, and the kilns of Ching-te-chen, which produced most of the porcelain exported to Europe, were destroyed. They would eventually be rebuilt, but European demand for porcelain would not wait.

European artisans tried without success to discover the jealously guarded secret of Chinese porcelain, or find an acceptable substitute. In 1575, Florentine artisans began producing soft-paste porcelain. True porcelain is non-porous, hard, and translucent; soft-paste porcelain is porous, soft, and opaque—though it is the right color. With the disruption in the China trade, the market for imitation porcelain expanded. By 1665, Delft became a major production center of a type of blue-and-white glazed pottery known as Delftware.

But the secret of making true porcelain eluded European chemists. The problem is that Chinese porcelain incorporates a type of clay called kaolin, which is abundant in China (it is also called "China clay") and uncommon elsewhere. It was not until the dawn of the 18th century that a deposit of kaolin was discovered near the city of Meissen.

Meissen was part of the territory ruled by the Duke of Saxony, Friedrich Augustus, who is said to have loved three things in life: power, porcelain, and gold. For power, he schemed to become the King of Poland in 1697 (reigning as Augustus II of Poland). This required him to become a Catholic, which he did, despite Saxony being the birthplace and heart of Lutheran Reformation; as a result, his wife left him. For porcelain, the Duke was an indefatigable collector, and owned more than 50,000 pieces by his death (the Duke's collection is now in the Saxon State Museum). For gold, the Duke employed an alchemist named Johann Friedrich Böttger, who claimed to be able to turn base metals into gold.

The Duke was no fool, and realized that most alchemists were frauds. Böttger was promised great rewards if he succeeded—and dire consequences if he failed. Since it is impossible to turn lead into gold using the means available in the 17th century, Böttger was in a precarious position, and searched desperately for a way out. Between 1702 and 1706, with the help of Tschirnhaus, Böttger discovered something even better than gold: he discovered how to make true porcelain. On June 6, 1710, the first European porcelain factory in Europe opened in Meissen.

7.2.6 The Great Northern War

Cannons dominated warfare beginning in the 17th century, and two things were required to make good cannon: iron (to make the cannon), and trees (to make the charcoal to smelt the iron). Sweden, with generous supplies of both, became the leading armaments producer in the 17th century, and a major military power. This was of grave concern to her neighbors: Denmark, Poland, and Russia. Thus in 1699, Augustus arranged a secret alliance between Denmark, Poland, and Russia, all of whom had longstanding grievances against Sweden. The next year, Augustus invaded Livonia (once part of Poland, but a Swedish possession since 1660) while the Danes invaded Schleswig, beginning the Great Northern War.

Unfortunately, the three allies faced Swedish King Charles XII. In the first year of the war, Charles forced the Danes out of the alliance, scattered the Polish troops, and drove the Russians from the Baltic, a series of successes that earned him the nickname of "the Swedish meteor." Charles invaded Saxony in 1706 and forced the porcelain collecting Duke to abdicate the Polish throne. This left Russia alone to stand against the Swedes.

Charles captured the city of Narva (now in Estonia) in 1701, but in 1704, the Russians retook the city: it was the first significant Swedish defeat of the war, and marked the beginning of an important Russian presence in the Baltic. The Russians dealt Charles a decisive defeat at the Battle of Poltava on July 8, 1709, and though the war would drag on for another twelve years, by the end, Sweden would be broken, Poland would be shattered, and Russia would emerge as a major European power.

7.3 Great Britain

The War of the Roses ended on August 22, 1485 when Lancastrian forces under Henry Tudor defeated the Yorkists at the Battle of Bosworth Field. Richard was killed, and Henry VII became the first king of the Tudor dynasty. He quickly consolidated his power and, to bring together the two sides of York and Lancaster (as well as eliminate one possible source of a rival to the throne) he married Edward IV's daughter Elizabeth.[10] The symbol of the Tudor dynasty thus became the red rose of Lancaster superimposed on the white rose of York.

Their first son Arthur was married to Catherine of Aragon in 1501 to cement an alliance with the growing power of Spain. Unfortunately Arthur died in 1502. Henry persuaded Catherine to marry Arthur's eleven-year-old brother Henry. The actual marriage did not occur until 1509, the year Henry became king, and required papal dispensation, since canon law deferred to the authority of Leviticus XX:21: "If a man shall take his brother's wife, it is an unclean thing ... they shall be childless."

Henry VIII cut a magnificent figure: young, handsome, athletic and well-educated. He was a skilled lutenist and wrote a number of songs (though "Greensleeves," often attributed to him, did not appear until more than thirty years after his death). He also received theological training, so when Luther began the Protestant Reformation in Germany, Henry responded with his *Declaration of the Seven Sacraments Against Martin Luther*. This earned him the title of Defender of the Faith in 1521, a title retained by all subsequent English monarchs.

Catherine gave birth to a daughter (Mary) in 1516, but the memories of the reign of Queen Matilda made Henry concerned over the viability of a reigning queen. Henry had long doubted the validity of his marriage to Catherine (he had filed a protest even in his father's time), and the fact that Catherine did not bear him a son added weight to his belief. Thus in 1527 he appealed to Pope Clement VII for an annulment. Legal opinion was split, and the Pope might have declared in favor of Henry. But he was reluctant to do so, since Catherine's nephew Charles V had just captured the city of Rome, and Clement was even then in the protective custody of the Emperor and his troops. Clement denied Henry's request, causing Henry to break with the Catholic Church and begin the English Reformation. First, Henry married Anne Boleyn in January 1533. Then he had Thomas Cranmer, the newly appointed Archbishop of Canterbury, annul Henry's marriage to Catherine on May 23, 1533. Finally in 1534, Parliament passed the Act of Supremacy that made the king the head of the Church of England, thereby legitimizing *ex post facto* all of Henry's actions.

If Henry began the English Reformation in the hopes of producing a male heir, he was disappointed. On September 7, 1533, Anne gave birth to another daughter (Elizabeth). Anne antagonized much of the court by her arrogance, and Henry by her inability to produce a male heir. She was brought up on charges of adultery (almost certainly false), tried, convicted, and executed on May 19, 1536. Eleven days later, Henry married Jane Seymour. On October 12, 1537, she gave Henry what no one else could: a son, Edward. Unfortunately the childbirth was so traumatic that she died 12 days later.

An unmarried king of England could enter into a political marriage, and in 1539, Thomas Cromwell, one of the architects of the English Reformation and Henry's chief advisor, suggested he marry Anne, the sister of Wilhelm, Duke of Cleves, a Protestant stronghold in Germany. Henry sent his court painter, Hans Holbein to paint her portrait. Holbein, used to painting the portraits of the rich and powerful, produced a flattering por-

---

[10]The importance of the Battle of Bosworth Field should not be misunderstood: Henry did not become king because he won the battle. He became king because he had a claim based on his ancestry; the battle simply established that his claim was better than those of the Yorkists.

trait of the woman Henry would call "the fat Flanders mare" when he met her on January 1, 1540. He had the marriage annulled a few months later, giving Anne a sizable pension on the condition she remain in England. For his part, Holbein would receive no more important commissions from the king, and Cromwell fell from power and was eventually executed.

7.3.1 Robert Recorde

Henry's last two marriages were to Catherine Howard, who was beheaded for adultery (Howard's guilt was incontestable), and Catherine Parr, who survived him. In 1543, the year Henry married Catherine Parr, ROBERT RECORDE (1510–1558) published *The Grounde of the Arts*, one of the most influential commercial arithmetics published in England. In dialog form, Recorde explained basic arithmetic operations and the rule of three. Recorde also wrote one of the first English algebras, *The Whetstone of Witte* (1557). The title is a pun: the Italians referred to the unknown as *cosa* ("thing" in Italian), and in Latin, *cos* is a whetstone. Thus algebra was the subject that sharpened one's mind. Recorde popularized Widmann's + and − symbols, as well as Stifel's $\sqrt{}$, and introduced the modern equals symbol: two parallel lines, "because no two things can be more equal."

On January 28, 1547, Henry VIII died, and the throne passed on to his son, Edward VI. Although Henry had grown rich through the seizure of monastic lands, most of these were sold almost immediately to pay off crown debts. Henry's failure to adequately manage England's economy left the monarchy in dire straits; Henry, notorious for his six wives, should be even more notorious for being the first English king to institute debasement of the currency.

Edward did what he could to restore the English currency. He issued a new coin: the silver crown, worth 5 shillings. It was the first English coin to use Arabic numerals to indicate the date. There was only one problem with issuing new silver coins: England's silver mines had been worked out, and new sources of silver had to be found. In 1551, Edward appointed Recorde to be the general surveyor of the mines and monies in Ireland, which included responsibility for the running of the Dublin mint and overseeing the newly opened mines in Wexford. Unfortunately Recorde could not make the Wexford mines profitable, and the operation was shut down in 1553.

The year was one of great troubles for England. Edward was dying, and his sister Mary was the heir apparent. However, Mary was staunchly Catholic, and Elizabeth, being the daughter of the executed Anne Boleyn, hardly seemed a suitable alternative. The powerful Dudley family put forward a reluctant fifteen-year-old as a candidate: Lady Jane Grey, the great-granddaughter of Henry VII. Before he died, Edward named her his successor, and on July 10, 1553, she was proclaimed queen. But the Dudleys neglected to build support for Jane, and Mary's legitimate claim outweighed her Catholicism. Jane abdicated the throne nine days later; she and her father were later executed.

As many feared, Mary attempted to restore England to Catholicism. She married his Most Catholic Majesty, Philip II of Spain, and with the help of Bishop Edmund Bonner of London, began a great persecution of Protestants. Bonner helped burn an estimated 300 heretics at the stake, earning him the epithet of "Bloody Bonner" (later transferred to Mary herself). What made Mary's executions particularly galling was their targets were rarely prominent Protestants, but instead tended to be drawn from the lower levels of society, particularly rural laborers. Mary's death in 1558 was greeted by great relief, and her half-sister Elizabeth became queen.

7.3.2   Elizabeth

Elizabeth reaffirmed the statues of Henry VIII in the Act of Supremacy (1559). But she had no desire to renew doctrinal disputes, and the Thirty-Nine Articles, which established doctrine for the Church of England, was deliberately ambiguous to be as inclusive as possible.

During Elizabeth's reign, English came into its own as a literary language. Inspired by the glories of the Elizabethan court, Edmund Spenser wrote the *Faerie Queen* (1590), and Christopher Marlowe told the story of a German alchemist who died around 1540: Faust. Marlowe's life is shrouded in mystery. In 1587, Cambridge University, noting Marlowe's many absences over the past three years, was about to deny him a Master's degree when a letter arrived from the Privy Council declaring that his absences were connected with "service to his country." This letter and the general mystery surrounding Marlowe's life has led to speculation that Marlowe worked for England's intelligence service.

Elizabeth's reign also saw British expansion into the New World. The early explorers and exploiters of the Americas were Spanish, Portuguese, and Italian; English interest was minimal until John Hawkins took a cargo of slaves from West Africa to the Spanish West Indies in 1563; this proved so profitable that on his second voyage (1564–1565), he had many backers, including the queen herself. However the slave trade was so lucrative that the Spanish were reluctant to share its profits with anyone else, and on Hawkins's third trip (1567–1569), the Spanish attacked his fleet; only two of his six ships escaped, one under the command of Hawkins himself, and the other commanded by a relative: Francis Drake. Drake's experience left him determined to wreak his vengeance on the Spanish, and in 1572, Drake began raiding Spanish commerce with the consent of the Queen.

In 1577, Drake began the voyage for which he was famous: the circumnavigation of the globe (the second, after Magellan's expedition in 1519–1522). The circumnavigation was more or less an accident: Drake intended to raid Spanish colonies on the Pacific Coast of North America and return to England via the Northwest Passage, a hypothetical waterway along the northern boundary of North America. Drake probably got as far north as the Vancouver area before the fog and cold caused him to give up; he sailed south, landed in the San Francisco area for provisions, then sailed across the Pacific. Drake returned to England on September, 26, 1580: of the five ships in Drake's fleet, two of the smaller ships were abandoned, one returned to England, one sank, and only Drake's flagship, the Golden Hind (originally called the *Pelican*) returned to England with 56 of its original crew of 100.

7.3.3   Gerardus Mercator

The exploits of Drake and other English adventurers caught the popular imagination, and *The Principal Navigations, Voyages and Discoveries of the English Nation* (1589 and greatly expanded second edition in 1598–1600) by Richard Hakluyt is one of our main sources of information about these early voyages. He gave public lectures on the exploits of the explorers, and became Oxford University's first lecturer in geography. Hakluyt used his renown to lobby for the exploitation of North America and the search for the northwest passage.

Hakluyt's keen interest in promoting navigation led him to meet merchants, sailors, mapmakers, and in general anyone interested in geography. One of his correspondents was GERARDUS MERCATOR (March 5, 1512–December 2, 1594), a Flemish engraver in the employ of Duke Wilhelm of Cleves (Anne of Cleves's brother). Drake's depredations on Spanish commerce notwithstanding, by far the greatest danger to the Spanish treasure ships carrying silver from the New World was bad navigation, and it was not unusual for an entire fleet to run aground because the navigator did not know where he was. The problem was creating a flat map that accurately represents the curved surface of the Earth.

The problem was compounded because mapmakers relied on sailors to provide them with navigational information. By the 1st century B.C., the Chinese knew that a needle made from a certain type of iron ore, if left free to rotate, would inevitably end by pointing south. By the 13th century, this discovery reached Europe, though the direction of the needle "changed" so that the compass pointed north.[11] The compass allowed for a new form of navigation: navigation by bearing. For example, a ship might sail in such a way that its course maintained a constant angle with magnetic north. Sailors assumed that this meant the ship was traveling in a straight line (along a great circle of the sphere), but while in the employ of the Duke of Cleves in the 1560s, Mercator realized that this was not in fact the case: the ship would in fact travel along a spiral centered at the magnetic pole. In 1880 this type of curve was named a loxodrome.

To reflect this fact of navigation by bearing, Mercator invented the Mercator projection, also called a cylindrical projection, which maps the surface of the sphere onto the surface of a tangent cylinder; the standard Mercator projection has the cylinder tangent to the surface of the Earth at the equator. The key virtue of the Mercator projection is that any straight line corresponds to a course of constant bearing, which made navigation easy. The main disadvantage is that the northern and southern extremities are greatly distorted in comparison to lands near the equator; thus Greenland appears larger than South America, though in reality South America is more than eight times as big.

7.3.4   Thomas Harriot

One of the main promoters of New World colonization was Walter Raleigh, who came to the Queen's attention in 1580 during a revolt in Ireland. Raleigh quickly became one of the queen's favorites, and worked hard to establish an English presence in the New World. In 1583, Raleigh, probably at Hakluyt's suggestion, hired THOMAS HARRIOT (1560–July 2, 1621) to teach sailors the principles of mathematical navigation. Among other things, Harriot proved that straight lines on Mercator projections corresponded precisely to paths of constant navigational bearings on a sphere. Unfortunately Harriot's notes, which were to be assembled into a textbook called the *Arcticon*, were lost, so the contents of his navigation course are a mystery.

Harriot is best known for his contributions to algebra, though since he published none of his mathematical works during his lifetime, his influence is hard to estimate. He was among the first to recognize the negative and imaginary roots of an equation, and that a

---

[11] Strictly speaking, the compass points to magnetic north, which is currently located in northern Canada.

cubic equation like $(x - a)(x - b)(x - c) = 0$ had no roots besides $x = a, b,$ or c. Ironically, his most widespread contribution to mathematics was not his own: the symbols $>$ and $<$, which were introduced by the editor of Harriot's *Analytic Art* (1631).

Harriot's contribution to New World colonization is easier to assess. In 1584, Raleigh sent an expedition commanded by the captains Philip Amadas and Arthur Barlowe to locate a site for a New World colony. They found one off the coast of what is now North Carolina: Roanoke Island. The next year 108 settlers, including Harriot (who was likely part of the initial survey), landed and attempted to found a colony under the command of Sir Richard Grenville. Shortly afterwards Grenville returned to England for more supplies, but when he returned in 1586, the colonists had vanished. The disappearance of the first Roanoke colony was no great mystery: two weeks before Grenville returned, the entire colony demanded they be returned to England, and Drake obliged them.

When he returned to England, Harriot published *A Brief and True Report of the New Found Land of Virginia* (1588). He went into great detail about the valuable commodities that might be obtained from the New World, including sassafras, which later English colonists would use as a substitute for hops in flavoring beer, and corn of various colors. Harriot also admits learning to smoke tobacco, whose virtues were so numerous that describing them "would require a volume by itself." Meanwhile Raleigh sent a second group of 150 colonists to Roanoke in 1587. The second colony passed a milestone with the birth of the first English child born in the New World, Virginia Dare (August 18, 1587), but it would take three years before an English ship returned to Roanoke. By then, the second colony had vanished, leaving an enduring mystery.

The delay was caused by the looming threat of a Spanish invasion. In 1588, in an attempt to deal once and for all with the nation that attacked his New World colonies and aided the Dutch rebels, Philip II of Spain sent an "invincible armada" to destroy the English fleet and invade England as a prelude to returning the country to Catholicism. The 200 ships of the English fleet (commanded by Drake, Raleigh, Grenville, and others) outnumbered the 130 ships of the Spanish fleet, many of which were support vessels (carrying troops or supplies, not weapons). Ironically, Philip helped build the English fleet as well: while Prince Consort of England, he convinced the Privy Council of the importance of a modern navy. While victory over the Spanish Armada was not a foregone conclusion, its defeat should not be considered too remarkable.[12]

7.3.5 Scotland

In 1567, Scottish Protestants, led by John Knox, forced the Catholic Queen Mary to abdicate in favor of her infant son, James VI.[13] However, many Catholic nobles remained in Scotland, and in 1593 a group of them sent eight signed but blank letters to Spain. The Spanish Blanks were intercepted and taken as evidence that the Catholic nobles were con-

---

[12]Regiomontanus, incidentally, predicted the year 1588 would see disaster on land and sea. As with the Oracle of Delphi and Nostradamus, vague predictions are invariably correct.

[13]To distinguish her from her contemporary, Mary I of England, she is usually known as Mary, Queen of Scots. After her abdication, she took refuge in the court of Elizabeth; however, the fact that she was a potential heir to the throne (her father's mother was Margaret Tudor, Henry VII's daughter) caused her to be involved in at least three attempts to assassinate Elizabeth and place Mary on the throne of England, eventually leading to Mary's execution on February 8, 1587.

spiring with Spain to overthrow James VI (the text was to be filled out by Jesuit priests and serve as an invitation to Philip II to invade Scotland). A group of Protestant nobles met with King James VI to demand measures be taken against the conspirators.

One of the petitioners was JOHN NAPIER (1550–April 4, 1617), whose father-in-law James Chisholm was one of the signers of the Spanish Blanks. But James VI did not want to antagonize the powerful Catholic nobles, and refused to take action. To persuade the king to purge his court of Papists, atheists, and anyone who did not support Protestantism, Napier wrote *A Plaine Discovery of the Whole Revelation of St. John* (1593), which contained thirty-six propositions that proved Catholicism would be all but extinct in Europe by 1639, and that the Second Coming of Christ would occur around 1698. Napier should have taken a cue from Regiomontanus, and left a vague prediction, but at least he followed Stifel's example and turned to mathematics proper. His invention of logarithms would, by greatly simplifying calculations, be instrumental in launching the scientific revolution.

A method of simplifying computations emerged around 1574 when the Danish astronomer TYCHO BRAHE (December 14, 1546–October 24, 1601) and his assistant, PAUL WITTICH (1555–January 9, 1587) discovered the formula:

$$2 \sin A \sin B = \cos(A - B) - \cos(A + B)$$

The importance of this is that it relates the product of two numbers (the sines of A and B) to the sum or difference of two numbers; thus it converts a difficult task (multiplication) into a sequence of simpler ones (addition, subtraction, and looking up numbers in a table).

For example, suppose we wish to find the product of 0.258819 and 0.342020. From a table of trigonometric values, we find that $0.258819 \approx \sin 20°$ and $0.34202 \approx \sin 15°$, so we have:

$$2 \sin 20° \sin 15° = \cos(20° - 15°) - \cos(20° + 15°),$$
$$2 \cdot 0.258819 \cdot 0.342020 = \cos 5° - \cos 35°$$
$$\approx 0.996195 - 0.819152$$
$$= 0.177043$$

Thus $0.258819 \cdot 0.342020 \approx 0.088522$.

This method of converting a product of trigonometric functions into a sum of trigonometric functions is referred to as prosthaphaeresis ("addition and subtraction" in Greek). Its usefulness lies in the fact that tables of trigonometric function values were widely available, so that finding (in this case) the inverse sine of a number, and the cosine of two angles, was easier than performing the multiplication of two six-digit numbers.[14]

The actual discoverer of the method of prosthaphaeresis is open to debate. Tycho was an inferior mathematician, so the discovery was probably made by Wittich. But Wittich did not publish his discovery until later, so the first published version of the prosthaphaeretic formulas appeared in *Foundations of Astronomy* (1588) by Nicolai Reymers Ursus. Since Ursus had visited Tycho in 1584, Brahe accused Ursus of plagiarism. Ursus in turn claimed that Wittich may have discovered the method, but the first *proof* of the validity of the

---

[14]The actual historical procedure would have been slightly different than in the example, since the tables of sines available were actually tables of chord lengths.

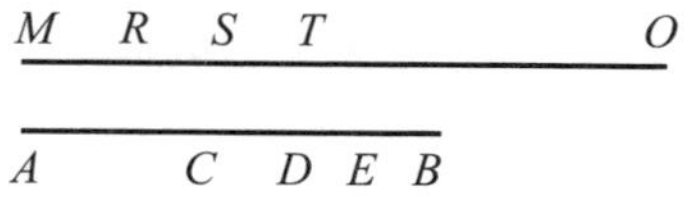

$$M \quad R \quad S \quad T \qquad\qquad\qquad O$$
$$A \qquad C \quad D \ E \ B$$

Figure 7.5. Napier's Logarithms: The logarithm of CB is MR.

method was due to JOOST BÜRGI (February 28, 1552–January 31, 1632). However, Bürgi was a student of Wittich at the University of Kassell, so he may have learned the method from Wittich in any case. To add to the confusion, the prosthaphaeretic rule was implicit in work done by JOHANNES WERNER (February 14, 1468–May 1522) between 1505 and 1513, and some scientists, notably Rheticus, had learned of it, so in 1611, Jacob Christman attributed the actual discovery to Werner; because of this, the method of prosthaphaeresis is sometimes called Werner's method. Finally, prosthaphaeretic methods were used by the eleventh century Islamic astronomer ibn-Yunis, though his discovery seems to have been unknown in the West until much later.

Regardless of who actually discovered the prosthaphaeretic formulas, they provided astronomers with a powerful tool: by converting a product into a sum, they would ease the computational difficulties. However, there was one flaw in the use of the prosthaphaeretic formulas: they required accurate tables of sines and cosines, and finding the sines or cosines of an arbitrary angle was itself a difficult computational task. Napier's contribution was recognizing that numbers in a geometric sequence could also help turn products into sums.

Nowadays this is easy to see, and we generally introduce logarithms as powers: since $a^3 a^2 = a^{3+2}$, it's clear that the addition of two exponents corresponds to a multiplication of two powers. But in the notation of Napier's time, this was far from obvious: for example, Viète would have written this product as A cub. A quad. aeq. A plani-cubo. Instead, Napier's *Description of the Wonderful Rules of Logarithms* (1614) defines the logarithm of a number in the following way: imagine a line AB with a length of $10,000,000$ units, and another line of infinite length MO. Suppose a particle moves along MO at a constant rate, passing through (in equal times) the points R, S, T, etc., while at the same time a particle moves along AB so that when the particle on MO is at R, S, T, etc., the particle on AB is at C, D, E, etc., where the velocity of the particle at A is equal to the velocity of the particle at M, and the remaining lengths CB, DB, EB are in geometric proportion. Then the lengths MR, MS, MT are the logarithms of the corresponding distances CB, DB, EB. In effect, Napier defined the logarithm of $10,000,000(0.9999999)^n$ to be n. Napier's terminology belies the origin of the logarithm in trigonometric ratios: he called the line AB the radius, which has a logarithm of zero, and the lengths CB, DB, EB, etc., were called the sines.

The idea of a logarithm was certainly "in the air;" in fact, Bürgi had a similar idea around the same time, though he used the powers of 1.0001, and did not publish his work until 1620. By then, logarithms had undergone a crucial change due to the work of HENRY BRIGGS (February 1561–January 26, 1630).

In 1615, Briggs visited Napier and suggested that logarithms would be more convenient if they were based on the successive powers of 10. Initially Briggs proposed $\log 10 = 10,000,000,000$ but after some discussion, Briggs and Napier decided that computations with logarithms would be simplest if $\log 10 = 1$ and $\log 1 = 0$. Napier declined to create a table of logarithms (he was in poor health and would die within two years), leaving it to Briggs to construct a new table of common logarithms.

There are two ways to construct such a table. Consider the problem of finding the logarithm to base ten of 2. One possibility is to find some power m so that $2^m \approx 10^n$, where n is a whole number; for example, $2^{63} \approx 10^{19}$. Thus $\log 2 \approx \frac{19}{63}$. A more tedious but generally more useful way to find logarithms is to find $\sqrt[n]{10}$. By taking successive square roots of 10, Briggs found a useful base that he could then raise to powers so that it would equal a given number; if $\left(\sqrt[n]{10} \right)^m = a$, then $\log a = \frac{m}{n}$. The construction of a table of logarithms established Briggs's scientific reputation. Thus it was natural that he would come to the attention of a theologian turned patron of science: Henry Savile.

7.3.6   Royalists and Parliamentarians

In 1603, Elizabeth died and James VI of Scotland became James I of a unified kingdom of England and Scotland. Those familiar with Shakespeare already know much of the life story of this prince of Denmark.[15] James's father, Lord Darnley, had been murdered, and his mother Mary subsequently married the chief suspect, James Hepburn. In addition his mother's chief councilor, David Rizzio, was stabbed to death in her presence.

James was probably the best-educated king of the time: he was literate in French, Greek, and Latin, and wrote treatises on many subjects. He denounced the "stinking suffumigation" in *A Counterblast to Tobacco* (1604). However James's real passion was theology and magic, and his blend of scholarship and superstition led Henry IV of France to call James "the wisest fool in Christendom." Indeed, James had written his own interpretation of the Book of Revelation (1585), which Napier used extensively in his *Plaine Discovery*.[16]

James handpicked Savile and 53 other scholars to prepare an English version of the Bible, now known as the *King James Version* (1611). Most of the translators worked in six groups in three locations over a period of seven years. The Oxford groups were entrusted to provide a new translation of the Gospels, the book of Acts, and the book of Revelation (the beginning and ending books of the New Testament).

Henry Savile was a member of one of the Oxford groups. Savile, who was personally wealthy, refused to be compensated monetarily for his work, so James knighted him instead. Sadly, Savile suffered a great personal blow shortly after his appointment as a translator: his only son, Henry, died at the age of 8, leaving Savile without heirs. In consequence, Savile chose to dedicate his enormous personal fortune to furthering education in England. Mathematics education in England was abysmal, leading the biographer John Aubrey to observe that soon, bartenders would be better at arithmetic than university professors; as late as 1573, it was a radical suggestion that candidates for the bachelor's degree know addition, subtraction, multiplication, division, and extraction of roots.

In 1619, Savile established two endowed chairs at Oxford University: the Savilian Professorships of Geometry and of Astronomy. The duties of the Savilian Professor of Ge-

---

[15] James's wife was Anne of Denmark, the daughter of Frederick II of Denmark; Frederick in turn was Tycho Brahe's patron.

[16] Shakespeare used one of James's treatises on witchcraft as source material for *Macbeth*. At the time, it was believed that James was descended from Banquo, hence the witch's prophecy that Banquo would be the ancestor of kings.

ometry included teaching the whole of Euclid's *Elements*, Apollonius's *Conics*, and the complete works of Archimedes; moreover, the students were to be taught the fundamentals necessary to study these works. The Savilian professor also had to provide instruction to students in the applications of mathematics to music, astronomy, and other fields; weather permitting, students were to be taken to the outdoors for a practicum in surveying and mensuration. Finally, the Savilian Professor was expected to make research contributions and place copies of his lecture notes in the University library.

Savile handpicked Briggs to be the first holder of the Savilian Professorship of Geometry; Briggs would also be the first to hold another prestigious post, the Gresham Professorship of Geometry. Upon Briggs's death, Peter Turner became the second Savilian Professor of Geometry. Turner was appointed not for his mathematical ability (which was minimal) but for his work codifying the statutes of Oxford University under the direction of the new Chancellor, Archbishop William Laud.

By then, England and Scotland had a new king, James's son Charles. Laud, as one of Charles's councilors, found it reprehensible that though England and Scotland were both ruled by the same person, they had radically different religious organizations. Both countries were Protestant, but the Church of England was Episcopalian (in which the ultimate authority in the Church lay with a single head—in this case, the King) while the Church of Scotland was Presbyterian (in which the ultimate authority in the Church lay with a council). With Archbishop Laud urging him on, Charles made the great mistake of trying to impose the *Book of Common Prayer* on the Scots, who rebelled in 1637.

Compromise might have been a possibility, and in 1640, Archbishop James Ussher, a friend of Briggs and one of Charles's advisers, suggested a way to iron out the differences between the two churches. Unfortunately Ussher's proposal fell on deaf ears, and Ussher, who might have gone down in history as the man who united the Churches of England and Scotland and stopped a civil war, is instead remembered as the person who calculated, on the basis of the genealogical account in the Old Testament, that creation occurred in 4004 B.C.[17]

On April 13, 1640 Charles summoned parliament to raise money to put down the rebellion, but parliament demanded that a generation of grievances be redressed. Charles refused to concede anything and less than three weeks later, dismissed the Short Parliament on May 5. Charles tried to prosecute the war using his own resources, ran out of money, and summoned parliament a second time on November 3, 1640. The Long Parliament would remain in session for twenty years, and began by impeaching two of Charles's most hated Councilors, including Laud (who was eventually tried and beheaded in 1645); disbanding the Star Chamber and the Court of High Commission (two of the main instruments of royal authority); and passing a bill requiring parliament to be summoned at least once every three years. Charles responded by ordering the arrest of several parliamentary leaders for treason, which came to nothing because upon hearing of the edict, they had taken refuge with the citizens of London, who refused to give them up.

The battle of wills continued until July 1642 when Parliament raised an army. Charles responded with an army of his own, and the English Civil War began between the Royalists (or Cavaliers), the supporters of the king, and the Parliamentarians (or Roundheads).

---

[17]Later accounts assert an exact date and time, but this was not part of Ussher's original work.

Roughly speaking, the Royalists held the north and west of England, while the Parliamentarians held the south and east, including London.

Cambridge University fell into a peculiar position. Since 1604, the University had the right to send two members to Parliament. But the town of Cambridge, too, could send members to Parliament, and here there was a conflict: the University supported the Royalists, while the town supported the Parliamentarians. In August, university officials began melting down the college's silver to help fund the Royalists. Before they could finish, they were interrupted when a small contingent of armed men arrived, led by one of the town's representatives to parliament: Oliver Cromwell.

In November 1642, Chichester (on the southern coast of England) fell to the Royalists. The Parliamentarians sent an army to recapture it under Sir William Waller; the city fell on December 27, 1642, after an eight day siege, and a Royalist letter in cipher was intercepted. The letter made its way into the hands of JOHN WALLIS (December 3, 1616–November 8, 1703) in 1643. Within two hours he decoded its content and endeared himself to the Parliamentary cause. Wallis subsequently decoded many Royalist communications.

Eventually the Parliamentarians overcame the Royalists, and on January 30, 1649 took the unprecedented step of executing the king, over the objections of Wallis and others. Wallis was rewarded for his work for the parliamentary cause by being appointed Savilian Professor at Oxford in 1649. Though Wallis's appointment was purely political, he soon proved that he deserved the position on merit alone.

Wallis's most important work was his *Arithmetic of the Infinite* (1656). By considering the partial sums of the powers of the whole numbers, Wallis came to a conjecture which, in modern terms, we would express as

$$\lim_{N \to \infty} \frac{\sum_{i=0}^{N} i^m}{\sum_{i=0}^{N} N^m} = \frac{1}{m+1}.$$

Then Wallis applied a crucial idea: by considering an area or volume as the sum of an infinite number of infinitely thin slices (whose width Wallis designated as $\frac{1}{\infty}$), he could compare the area of a figure to the area of an enclosing figure whose area could be easily calculated. For example, consider the region under the parabola $y = x^2$ above the x-axis over $0 \leq x \leq 1$. A comparison of the infinitesimal rectangles that make up the parabolic region and the infinitesimal rectangles that make up the rectangle leads to a limit of the preceding form, where $m = 2$; hence the region under the parabola is one-third the area of the enclosing figure.

Using incomplete (scientific) induction and interpolation, Wallis then conjectured corresponding formulas for other areas. In the end, he considered the area under the curve $y = \sqrt{R^2 - x^2}$, where R is a constant, and compared it to the square whose side length is R. Wallis found the ratio between the square and the region under the curve (which is a quarter of a circle); in effect, he found that

$$\frac{4}{\pi} = \frac{3 \cdot 3 \cdot 5 \cdot 5 \cdot 7 \cdot 7 \cdots}{2 \cdot 4 \cdot 4 \cdot 6 \cdot 6 \cdot 8 \cdots}.$$

After Charles's execution, Cromwell assumed the title of Lord Protector, and England became a Republic. Cromwell's main goal during the years of his rule as Lord Protector was to establish a stable, secure realm. Royalists represented a clear and present danger, for Charles's son Charles was still alive and with the help of foreign powers, might reclaim the throne. To reduce this danger, Cromwell passed a number of statutes depriving Royalists of position and power.

One victim of the statutes passed by Parliament to punish supporters of the executed king was ISAAC BARROW (October 1630–May 4, 1677). Barrow's early life was full of trouble, mostly brought on by himself: he gained an early reputation as a bully and a troublemaker, and his father, a linen draper, was said to have prayed that if God had to take one of his children, he would prefer that it be Isaac. It took a stern headmaster at Felstead, in Essex, to turn Isaac towards the path of scholarship.

In 1655, the Regius Professor of Greek at Cambridge was forced to resign because of his support of the Royalist cause during the civil war. Barrow was the logical candidate for the position, but he, too, had been a Royalist, and was passed over in favor of the Parliamentary candidate. Barrow realized that so long as the political climate made Royalists unpopular, he would never achieve any high position, so he left England to travel through Europe.

Cromwell was a Puritan, and while the name today conjures up an image of a stern moralist who seeks to ban all outward expressions of frivolity, Cromwell himself was a moderate. He banned races and cockfights, but this because he feared that these public meetings might serve as a cover for seditious activities, and not from any considerations of morality. Indeed, Cromwell showed exemplary tolerance: Catholicism was still technically illegal, but the laws were rarely enforced, and even the Jews were invited to return to England.

However, some of Cromwell's local governors were extremists who closed down taverns and banned games and other recreations. Cromwell's tolerance of Protestant sects like Baptists and Quakers, not to mention his tolerance of the Catholics, led to further confrontations with the radicals. Like the Stuart kings he replaced, Cromwell resorted to dissolving Parliament on several occasions, summoning it only when it was necessary to raise money. Unlike the Stuart kings, Cromwell was able to maintain his hold on power thanks to his good relationship with the army. His son Richard, who became the new Lord Protector upon his father's death in 1658, was not so adept at controlling the army, and within a year he was forced to resign.

7.3.7 The Restoration

In 1660, Thomas Scott, Parliament's director of intelligence, obtained some letters whose contents were encrypted; he passed these onto Wallis for analysis. The decrypted messages indicated that Charles, the son of the executed king, was negotiating with Presbyterian ministers in London for a return to England. Parliament offered Charles the throne on May 8, under certain conditions (essentially, that Charles respect the authority of Parliament), and on May 25, Charles landed in Dover, beginning the Restoration Era in England. Charles II conceded many things, but insisted on the right to prosecute and execute those directly responsible for his father's death. The 57th signature on the death warrant was Scott's, who subsequently fled to Brussels. Somehow he was persuaded to return to England, whereupon he was tried, found guilty of treason, then hung, drawn, and quartered.

The return of the king meant that Barrow and other Royalists, who had been living in continental Europe, could come back to England. Barrow returned and took up the position of Regius Professorship of Greek. However, the terms of the Regius Professorship required him to forego any other paid university position, and at 40 pounds a year, the income of the professorship was insufficient for his needs. He applied for the Professorship of Geometry at Gresham College in London, obtaining the position (and the salary of 50 pounds a year) in mid-1662.

While at Gresham College, Barrow joined a group of scientists who had met on a regular basis since November 28, 1660 for the promotion of "physico-mathematical experimental learning"—what we might call quantitative science. The original group included Christopher Wren, Robert Boyle, and William Brouncker. On July 15, 1662, the Royal Society of London received a charter from Charles II. The charter included the right to operate a printing press, and it began to print a new type of material: a periodical.

During the Thirty Years' War, the first periodicals, known as *corantos*, appeared. As the name suggests, they contained reports of current events, and were the precursor to the modern newspaper. On December 2, 1620, the first English translation of a Dutch coranto appeared in Amsterdam, and on May 23, 1623, Nicholas Bourne and Thomas Archer began publishing the *Weekly Newes from Italy*, the first English newspaper. The word news, incidentally, refers to the "newness" of the information (we still ask each other "What's new?"), and the claim that it is an acronym for "North, East, West, and South" is spurious. Though the early newspapers claimed to be perfect (i.e., complete) and impartial, they were, of course, used by both sides during the Civil War for propaganda purposes. Thus, beginning with Cromwell, newspapers were subject to strict licensing regulations.

Newspapers helped printers solve the financial problems created by copyright infringement. A large, popular work could be copied relatively easily by other printers, but pirating a newspaper was pointless, for by the time a print run could be organized, the contents were literally old news. But aside from journalistic bias, there was little difference between one newspaper and another: they all reported the same stories. What was needed was a regular publication that contained a variety of stories of interest. In 1663, Johann Rist, a pastor and hymn-writer, began publishing *Edifying Monthly Discussions* in Hamburg: it was the first periodical (though its period was irregular). In 1665, the first scientific journal, the *Journal des Scavans* appeared in France under the editorship of Denys de Sallo to "make known that which is new in the Republic of Letters."

Under the guidance of HENRY OLDENBURG (1618–September 5, 1677), the Royal Society began publishing the *Philosophical Transactions of the Royal Society* in March 1665. Since the Royal Society's primary interest was experimental science, very few mathematical contributions made their way into the *Philosophical Transactions*. Indeed, there were very few mathematical publications of any kind in any journal, for mathematics suffers from a unique feature: it is possible to solve a problem and show to others that your solution is correct *without* revealing your methods.

For example, one of the first books published by the Royal Society was *Micrographia*, by Robert Hooke, in which Hooke described objects he viewed under a microscope (including compartments in cork slices which he dubbed cells). But since the only way to convince others that the objects actually existed was to have them look through the micro-scope itself, it was necessary to include instructions on how to make a microscope: even had he wanted to, Hooke could not keep the *method* of discovery secret. Thus scientists honor the discovery, not the method, and the best way to ensure prestige was to be the *first* to discover—which encouraged rapid publication of results.

Mathematicians, on the other hand, could solve problems and present their solutions without presenting their methods, and prestige came from solving problems others could not. Thus only amateurs and professors in endowed positions published original mathematical research. Barrow was instrumental in changing the culture of secrecy, for he argued that academics had the responsibility of publishing their works to further advance scientific progress.

In 1663, Barrow was part of a commission deciding how to administer the will of Henry Lucas. Lucas was one of Cambridge University's two representatives to the Short Parliament, and had left land with an annual income of 100 pounds to fund a professorship in mathematics. Barrow and others drafted out the details of the Lucasian Professorship, and Barrow himself became the first Lucasian Professor. Between 1664 and 1666, Barrow delivered a series of lectures covering a wide variety of mathematical topics, and though he was required to deposit copies of his lectures in the university library, he went further and published his *Geometrical Lectures* in 1670.

Barrow was the first to explicitly recognize the connection between the tangent and area problems, one version of the Fundamental Theorem of Calculus. In modern terms, we would state Barrow's result as follows:

Theorem 7.3. *Let $y = f(x)$ be a curve that is continuous, positive and increasing over the interval $0 \leq x \leq M$, and let $y = F(x)$ be such that $F(a)$ is the area under $y = f(x)$ and above the x-axis over the interval $0 \leq x \leq a$. Then the line passing through the point on $y = F(x)$ where $x = a$ with slope equal to $f(a)$ will be tangent to the curve.*

Barrow was a mathematical conservative, and preferred to avoid the use of infinitesimal methods whenever possible (his proof of the fundamental theorem, for example, is a geometric one that relies on the algebra of inequalities). However, one of his colleagues—later determined to be Isaac Newton—convinced Barrow that infinitesimal methods were valuable, even if they were not rigorous, and Barrow relied on them to prove the converse of Theorem 7.3, using essentially the modern proof.

Barrow held the position of Lucasian Professor for several years, and continued to hold the Chair of Geometry at Gresham College, in London. Today, Cambridge is just a short train ride from London, but in Barrow's time, it was a two-day trip. Naturally questions arose about Barrow's ability to hold both posts. Since the terms of the Lucasian Professorship forbade Barrow from holding any other position at Cambridge, and in particular prevented him from advancement to the higher, administrative ranks, he resigned the professorship in 1669 and rose rapidly, becoming chaplain to Charles II in 1670, the Master of Trinity College in 1674, and then Vice Chancellor of the University in 1675.

Barrow believed that mathematicians ought to publish their work, but at the same time did not feel it was their duty to be editors; his original intention was to publish the *Geometrical Lectures* "just as they were born." However, Barrow was amenable to letting one of his students edit the lectures to put them into readable form. The editor was Barrow's handpicked successor as Lucasian Professor: ISAAC NEWTON (December 25, 1642–March 20, 1727).

Newton was born a few months after the death of his father; his mother remarried. The relationship between Newton and his mother Hannah Ayscough and stepfather Barnabas Smith was strained at best, and Newton even threatened to burn down the house with his parents inside. Upon his stepfather's death in 1653, his mother sent Newton to look after the family estate (the Ayscoughs were relatively prosperous farmers). However, Newton showed neither ability nor interest in estate management, and an uncle, William Ayscough, persuaded Newton's mother that the boy should be sent to a preparatory school for university. Newton entered Trinity College (part of Cambridge University), his uncle's school, on June 5, 1661.

In the 1660s, plague struck England. It devastated London and might have had catastrophic effects but for the Great Fire of London, which started when a baker's shop accidentally caught fire on September 2, 1666. The fire burned for three days and destroyed much of the city of London, including the original St. Paul's Cathedral and 13,000 houses (as well as the entire unsold stock of the *Philosophical Transactions*, a loss that nearly bankrupted the publishers). Miraculously, only eight people are known to have died, and the fire burned the thatch that housed the rats that had the fleas that carried the plague, helping bring an end to the epidemic. Moreover the devastation allowed much of the city to be rebuilt according to an overall plan, and Christopher Wren oversaw the building of a newer and bigger St. Paul's Cathedral.

In 1665, the plague spread to Cambridge, forcing the closure of Trinity College. Newton returned to his family's farm in Lincolnshire. The next few months were, perhaps, the most significant few months in the entire history of science, for Newton's four major contributions to mathematics and physics were made during this self-imposed exile from the plague: the binomial theorem; calculus; the theory of gravitation; and the discovery of the nature of light and colors.

Newton claimed to have discovered the generalized binomial theorem before he was acquainted with the procedures of root extraction, suggesting a date of around 1664 or 1665, but the first written appearance of the binomial theorem was in a letter from Newton to Henry Oldenburg, Secretary of the Royal Society, on June 13, 1676. Around 1669, Newton first put forth the ideas expressed in his *Analysis by Equations of an Infinite Number of Terms*, where he developed what we would now call the series expansion of a function, and around 1671, he wrote *On the Method of Fluxions and Infinite Series* which developed what we would now call differential calculus.

However, Newton did not publish his methods of calculus for many years. Part of the reason was the contentious Robert Hooke. Hooke anticipated (in rough outline) both Newton's work on gravitation and Huygens's theory of light, and when Newton presented his own theory of light in 1672, Hooke said what was correct in the theory was stolen from him, and what was original was wrong. As a result Newton became reticent and reluctant to publish other results that Hooke might criticize or claim he had discovered earlier.

7.3.8 The Glorious Revolution

Charles had no heirs, and it seemed certain that, upon his death, his brother James would succeed him. However, this posed a problem. It was an open secret that James was Catholic, and the radicals sought every means at their disposal to pass a bill of exclusion, keeping the unpopular James from the throne.[18] But Charles refused to consider disinheriting James, for it would call into question the very legitimacy of hereditary succession.

Parliament was dominated by Charles's supporters, but in 1678, a powerful weapon was dropped into the hands of the exclusionists by Titus Oates, who claimed to have evidence of an international conspiracy: the Popish plot. According to Oates, the Jesuits, with the help of French troops, were going to kill Charles II, place his brother James on the throne, and massacre English Protestants. Oates was hardly a credible witness: he was expelled from the Merchant Taylors School in London in 1665, jailed for perjury in 1674, and dismissed from the navy for misconduct. But his tall tale caught the ear of one of the leading exclusionists, Anthony Ashley Cooper, the Earl of Shaftesbury, who used Oates's accusations to fan the flames of paranoia, and began organizing a national campaign to push through a bill of exclusion.

One of the reasons the Popish plot gained momentum was that Oates presented his "evidence" to an English magistrate, Edmund Berry Godfrey, who was murdered shortly afterwards. To the paranoid, this proved the reality of the conspiracy. Shaftesbury took the opportunity to blame the murder on his political enemy: Samuel Pepys (pronounced "Peeps"). There was no reality whatsoever to the conspiracy, but such was the climate of paranoia that Pepys, unjustly and falsely accused, might have been executed had not Charles dismissed Parliament, delaying the trial indefinitely.

Pepys is best known for a diary he kept between January 1, 1660 and May 31, 1669, which gives us an incredibly rich description of life in Restoration London. But he was no passive observer of Restoration society. As Secretary of the Admiralty, he was almost single-handedly responsible for tripling the size of the navy, turning it from an under-gunned, undermanned, underfunded force to a modern, powerful fleet. In 1684, Pepys became the President of the Royal Society (he had been a member since February 14, 1664). Thus Newton's master work *Mathematical Principles of Natural Philosophy* (1687) appeared under Pepys's auspices on July 5, 1686.

In the *Principia*, Newton introduced limits (though not in the modern form), and gave a lemma equivalent to noting that the Riemann sum converges to the area under a curve as the width of the widest rectangle tends to zero. Once again Hooke attempted to claim credit (this time for the inverse-square law of gravitation), and Newton vowed to publish no more until after "that horrible man" was dead. As a result Newton's first systematic account of his calculus did not appear until 1704, when "Treatise on the Quadrature of Curves" appeared as one of two appendices to his *Optics*. Newton's publication might have been delayed in any case, for shortly afterwards he became active in national politics.

For trying to steal the throne from James, the exclusionists were called Whigs, a Scottish slang term for a horse thief. In turn the Whigs labeled their opponents Tories, an Irish slang term for a Catholic outlaw. Three times Parliament met, and the Whigs attempted to push through an exclusion bill; three times Charles dissolved Parliament before the bill could pass both houses. It was only by such drastic actions that Charles was able to save the throne for his brother James. On February 6, 1685, Charles died and James succeeded him.

---

[18]There is an apocryphal story that, one day, the brothers were walking along, and James expressed concern that someone might try to assassinate Charles. "Don't worry; no one would assassinate *me* to make *you* king" was Charles's response.

James believed that most Englishmen were naturally Catholics, and remained Protestant only because of legal penalties imposed on Catholics. Thus, he planned to eliminate the anti-Catholic laws. He found a friend in William Penn, who just prior to James's accession established a Quaker colony in the New World: Pennsylvania (named after Penn's father, the Admiral William Penn). With Penn's support, James drafted the Declaration of Indulgence (April 1687), which condemned compulsory religious conversion of any sort, and suspended all laws regulating ecclesiastical matters: in effect, James declared that no person could be legally prosecuted for their religious practices.

One place where Catholics faced discrimination was at the major universities, which required professions of faith that conflicted with Catholic beliefs; effectively Catholics were prevented from attending Cambridge or Oxford. James pressured the universities to admit Catholics, and when Cambridge refused to give a degree to a Benedictine monk, James dismissed the Vice Chancellor of the University.

Such high handed behavior incurred the wrath of parliament, and might have led to another civil war. But James only had daughters, who had married Protestant princes, so upon his death England would become Protestant again. The situation changed when Queen Mary gave birth to a son, James Edward, in June 1688, and England looked as if it was doomed to have a Catholic monarchy. To forestall this, parliament called on the husband of James's Protestant daughter Mary: William of Orange. William arrived in England with 15,000 men on November 5. When James's supporters melted away, he fled to France, landing on Christmas Day, 1688.

The result was a legal quagmire. William had no claim to the throne, though his wife Mary did. Mary had no intention of ruling separately from William. James was still technically king. Parliament could declare James deposed—but only James could summon parliament. Fortunately the Civil War had established the precedent whereby a Convention, consisting of members of parliament, could meet. On January 15, 1689, Newton was elected to represent Cambridge University at the Convention, and three days later, the Convention offered the throne jointly to William and Mary, declaring that by fleeing to France, James had abdicated the throne. The Glorious Revolution had overthrown a reigning king with almost no bloodshed, though for more than half a century, James's supporters (known as Jacobites, after the Latinization of James) would try to place a Stuart on the throne.

Newton's involvement in the government lasted past the Glorious Revolution itself. In 1693, Charles Montagu helped the government borrow 1,000,000 pounds at an annual interest rate of 10%: it was the beginning of the English national debt. The next year Montagu arranged for the creation of the Bank of England. Then in 1695, Montagu addressed a key problem with English currency: clipping. If a coin is made out of a precious metal, it is possible to "clip" the rim of the coin and collect the shavings. One way to prevent this sort of fraud is to mill the edges of the coin: a clipped coin will have a smooth edge, and thus be easily detectable.

The problem was that most of the coins in circulation were unmilled, mainly because they had been minted long before. Thus Montagu called for a great recoinage: older coins were to be taken out of circulation, melted down, and reissued—this time with milled edges to prevent clipping. To oversee the recoinage, Montagu had Newton, a close personal friend, appointed to the Mint. Newton threw himself so assiduously into the details of running the mint that his original scientific work effectively ceased by 1699, when he became Master of the Mint. Newton gained a reputation for being almost obsessed with the quality of the coins produced, and rejected several consignments on the grounds that the coins produced were improperly made.

For Further Reading

For the history of early modern Europe, see [13, 14, 16, 19, 20, 29, 78, 50, 91, 133]. For the mathematics and science of early modern Europe, see [94, 98, 111].

8

The Eighteenth Century

8.1 Great Britain

Neither William nor Anne (Mary's sister, and William's successor) produced an heir to the throne, so on June 12, 1701, Parliament passed the Act of Settlement which expressly stated that the crown of England would pass to Sophia of Hanover, the granddaughter of James I (and daughter of Frederick V the "Winter King"), or any Protestant children she might have, and *not* to ex-king James or his children. Thus on William's death in 1702, Anne became Queen of England and Scotland (Mary predeceased her husband).

Anne knighted Newton in 1705 (the first scientist to be honored for his scientific work), and became the first monarch of Great Britain after Parliament passed the Act of Union on May 1, 1707, formally recognizing the unification of Scotland and England. A new nation needed a new flag, and the St. Andrew's cross (a white ×, symbol of Scotland) combined with the St. George's cross (a red +, symbol of England) to form the "Union Flag", a precursor to the modern British flag.

In addition to a new flag, England received a new symbol. In 1712, Anne's personal physician JOHN ARBUTHNOT (April 29, 1667–February 27, 1735) published *The Law is a Bottomless Pit*, usually known by its subtitle, *The History of John Bull*. Published as a series of five pamphlets, the satire told the tale of how an honest clothier, John Bull, and a linen-draper, Nicholas Frog, spent all their money in a lawsuit against Louis Baboon. The personifications were transparent to anyone: John Bull was England, Nicholas Frog the Netherlands, and Louis Baboon was France. The pamphlet proved so popular that John Bull has been the personification of England ever since.

The History of John Bull was written to support the peace initiative to end the War of the Spanish Succession (1701–1714), which began when Charles II of Spain willed his kingdom to Louis XIV's grandson Philip. Almost immediately France found itself facing an alliance consisting of England, the German Empire, the Dutch Republic, Prussia, Hanover, Portugal, and (after initially supporting France) Savoy. John Bull spent all his money on the lawsuit; the British were more fortunate and British participation ended with the Treaty of Utrecht (April 11, 1713), an event commemorated by Georg Friedrich Handel, then living in London. Handel's *Utrecht te Deum* and *Jubilate* caught the royal fancy, and Queen Anne gave him an annual pension of 200 pounds. The Treaty allowed Louis's grandson Philip to retain the Spanish throne, in exchange for renouncing his claim to the French throne, but this was a consolation prize: both Spain and France lost much of their overseas possessions to Britain, and in particular, the English gained the *asiento*: the right to sell African slaves in Spanish colonies.

8.1.1 Probability

Insurance is as old as civilization, though with the development of the theory of probability in the 17th century, it became possible to use mathematics instead of intuition to determine premiums. One special type of insurance is the annuity, whereby a premium is paid on a regular basis and if the beneficiary reaches some specified age, he or she will receive a payment (retirement pensions are one type of annuity). The advantage to the policy-holder is that the sum of the premiums is less than the value of the annuity; the advantage to the insurer is that there is a chance the policy-holder will die before being able to collect. But in order to determine the correct price for the annuity, it was necessary to have accurate tables of mortality, and while bills of mortality existed, they frequently omitted key information, such as the age of death; moreover, these bills of mortality were generally for large cities, like London or Dublin, and in this case, many of those who died were not actually born in the city, further obscuring the true death rate.

EDMOND HALLEY (October 29, 1656–January 14, 1743), who paid the printing costs associated with the *Principia*, was among the first to publish a useful table of mortality, and deserves the accolade "Father of Actuarial Science" for being one of the first to apply probability to insurance. In "An Estimate of the Degrees of Mortality of Mankind, Drawn from Curious Tables of the Births and Funerals at the City of Breslaw; With an Attempt to Ascertain the Price of Annuities upon Lives" (1691), Halley used demographic information from Breslau (the capital of Silesia) for the years 1687–1691. Unlike London or Dublin, Breslau had no large-scale immigration, and the data presented to the Royal Society included the age of death and sex of the decedent. The mortality tables are sobering. Over the five year period, there was an average of 1238 births per year. Of these children, 348 would die in their first year, and 198 more would die before their sixth birthday. Halley then used the mortality table to determine the premium to be charged for a given annuity payable in three years if the beneficiary survived.

Probability appeared in other contexts as well. A number of leading scientists and intellectuals of the time (including Arbuthnot and Newton) were Deists, who believed that God's characteristics are amenable to rational analysis. Then as now, probability appeared in arguments that "proved" the existence of God. Arbuthnot, who translated Huygens's treatise on probability into English in 1692, presented "An Argument for Divine Providence, Taken from the Constant Regularity Observed in the Births of Both Sexes" (1710). In it, Arbuthnot noted that the number of boys born in London exceeded the number of girls born for the past 82 years. Since (Arbuthnot said) the probability of this was $\left(\frac{1}{2}\right)^{82}$, the extreme improbability of the event was evidence of the existence of God.

Newton himself gave two mathematical "proofs" of the existence of God. In the 1726 edition of the *Principia* and in the 1717 edition of *Optics* Newton argued that "blind fate" could never have made all the planetary orbits very nearly circular and very nearly in the same plane: this required divine intervention.[1] Also in the *Optics* Newton argued that the

---

[1]Laplace used the same argument a century later, but substituted "physical cause" for "divine intervention."

combined gravitational influences of the planets on one another might be disruptive over the long term, so the solar system might occasionally need divine intervention to be set straight.

Alternatively, probability might be applied to another problem: the accuracy of religious texts. JOHN CRAIG (1663–October 11, 1731), in his *Mathematical Principles of Christian Theology* (1699), concluded that the evidence of the truth of the Gospels declines with time, reaching 0 in the year 3144; hence the Second Coming had to occur before then. A similar sentiment (minus eschatology) appeared in "A Calculation of the Credibility of Human Testimony" (1699), an anonymous work (possibly written by Halley) which calculated that a written source might be recopied for 7000 years before the probability exceeded $\frac{1}{2}$ that it was incorrectly copied.

More profane uses of probability existed as well. On November 22, 1693 Pepys wrote to Newton asking him to resolve a problem posed by Thomas Neale, William III's Groom Porter. As Groom Porter, Neale's responsibilities included supplying the king and company with cards, dice, and settling rules disputes. Neale, who would be Newton's predecessor as Master of the Mint, posed the following problem. Suppose A has 6 dice and needs to throw at least one 6; B has 12 dice and needs to throw at least two 6s; C has 18 dice and needs to throw at least three 6s. Who had the "easiest" task? Newton's response to Pepys extended over several letters: probability was still in its infancy, so the problem, as stated by Pepys, was ambiguous (for example, the stakes were unspecified). Newton's letter of December 16, 1693 gave a detailed solution to the question of who had the greatest probability of winning, calculated using a binomial probability.

More general problems in probability were addressed in "On the Measurement of Chance" (1711), by ABRAHAM DE MOIVRE (May 26, 1667–November 17, 1754). De Moivre was a French Huguenot, but on October 18, 1685, Louis XIV revoked the Edict of Nantes, effectively depriving French Protestants of their legal rights. More than 400,000 Protestants fled France after the revocation. Most went to England, the Americas, or the Netherlands. After spending two years in a French prison, de Moivre arrived in London in 1688, where his mathematical prowess became quickly apparent; by the end of his life, Newton was telling prospective students, "Go to Mr. de Moivre; he knows these things better than I."

"On the Measurement of Chance" broke new grounds in the investigation of probability. Previous authors had only considered games where the two players had an equal chance of winning, but de Moivre examined cases where the two players had unequal chances of winning. In addition, de Moivre made extensive use of the generating function, which has since become a key component of the theory of combinatorics. De Moivre greatly expanded his investigations into probability in *The Doctrine of Chance* (1718).

Though de Moivre's contributions to probability were important, he is commemorated for a theorem stated in his *Miscellaneous Analysis* (1730): If $l = \cos nB$ and $x = \cos B$, then according to de Moivre:

$$x = \frac{1}{2} \sqrt[n]{l + \sqrt{l^2 - 1}} + \frac{\frac{1}{2}}{\sqrt[n]{l + \sqrt{l^2 - 1}}}$$

This points to the fundamental flaw in probability based ontological arguments: they assume randomness for events that might or might not be truly random.

which is equivalent to

$$\cos B = \frac{1}{2} (\cos nB + i \sin nB)^{1/n} + \frac{1}{2} (\cos nB + i \sin nB)^{-1/n} ,$$

a form of what is now called de Moivre's Theorem (it was Euler who put de Moivre's theorem in its familiar form).

De Moivre played a small role in the priority dispute over the inventor of calculus. The dispute smoldered for many years after the publication of Newton's works, but exploded in 1710, when JOHN KEILL (December 1, 1671–August 31, 1721) openly accused Leibniz of plagiarism. In 1712, the Royal Society appointed a commission to review the claims of Newton and Leibniz. The members of the committee (not known for a century afterward) included Halley, de Moivre, and Arbuthnot; and two of Keill's friends, JOHN MACHIN (1680–June 9, 1751) and BROOK TAYLOR (August 18, 1685–December 29, 1731), though in fact the report published by the commission was authored by Newton. Not surprisingly the commission concluded that Leibniz did indeed plagiarize Newton's work.

8.1.2   The Theory of the Moon

The priority dispute would have far-reaching consequences. It is not true, as is frequently claimed, that Leibniz's notation was so superior to Newton's that English progress was retarded (physicists, for example, still use Newtonian notation). However, Leibnizian calculus is primarily algebraic, while Newtonian calculus is primarily geometric, making it more difficult to apply. Moreover, the fact that the foreigner Leibniz was so badly treated by the English turned what should have been a purely academic dispute into one involving national pride. This was critical, for in the early 18th century, there were two systems of physics in existence. The first was the system elucidated by Newton in the *Principia*, and the second was that of Descartes.

The fundamental problem was action at a distance. How could the sun's gravity make itself felt across a vacuum, when every known force was one of contact: you had to touch the object you were affecting, or touch it with something that you were touching or had touched. Indeed, the fact that gravity was an attractive force made matters more difficult: you could push something away from you without touching it by throwing something at it, but how could you pull something *towards* you without making direct contact?

Even before Newton, Descartes offered an explanation. According to Descartes, the universe was at every point filled with particles, imbued with certain properties. One of these properties was that they swept eternally in a vortex around the sun. The planets, surrounded by the vortex particles, were swept along like leaves in a stream, accounting for the orbits (and, incidentally, answering Newton's question of how the planets could all move in the same direction and in very nearly the same plane: they had to, since to move against the vortex or out of the plane of the vortex would be like trying to swim upstream). Leibniz's followers, particularly Johann Bernoulli, tended to be Cartesians, and filled the universe with particles.

Newton chose to let the predictive power of universal gravitation speak for itself. In the *Principia* Newton showed how universal gravitation accurately accounted for Kepler's three laws and numerous other phenomena besides; as for how gravity made itself felt across a vacuum, "I make no hypotheses" was Newton's response. In the third edition of the *Principia* Newton added a General Scholium that declared that "The hypothesis of vortices is pressed with many difficulties." But Newton obscured an equally important fact: universal gravitation could not accurately account for the motions of the Moon. Combined with the conceptual difficulties of action at a distance, this observational fact prevented widespread acceptance of Newtonian physics until mid-century. By then, the motion of the Moon became a matter of paramount importance.

During the War of the Spanish Succession, an English fleet under the command of Cloudesley Shovel seized Gibraltar in 1704. On his way back from Gibraltar in 1707, Shovel's fleet encountered a severe storm that caused it to lose its bearings, and when the weather cleared, the lost fleet sailed into destruction on the rocks off the coast of the Scilly Islands near Cornwall. Four ships and two thousand men perished in the disaster. With Britain suddenly a global power, it was absolutely vital that English ships have some way of determining their position at sea, so in 1714, Parliament announced the Admiralty Prize: 10,000 pounds for a device that could accurately determine longitude within 1° (with progressively higher amounts for more accurate determinations).

In principle the problem of longitude is easy to solve: the difference between local time and time at a fixed meridian gives the displacement between the two points. For example, if it is noon in London and 11 AM aboard a ship in the Atlantic, then the ship is 1/24th of the circumference of the world, or 15° of longitude, to the west of London. In practice, it is easy to find local time (the height of the sun above the horizon determines this), so it is only necessary to determine the "at home" time. Thus, the solution to the problem of longitude comes down to finding an accurate timepiece.[2]

Newton (a member of the commission) doubted that a mechanical clock could be made accurate enough to determine the time, and preferred an astronomical method: a theory of the Moon that could be used to prepare tables giving the Moon's position years in advance (in this context, a "theory" is a mathematical model that can be used to determine the Moon's position). By comparing the Moon's observed position to that listed on a table of ephemerides (prepared for the home longitude), the "at home" time could be determined. As early as 1700, Newton derived equations for a theory of the Moon that gave the lunar position accurately enough to find the longitude within 3°; Newton's *Theory of the Moon's Motion* was reprinted in 1702 by DAVID GREGORY (June 3, 1659–October 10, 1708). However, the final resolution to the problem of longitude would be by the invention of an accurate nautical chronometer by John Harrison. Harrison finished his first clock in 1737, and though he satisfied the terms of the Admiralty Prize, he never received the full reward; in fact, no one ever would.

If universal gravitation accurately represented the physics of the universe, then in principle it should be possible to calculate the future position of the Moon. Gravitational perturbations of the sun could explain all the variations in the Moon's orbit save one: the motion of the apogee (the point of the Moon's orbit farthest from the Earth). The apogee advances by about 3° per orbit, and Newton's calculation in the *Principia* could only account for half this motion. Despite various attempts, Newton and his contemporaries were never able to explain the full amount, leading him to declare that the problem of the Moon's motion was the only one that ever made his "head ache." More importantly, the inability of uni-

---

[2] Arbuthnot's friend, Jonathan Swift, referred to the problem of longitude in *Gulliver's Travels*.

versal gravitation to account for the full motions of the Moon delayed its acceptance by continental scientists for half a century.

8.1.3 Expatriates in Venice

Ironically, at the same time the mathematical followers of Leibniz and Newton were drawing apart from each other, their home states of Hanover and England were pulling closer together. Anne died on August 1, 1714, and the crown passed to George I of Hanover. George spoke no English, and communicated with his ministers in French. As with most newly enthroned monarchs, George dismissed many of Anne's ministers and appointed his own favorites.

One of those dismissed was John Erskine, the Earl of Mar. Disgruntled, the Earl of Mar took up the Jacobite cause, and on September 1715, Mar proclaimed James Edward (James II's son) the king of Great Britain. Mar's troops moved from Perth towards Stirling, guided by a highland outlaw, Rob Roy MacGregor. MacGregor had divided loyalties, for leading the government forces was the Duke of Argyll, who supported MacGregor in his campaign against the Duke of Montrose. At the Battle of Sheriffmuir (November 13, 1715) the Jacobite forces were routed. "The Fifteen" was quickly suppressed, and Erskine ended his life in exile. For his part in helping the Jacobites, MacGregor had a price put on his head for treason, though his life and cause were romanticized in *Highland Rogue* (1723), an anonymous tract probably written by Daniel Defoe. The publicity garnered by the account led to a Royal Pardon for MacGregor in 1727.

Many changes could be expected in the wake of the Jacobite uprising. After the Civil War, Parliament was obliged to hold elections every three years, but the Whig-dominated Parliament, sensing a strong Tory sentiment among the electorate, pushed through the Septennial Act (1716), extending the time between election to seven years (the current British system).

Another consequence was the restoration of loyalty oaths. When JAMES STIRLING (May 1692–December 5, 1770), entered Oxford University in 1711, he was excused from the loyalty oath because of the prominence of his family. But in the aftermath of the failed Jacobite rebellion, his exemption was eliminated and when he refused to swear the oath, he lost his scholarship and was later accused of treasonable correspondence with other Jacobites, a charge of which he was acquitted. He stayed in England for a few more years, but in 1717 Stirling went to Venice with the Venetian ambassador to England, Nicholas Tron. Stirling stayed in Venice for a few years, but very little is known about what he did there.

Venice was an important cultural center, and dominating the scene were the *opera in musica*, "works in music." Though Venice was not the birthplace of the opera (that honor belongs to Florence, where Ottavio Rinuccini's *Daphne* was performed in 1597), it was home to the *Teatro San Cassiano*, the first opera house opened to the public, sponsored by the Tron family in 1637. Perhaps the best-known musician in 1720s Venice was a man whose hair gave him the nickname of "the Red Priest." Antonio Vivaldi taught music to young girls at an orphanage, the Pietà, which staged concerts as a way to raise money.

Stirling eventually returned to Britain, and by the time of the publication of *Method of Differentials* (1730) he was in London. *Method of Differentials* was by far his most important work, discussing infinite series, interpolation, and quadrature, and including the approximation formula for $n!$ known as Stirling's formula:

$$n! \approx \left(\frac{n}{e}\right)^n \sqrt{2\pi n}.$$

The result was also included in de Moivre's *Doctrine of Chance*.

John Law, a Scotsman, also called Venice his home during the 1720s. Law studied mathematics in London in the 1690s. It is tempting to think that he studied with Arbuthnot, the translator of Huygens's treatise on probability, for Law reputedly had some mathematical system for winning games of chance; certainly, Law made a living as a successful gambler.

In 1705, Law published *Money and Trade Considered*. Law's thesis was that paper money, backed by a central bank, would promote trade and the economy. The idea had already been implemented in New England when the Massachusetts Bay Company began issuing paper money in 1690, but in Old England, the government was skeptical: how could anyone have confidence in mere paper, unless it were backed by silver or gold specie?

France, however, was a different story, since no one had confidence in the debased coins issued by the monarchy. Law went to France in 1713, where he convinced the French government to implement his scheme. A national bank, Law and Company, was established in 1715. Law's bank notes were backed by real bullion, unlike the nearly worthless French currency. Law convinced the Regent, the Duke of Orleans, to grant near-total control over the Mississippi territory in New France to the Mississippi Company. The Mississippi scheme allowed those who held nearly worthless *assignats* (promissory notes that the government would pay the holder a specified sum) to redeem them for paper money issued by Law's bank; the paper money could then be used to purchase stock in the Mississippi Company. The great influx of capital promised to restore the French crown to solvency.

Unfortunately it was not long before the French government deluded itself into thinking that it could simply print money. When Law ran the bank, he never issued more than 60 million livres (that being the amount he could redeem). When Law's bank was reconstituted as the National Bank of France, the government almost immediately printed one *billion* livres, and printed even more in the years to follow. In 1720, it was discovered that government had insufficient specie to cover the nearly 2.6 billion livres in circulation, causing widespread panic and a collapse of the economy. Law, though blameless, was the obvious scapegoat, and he fled to Venice. About the only consolation was that largely worthless stock in the Mississippi Company was bought with largely worthless banknotes, which in turn were bought with largely worthless *assignats*. Thus, the actual consequences of the Mississippi scheme to the French government were minimal: some unpopular ministers fell from power, and others absented themselves from Paris, but the government itself underwent little change.

Coincidentally, England was undergoing her own market crash, though for entirely different reasons. In January 1720, the South Sea Company announced it would assume the management of England's public debt. The value of company shares increased rapidly. Newton had owned South Sea Company stock since 1713, and sold £3000 of the stock (a portion of his holdings) in April 1720. The shares continued to rise, and Newton purchased an additional £1000 in June. By August the shares were selling for seven times what they were in January, but then the market collapsed and the shares returned to their starting value by December. The South Sea Bubble bankrupted thousands who had borrowed heavily to buy the overpriced stock. Newton's niece, Catherine Conduitt, reported that he lost £20, 000 in the collapse, though it is not clear if this was a realized loss or simply a loss in the paper value of his holdings.

The House of Commons demanded an inquiry, and it soon became clear that part of the cause of the collapse was corruption within the government: in modern terms, several members of the government, including the king and some of his mistresses, traded on inside information right before the collapse of the stock prices. The foreign-born royal family might have suffered a great blow to its prestige were it not for the clever handling of the House of Commons by an up-and-coming politician: Robert Walpole. In gratitude, George allowed Walpole to sacrifice some of his German favorites (someone had to be blamed), and Walpole began a long period as the true leader of the government and the first modern Prime Minister of Great Britain (though he himself rejected the title).

George Berkeley also resided in Venice in the years following the Jacobite uprising. Berkeley's philosophy included the tenet that an object has no existence unless it is capable of being perceived. At one point, Berkeley attempted to begin a university in the New World (in Bermuda); he lived in Newport, Rhode Island between 1728 and 1731 while trying to raise money. The venture failed, and he returned to England, leaving his plantation and thousand volume library to Yale College (now University).

In 1734, Berkeley became the Bishop of Cloyne (in Dublin); in the same year, he published *The Analyst; or, a Discourse Addressed to an Infidel Mathematician* (generally believed to be aimed at Halley). Berkeley's contention was that if mathematicians could accept Newton's fluxions, then they should have no problems accepting the mysteries of theology. For example, in finding the ratio between the fluxion of x^n to the fluxion of x (the equivalent to finding the derivative of $y = x^n$), Newton supposed x increased by an amount equal to o, in which case the ratio of the increments was $nx^{n-1} + o(\cdots)$. By then letting o become nothing, the ratio was determined to be nx^{n-1}. But, Berkeley objected, if o was nothing, then neither x nor x^n increased at all, and the ratio was meaningless. If this "mystery" could be accepted, Berkeley argued, then surely a mathematician would not quail at the mysteries of religion.

Berkeley was quick to point out that he had no questions as to the *results* of calculus, but he believed this was fortuitous: calculus was correct because the logical errors in the analysis canceled each other out. The criticisms stung English mathematicians. By mid-century English mathematicians were making very few contributions to analysis, instead concentrating their efforts on areas of mathematics, such as probability and algebra, whose foundations were considered rigorous.

8.1.4 Maclaurin

The forces the Duke of Argyll used to win the Battle of Sheriffmuir included the Berwick militia, under the command of Alexander Hume, Baron Polwarth. Polwarth became a diplomatic envoy of George I, and sent his son, also named George, on a Grand Tour of Europe, accompanied by COLIN MACLAURIN (February 1698–January 14, 1746). Unfortunately George Polwarth the younger died in Montpellier, France in 1724, and Maclaurin returned home.

Maclaurin's major mathematical work was *A Treatise on Fluxions* (1742), the first systematic exposition on Newton's calculus, which contained a detailed discussion of infinite series, including Taylor's theorem (credited to Taylor). In Maclaurin's *Treatise on Algebra* (1748), Cramer's Rule appeared for the first time, though Maclaurin's description, given entirely in words, was hard to follow and difficult to generalize.

In 1745, the Jacobites made one last bid to restore the Stuarts by rallying around James Edward's son, Charles Edward, known as "Bonnie Prince Charlie" or, to distinguish him from his father, the Young Pretender. The Forty-Five was the last and, in some ways, the most dangerous threat to the crown, which could not devote its full effort to suppressing the rebellion because of its involvement in continental conflicts.

In 1738, Captain Robert Jenkins appeared before parliament and displayed what he said was his ear, cut off by Spanish coast guards as he sailed into the West Indies. The War of Jenkins Ear against Spain erupted in 1739, over Walpole's objections. When the War of the Austrian Succession began in 1740, the British, realizing that Austria's defeat would enhance the power of France, entered the war on the side of Austria. But the war was not going well by 1745, and since France was fighting Britain, the French were willing to sponsor an expedition to restore the Stuarts.

Charles Edward landed in Scotland in July, and won a string of easy victories. By September they were marching towards Edinburgh. Maclaurin helped prepare the defenses for the city, but the resources available to him were so limited that he could do little, and the city fell to the Jacobites. Maclaurin never recovered from an illness he acquired during the subsequent events, and died in January of the next year. Meanwhile, the Young Pretender failed to gain much of a following in England proper, and on April 16, 1746, his army was defeated at the battle of Culloden Moor, the last land battle ever fought on the island of Great Britain. The British forces were commanded by the Duke of Cumberland, who remained in Scotland to round up Jacobite sympathizers; his subsequent executions (about 120 in all) earned him the nickname of "Butcher Cumberland." Bonnie Prince Charlie fled to the island of Skye in the Hebrides, where Flora MacDonald disguised him as a woman, "Betty Burke", and helped him escape overseas.

An air of nostalgia overhangs Bonnie Prince Charlie, for the Jacobite invasion was seen as a bid for Scottish independence. In the aftermath of the Forty Five, the British government broke up the highland clans, banned the wearing of kilts and tartans, confiscated arms, forbade the playing of bagpipes and, since the relatively infertile soil of Scotland made crop-raising unprofitable, forced the clans off their lands to make way for large herds of sheep. The failed Jacobite invasion would be commemorated in songs like "Twa [Two] Bonnie Maids" (1819) and "The Skye Boat Song" (1884).

8.1.5 Mathematics and Coffee

One of the most unusual features of the pre-modern era is the constant consumption of alcoholic beverages. Pepys's diary shows that ale or beer was consumed at breakfast, lunch, and dinner, and often in large quantities. Though 17th century beer would be fairly weak by today's standards (averaging less than 1% alcoholic content), the constant state of partial inebriation must have had a profound effect on how the world was viewed. When soldiers returning from the Netherlands brought gin to England, people consumed it as if it was

Figure 8.1. William Hogarth's *Gin Lane*.

beer, with predictable results. In 1751 William Hogarth produced the contrasting prints "Beer Street," whose inhabitants are tranquil, happy, well-fed and well-dressed, and "Gin Lane", where half-starved men fight with a dog over a bone, a neglectful mother allows her baby to fall off a railing, and a building collapses in the distance from ill-repair.

Coffee provided an alternative to alcohol; the first coffeehouse in London opened in 1652 (in a building that is now a pub on St. Michael's Alley). Within a few years, coffee-houses began serving tea as well.

Then as now coffeehouses were more than places to buy coffee or tea. They were places to meet and discuss the latest news, and to conduct business. For example, as early as 1688, insurance brokers, businessmen, and ship owners met at a coffee house owned by Edward

Figure 8.2. William Hogarth's *Beer Street*.

Lloyd. Lloyd's main function, besides providing coffee and a meeting place, was to provide his customers with shipping lists and cargo manifests, and in 1696 he began publishing Lloyd's List, still being published on a regular basis. Generally speaking no single insurer was willing to be responsible for an entire ship and its cargo; instead, they would underwrite the amount they were willing to be responsible for beneath the ship's name.[3]

A somewhat more unusual activity occurred at the coffeehouses: the teaching of mathematics. Nor were the mathematicians second-rate amateurs: WILLIAM WHISTON (December 9, 1667–August 22, 1752), who would become Newton's successor as Lucasian

---

[3]Lloyd himself did not sell insurance; he simply provided a place for people to buy it. This is still one of the main functions of Lloyd's of London.

Professor, was a coffeehouse mathematician who introduced Samuel Clarke to Newton's mathematics in a Norwich coffeehouse around 1700. Whiston would eventually be dismissed from his post in 1710 (he subscribed to Arianism, the belief that denied the divinity of Christ, which would influence the formation of the Unitarian churches), and returned to his coffeehouse roots, this time in London. He would be one of the first to perform public scientific demonstrations.

Meanwhile Clarke went on to be one of the staunchest defenders of Newton's physics, entering into a famous debate with Leibniz over the nature of space, time, and motion. Leibniz adhered to Cartesian physics, and in a letter to Clarke dated August 18, 1716, derided the "miraculous and imaginary" action of gravity over an intervening vacuum. Clarke's reply on October 29, 1716 was quintessentially Newtonian: gravity was an observed fact, and the fact that scientists were not able to explain *how* it worked should not be used as a reason to reject the fact of universal gravitation.

Perhaps the most distinguished of the coffee-house teachers was a self-taught mathematician, THOMAS SIMPSON (August 20, 1710–May 14, 1761), who worked by day as a weaver. In 1740, Simpson published *The Nature and Laws of Chance*, a text largely based on the work of de Moivre. Simpson incurred de Moivre's wrath, for his cheaper editions of de Moivre's work threatened de Movire's sales. In retaliation, de Moivre warned his readers not to be taken in by other editions that "mutilated" his propositions and "confounded" them with a "crowd of useless symbols."

One of the early debates in statistics was whether a single, accurate measurement was better than the average of a number of measurements. On April 10, 1755, Simpson presented the first paper on the theory of errors. In effect, Simpson assigned a probability distribution to the size of the error, and showed that the probability that the mean would be within a certain distance of the true value was significantly greater than the probability than any single observation would be within the same distance.

Simpson was one of a very few English mathematicians who decried slavish adherence to Newton's mathematics; indeed, in a collection of his works, *Miscellaneous Tracts on Some Curious, and Very Interesting, Subjects in Mechanics, Physical Astronomy, and Speculative Mathematics* (1757), he wrote, "it appears clear to me, that, it is by a diligent cultivation of the Modern Analysis, that foreign Mathematicians have, of late, been able to push their Researches farther, in many particulars, than Sir Isaac Newton and his followers here, have done."[4] Unfortunately, English mathematicians continued to apply the geometrically based Newtonian calculus for generations after Simpson, and as a result, English progress in mathematics languished for a century.

8.2 Central and Eastern Europe

Between 1697 and 1698, a six-and-a-half foot tall "Sergeant Peter Mikhailov" traveled with a Russian delegation through Europe. The "Sergeant" studied carpentry and western methods of naval construction in Amsterdam and in the Royal Naval yard in Deptford, England. In his free time, he visited schools, museums, and Parliament. It was an open secret that the

---

[4]Simpson [1757], n.p.

"Sergeant" was Peter I absolute ruler of Russia; he would soon put his observations into practice as the first of the Enlightened Despots.

Peter intended to use all of his autocratic power to drag his country into the eighteenth century. Some of his changes were literally cosmetic. For example, western European nobility were clean-shaven. Thus Peter ordered the Russian boyars to shave, and walked the streets himself, shears in hand, looking for nobles who disobeyed his orders; if he found one, he gave the offender a none-too-careful shave. Other changes were more fundamental. He replaced the Old Church Slavonic alphabet with a simplified script, and reorganized the Russian army by introducing flintlocks, bayonets, and a merit-based system of promotions. The tireless Peter led the new Russian army to victory against the Swedes at the Battle of Poltava (July 8, 1709 O.S.), effectively ending the Great Northern War (though it would drag on for several years after).

During the war, Peter secured a spot of land at the mouth of the river Neva on the Gulf of Finland, and established a new capital there: St. Petersburg (ostensibly named after St. Peter). Under the terms of the Treaty of Nystad (August 30, 1721), Russia returned Finland and paid an indemnity to Sweden. But Sweden gave up Livonia and Estonia (now Latvia and Estonia), which gave Russia a presence on the shores of the Baltic and signaled the establishment of Russia as a major power; this was commemorated when the Russian Senate voted to make Peter not only Tsar but "Emperor of all the Russias."

In the autumn of 1724, Peter plunged into the Gulf of Finland to help save some soldiers in danger of drowning aboard a grounded ship; he caught a chill and died a few months later in the depths of a northern winter. In a suspiciously familiar story, Peter, unable to speak in his last moments, gestured for some paper so he could designate his successor, but only wrote as far as "Everything should be given to" before he died. The palace guards helped make his second wife Catherine Tsarina.

8.2.1   The Bernoullis

To emulate the French and British academies of science, Peter chartered the Russian Academy of Sciences in 1724. CHRISTIAN GOLDBACH (March 18, 1690–November 20, 1764) became a professor of mathematics at the Academy in 1725, and toured Europe recruiting capable scientists and mathematicians willing to relocate to distant Russia. His most important recruits all came through the University of Basel, in Switzerland.

Basel was the home to the Bernoulli family, wealthy merchants who fled Antwerp during the persecution of the Protestants during the Eighty Years War. The father of JAKOB BERNOULLI (December 27, 1654–August 16, 1705) wanted him to become a pharmacist, but Jakob developed an interest in mathematics and between 1676 and 1683, studied mathematics in Geneva, Paris, the Netherlands, and England. In 1683 he returned home and accepted the chair of mathematics at the University of Basel.

Jakob's great work was the *Art of Conjecturing*, which included a proof of the Law of Large Numbers. Unfortunately it was unpublished at the time of his death in 1705, and the manuscript languished until his nephew NIKOLAUS BERNOULLI (October 21, 1687–November 29, 1759) edited and published it in 1713.[5] By then *Analytical Essay on Games*

---

[5] Since Nikolaus was the name of the father of Johann and Jacob Bernoulli, the name Nikolaus was a common one in the family. This Nikolaus is usually referred to as Nikolaus (I) Bernoulli.

of Chance (1708) by PIERRE RÉMOND DE MONTMORT (October 27, 1678–October 7, 1719) had appeared, which duplicated many of the key results of the *Art of Conjecturing*, so Jakob Bernoulli's work had less impact than it might have otherwise.

On September 9, 1713 Nikolaus posed a series of problems to Montmort. The fifth problem raised a peculiar paradox involving expected value. Consider the following game: Suppose *B* rolls a die, and *A* will pay him 1 coin if he obtains a 6 on the first throw; 2 coins if he obtains a 6 on the second throw; 4 coins if he obtains a 6 on the third throw, and so on. It is easy to show that the expected value of this game is infinite, so a rational gambler should be willing to pay any price to enter such a game. On the other hand, even a mathematician confident in the theory of probability would hesitate to pay even a modest amount (say twenty dollars) to play the game.

The discrepancy between the rational and the actual decision became known as the St. Petersburg Paradox because it was resolved by DANIEL BERNOULLI (February 8, 1700–March 17, 1782) in a paper published by the St. Petersburg Academy in 1738.

Actually, the problem had been solved ten years before by GABRIEL CRAMER (July 31, 1704–January 4, 1752). Since 1724, Cramer had been teaching at the Academy of Calvin, and knew the Bernoullis and their mathematics well: he would be entrusted with the preparation of the complete works of both Jakob Bernoulli (1744) and his young brother Johann (1742). Cramer is best known for Cramer's Rule, which appeared in his *Introduction to the Analysis of Algebraic Curves* (1750) as a means of solving five equations in five unknowns.

In a letter to Nikolaus Bernoulli on May 21, 1728, Cramer not only solved the problem, but rephrased it in its modern form: Suppose *B* were to flip a coin, and *A* were to pay *B* 2^{n-1} coins if the first "Heads" result occurs on the nth toss. Cramer resolved the paradox by noting that mathematicians value money in proportion to its quantity, while "men of good sense" value money in proportion to the use they may make of it. Daniel elaborated on this idea and developed the concept now known as decreasing marginal utility: the value of a commodity is dependent on how much of that commodity you already possess. For example, a billionaire might not hesitate to wager a thousand dollars on a game of this sort, whereas someone of more modest means might quail at even a ten dollar ante.

If Jakob Bernoulli made profound discoveries in probability, his young brother JOHANN BERNOULLI (August 6, 1667–January 1, 1748) made equally profound discoveries in calculus. Jakob taught Johann mathematics (over their father's objections), and the brothers corresponded with Leibniz. On the whole, Jakob was a more rigorous mathematician, while Johann was more innovative. For example, Johann invented the technique of solving differential equations using integrating factors, partly in response to Jakob's more complicated method of solution. Johann also discovered an integral version of the Taylor series. In modern terms if n is a function of z, then

$$\int n \, dz = nz - \frac{z^2}{1 \cdot 2}\frac{dn}{dz} + \frac{z^3}{1 \cdot 2 \cdot 3}\frac{d^2 n}{dz^2} - \frac{z^4}{1 \cdot 2 \cdot 3 \cdot 4}\frac{d^3 n}{dz^3} + \cdots .$$

Johann's best-known discovery was made in connection with GUILLAUME FRANÇOIS ANTOINE (1661–February 2, 1704), the Marquis de l'Hôpital. The Marquis came from a distinguished family that produced two Marshals of France and several lieutenant generals;

naturally, l'Hôpital entered the military and became a cavalry captain. Unfortunately, he was terribly nearsighted, and unable to see clearly beyond about ten feet. Thus, l'Hôpital was kept from active service, and eventually resigned his commission. He turned his attention to mathematics.

On March 17, 1694, l'Hôpital wrote to Johann Bernoulli, offering to pay him 300 livres a year on the condition that Bernoulli work on problems posed by l'Hôpital, and that Bernoulli not reveal his solutions to anyone else. On June 7, 1694, l'Hôpital asked Johann to evaluate

$$y = \frac{\sqrt{2a^3x - x^4} - a\sqrt[3]{a^2x}}{a - \sqrt[4]{ax^3}}$$

when $x = a$ (in modern terms, he wanted a value that would make y continuous at $x = a$, and thus needed to find the limit as x approached a). On July 22, 1694, Bernoulli replied with an explanation equivalent to what we now call l'Hôpital's rule. L'Hôpital included it in his book, *Analysis of the Infinitely Small For the Investigation of Curves* (1696), the first differential calculus text. He made no claim to originality, and gave generous credit to both Johann and Jakob Bernoulli for their contributions, but Johann felt slighted nonetheless.

After Jakob's death in 1705, Johann received the chair in mathematics at the University of Basel. Despite being in existence for more than two centuries, the university remained small, boasting of a dozen or so teachers and less than 100 students. But those hundred included Johann's sons Daniel and NIKOLAUS (II) BERNOULLI (February 6, 1695–July 31, 1726).

Daniel studied medicine as well as mathematics, and had two opportunities to obtain a chair at the University of Basel. Though he was qualified, the final decision among the qualified candidates was by lottery, and Daniel lost both times. Thus he and his brother Nikolaus went to Venice to continue his study of medicine, arriving in 1723. Fortunately Daniel became too ill to continue his medical studies, and wrote four papers on mathematical and physical topics during his convalescence. These attracted the attention of Goldbach, who (like Law, Stirling, Berkeley, and others) was also in Venice; Goldbach invited the brothers to St. Petersburg.

Daniel was one of the first continental physicists to wholeheartedly embrace Newton's mechanics, and made great strides developing fluid mechanics (the weakest part of the *Principia*). One of Daniel's main contributions to mathematics concerned the theory of recurrent sequences; he presented some of his main results in 1728. To find a closed form expression for a recurrent sequence, he assumed that the terms of a recurrent sequence were given by a geometric sequence (de Moivre used the same idea, though Bernoulli gave a proof as well). For example, Bernoulli analyzed the Fibonacci sequence (though he did not call it that), where $a_{n+2} = a_{n+1} + a_n$. Next he supposed that $a_n = a_0 r^n$. Hence:

$$a_0 r^{n+2} = a_0 r^{n+1} + a_0 r^n,$$
$$r^2 = r + 1.$$

Thus $r = \frac{1 \pm \sqrt{5}}{2}$. Since there are two roots, the general term of the recurrent sequence will be

$$a_n = \alpha \left(\frac{1 + \sqrt{5}}{2}\right)^n + \beta \left(\frac{1 - \sqrt{5}}{2}\right)^n$$

where α and β are constants to be determined. Bernoulli let the first term of the sequence be 0 (which is consistent with Leonardo of Pisa's original problem); hence $a_0 = 0$ and $a_1 = 1$. This gives the system of equations:

$$0 = \alpha + \beta,$$

$$1 = \alpha \left(\frac{1 + \sqrt{5}}{2} \right) + \beta \left(\frac{1 - \sqrt{5}}{2} \right).$$

Hence $\alpha = \frac{1}{\sqrt{5}}$ and $\beta = -\frac{1}{\sqrt{5}}$.

Like his uncle Jakob and cousin Nikolaus, Daniel also contributed to the development of the theory of probability. He was unaware of Cramer's solution to the St. Petersburg problem, so he presented essentially the same solution in a paper published by the Academy in 1738. He also applied probability and expected value to a more serious issue: epidemiology and public health.

Smallpox was one of the most dreaded diseases of the era. Not only did it kill a sizable percentage of its victims, those who survived were often scarred for life. A 1713 letter to the Royal Society described the practice of inoculation, which originated in the Ottoman Empire: pus from the sores of someone already infected with smallpox would be injected into someone not yet infected. The result was a mild, non-scarring, non-lethal case of smallpox that would protect the person from future outbreaks. When a smallpox epidemic hit New England in 1721, Cotton Mather of the Massachusetts Bay Colony encouraged the Boston physician Zabdiel Boylston to begin widespread inoculation. On June 26, 1721 Boylston was the first person to inoculate patients in the New World. The first patients were Boylston's own son Thomas—and two slaves. Despite threats, Boylston inoculated some 240 persons, only six of whom died of smallpox.

The problem was that, while inoculation worked in theory, in practice there was no way to control the severity of the resulting disease. The recipient *usually* come down with a mild case, as desired, but they *might* come down with a full blown case, with potentially fatal consequences. In 1760 Daniel analyzed the practice of inoculation by determining the expected value of the number of years of life added to a person's lifetime *if* they were inoculated. In order to complete the analysis he had to develop a model for the number of persons in a population who had not yet caught the disease; his differential equations led him to one of the first examples of the logistic model in epidemiological studies. Based on empirical evidence, Daniel assumed that a person had one chance in eight of contracting smallpox, and one chance in eight of dying from it; he concluded that on average inoculation would extend a person's life by three years.

8.2.2 Leonhard Euler

Nikolaus (II) Bernoulli might have had a glorious career as a mathematician. However he died shortly after his arrival in St. Petersburg of a hectic fever (a recurrent fever with a daily spike in temperature, followed by chills and sweating). This left a vacancy in the academy, and on Daniel's recommendation, Goldbach offered the position to another graduate of the University of Basel: LEONHARD EULER (April 15, 1707–September 18, 1783). Euler arrived in St. Petersburg on May 24, 1727, bringing (at Daniel's request) some essentials of life unavailable in Russia: tea, coffee, and brandy. Daniel eventually tired of life in Russia and returned to Basel in 1733.

Euler took the tools created by Newton, Leibniz, and the Bernoullis and applied them to every area of mathematics; indeed, it is not much of an exaggeration to say that Euler created modern mathematics. He published an average of 800 pages a year during his productive life; a complete bibliography of Euler lists 866 items, including some very lengthy books, not to mention several thousand letters by which he communicated other results to the mathematicians of Europe.

Many of Euler's results are obtained through the clever use of infinite series, though at the time a rigorous theory of series had not yet been developed, and many of Euler's techniques (though not his results) are no longer considered valid. For example, in "On the Sums of Many Series of Reciprocals", finished around 1736 but not published until 1740, Euler showed

$$\frac{\pi^2}{6} = 1 + \frac{1}{4} + \frac{1}{9} + \dots.$$

The summation of the series had eluded the best efforts of Leibniz and the elder Bernoullis.

Goldbach suggested that Euler examine some of Fermat's results in number theory. Euler, like most of Fermat's correspondents, was at first uninterested, but Goldbach was persistent (and, as Euler's superior at the Academy, was in a position to be insistent as well). Euler soon found that number theory was a rich field worthy of investigation. He began by disproving some of Fermat's claims, including Fermat's conjecture that numbers of the form $2^{2^n} + 1$ were always prime. Euler also proved Fermat's Lesser Theorem: If p is prime, then $a^{p-1} - 1$ is divisible by p. In fact, he proved this theorem four times, extending and generalizing it into the Euler-Fermat Theorem: If k is the number of numbers relatively prime to n, then n divides $a^k - 1$. This was the first appearance of what is now known as the Euler-ϕ function (where $\phi(n)$ is the number of numbers relatively prime to n).

Euler and Goldbach were representative of a peculiar situation in 18th century Russia: the domination of Russian cultural and political life by non-Russians. Germans were particularly well-represented: during the reign of Tsarina Anna, foreign affairs were handled by Andrey Osterman (from Westphalia), the army was controlled by Burkhard Münnich (from Saxony), and the court favorite was her lover, Ernst Johann Biron, the Duke of Courland.

Upon Anna's death, Biron became the regent for her two-month-old grand nephew Ivan (who was only one-quarter Russian). Biron was universally hated: his purges resulted in the execution of several thousand people and the exile of more than 30,000 to Siberia. Within a month he was replaced by Münnich, but to the Russian nobles, this merely meant the replacement of one German with another. Anti-foreign sentiment was running high, and Euler gladly accepted an offer from Frederick II, the King of Prussia, at the Berlin Academy of the Sciences. A few months after Euler's departure, the Russian nobles revolted, and there was a great purge of the Germans in the government: Biron, Osterman, Münnich and Ivan's mother Anna were sent to exile in Siberia, and Ivan was imprisoned for the rest of his life. Euler avoided the purge, but entered a war.

Frederick II inherited the third largest army in Europe from his father, Frederick William I; only the armies of France and Russia were larger than that of Prussia. This was an incredible achievement, considering that these countries were ten or more times the size of Prussia in population and wealth, but Frederick William I did something even more amazing: to pay for his massive military buildup, he instituted a system of compulsory primary education and encouraged commerce and industry. The taxes generated by the resurgent Prussian economy allowed him to end his reign with a budget surplus. To maintain the growing Prussian economy and balanced budget, Frederick William I avoided going to war. His son was not so scrupulous.

On April 19, 1713, the Austrian Emperor Charles VI proclaimed the Pragmatic Sanction, which declared the indivisibility of the Habsburg lands: they would pass intact to his eldest son; in the event that he had no sons, they would pass to his eldest daughter. In 1720, after the death of his only son, Charles reaffirmed the pragmatic sanction, trying to guarantee the inheritance of his daughter Maria Theresa. It did no good, and on Charles's death on October 20, 1740, several powers claimed part or all of the Habsburg lands.

Prussia had no claim to any Habsburg territory, but it did have an enormous army. Frederick II offered his help, in exchange for Silesia (one of the richest provinces of the Empire). Maria Theresa refused, so Frederick invaded Silesia hoping to seize by force what he could not obtain by negotiation. He defeated an Austrian army at the Battle of Mollwitz (April 10, 1741), though the Austrians captured a number of prisoners, including Maupertuis, another scholar recruited by Frederick for Berlin. Maupertuis would not arrive in Berlin until 1744, and when he accepted the Presidency of the Berlin Academy of Sciences, he was expelled from the Paris Academy.

The victory of Prussia over Austria led to the formation of an alliance including France, Spain, Bavaria, and Saxony, all hoping to dismember the Habsburg empire in the War of the Austrian Succession. But Maria Theresa proved a capable ruler, and by 1748, Habsburg territory remained mostly intact, with one exception: Silesia.

8.2.3 Maria Agnesi

At the start of the war the Austrian empire included Milan, the birthplace of MARIA GAETANA AGNESI (May 16, 1718–January 9, 1799). Agnesi's father Pietro was a wealthy silk merchant who recognized his daughter's talents early and arranged for her to have the best available tutors. He also arranged for disputations at his home where the young Maria would debate with notable scholars, frequently in their own native language: by the time she was eleven, Agnesi could speak Latin, Greek, Hebrew, Italian; French, Spanish, and German.

One of the frequent visitors to the Agnesi residence, Ramiro Rampinelli, encouraged Agnesi to write a textbook on differential calculus. The result was the remarkable *Foundations of Analysis for the Use of Italian Youth* (1748, 1749), dedicated to the Empress Maria Theresa. In two volumes and 1000 pages, Agnesi covered nearly the whole of mathematics, from algebra to infinitesimal analysis. Her best-known contribution is perhaps her least important: the description of a curve invented by LUIGI GUIDO GRANDI (October 1, 1671–July 4, 1742) and which was called the "turning," *la versiera* in Italian. John Colson, who had learned Italian so he could translate *Foundations of Analysis* into English, misread this as *l'avversiera*, "she-devil" or "witch," and consequently the curve is called the witch of Agnesi. *Foundations of Analysis* was received so well that Pope Benedict XIV praised her work, and appointed her to the Chair of Mathematics and Natural Philosophy at the University of Bologna in September 1750. She accepted, but did not teach.

One of the interesting features of *Foundations of Analysis* is that it was written in Italian *for* Italians (Euler's major works, in contrast, were written in Latin). Most of the Italian peninsula was ruled by Spain, Austria, France, or the Papacy. The dominant *Italian* power on the peninsula was the Kingdom of Sardinia-Piedmont, created from the Duchy of Savoy as a result of the War of the Spanish Succession and ruled by the ambitious Charles Emmanuel III.

During the War of the Polish Succession (1733–1735), Charles Emmanuel joined France and Spain against Austria, hoping to be rewarded with Austria's holdings in southern Italy. Unfortunately, these territories (Naples and Sicily) were given to Spain, though he received Novarra and Tortona. Thus he entered the War of the Austrian Succession on the side of Austria, hoping to expel Spain from Italy. This time he was more successful, and Sardinia-Piedmont received part of the Duchy of Milan. The growing power of Sardinia-Piedmont created hope that a Kingdom of Italy might soon be realized.

8.2.4 Berlin

Around 1755, Euler began communicating with a young mathematician living in Turin, the capital of Sardinia-Piedmont: JOSEPH LOUIS LAGRANGE (January 25, 1736–April 10, 1813), the great-grandson of a French cavalry captain in the service of the Duke of Savoy, Charles Emmanuel II.

Euler was impressed with Lagrange's work on the calculus of variations, and allowed Lagrange to publish first so that the young mathematician might be able to establish his reputation. Indeed, Lagrange's treatment has since become the foundation for the current theory. On September 2, 1756, with Euler's sponsorship, Lagrange became a member of the Berlin Academy.

Lagrange quickly became one of the most respected mathematicians in Europe. In a different time or place, Lagrange might have received royal patronage from the king. However Charles Emmanuel's quest to turn Sardinia-Piedmont into a major power meant that little of the state's resources could be spared to support culture or the arts. D'Alembert, who met Lagrange when he visited Paris in 1763, tried to find him a more suitable position. He convinced Frederick II to invite Lagrange to Berlin, so that (in Frederick's own words) "the greatest king in Europe" would have "the greatest mathematician" at his court. This was the worst possible thing for Frederick to say, because Euler was already in Berlin. Lagrange, not wanting to give the impression that he considered himself Euler's superior, declined Frederick's offer and stayed in Turin.

Frederick had just survived the Seven Years' War, which began in August 1756 with a pre-emptive strike by Prussia against Saxony to forestall an invasion by Austria and Russia. France joined the others in an alliance to crush Prussia once and for all; many of the smaller German states, mindful that if Prussia went, they were likely to be next, allied with Frederick. In quick succession Frederick and his allies defeated the Austrians, Saxons, French, and Russians. Since military victories are more glamorous than responsible government, it is the son and not the father who became known as Frederick the Great.

Unfortunately Frederick was winning battles but losing the war. While Frederick defeated the Austrians at Liegnitz, the Russians under the command of General Tottleben entered Berlin and on October 9–12, 1760, burned much of it, including Euler's farm in nearby Charlottenburg. When Tottleben learned of the owner of the farm, he declared that he was not waging war on mathematicians, and indemnified Euler for damages; the Tsarina Elizabeth sent Euler an additional 4000 florins.

Prussia might have been wiped from the map of Europe but for a fantastic stroke of luck. In 1762, the Russian Tsarina Elizabeth died, leaving the throne to her nephew Peter, the Duke of Holstein-Gottorp. Peter detested the Russians and hero-worshipped Frederick, so he withdrew from the alliance against Prussia. Subsequently Peter was deposed and assassinated. His wife Catherine II, eventually known as Catherine the Great, succeeded him. Though she had no love of Frederick, she saw no point in continuing the war. Without Russia's help, Austria and France could not inflict a decisive defeat on Frederick, so in 1763, the war ended with the signing of the Treaty of Hubertusburg, restoring the central European situation to what it was before the beginning of the war.

The immense cost of the war made Frederick reluctant to enter into any other conflicts, and for the rest of his reign, the Prussian war machine remained at peace. Frederick turned to the finer things in life and tried to make Berlin a center of culture. He furnished his palace *Sans Souci* ("without a care" in French) with the finest artwork he could buy. He wrote music, scholarly treatises on history and politics, and mediocre poetry. Unfortunately for those around him, Frederick believed in personal rule (he despised kings who ruled through their ministers).

Euler found himself an unwilling recipient of Frederick's advice. The king's liberal tendencies, particularly with regards to religion, further annoyed the pious Euler. By 1766, Euler had had enough, and accepted an offer from Catherine the Great to return to St. Petersburg, where he stayed for the rest of his life.

A few years later Denis Diderot arrived at Catherine's court. Diderot's work on the encyclopedia placed him in dire financial straits, so to help him Catherine bought Diderot's library and paid him to organize it; he accepted and in 1773, visited Catherine in Russia to thank her and draw up a plan for an Imperial Russian University. He left a few months later.

The nineteenth century English logician Augustus de Morgan gave the following explanation for Diderot's departure: Diderot, a notorious atheist, was informed that Euler had a mathematical proof of the existence of God. When pressed, Euler announced, "Sir: $\frac{a+b_n}{n} = x$, therefore God exists; answer that!" The story is extremely improbable: Diderot never intended to stay in Russia, so no reason must be invented for his departure, and Euler's reverence for both mathematics and God make it unlikely that he would use them so frivolously.

Frederick was furious at Euler's departure, though the fact that this paved the way for Lagrange's acceptance (on November 6, 1766) helped mollify him. Lagrange joined a distinguished group of scientists and mathematicians that included JOHANN HEINRICH LAMBERT (August 26, 1728–September 25, 1777). In 1768 Lambert proved that π was irrational by proving that if $\tan v$ was rational, then v irrational; since $\tan \frac{\pi}{4} = 1$, this proved that $\frac{\pi}{4}$ was irrational, and hence π was irrational.

Lambert was also involved in the creation of non-Euclidean geometry through his *Theory of Parallel Lines*, written in 1766 though not published until twenty years later. Unbeknownst to him much of his work was anticipated by that of the Jesuit priest GIOVANNI GIROLAMO SACCHERI (September 5, 1667–October 25, 1733), who attempted to prove

Euclid's fifth postulate in *Euclid Vindicated From All Faults* (1733). Both Lambert and Saccheri considered a quadrilateral $ABCD$, and AD, BC perpendicular to AB, then there are three possibilities for the angles $\angle ADC$ and $\angle DCB$:

1. The hypothesis of the right angle: the two angles are both right angles. This corresponds to Euclid's fifth postulate.

2. The hypothesis of the obtuse angle: the two angles are obtuse. In this case, parallel lines will always intersect.

3. The hypothesis of the acute angle: the two angles are acute. In this case, there will be non-parallel lines that never intersect.

Saccheri easily showed that the hypothesis of the obtuse angle led to a contradiction. But he was unable to find a corresponding contradiction for the hypothesis of the acute angle. What he found was that, under this hypothesis, two lines perpendicular to the same line would approach each other asymptotically.

Lambert, following the same line of argument, arrived at the same conclusion. But whereas Saccheri believed his conclusion was contradictory (and thus Euclid was "vindicated from all faults") Lambert considered further effects of the hypothesis of the acute angle, leading him to the first mathematical investigation of non-Euclidean geometry and the discovery of the hyperbolic functions.

8.3   France

The eighteenth century cannot be understood without taking into account the influence of François-Marie Arouet, better known as Voltaire. On May 5, 1726, after a quarrel with the Chevalier du Rohan (who had made fun of his pseudonym), Voltaire went to England, where he would meet Berkeley, Clarke, and others, and be introduced to the discoveries of Newton. Voltaire would later contrast the relative freedom in England with the rigid strictures in France, though he did note that the case of Admiral John Byng, court-martialed and executed by firing squad for failing to relieve Minorca during the opening phase of the Seven Years' War, showed that the English sometimes found it necessary to shoot one admiral "to encourage the others."

When he returned to France, Voltaire published *Philosophical Letters* (1734), which concluded with an attack on Pascal's belief that man's purpose was to reach heaven through penitence. Instead, Voltaire argued, man's goal should be to increase the happiness of his fellows through the rational application of science and technology, the very goals and methods espoused by the philosophers of the Enlightenment.

The attack on religion and the monarchy implicit in the *Philosophical Letters* led to an arrest warrant being issued for Voltaire, who sought refuge in the home of the Marquise of Châtelet, ÉMILIE LE TONNELIER DE BRETEUIL (December 17, 1706–September 10, 1749). In many ways the Marquise was Voltaire's protégé. Even her name shows traces of Voltaire's influence: it was originally spelled Chastelet, but Voltaire (a member of the French Academy, dedicated to maintaining the integrity of the French language) eulogized her as Châtelet, reflecting the growing custom of using the circumflex to indicate a silent (and eventually omitted) "s."

Émilie's parents arranged for her to learn fencing, riding, gymnastics, as well as Latin and mathematics. She met Voltaire at a party hosted at her parents house, and after the birth of her second son with the Marquis, she and Voltaire became lovers around 1733. Voltaire and Châtelet collaborated to publish *Elements of the Philosophy of Newton* (1736), where "philosophy" refers to physics. Châtelet and Voltaire submitted separate entries for the Paris Academy Prize of 1737 on the nature and propagation of fire; unfortunately, they competed with Euler and lost.[6] In 1740, she tried to reconcile Newtonian and Cartesian physics in *Concepts of Physics*, and went on to prepare a complete translation of the third edition of the *Principia* into French.

Châtelet's achievement in obtaining a first class scientific education was impressive, for it was difficult for a woman to enter into scientific debate at the time. The proceedings of the French Academy of Sciences were closed to women, as were the universities. They could not even attend the informal scientific lectures in the coffeeshops, for the cafés limited their clientele to men in the belief that women should not drink coffee, a belief satirized by Johann Sebastian Bach in his *Coffee Cantata* (1732). In order to gain entry to the coffeehouses, Châtelet resorted to the expedient of dressing as a man.

At the Café Gradot, Châtelet met with PIERRE LOUIS MOREAU DE MAUPERTUIS (September 28, 1698–July 27, 1759), who was one of her mathematics teachers (and lovers). Maupertuis would play a key role in the acceptance of Newtonian physics outside of Britain. In the *Principia*, Newton made the prediction that the Earth was an oblate spheroid, flatter at the poles than at the equator. If this was so, then a degree of latitude would decrease in length as one moved towards the equator. But in 1713 Jacques Cassini, using data collected by his father Gian-Domenico Cassini, suggested instead that the Earth was a prolate spheroid, flatter at the equator than the poles. Jacques Cassini and his son, César-François Cassini carried out additional measurements and in 1733, announced a confirmation of the earlier results.

However the measurements of the Cassinis had been made in the north and south of France, and the difference in length between a degree of latitude was small enough that errors in the measurements would overwhelm any actual differences. Moreover, both Saturn and Jupiter *were* oblate spheroids (a fact discovered by Gian-Domenico himself), so it was plausible that the Earth was as well. Additional evidence was needed to settle the question. In particular, scientists needed the measurement of the length of a degree of latitude at points widely separated on the Earth's surface. Thus in 1735, the Academy of Sciences sent out two expeditions: one to Lapland, under the command of Maupertuis, and the other to Peru, under the command of Charles-Marie de la Condamine. The Lapland expedition returned in 1737, and Maupertuis announced the results: Newton was correct.

The Lapland expedition included ALEXIS-CLAUDE CLAIRAUT (May 7, 1713–May 17, 1765), a child prodigy who presented his first paper to the Academy shortly after he turned 12. On September 4, 1729 the sixteen-year-old Clairaut became the youngest person ever proposed for membership in the Academy, and on July 11, 1731, he became the youngest person ever elected. Clairaut became another of Châtelet's lovers, and later collaborated with her in the translation of Newton's *Principia*. Clairaut was instrumental in solving one of the last obstacles to the acceptance of universal gravitation: the motion of the Moon's apogee.

---

[6]Euler would win ten Academy Prizes altogether, setting a record.

In the *Principia*, Newton showed that if the sun and a planet were alone in the universe, then the planet's orbit would be an invariable ellipse with the sun at one focus. The gravitational perturbations of the other planets should cause this ellipse to change in size, shape, and spatial orientation over time; these changes are collectively referred to as "inequalities" or "variations." Newton derived all of the Moon's inequalities *except* the motion of the apogee by universal gravitation, but this last eluded Newton and his immediate successors.

One possibility centered around the inverse square law. Newton claimed it for point masses, and proved it for perfectly homogeneous spheres. But Clairaut knew from firsthand experience that the Earth was neither. In a paper presented to the French academy on November 15, 1747 he suggested that the inverse square law might not be applicable to the planets. His conclusion was almost immediately attacked by Georges-Louis Leclerc de Buffon. Buffon's main argument was that the inverse-square law is so aesthetically appealing that, without great evidence to the contrary, it would be difficult to justify abandoning it.

Another explanation arose from the fact that the equations of motion cannot be solved exactly, so approximations must be used. Thus, the approximations used might be insufficiently accurate. On January 21, 1749, Clairaut deposited (but did not read) a paper with the French Academy, and on May 17, 1749 Clairaut announced that full motion of the Moon's apogee *could* be explained by the simple inverse-square law alone. However he declined to present his work. To draw him into an actual publication, Euler proposed (anonymously) that the first prize question of the Russian Academy of Sciences ought to be the explanation of the motion of the Moon's apogee, either by Newtonian mechanics or by some other means. Clairaut, with paper ready, easily won the prize by showing that the nonlinear terms, which previous mathematicians had ignored as insignificant, accounted for the remaining unexplained motion of the apogee.

8.3.1 D'Alembert

Châtelet was very liberated for a woman of her time, though it was her intellectual and not her sexual freedom that was unusual. Indeed, the practice of taking multiple lovers was common among the French upper class at the time, and would later be one of the main themes of the novel *Dangerous Liaisons* (1782) by Pierre Choderlos de Laclos. Illegitimate children and disputed parentage were inevitable consequences of this practice (indeed, Châtelet's death in 1749 was due to complications giving birth to the illegitimate daughter of the poet Saint-Lambert).

In 1717, the unmarried Claudine-Alexandrine Guérin de Tencin gave birth to a son by one of her lovers, the Chevalier Louis-Camus Destouches. Tencin abandoned the boy on the steps of the church of St. Jean le Rond, from which the boy took his name: JEAN LE ROND D'ALEMBERT (November 17, 1717–October 29, 1783). Destouches arranged to have the boy raised by a glazier and his wife, the Rousseaus, and paid for d'Alembert's education. According to legend, Tencin later tried to claim d'Alembert as her son, and d'Alembert responded, "The glazier's wife is my mother."

D'Alembert helped found the theory of partial differential equations, but in some sense, his disputes with other mathematicians were far more important. His criticisms of Clairaut's work led him to the discovery of what came to be known as the Cauchy-Riemann equations. When Euler claimed a proof of the fundamental theorem of algebra, d'Alembert remarked, "Mr. Euler hasn't yet published any of his work on the subject" and presented his own proof (1748), using ideas from complex analysis. When considering the shape of a vibrating string fixed at both ends, d'Alembert argued that the function that gave the string's shape at any time t had to be twice-differentiable, which had to be reconciled with Euler's argument that the initial position of the string might not correspond to a such a function (for example, a $\wedge$ would not even be once-differentiable at the summit); this forced a reconsideration of what mathematicians then called a continuous function.

D'Alembert was a popular guest at the salons, regular meetings among the intelligentsia at the house of some wealthy benefactor. D'Alembert's mother, the Mademoiselle de Tencin, hosted a salon whose guests included Charles Louis de Secondat, the Baron of Montesquieu, advocate of separating the functions of the government into independent executive, legislative, and judicial branches so no branch could dominate the others; Bernard le Bouyer de Fontenelle, the Secretary of the Academy of Sciences whose *Eulogies of the Academicians* (1740) provides us with biographical information of many of the key scientists and mathematicians of the early 18th century; and Pierre Marivaux, the French comedic playwright second only to Molière in popularity.

After de Tencin's death in 1749, Fontenelle and Montesqieu began to attend the salon of Marie Anne de Vichy-Chamrond, the Marquise du Deffand. D'Alembert also spent time there, where he met Julie de Lespinasse , who like d'Alembert himself was the illegitimate child of nobility (she was the daughter of the Countess of Albon). When the Marquise lost her sight in 1754, Lespinasse began taking on a more prominent position as hostess. This incurred the jealousy of the Marquise, who dismissed her in 1764. In response she established her own salon and, to the Marquise's dismay, drew away most of her attendees. D'Alembert and Lespinasse bonded closely and d'Alembert even took up residence in her home, but there is no indication that the two were ever more than friends.

In the late 1740s Diderot recruited d'Alembert and others to help write an encyclopedia covering the whole of human knowledge, producing what is considered the first modern encyclopedia. The first volume, with a preface by d'Alembert, appeared in 1751.

D'Alembert wrote most of the scientific and mathematical articles, and his work shows considerable insight. For example, his article on "Limit" begins:

> Limit, s.f. (mathematics) One says that one quantity is the *limit* of another quantity, when the second can approach the first nearer than any given amount, however small it may be supposed, without surpassing the quantity being approached...

Elsewhere d'Alembert defined the derivative in terms of the limit of the difference quotient, again anticipating the modern definition.

D'Alembert's articles on probability were interesting because of their errors. In "Cross or Tails" (French coins had a cross on one side), d'Alembert questioned the standard theory of probability, and claimed that the probability of throwing a cross on two throws was 2 in 3, because there were only three possible outcomes: cross on the first throw; tails on the first and cross on the second; and tails on both throws.

Altogether d'Alembert wrote 1309 articles, including Bibliomania ("the passion to have books, or to collect them"), Depopulation ("strictly speaking the act of depopulating a place or a country"), and others. A 1756 visit to Voltaire in Geneva provided the background for d'Alembert's article on that city. About the ministers, d'Alembert wrote: "The clergy of Geneva have exemplary morals... at no point does one see, as in other countries, bitter arguments with one another over unintelligible matters..." However, he adds, "Many no longer believe in the divinity of Jesus Christ", and (probably at Voltaire's instigation) urged Geneva to lift the "barbaric prejudice against theater." This prompted Jean-Jacques Rousseau to write "Letter to M. d'Alembert on the Theater," arguing that the theater invariably corrupted communities it infiltrated.

Rousseau (no relation to d'Alembert's foster parents) was born in Geneva, but spent many years in Paris. Rousseau's early interest was music, and Diderot enlisted him to write most of the music articles in the *Encyclopedia*. In 1752, Rousseau wrote and produced *The Village Soothsayer*. It proved immensely popular, and was later used by a twelve-year-old Wolfgang Amadeus Mozart as the basis for his *Bastien and Bastienne*.

The popularity of *The Village Soothsayer* repelled Rousseau, and thereafter he focused on philosophical writings. His most important work was *The Social Contract* (1762), where he argued that a just society is based on subordinating the needs of the individual to the general will, which might be viewed as the needs of society as a whole. For example, consider the problem of taxation. Everyone benefits when the government can pay its debts, but ultimately taxes are paid by individuals. In a just society, the needs of individuals (to keep all their wealth) must be subordinated to the needs of the society (funds to pay for projects that benefit everyone).

Unfortunately 18th century France was far from just. The government was nearly bankrupt because those who could most easily afford to pay taxes (the nobles) were also the ones who decided who should pay taxes. Through the *Parlements*, they blocked any legislation that required them to pay taxes, passing the burden on to the middle and lower classes.

With French finances heading towards disaster, something had to be done. Joseph-Marie Terray became the Controller General of France in December 1769, and imposed drastic measures to restore the state to solvency. He repudiated part of the debt, suspended interest payments on the remainder, and required forced loans from the nobility. His proposals, particularly those that required the nobility to pay taxes to the government, met with vigorous opposition from the *Parlements*. Thus, with the help of René de Maupeou, two of the seven *Parlements* were disbanded in early 1771, and the Paris *Parlement* (the most powerful) lost its power to veto royal decrees. Terray began a massive reform of the royal finances, and it looked as if financial disaster might be averted.

8.3.2 The Stability of the Solar System

The year 1749 was a very good year for Newtonian physics. Not only did Clairaut explain the motion of the Moon's apogee using universal gravitation, but Euler published a paper (for a French Academy Prize) that explained, again using Newtonian gravitation, an inequality in the motion of Jupiter and Saturn. Finally, d'Alembert showed how the precession of the equinoxes could be explained by considering the effect of the Moon's gravitational force on an oblate spheroid like the Earth. The death of Marquise Châtelet, Newton's great supporter in France, was tragic, but balanced by the birth of PIERRE SIMON DE LAPLACE (March 23, 1749–March 5, 1827).

In 1768 Laplace approached d'Alembert in Paris, seeking a position to support himself; he had with him a letter of recommendation from his teacher, Pierre le Canu. D'Alembert sent Laplace away with a mathematical problem, telling him to return in a week. According to legend, Laplace solved the problem overnight. This impressed d'Alembert more than any letter of recommendation, so he arranged to have Laplace given a teaching position at the Military School in Paris. Among Laplace's duties were examining cadets for entry into the artillery school; many years later, in September 1785, Laplace would pass a young 16-year-old Corsican named Napoleon Bonaparte.

In 1771 Laplace applied for membership to the French Academy of Sciences. Despite Laplace's obvious promise, he lost to ALEXANDRE-THÉOPHILE VANDERMONDE (February 28, 1735–January 1, 1796), known for the Vandermonde determinant, which appears in approximation theory but nowhere in Vandermonde's own work. Vandermonde was elected to the Academy on the strength of his first paper, which investigated the solvability of equations by means of radicals.

The next year Laplace applied again, but this time he was passed over in favor of Jacques Antoine Joseph Cousin, a professor at the Royal College. Vandermonde had at least published some important mathematical papers, but Cousin was a relative nobody. Laplace took offense, and prepared to leave Paris. At Laplace's request, d'Alembert asked Lagrange if there were any openings at the Berlin Academy of Sciences. Fortunately for the honor of French mathematics, Laplace was elected as an adjoint in the Academy of Sciences on March 31, 1773. Shortly thereafter Laplace began to work on the problem for which he would become famous: the proof of the stability of the solar system.

The triple triumph of universal gravitation in 1749 left unresolved the question of how gravity made itself felt across a distance. Many scientists held out for the existence of some sort of interplanetary medium. Such a medium, however, would slow the planets as they moved through it, causing them to spiral into the sun; this in turn would cause their mean motions to accelerate and manifest as a discrepancy between the predicted and actual position of a planet.

As early as 1695, Halley noted that the Moon seemed to be undergoing an acceleration in its motion, and in a letter presented to the Royal Society of London on June 1, 1749, the Reverend Richard Dunthorne brought Halley's discovery to the forefront. By an exhaustive comparison of eclipse data from ancient times to the present, Dunthorne concluded that the Moon's mean motion had undergone an acceleration in historic times. On June 28, 1749, Leonhard Euler wrote a letter to the Royal Society of London that suggested the length of the solar year was shortening: evidence that the Earth had also undergone an acceleration in its mean motion.

In his 1749 letter, Euler pointed out that the planets, as they moved through the resisting medium, would spiral in towards the sun and eventually be consumed by it. Moreover, the planets had to be farther from the sun in the past, and at some point they would have been closer to some star other than the sun, so they would have fallen into orbit about that star instead of the sun. Thus the resisting medium implied the solar system had both a beginning and an end: the first time such a suggestion had been made based on a scientific model of the universe rather than a theological one.

In 1762 the French Academy of Sciences posed as its prize question the problem of the motion of the planets through a resisting medium. Euler's son JOHANN ALBRECHT

EULER (November 27, 1734–September 18, 1800) submitted an entry that suggested that the interplanetary medium had little effect, though significantly, Euler only considered the effects over what we would call "historic" time scales, several thousand years, for good reason: the idea that the Earth was more than a few thousand years old was a generation in the future. Johann Albrecht Euler won an Award of Merit for his entry.

The winning entry was by CHARLES BOSSUT (August 11, 1730–January 14, 1814), who began by noting that universal gravitation was an "incontestable truth," a remarkable change of heart from as little as fifteen years before. An exhaustive mathematical analysis led Bossut to conclude that the acceleration of the mean motion of the Moon and the Earth were real phenomena that could be explained by the presence of a resisting interplanetary medium; however, Bossut concluded, universal gravitation could explain all the remaining variations in the planetary motion.

However, the acceleration of the Moon's mean motion was the *only* evidence in favor of an interplanetary medium. In 1772 the Paris Academy offered a prize for a new model of the Moon's motion. Lagrange and Euler won for their contributions. After showing that universal gravitation could *not* create a secular acceleration in the Moon's mean motion, Lagrange questioned the evidence itself. If a secular acceleration existed, then the second differences between the observed and predicted positions of the Moon ought to be constant.[7] However, Lagrange noted, this was not the case; he concluded that the "secular acceleration" was likely a result of imperfect observation. Shortly afterwards Lagrange, and somewhat later Laplace, began a mathematical analysis of observational error (though both were anticipated by Simpson).

Even without an interplanetary medium, the mutual gravitation of the planets might cause drastic changes in the planetary orbits, leading to disruption of the solar system. The stability question focuses on the gravitationally induced changes in the parameters of the elliptical orbits of the planets. Three of the parameters are of importance: the length a of the semimajor axis of the ellipse; the eccentricity e of the orbit; and the inclination θ of the planet of the orbit to the planet of the ecliptic. If these parameters were subject to periodic variations only, then it could be tentatively declared that the solar system was stable. In 1760 the Paris Academy posed the question of whether the solar system was gravitationally stable over long periods of time; the entry by Euler's son CHARLES EULER (July 26, 1740–July 3, 1790) attempted to show that the semimajor axis of the elliptical orbits underwent only periodic variations, but the analysis was flawed and no prize was awarded.

Between 1773 and 1785 Lagrange and Laplace would answer Newton's question about the long term stability of the solar system. In 1773, Lagrange submitted a paper to the French academy that showed the inclinations of the planetary orbits to the ecliptic underwent periodic variations only. The paper was submitted to Laplace to referee. Laplace's behavior in this matter was not admirable: he used Lagrange's methods to show that the eccentricities and the position of the aphelia of the planetary orbits, too, were subject only to periodic variations. He then presented his own results to the Academy on December 17, 1774, *before* submitting the report on Lagrange's method. Lagrange's paper appeared somewhat later; it was the last paper on the subject he submitted to the Paris Academy.

---

[7] Assuming a constant acceleration.

In 1776 Lagrange presented a paper to the Berlin Academy showing that if a is the length of the semimajor axis, then in general, $\dfrac{d\left(\frac{1}{2a}\right)}{dt}$ could be written as a sum of sines and cosines; hence the variations were periodic. Lagrange refined his proof in 1783 and 1784, and his contemporaries hailed him as the man who proved the stability of the solar system.

There remained the question of the size of the periodic variations; this was especially problematic since Lagrange's expression for $\dfrac{d\left(\frac{1}{2a}\right)}{dt}$ included terms of the form $\cos(nt + k)$, where n could be a very small number; upon integration, this would yield a term in the expression for $\frac{1}{2a}$ of the form $\frac{1}{n}\sin(nt + k)$, which could potentially be a very large number. Thus Laplace's main contribution to the stability problem was a 1785 paper where he showed that the periodic variations were necessarily small. If m, m', m'', ...are the masses of the planets whose respective orbits have eccentricities e, e', e'', ..., inclinations θ, θ', θ'', ...and semimajor axes a, a', a'', ..., then:

$$m\left(e^2 + \theta^2\right)\sqrt{a} + m'\left(e'^2 + \theta'^2\right)\sqrt{a'} + m''\left(e''^2 + \theta''^2\right)\sqrt{a''} + \ldots = \text{constant}.$$

Since all the terms are positive, no single term can be too large, and thus no single element of any orbit can become too large.[8]

The work of Laplace and Lagrange made it possible to suppose the solar system could exist for indefinitely long periods of time. In 1785, shortly after Laplace's work appeared, the Scottish geologist James Hutton published the first of a series of papers suggesting that the geological features of the Earth were the result of very slow physical processes operating over very long periods of time; it was the birth of uniformitarianism, and the beginning of the road that would eventually lead to Darwin's theory of evolution. Unfortunately Hutton wrote in a very turgid style, so his ideas remained relatively unknown until a fellow geologist JOHN PLAYFAIR (March 10, 1748–July 20, 1819) presented them in a more readable form in *Illustrations on the Huttonian Theory of the Earth* (1802).

Playfair excelled at presenting difficult ideas in an easily understood form. This led him to make an important contribution to mathematics. Since the time of Euclid, attempts had been made to replace the fifth postulate with something simpler. However, none of the alternate fifth postulates proved particularly successful until Playfair, whose 1795 *Elements of Geometry* introduced what is now called Playfair's Axiom:

> Two straight lines cannot be drawn through the same point, parallel to the same straight line, without coinciding with one another.

8.3.3   The Revolution

Louis XV died in 1774, to be succeeded by his weak-willed, indecisive and emotionally immature grandson Louis XVI. Terray's reforms of French finances were possible only because Maupeou had hamstrung Parlement; further reform might have been possible if

---

[8]The stability question is still open, because in general the problem cannot be solved exactly, and can only be solved using approximations.

the *Parlements* remained weak. But Maupeou's enemies persuaded Louis XVI to dismiss both Maupeou and Terray, and restore the power of the *Parlements*. Parlement immediately reasserted the privileged position of the nobility and clergy, and thwarted every attempt to stabilize French finances. Events in the Americas worsened the situation.

When the American Revolutionary War broke out, some saw it as an opportunity to take back the empire (mainly North America and India) they lost to Britain in the aftermath of the Seven Years' War. But Louis's Comptroller General, the Baron Turgot, opposed intervention, pointing out that the monarchy was essentially bankrupt. Turgot was dismissed on May 12, 1776 after trying to reform French finances by eliminating the special privileges of the nobility. Afterwards, the Foreign Minister, the Count of Vergennes, called for covert aid to be sent to the colonists. Many Frenchmen, including the Marquis de Lafayette, went to the Americas to help the rebels, and an official treaty of assistance was signed on February 6, 1778.

At Yorktown, the French fleet under Admiral François-Joseph Paul de Grasse blockaded the harbor and prevented British reinforcements from reaching Lord Charles Cornwallis, forcing his surrender on October 19, 1781. The French went on to capture St. Kitts in January 1782. De Grasse dispatched the *Solitaire*, a 74-cannon ship under the command of JEAN CHARLES DE BORDA (May 4, 1733–February 19, 1799), to protect a six-ship squadron carrying troops to reinforce the garrison at Martinique. The troops were delivered, but as the *Solitaire* rejoined the fleet off the coast of Dominica, a 36-ship British fleet under the command of George Rodney appeared. At the Battle of the Saintes (April 9–12, 1782) the French fleet was defeated, and de Grasse and Borda were captured.

Borda had a unique background for a naval commander: he was a cavalry officer *and* a member of the Academy of Sciences. Borda was appointed mathematician to a cavalry corps, and was responsible for performing ballistic computations. His mathematical investigations of projectile motion led to his election to the Academy in 1756. When the Seven Years' War broke out Borda served with distinction. Borda, however, soon realized the importance of sea power, so on September 4, 1758 he enrolled in the naval school at Mézière (where Bossut was teaching).

The American Revolutionary War ended with several treaties collectively referred to as the Peace of Paris (1783). The United States gained its independence from Britain. France obtained Senegal, the island of Tobago—and 2,000,000,000 livres more debt.

Time after time, finance ministers came to the same conclusion: French finances could be restored by taxing the nobles and clergy. But the nobles and clergy controlled the *Parlements*, and refused to ratify any legislation that required them to pay taxes. In this environment where the will of a minority could thwart the will of the majority, Borda and others began to consider voting systems in general. In 1770 Borda read a memoir to the Academy of Sciences on some of the problems of existing voting systems. Criticizing the French system was politically inexpedient; fortunately, Borda had the Polish parliament to point to as a model of a disastrous voting system.

In Poland, any noble could become king, so in principle, all nobles were equal politically. This equality was institutionalized in the liberum veto, whereby *any* noble could defeat *any* measure before the *Sejm* (parliament) by his veto; indeed, the liberum veto could be used to force the dissolution of the *Sejm* and the nullification of *all* the acts passed during its tenure. For example, Augustus III presided over 15 *Sejms*. Five of them were dissolved, and the remainder accomplished nothing, thanks to judicious use of the liberum veto by nobles supported by Austria or, more frequently, Russia.

Stanislaw Poniatowski became King of Poland in 1764. He faced the same problems Louis XVI did: every attempt at reform was blocked by his own government. Unfortunately, Stanislaw also had powerful neighbors. On August 5, 1772, Russia, Prussia, and Austria orchestrated the First Partition of Poland. Poland lost about one-third of its territory and nearly half of its population. Despite the looming threat of national annihilation, it took nineteen years before the *Sejm* adopted a new constitution that eliminated the *liberum veto*.

On June 16, 1770, Borda read *Memoir on Elections by Means of Ballots* to the Academy, though the paper would not be published until 1784. Borda offered an alternative to the traditional plurality system, where the candidate with the most votes wins: a system of voting he called "election by order of merit," though it is now called a Borda count. In a Borda count, the electors rank the candidates, from first choice to last choice. A last place vote was worth a points; a next-to-last place vote was worth $a + b$ points; a third-to-last place vote was worth $a + 2b$ points, and so on. In the simplest case, if there are n candidates, a last place vote is worth 1 point, a next-to-last place vote is worth 2 points, and so on. The candidate with the most points is declared the winner.

For example, suppose there are seven electors and three candidates A, B, and C, and that Electors 1, 2, and 3 rank the candidates ABC (i.e., A is their first choice, B is their second, and C is their last), Electors 4 and 5 rank the candidates BCA, and Electors 6 and 7 rank the candidates CBA. With plurality voting, each elector votes only for the candidate they rank first. Thus, Electors 1, 2, and 3 vote for A, Electors 4 and 5 vote for B, and Electors 6 and 7 vote for C; candidate A has the most votes, so he or she wins the election.

On the other hand, note that candidate A was actually the last choice of four of the seven electors. A Borda count would take this into account; if a third place vote was worth 1 point, a second place vote worth 2 and a first place vote worth 3, then candidate A would receive 13 points, candidate B would receive 16 points, and candidate C would receive 13 points, and B (who was the first or second choice of every elector) would be the winner.

An objection to Borda's method was raised by MARIE JEAN ANTOINE NICOLAS DE CARITAT CONDORCET (September 17, 1743–March 29, 1794). Condorcet had been appointed the Inspector General of the Mint by Turgot, but when Turgot was dismissed in 1776, Condorcet offered his resignation (which was refused until 1791). In 1785, Condorcet published his most important work, *Essay on the Application of Probability to Decisions Rendered by Plurality Voting*, which gave a detailed analysis of the probability that an electorate would come to the "right" decision by plurality vote.

In *Essay*, Condorcet voiced an objection to the Borda count and presented another system of voting now called the Pairwise Comparison Method. Condorcet considered a case with 81 voters and three candidates, A, B, and C. Suppose 30 voters ranked the candidates ABC; 1 ranked the candidates ACB; 10 CAB, 29 BAC, 10 BCA, and 1 CBA. A Borda count would give A 182 points; B 190 points, and C 114 points; thus B would win. However, Condorcet pointed out, 41 voters agreed with the proposition, "A is better than B", whereas only 40 voters felt the opposite; likewise, 60 voters felt "A is better than C". Had the election been held between A and B alone, or A and C alone, then A would have won. Condorcet's proposal was to compare the candidates pairwise in just this fashion: the candidate who would have won the most elections in a two-way race would be declared the winner.

Meanwhile, the French financial situation continued to deteriorate. In 1788 Jacques Necker, who had been finance minister between 1771 and 1781, was called out of retirement to restore the crown to solvency. Necker was hardly a good choice: in 1781, he claimed that the French government had an annual surplus of 10,000,000 livres (in fact there was an annual *deficit* of 46,000,000 livres), and helped finance French participation in the American Revolutionary War through loans. Between the refusal of the *parlements* to allow the privileged classes to be taxed, and poor fiscal practices, the expected states revenues for 1788 were 146,000,000 livres, while *interest* on the public debt stood at 318,000,000 livres, and state expenditures for the year were expected to be on the order of 600,000,000 livres. Enough was enough: to override the influence of parlement, Louis XVI summoned the Estates-General. On May 5, 1789, the Estates-General met for the first time since 1614.

The problem was that the Estates-General consisted of three groups: the First Estate (the nobility); the Second Estate (the clergy); and the Third Estate (everyone else). In *Parlements*, the nobility and clergy had joined forces to block every attempt at reform. What was to keep the privileged classes from doing the same thing in the Estates-General? Louis XVI granted the Third Estate the right to have twice as many members (600) as the First and Second estates. If the three estates voted at the same time, then the Third Estate's greater numbers would be able to force through necessary reforms. But then Louis nullified the numerical advantage by requiring the estates to vote separately.

This infuriated the Third Estate. In January 1789, the churchman Emmanuel-Joseph Sieyès published *What is the Third Estate?*, a pamphlet that argued that the Third Estate *is* France, and that it alone had the right to draft a new constitution. Taking Sieyès's lead, the Third Estate broke away and announced the formation of the National Assembly on June 17, 1789, and invited the First and Second Estate to join them. Louis was furious at this action, suspended the meeting of the Estates-General, and locked the meeting hall, forcing the delegates of the National Assembly to meet at a nearby tennis court. On June 20, 1789, they swore the Oath of the Tennis Court: they would not disband until they had written a constitution for France. Louis vacillated, and ordered the First and Second Estates to join the National Assembly. The French Revolution had begun.

The French Revolution might be summarized as both the ultimate expression and the ultimate perversion of the Enlightenment. On the side of reason and rationality, the Committee of Weights and Measures was established on June 22, 1789, with a mandate to reform the complicated system of weights and measures then in existence; the result was the metric system, adopted by the Assembly on March 25, 1791.

On the other hand, reason took a back seat to passion when a mob stormed the Bastille on July 14, 1789, intent on freeing political prisoners. After murdering the warden and going to the cells, the mob found only some thieves, drunkards, and prostitutes.

Meanwhile, the National Assembly began to write a constitution for France. In the legislative hall, the representatives of the nobility (whose main interest was preserving the status quo with as few changes as possible) sat in the place of honor, on the right-hand side of the president of the assembly, while the representatives of the Third Estate (whose main interest was in an overall reform of the entire political system) sat on the left-hand side; ever since then, conservative politics has been referred to as right wing, while liberal reformers have been labeled left wing.

The queen, Maria Theresa's daughter Marie Antoinette, urged Louis to flee to safety, so on June 20, 1791 the royal family left for Metz. They got as far as Varennes, near Verdun, before they were intercepted and brought back to Paris. Louis was forced to accept the new constitution, which severely limited the power of the king and placed nearly all power in the hands of the elected assemblies. Its work done, the National Assembly disbanded itself and prepared the way for its successors, first the short-lived Legislative Assembly, and then the National Convention, which first met on September 20, 1792.

Condorcet was a member of both the Legislative Assembly (representing Paris) and the National Convention (representing Aisne). On September 21, 1792 the Convention announced the abolition of the monarchy: France was now a Republic (later referred to as the First Republic). But what should be done with the king? The radicals, known as the the Jacobins (because their meeting place was once the Church of St. Jacques), wanted to execute the king. Condorcet sided with the moderate Girondists (because many of their deputies came from the region of Gironde), who argued that the king had broken no law that would warrant his death. The radicals, backed by a diverse group of parties known collectively as the Mountain (because their deputies sat in the balcony seats of the assembly) won the day.

Louis was tried, condemned to death, and executed on January 21, 1793. The War of the First Coalition (consisting of Austria, Prussia, Spain, the Netherlands, and Britain) began soon after. On March 18, the Austrian army entered Belgium, and the French general commanding the armies in Belgium, Charles Dumouriez, defected to the Austrians. On April 6, with military reverses all around them, the National Convention gave nearly dictatorial powers to one of the most notorious groups in history: the Committee of Public Safety.

Since Dumouriez was a Girondist, the Committee ordered the arrest of 31 Girondists on June 2, and the radicals presented a new constitution to the Convention. Condorcet urged the Convention to reject the new constitution as unnecessary. This sealed his fate, and on July 8, 1793, the Convention issued a warrant for his arrest on the charge of opposing the unity of the Republic. The penalty for this crime was death, so Condorcet fled. According to one account he was caught when he ordered an omelet at an inn in Clamart. When asked how many eggs he wanted in it, Condorcet replied, "A dozen." Condorcet then tried to claim his profession as a carpenter, but the lack of calluses on his hands gave him away. He was arrested, but before he could be taken away to Paris for execution, took poison on the night of March 26, 1794.

The ordering of an omelet is a commonplace thing to us, but in Condorcet's time it was very new. Inns might serve food to guests, but guests had the choice of eating at the host's table, or cooking their own food. Then in 1762 a Parisian soup-vendor, A. Boulanger (the name means "baker") opened the first restaurant, named because a diner could choose from a menu of "restorative" dishes. The first luxury restaurant in Paris was the Great London Tavern, opened by Antoine Beauvilliers in 1782. A frequent diner there was Anthelme Brillat-Savarin, a lawyer who served in the Third Estate and later wrote the eight-volume *The Physiology of Taste* (1825), which includes the immortal phrase, "Tell me what you eat and I'll tell you what you are." The number of restaurants increased rapidly, and by

1804, Paris boasted of more than 500 restaurants. The proliferation of the restaurant is sometimes attributed to the flight of the nobles, who left their chefs and cooks without a means of support, though in fact many restaurateurs (such as Beuavilliers) opened their establishments before the Revolution to capitalize on the middle class desire to act, dress— and eat—like the nobles.

Condorcet's suicide occurred in the middle of the Reign of Terror, instituted by the Committee for Public Safety to ferret out the enemies of the Republic. But one who believes enemies can be anywhere will invariably find them everywhere. Institutions established by the monarchy were viewed with great suspicion. The Academy of Sciences was disbanded on August 8, 1793. Borda (member of an ancient noble family), Laplace (son of minor nobility) and others were dismissed from the Committee of Weights and Measures.

Oddly enough Lagrange, who had returned to Paris in 1787, was appointed as a replacement despite the fact that he was technically an enemy alien. Lagrange soon found himself threatened by a law of September 1793 subjecting all enemy aliens to arrest and confiscation of their property; Antoine-Laurent Lavoisier, the founder of quantitative chemistry (and another one of those dismissed from the Committee of Weights and Measures) intervened to protect Lagrange. Unfortunately Lavoisier could not protect himself, and he would be guillotined for being a tax farmer; Lagrange would comment, "It took but a moment to sever that head, and a hundred years may not produce another like it."

Maximilian Robespierre was appointed to the Committee on July 27, 1793. Robespierre intensified the Reign of Terror, but arresting and executing losing generals failed to produce victorious ones. More useful were the actions of Lazare Carnot, who joined the Committee on August 14, 1793. Carnot almost immediately issued a decree that (on paper, at least) called for universal, compulsory military service for all French men, though in practice only unmarried men between 18 and 25 were organized into battalions. Several hundred thousand troops were raised in this way. But Carnot realized that modern warfare required more than bodies, and issued other decrees that mobilized the resources of the entire nation to support the army.

On October 5, 1793, the Convention introduced its most universally despised measure: the decimal calendar. The French Republican Calendar consisted of 12 months, each consisting of 3 ten-day weeks, with five (or six) extra days added at the end of the year. The old month names were discarded, and new names evocative of the season were instituted; September 22, 1792 (the date of the autumnal equinox and coincidentally, the date of the abolition of the monarchy) became 1 Vendémiaire of Year I (with the years designated by Roman numerals).

On 4 Frimaire, Year II (November 24, 1793) Borda proposed and the Convention imposed the next logical step: decimal time. Each day was divided into 10 hours, and each hour was divided into 100 decimal minutes consisting of 100 decimal seconds. Borda also suggested dividing a right angle into 100 degrees, each divided into 100 minutes of 100 seconds.

Slowly the tide of battle began to turn. Carnot himself participated in the Battle of Wattignies (October 16, 1793), helping lift the siege of Maubeuge by the Austrians, before returning to Paris. Then in December, Toulon, captured by a British and Spanish fleet in August, was recaptured. Augustin Robespierre wrote to his brother praising the "transcendent merits" of a young artillery officer instrumental in the defeat of the British: Napoleon Bonaparte.

Robespierre and his supporters espoused a program of political *and* social equality, while Carnot and others felt that the revolution's impact ought to be limited to establishing political equality. On 22 Prarial, Year II (June 10, 1794), Robespierre wrested from the Convention a new law that revoked the right of the accused to a public trial, and gave the jury only two choices: acquittal or death. Some 1400 people were executed over the next month, most through the use of a new device: the guillotine, introduced as a merciful alternative to the usual form of execution (hanging).

Robespierre finally went too far by establishing a new, secular religion, threatening to overturn the existing social order in Catholic France. On 9 Thermidor, Year II (July 27, 1794—one year to the day after becoming a member of the Committee of Public Safety) Robespierre and his supporters in the Committee were arrested; they were executed the next day. Though the intention was simply to eliminate Robespierre, the Thermidorian Reaction ended the Reign of Terror. An estimated 300,000 people were arrested during the Reign of Terror, and more than 15,000 were executed, not counting those, like Condorcet, who cheated the executioner by committing suicide.

8.3.4 Napoleon and his Mathematicians

Before the Revolution, GASPARD MONGE (May 9, 1746–July 28, 1818) took a post examining prospective naval officers. At the time, the most important qualification for becoming an officer was noble birth, but Monge took the unpopular view that the commander of an expensive ship should know more than his ancestry. Somehow he avoided dismissal by the patronage-riddled government, and his humble birth—his father was a peddler—endeared him to the revolutionaries, who made him Minister of the Navy in late 1792. But Monge was no more willing to put up with interfering Republicans than he was with interfering nobles, and resigned on February 13, 1793, only to be reappointed five days later. It was not until April 10, 1793 that his resignation was finally accepted.

Monge and Carnot both served on the Commission of Public Works, which had a mandate to organize new schools. On March 11, 1794, the Commission established the Central School for Public Works, later changed to the *École Polytechnique* ("Polytechnical School"). The school opened in December with a rigorous curriculum, which included 800 hours of mathematics lectures, delivered by *Professors*, in addition to time spent with *Repeaters* who explained the lectures and helped students prepare for tests given by the *Examiners*. Lagrange was the first professor of analysis there, while Monge lectured on a new branch of mathematics he pioneered: descriptive geometry.

For centuries, engineers and mapmakers had been struggling with the same problem that plagued artists: presenting a three-dimensional object as a two-dimensional drawing. The medieval solution had been to show 3/4 of the object, and Renaissance artists began to use perspective, but both would distort key details and omit vital information. Monge's solution was brilliant in its simplicity: the object was *not* shown in perspective, but rather as a projection onto a set of two perpendicular planes.

A geographic feature, like a mountain, poses a different problem. Using Monge's system, the mountain would be shown first in profile, and then from a vantage point directly above it. Neither would allow the elevations of the mountain to be accurately presented. Various methods were invented to deal with representing geographical features, including contour lines and slope hachuring; one of the most successful practitioners was GUILLAUME-HENRI DUFOUR (September 15, 1787–July 14, 1875), who gave his name to the Dufour map, which incorporated the use of contour lines to show elevations.

The Commission of Public Works established a second school to train teachers: the École Normale Supérieure, which opened its doors in October 1794 and enrolled 1200 students. "Superior" indicated it was what we would call a college, while "Normal" distinguished it from the universities; it gave rise to the use of the term Normal School to refer to teacher's colleges during the 19th century. Laplace, Vandermonde, Monge, and Lagrange taught and wrote textbooks for their courses.

Lagrange taught calculus, but one of his students complained that he spoke too softly with too thick an accent, which made little difference since most of the students were incapable of appreciating his work anyway. The critic was JEAN-BAPTISTE FOURIER (March 21, 1768–May 16, 1830), whose life suffered from an excess of luck, both good and bad. Fourier arrived in Paris on the eve of the Revolution; shortly thereafter, his school closed. He presented a paper on finding bounds on the roots of an algebraic equation, which was lost; consequently the theorem was rediscovered by JACQUES CHARLES FRANÇOIS STURM (September 29, 1803–December 18, 1855) and named after the latter. During the Terror, Fourier defended some of those accused by the Committee for Public Safety, and was imprisoned. He appealed to Robespierre, but his appeal was rejected and Fourier would have been executed. Fortunately Robespierre fell and Fourier was released.

Fourier then enrolled in the Normal School, which closed. But Fourier impressed his teachers, who arranged for him to obtain a position teaching at the École Polytechnique in 1795. Later that year, some members of the National Convention called for the arrest of the remaining supporters of Robespierre and the Committee for Public Safety. Since Fourier wrote to Robespierre, he was arrested; only sworn testimony from his friends saved him from a lengthy prison sentence.

Meanwhile, Napoleon's star continued to rise. On October 5, 1795, a revolt began in Paris under the auspices of royalists hoping to restore the monarchy. When a mob marched on the Tuileries, Napoleon ordered his regiment to fire their cannons into the crowd. The whiff of grapeshot saved the Republic, and sent a clear message: revolt would not be tolerated. Shortly thereafter the National Convention dissolved itself, and a new government was constituted: the Directory, so-named because executive power in the government was vested in the hands of five Directors, drawn from the legislature.

The First Coalition might have been able to restore the monarchy during this time, but two of its members, Austria and Prussia, were preoccupied with Poland. On January 23, 1793, Russia and Prussia agreed to the Second Partition of Poland, annexing half of what remained of Poland after the First Partition. A revolt began under the leadership of Thaddeus Kosciuszko, veteran of the American Revolution.[9] Despite his successes in America, he could not prevent Russia, Prussia, and Austria from dividing up the remnants of the kingdom in the Third Partition of Poland (October 24, 1795).

---

[9] Kosciuszko played a key role in defeating a British army at Saratoga (October 17, 1777), and began building fortifications at West Point on the Hudson River.

Digesting Poland turned the eyes of the great powers away from France, giving the new Republic time to prepare for invasion. Carnot suggested a three-pronged attack against the First Coalition. Austria, the greatest danger, would be met in Italy by an army under Napoleon's command; by the end of 1797, Napoleon was in control of northern Italy. Monge accompanied Napoleon's army to Italy as part of a commission to select works of Italian art to be looted and brought back to fill the newly opened Central Museum of the Arts, housed in a former royal palace: the Louvre.

A public museum of art had been the dream of the Comte d'Angiviller, Louis XVI's Director of Buildings. The revolution forced d'Angiviller to flee France in 1791, but the project continued without him, and the Central Museum of the Arts opened in 1793. After France annexed Belgium, Flemish paintings, including some by van Eyck and many by Rubens, were sent back to the Louvre, expanding its collections; Monge added the works of Italian masters.

In 1798, Napoleon made a fateful decision. The Directors wanted an invasion of Britain, France's greatest adversary, but Napoleon realized that to invade without mastery of the sea was suicide. But there was a second option: economic warfare could be waged on the "nation of shopkeepers" by striking Egypt and threatening the route to India. To oversee the looting of antiquities that would surely fall into French hands after the conquest of Egypt, Napoleon sent for Monge, who sent for Fourier.

The Egyptian campaign began well, but on August 1, 1798 at Abukir, just east of Alexandria, the British fleet under the command of Admiral Horatio Nelson destroyed the French fleet, trapping the French army in Egypt. The victory of the British encouraged the allies, who formed the Second Coalition on December 24, 1798. Two new allies were Russia and the Ottoman Empire (the nominal rulers of Egypt), and with Napoleon stuck in Egypt, the armies of the Second Coalition scored great victories against the French. Sieyès (who was by then a Director) insisted that only a military dictatorship could save the Republic: "I am looking for a saber," he declared. Napoleon got the hint, and sneaked out of Egypt on August 22, 1799, abandoning his army.

Fourier and others remained. A few days after Napoleon's flight a French soldier named Boussard (or Bouchard) discovered a slab of basalt on which a royal decree was written in three languages: Greek, Demotic (a form of Egyptian), and Hieroglyphic. Since scholars knew how to read Greek and Demotic, the Rosetta stone (named after a town near where it was found) could be used to decipher the hitherto unreadable hieroglyphics. Fourier himself would help write the *Description of Egypt*, founding modern Egyptology in the process.

Napoleon landed in France on October 8. By then, French armies had scored some victories against the Second Coalition, and the Republic was no longer in need of saving. However, Napoleon and Sieyès continued their plans. On October 19, 1799, Napoleon wrote to Laplace, who had just sent him the first two volumes of his *Celestial Mechanics*. Napoleon promised to read them the first six months he had free, and invited Laplace to dinner the next day. Subsequent events lead us to conclude that they spoke of matters here on Earth rather than in the skies, for Napoleon overthrew the Directory of November 9 and, as First Consul, made Laplace the Minister of the Interior on November 12, 1799.

As First Consul, Napoleon was effectively the dictator of France, and as the Minister of the Interior, Laplace oversaw all domestic affairs except for finance and the secret police.

Among other things, Laplace recommended that July 14 (the date of the storming of the Bastille) become a national holiday. Napoleon was less than pleased with Laplace's performance, and criticized him for bringing "the spirit of the infinitely small" to the government. Thus Laplace was replaced a few weeks later by Napoleon's brother Lucien. However Laplace continued to receive honors from his former student, becoming a Grand Officer in the newly formed Legion of Honor (1802), and Chancellor of the Senate (1803).

In 1802, Napoleon became Consul for life, with the right to appoint his successor. This effectively restored the monarchy, but his popularity was so great that few cared that the net effect of the Revolution was a replacement of the Bourbons with the Bonapartes. Indeed, since France now ruled over many non-French territories, Napoleon made ready to announce the formation of the Empire of France.

Students at the Polytechnical School made ready to publish an attack on Napoleon's plans for an empire, but they were talked out of this potentially dangerous course of action by one of the professors: SIMÉON-DENIS POISSON (June 21, 1781–April 25, 1850). Poisson feared that Napoleon would take action against the school (which was easy to find) and not the students (who could melt into the larger society). Poisson himself had no interest in politics, and once said, "Life is good for only two things: to study mathematics, and to teach it." In both he was successful, and few indeed are the branches of mathematics that do not have a Poisson object, method, or theorem.

On May 28, 1804, Napoleon was proclaimed Emperor of the French. He crowned himself on December 2, 1804, and within a few years, restored the pre-revolutionary system of nobility. Fourier became a Baron; Laplace and Monge became Counts. By then Napoleon was fighting yet another war with the British and a coalition of European powers.

At the Battle of Trafalgar on October 21, 1805, a British fleet destroyed French naval power and England's mastery of the seas was assured. But once again Napoleon proved invincible on land. At Austerlitz he crushed a combined Austro-Russian army (December 2, 1805); since the emperors of Austria, Russia, and France were all present, the battle is usually called the Battle of the Three Emperors. On July 12, 1806, Napoleon declared the existence of the Confederation of the Rhine, reorganizing the non-existent Holy Roman Empire into a French dominated political unit. Austria, Prussia, Brunswick, and Hesse declined membership and continued the fight against Napoleon. But the Prussian army that once held off Austria, Russia, and France found that past military glories were no substitute for current military preparedness, and was crushed at Jena and Auerstädt (October 14, 1806).

8.3.5   Gauss

The Duke of Brunswick, Carl Wilhelm Ferdinand, died from wounds received in the battle of Auerstädt. The Duke, Frederick the Great's nephew, hoped to emulate his uncle's success at establishing an enlightened despotism; he reformed the duchy's finances and educational system and called on all the teachers in the duchy to watch for especially promising students.

CARL FRIEDRICH GAUSS (April 30, 1777–February 23, 1855) came to the attention of the Duke as a result of an incident that occurred around 1785. Gauss's teacher (named Büttner) assigned his class the problem of summing an arithmetic sequence; Gauss almost immediately wrote down the correct answer. In the oldest version (told by a colleague of

Gauss, who claimed that Gauss often recounted it), the sequence is not identified, nor is Gauss's method explained.[10]

Büttner assigned Johann Martin Bartels, to work with Gauss on more advanced mathematical topics.[11] At 14, Gauss began receiving a yearly stipend of 158 thalers from the Duke to continue his studies. Gauss graduated from the University of Brunswick in 1795 and continued his studies at the University of Göttingen.

Göttingen, in Hanover, was one of the newest universities in Europe. It had been established in 1737 by George Augustus, the Elector of Hanover (also George II, king of England), and rapidly became the center of a new German humanism. Between 1740 and 1780, Göttingen was the home of a group of poets calling themselves the Göttingen Grove, who wanted to free poetry from the rationalism of the Enlightenment (and purge it of French influences): they were one of the precursors to the Romantics.

More central to the new German humanism and the Romantic movement was Christian Gottlieb Heyne, a philologist at Göttingen who specialized in mythology. Heyne noted that our knowledge of the ancient myths is invariably through literature from a much later and more sophisticated period, so the sense of fear and awe felt by the original mythmakers is absent. Thus Heyne attempted to trace the development of poetry in an effort to recover the original myths and regain the primordial sense of wonder.

Mathematics at Göttingen was represented by ABRAHAM GOTTHELF KÄSTNER (September 27, 1719–June 20, 1800). Kästner had compiled monumental encyclopedias on mathematics, but he was a poor teacher. Gauss later called Kästner (who composed epigrams) the best mathematician among poets, and the best poet among mathematicians. Thus Gauss had a difficult choice to make: whether to pursue philology and follow in the footsteps of the brilliant Heyne, or mathematics and be associated with the mediocre Kästner.

Fortunately for mathematics, Gauss made a profound discovery in 1796. Euclid's *Elements* included compass and straightedge constructions for regular polygons of 3, 4, 5, and 15 sides. Since bisecting an angle is also possible using compass and straightedge, then polygons with $3 \cdot 2^n$, $4 \cdot 2^n$, $5 \cdot 2^n$, and $15 \cdot 2^n$ sides were also possible. Gauss discovered a means for constructing a regular 17-sided polygon (a heptadecagon). Kästner's reaction was, apparently, muted: he dismissed the importance of the discovery.[12] Gauss quit the University of Göttingen before taking a degree, and returned to the University of Brunswick where in 1799, he completed his dissertation, giving a new proof of the fundamental theorem of algebra (the first of four proofs he would give over the next fifty years).

In 1801, Gauss published *Arithmetical Investigations*, which included the algebraic basis of his construction of a regular heptadecagon. Consider a line segment AB. If we can, using compass and straightedge only, construct a line CD of length n times AB, then n is said to be a constructible number. It is clear that the constructible numbers include all whole numbers; since we can divide a line into any number of equal segments, then all positive rational numbers are also constructible. If A is one endpoint of the diameter of a circle, B on

---

[10]The story that Gauss summed the sequence $1 + 2 + 3 + \cdots + 98 + 99 + 100$ first appeared in 1938, in a paean to Gauss by Ludwig Bieberbach.

[11]Later, Bartels would have another famous student: Nikolai Ivanovich Lobachevsky.

[12]To be fair, Gauss's discovery was a straightforward modification of the methods of Lagrange and Vandermonde.

the diameter, and C at the other endpoint of the diameter, then the perpendicular through B to a point D on the circle will satisfy $BD = \sqrt{AB \cdot BC}$; so quadratic irrationals (one that is the solution to a quadratic equation with integral coefficients) are also constructible. In general, a number N is constructible if it is a positive rational number, or if it is a solution to a quadratic equation whose coefficients are constructible numbers.

Gauss knew that the roots of $x^n - 1 = 0$ could be expressed as $\cos \frac{2\pi k}{n} + i \sin \frac{2\pi k}{n}$ for $k = 0, 1, 2, \ldots, n - 1$, which allowed the circle to be divided into n equal arcs as a prelude to the construction of a regular n-gon. Thus if the roots of $x^n - 1 = 0$ are constructible, so is the regular n-gon.

Suppose r is a root of this equation. Obviously $r^n = 1$. If $r^m \neq 1$ for $m < n$, and all roots are powers of r, then r is said to be a primitive root. For example, the equation $x^4 - 1 = 0$ has roots $1, -1, i$, and $-i$. The roots i and $-i$ are primitive, since the powers of i (or $-i$) generate all roots; meanwhile 1 and -1 are not primitive, since their powers do not generate all the roots. Thus the constructibility of a regular n-gon hinges on the constructibility of a primitive root of $x^n - 1 = 0$.

Since $x^n - 1 = 0$ has one (non-primitive) root $x = 1$, we may factor out $x - 1$. This gives us the cyclotomic equation

$$x^{n-1} + x^{n-2} + x^{n-3} + \ldots + x^2 + x + 1 = 0$$

(named because it originates in the cutting of a circle). Lagrange and Vandermonde realized that the roots had to satisfy certain equations, which allowed them to find the nth roots of unity. Gauss's contribution to the problem of cyclotomy was finding a simple means of constructing equations satisfied by some of the roots.

Gauss's method can be illustrated using the equation $x^5 - 1 = 0$, which gives rise to the cyclotomic equation $x^4 + x^3 + x^2 + x + 1 = 0$. Take a primitive root r, and consider an arbitrary power of r, say r^λ. Then consider the roots r^{λ^n}. For example if $\lambda = 3$, then the roots represented by r^{3^n} are:

$$r^3, r^9 = r^4, r^{27} = r^2, r^{81} = r \ldots$$

where we use the fact that $r^5 = 1$ to simplify r^9, r^{27}, and so on. In this case, the powers of r of the form r^{3^n} generate all roots.

On the other hand, consider r^{4^n}:

$$r^4, r^{16} = r, r^{64} = r^4, \ldots.$$

This separates the roots into two sets, r and r^4 (which are formed from the powers of r^4) and r^2 and r^3. In the general case, if one group of roots is $r^m, r^k, r^p, \ldots$, the other sets will be $r^{jm}, r^{jk}, r^{jp}, \ldots$ for some value or values of j.

Now let the two sums $r + r^4$ and $r^2 + r^3$ be the solutions to an equation; since there will be two solutions, the equation will be of degree 2, and the simplest such equation will be:

$$
\begin{aligned}
(y - [r + r^4])(y - [r^2 + r^3]) &= y^2 - [r + r^4 + r^2 + r^3]y + (r + r^4)(r^2 + r^3) \\
&= y^2 - [r + r^2 + r^3 + r^4]y + (r^3 + r^4 + r^6 + r^7) \\
&= y^2 - [r + r^2 + r^3 + r^4]y + (r^3 + r^4 + r + r^2) \\
&= y^2 - [r + r^2 + r^3 + r^4]y + [r + r^2 + r^3 + r^4].
\end{aligned}
$$

Since r solved the cyclotomic equation $x^4+x^3+x^2+x+1 = 0$, then $r^4+r^3+r^2+r+1 = 0$, or $r^4 + r^3 + r^2 + r = -1$. Hence our equation can be simplified as:

$$(y - [r + r^4])(y - [r^2 + r^3]) = y^2 + y - 1.$$

Since this equation is quadratic, we can solve it easily and find $y = \frac{-1\pm\sqrt{5}}{2}$. One of the roots corresponds to $r + r^4$, and the other corresponds to $r^2 + r^3$.

How can we determine which of the two solutions is $r + r^4$ and which is $r^2 + r^3$? Gauss suggested that the approximate values of the two solutions could be determined and compared with the known values of the roots. For example, if we suppose r is the principal root $\cos\frac{2\pi}{5} + i \sin\frac{2\pi}{5} \approx 0.309 + 0.951i$, then $r + r^4 \approx 0.618$. Hence $r + r^4 = \frac{1+\sqrt{5}}{2}$.[13]

To disentangle the two roots we can construct another quadratic equation with r and r^4 as roots:

$$(z - r)(z - r^4) = z^2 - (r + r^4)z + r^5$$

$$= z^2 - \left(\frac{-1 + \sqrt{5}}{2}\right)z - 1.$$

Solving this equation will give us r and r^4, and one of these will be the primitive root; from the primitive root, all the others can be constructed.

Note that in this case the primitive root was found by solving a quadratic equation whose coefficients included a number found by solving another quadratic equation whose coefficients were rational: thus a primitive root of $x^5 - 1 = 0$ is constructible, so the regular pentagon is also constructible. For a regular heptadecagon, Gauss noted that the roots could be split first into two groups of eight (so the sum of each eight could be found by solving a quadratic equation); each group of eight could then be split into two groups of four (and again the sum of the four could be found by solving a quadratic equation); each group of four could be split into two groups of two, and finally the groups of two could then be solved using a quadratic equation.

On the other hand, consider the problem of constructing a regular heptagon. This corresponds to the problem of finding the roots of $x^7 - 1 = 0$. Let r be one of the primitive roots. Then r^{2^λ} will generate a set of three of the roots:

$$r^2, r^4, r^8 = r$$

and the roots will split into two groups, r^2, r^4, r, and r^6, r^5, r^3. As before, let the sum of the roots in each group be the roots of an equation. In this case the equation will be

$$(y - [r^2 + r^4 + r])(y - [r^6 + r^5 + r^3]) = y^2 + y + 2$$

with solutions $y = \frac{-1\pm\sqrt{-3}}{2}$. From this we can determine that

$$r^2 + r^4 + r = \frac{-1 + \sqrt{-7}}{2} \quad \text{and} \quad r^6 + r^5 + r^3 = \frac{-1 - \sqrt{-7}}{2}$$

---

[13] Actually, we can determine this directly, since if $r = \cos\left(\frac{2\pi}{5}\right) + i \sin\left(\frac{2\pi}{5}\right)$, then $r + r^4 = 2\cos\left(\frac{2\pi}{5}\right)$, which is a positive real number.

The problem arises in trying to disentangle the three roots $r^2 + r^4 + r$. In order to do so, we let r^2, r^4, and r be the roots of an equation:

$$(z - r)(z - r^2)(z - r^4) = z^3 - (r + r^2 + r^4)z^2 + (r^3 + r^6 + r^5)z - r^7$$

$$= z^3 - \left(\frac{-1 + \sqrt{-7}}{2}\right)z^2 + \left(\frac{-1 - \sqrt{-7}}{2}\right)z - 1.$$

Solving this cubic requires extracting a cube root, which is geometrically inconstructible. This suggests that the regular heptagon cannot be constructed using compass and straight-edge alone, though it was not until 1837 that PIERRE WANTZEL (June 5, 1814–May 21, 1848) proved the impossibility of certain constructions.

Gauss's analysis suggests the following: Suppose p is prime and $p-1$ has a prime factor k. Then at some point in the process of separating the roots into sets, we must separate the roots into k sets, which will require the solution to a kth degree equation. But if $k > 2$, the solutions do not correspond to constructible numbers. Thus the only regular p-gons (where p is prime) that can be constructed are those where $p = 2^q + 1$. Furthermore, if q itself has any odd factors, then $2^q + 1$ will be composite. Thus the constructible polygons correspond to the Fermat primes $F_n = 2^{2^n} + 1$, a remarkable link between number theory and geometry, two seemingly unrelated branches of mathematics.

8.3.6 Germain

The French army besieged Hanover after the defeat of the Prussians, and during the attack, the commander of the French forces, Joseph-Marie Pernety, received a letter from a friend's daughter, SOPHIE GERMAIN (April 1, 1776–June 27, 1831). Germain requested that Pernety look after Gauss and make sure that no harm came to him. Perhaps mindful that Marcellus, the Roman general, is best-known as the general whose troops killed Archimedes, Pernety dispatched some troops with strict orders to protect Gauss.

Finding himself under French protection surprised Gauss, but solved a mystery for him. For many years Gauss had been in communication with "Monsieur le Blanc," discussing various results from *Arithmetical Investigations*; though Gauss was probably aware that the name was a pseudonym, he had no idea until the siege of Hanover that the mysterious Monsieur le Blanc was a woman. Germain had sent many results to Gauss, including proofs of restricted forms of Fermat's Last Theorem. Her greatest result was showing that if n is an odd prime less than 100, then $x^n + y^n = z^n$ has no integer solutions x, y, z, where x, y, z are all relatively prime to n. Germain also corresponded with ADRIEN-MARIE LEGENDRE (September 18, 1752–January 10, 1833), and Legendre's second edition of *Theory of Numbers* included a supplement, largely written by Germain.

Germain had used the "Monsieur le Blanc" pseudonym before. At the Polytechnical School, students were expected to write end-of-year reports, summarizing their accomplishments. Lagrange found the report of one "Monsieur le Blanc" especially good, and inquired to "his" identity. When Lagrange found that "Monsieur le Blanc" was in fact a woman, he was impressed: women were not allowed to study at the Polytechnical School, which meant that Sophie Germain was self-taught. Lagrange became one of her greatest supporters.

8.3.7   Projective Geometry

After the crushing of the Prussian army at the battles of Jena and Auerstädt, Napoleon issued the Berlin Decree (November 21, 1806) to strike at the British through economic warfare. The Continental System prohibited trade with Britain, but Portugal, who would be economically ruined by the blockade, and Russia, where Tsar Alexander I distrusted Napoleon's motives, refused to join the boycott.

With the cooperation of Spain's Charles IV, French troops used Spain as a staging area for an invasion of Portugal, beginning the Peninsular War. By the end of November 1807 the conquest of Portugal was complete and the Portuguese royal family fled to Brazil. Though their mission in Portugal was accomplished, French troops remained in Spain, clearly intent on staying. Thus on March 17, 1808 a palace revolution deposed the complacent Charles and placed his son Ferdinand VII on the throne in the hopes that he would lead the nation against the invaders. Napoleon forced Ferdinand to return the throne to his father, who then abdicated in favor of Napoleon's brother Joseph. More French troops came to Spain to support the new regime.

The troops included CHARLES-JULIEN BRIANCHON (December 19, 1783–April 29, 1864), an artillery lieutenant who had just graduated first in his class at the Polytechnical School. While there, Brianchon studied under Monge and made one of his most important discoveries:

Theorem 8.1 (Brianchon's Theorem). *Suppose six lines are tangent to a conic (and thus form a hexagon circumscribed about the conic section). Then the lines joining the opposite vertices intersect at a point.*

This theorem is even more remarkable because it is one of the first examples of the principle of duality. Consider the two statements:

 1. A unique line is common to any two non-coincident points.

 2. A unique point is common to any two non-parallel lines.

By exchanging "point" and "line" in the first statement (and "parallel" for "non-coincident"), we obtain the second statement. Because the two statements are related in this way, they are said to be duals of each other; because points and lines are related in this way, many theorems about lines have an equally valid dual theorem about points, and conversely. For example, "Any three points define a unique triangle" and "Any three lines define a unique triangle" are dual statements. Moreover, since the points that define a triangle correspond to its vertices, while the lines correspond to its sides, then the vertices and sides of a triangle are duals; finally, since both figures are triangles, then a triangle is its own dual.

What is the dual of Brianchon's theorem? First, we can exchange "line" for "point," and begin with six points on a conic (which will form a hexagon inscribed in a conic). A line joining opposite vertices may be viewed as the line that both points have in common; thus the dual is a point that two lines have in common. Hence the dual of the "line joining opposite vertices" is the "point at the intersection of opposite sides." Finally the intersection point all three lines have in common becomes a line that all three points have in common. Thus the dual of Brianchon's theorem is that, given a hexagon inscribed in a conic section and the opposite sides extended until they meet at points A, B, and C, then the three points

Figure 8.3. Goya's *The Third of May*.

will be on the same line. This is Pascal's hexagon theorem, discovered by Pascal when he was only 16.[14]

It would be some years before Brianchon could return to mathematics, because he and the rest of the French army were busy keeping Joseph Bonaparte on the throne. On May 2, 1808 the citizens of Madrid rose up against the French invaders; the events were commemorated in 1814 by Francisco Goya in two paintings, *The Second of May* (showing a group of citizens attacking imperial soldiers) and the more iconic *The Third of May* (showing imperial soldiers massacring a group of citizens). Joseph Bonaparte was eventually crowned King of Spain on June 15. Though more and more French troops entered Spain to support him, they could make little headway against small bands of Spanish patriots fighting a "little war" (*guerrilla*) against the French invaders. The British took advantage of the situation and eventually sent troops under the command of Arthur Wellesley, the future Duke of Wellington. As is often the case, the guerillas could not inflict serious damage on the invaders, but neither could the invaders win a decisive victory over the partisans.

To deal with Russia, Napoleon organized the largest army yet assembled. The 600,000 troops of the *Grand Armée* entered Russia in June 1812. Marshal Mikhail Illarionovich Kutuzov, commander of the Russian armies, intended to trade time for territory, but public pressure forced him to fight one of the bloodiest battles of the era at Borodino on Septem-

---

[14]It should be pointed out that it is possible to interpret Pascal's original statement of the theorem as Brianchon's theorem.

ber 7. The Russians lost 45,000 men, while the French lost 30,000. Kutuzov continued his Fabian tactics, and Napoleon arrived in Moscow on September 14, 1812. To prevent the French from using the city for winter quarters, the city was burned between September 15 and 19. After five weeks of waiting for the Russians to surrender, Napoleon realized the harsh Russian winter was fast approaching, and began a retreat from Moscow on October 19.

This time, the Russians harried Napoleon's forces with the aid of their greatest ally: "General Winter." Kutuzov attempted to block Napoleon's retreat near Smolensk at the Battle of Krasnoii (or Krasnoe) on November 14–18, 1812.[15] A bold (and suicidal) attack by Marshal Michel Ney on November 18 helped save part of the French army, but some 5000 French soldiers were killed and 8000 captured or missing.

Lieutenant JEAN VICTOR PONCELET (July 1, 1788–December 22, 1867), a graduate of the Polytechnical School, was among those captured. Poncelet was fortunate: Of the 600,000 men who went into Russia, fewer than 100,000 returned alive. Most of the casualties were from disease and cold, not enemy fire. To pass the time while prisoner, Poncelet taught geometry to fellow captives, and took the first steps towards developing projective geometry, a powerful tool for geometrical investigations. For example, the projection of a line in one plane onto another plane is another line, and the projection of a conic section onto another plane is another conic section. Thus the theorem of Pascal (or Brianchon) could be proven for one type of conic section (for example, the circle), and be valid for all conic sections via projection.

Projective geometry was clearly "in the air," for similar ideas were raised by one of Monge's students, JOSEPH DIAZ GERGONNE (June 19, 1771–May 4, 1859). Gergonne reputedly said that "No mathematical theory is complete until it can be explained to any passer-by on the street" (though this is frequently attributed to other mathematicians as well). Gergonne was also concerned that both the *Memoirs of the Academy of Sciences* and the *Journal of the Polytechnical School* were published only irregularly; he tried to encourage the publication of a more regular journal to inspire a new generation of geometers, and when that failed he himself began in 1810 to publish *Annals of Pure and Applied Mathematics*, which would last for twenty-two years. Because of his position he published many of his own papers (245 altogether), many of which were on projective geometry. This led to a priority dispute with Poncelet over the principle of duality (though in fact both were anticipated by Brianchon).

The allies chased Napoleon back to Paris. Hasty defenses had been constructed, and the students at the Polytechnic School (who were destined for the army in any case) formed a battalion. One of the students was MICHEL CHASLES (November 15, 1793–December 18, 1880), originally named Floréal.[16] Fortunately for Chasles, Marshal Auguste de Marmont realized his position was impossible and surrendered the city at the end of March 1814. Chasles would go on to further develop projective geometry and begin the study of

---

[15]Tolstoy's epic *War and Peace* (1865–9), which opens Part 11 with a discourse on the use of infinitesimal methods in solving problems in dynamics, lists the battle as beginning on November 5. The Russian Empire had not yet adopted the Gregorian calendar, so November 5 in Russia was November 14 in the rest of Europe—even France, for Napoleon abolished the Republican calendar on January 1, 1806.

[16]Floréal was one of the months in the decimal calendar, though it is not clear if there is any connection. He changed his name to Michel on November 22, 1809.

enumerative geometry (now a branch of algebraic geometry), which in essence sought to determine the number of figures of a given type that satisfy a given relationship.

Charles Talleyrand, the French foreign minister, had the Senate declare that Napoleon and his family had forfeited the throne. On April 13, 1814, Napoleon agreed to abdicate. As Chancellor of the Senate, Laplace could either sign the decree ratifying Napoleon's abdication, or resign, showing solidarity with his patron. Laplace signed. The terms of abdication were generous: Napoleon was given the island of Elba, just off the Italian coast, as his own Principality, and an annual stipend of 2,000,000 francs. The victorious allies installed Louis XVI's brother as Louis XVIII.[17] There was no Louis XVII, since Louis XVI's son Louis-Charles died in 1795.

Louis XVIII's reign was interrupted in 1815, when Napoleon escaped from Elba, rallied his supporters, and marched on Paris. Napoleon was tired of sycophants, and even rewarded some of those who opposed him on principle: Fourier, whose support had been lukewarm after the disastrous Egyptian expedition, became the Prefect of the Rhône, and Carnot, who voted against Napoleon becoming Consul for life and the establishment of the Empire, became the Minister of the Interior. Monge, who had helped in the defense of Lìege against the allies and whose support had never wavered, rejoined Napoleon enthusiastically. Laplace left Paris before Napoleon arrived. So did Louis XVIII, though he returned with the armies of Austria, England, Prussia, and Russia: over a million men marched against Napoleon. At the Battle of Waterloo (June 18, 1815), the Duke of Wellington gained renown by crushing the Bonapartists for the last time, and afterwards, Wellington summarized the hopes of a continent by saying, "I hope to God I have fought my last battle."

For Further Reading

For the Enlightenment and French Revolutionary era, see [43, 51, 53, 76, 112, 132]. The mathematics of this and all subsequent eras are readily available, either in the form of collected works or compiled journals; in many cases, these are also available through digital libraries.

---

[17]Louis XVI was Louis-Auguste, while his brother was Louis-Stanislas-Xavier.

9

The Nineteenth Century

9.1 France

In the aftermath of the Napoleonic Wars, the delegates at the Congress of Vienna sought to establish the balance of power as the key theme in European politics. The delegates joined small nations together to form larger, and theoretically more powerful, states. Belgium and the Netherlands, different in religion, language, and culture, became Kingdom of the Netherlands; the German states, different in religion and political orientation, were welded together into the German Confederation; and Sweden's claim to Norway was recognized. It is perhaps no coincidence that Mary Shelley's *Frankenstein* (1818), the story of a man who created a monster by joining together body parts from several corpses, was written only a few years after the Congress of Vienna.

The restored Louis XVIII wasted little time punishing Napoleon's supporters. Fourier lost his baronial title, and Carnot and Monge were expelled from the French Academy and replaced with AUGUSTIN-LOUIS CAUCHY (August 21, 1789–May 23, 1857). Cauchy, through the lessons he taught at the Royal Polytechnical School (formerly the Polytechnical School), would be largely responsible for establishing calculus on a rigorous basis, giving what amounts to the modern definition of limit, derivative, continuity, convergence, and the definite integral; Cauchy's publication record (789 items) is second only to Euler's (866 items).

On June 4, 1814, Louis issued a constitution that established a bicameral legislature consisting of an elected Chamber of Deputies and a hereditary Chamber of Peers; he also gave guarantees of religious and civil liberty. In August 1815, the Ultras, a group of conservatives who wanted to restore the nobility to the privileges and positions they held before the Revolution, gained a majority in the Chamber of Deputies. Fearing that this would simply lead to another Revolution, Louis dissolved the Chamber in September 1816. The new Chamber of Deputies had a moderate majority, though the Ultras maintained a significant presence, and the independents (or liberals), a hodgepodge of Bonapartists, Republicans, and constitutional monarchists united solely by their distaste for the Bourbons and their desire to maintain the advances made during the Revolution, gained steadily.

In 1819, Maximilien Foy was elected to the Chamber of Deputies. Foy was a liberal who had served with distinction under Napoleon in campaigns from Constantinople to Spain, though he voted against the Consulate and the Empire. He quickly became the spokesman for the liberal opposition to the Bourbons, and one of the leading political figures in France.

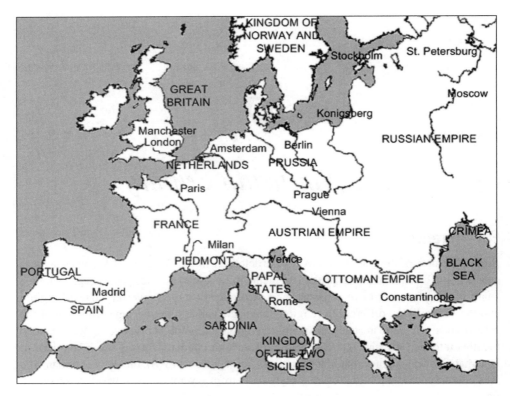

Figure 9.1. Europe after the Congress of Vienna

A political crisis soon arose. Spanish South America had been in revolt since 1810, and on August 7, 1819 the rebels under Simón Bolívar defeated a Spanish army stationed near Bogata at the Battle of Boyacá and freed New Granada (now Colombia and Venezuela) from Spanish control. Back in Spain, the recently restored Ferdinand VII assembled an army to retake New Granada. While in exile, Ferdinand declared himself a liberal and a supporter of the 1812 constitution, but as soon as he regained the throne he repudiated the constitution (May 4, 1814) and six days later, arrested many of the liberal leaders, prelude to restoring the absolute monarchy. Dissatisfaction with the reactionary Ferdinand grew, and Ferdinand's army, intended for South America, instead marched against him under the banner of a major, Rafael de Riego.

In 1823, delegates at the Congress of Verona sanctioned French intervention in Spain to suppress the revolt. With French support, Ferdinand crushed the rebels; the "Ominous Decade," marked by a persecution of the liberals in Spain, began.

The restoration of the absolute monarchy raised the possibility that Spain would attempt to reconquer South America (by then, Bolívar's forces had liberated Ecuador and were well on their way to freeing Peru as well). Consequently the American President, James Monroe, announced the Monroe doctrine on December 2, 1823: it declared the Americas "off limits" to European intervention. Since the British found it convenient to keep *other* European powers out of the Americas, the Royal Navy played a major role in enforcing the Monroe doctrine.

Around this time, Foy hired GUSTAV PETER LEJEUNE DIRICHLET (February 13, 1805–May 5, 1859) to teach German to his children. Dirichlet's grandfather came from the Belgian town of Richelet (now Richelle, a few miles northeast of Liège) and was known as "the youth [*le jeune*] from Richelet [*de Richelet*]." Dirichlet himself was born in Düren, then part of the French Empire, though the Congress of Vienna included it in the German Confederation (it is currently in Germany). Though his family name was probably pronounced "dee-ruh-shuh-lay", Dirichlet himself used the pronunciation "deer-uh-clay."

Dirichlet came to Paris to study mathematics. Number theory was his great interest, and one of the few possessions he brought with him was Gauss's *Arithmetical Investigations*. Dirichlet made some progress in the proof of Fermat's Last Conjecture.

The first appearance of Fermat's Last Conjecture occurred in 1670, when Fermat's son Samuel published his father's notes on Diophantus's *Arithmetic*. Frenicle de Bessy, based on work Fermat sent him, proved the $n = 4$ case in a book published posthumously in 1676, and Euler proved the $n = 4$ case in 1738. As early as 1753, Euler also claimed to have a proof for the $n = 3$ case, but the first appearance of this proof was in his *Elements of Algebra* (1770); as written, the proof is incomplete.

In a paper presented to the French Academy of Sciences on July 11, 1825 Dirichlet examined the $n = 5$ case of Fermat's Last Theorem. Germain's work proved that if a solution to the $n = 5$ case existed, at least one of the numbers must be divisible by 5. But if two of the numbers are divisible by 5, then all three are; hence any primitive solution must have exactly one number divisible by 5. Dirichlet showed that the number divisible by 5 could not be even. "Hence it remains only to treat the case where the number divisible by 5 is odd; but the method presented in this Memoir appears insufficient for this case, and I do not see how one can complete the demonstration of this particular case of this theorem of Fermat."[1] Legendre supplied the missing pieces, and a complete proof of the $n = 5$ case was announced on November 14, 1825.

9.1.1 The Revolution of 1830

Louis XVIII died in 1824 and his brother became King Charles X. Charles tried to restore the monarchy and nobility to the privileges and positions they held before the Revolution, with predictable results. On July 27, 1830, the mobs of Paris took to the streets. Charles abdicated on August 2, eventually settling in Prague. Louis-Philippe became the king of the *French* (as opposed to king of France) on August 9. He was known as the Citizen King. Cauchy refused to swear an oath of loyalty to the new king and emigrated to Turin, where he served the King of Sardinia, Charles Albert for a time. However in 1833 ex-King Charles asked Cauchy to join him in Prague and tutor his grandson Henry, the Duke of Bordeaux; Cauchy accepted. Cauchy tutored the Duke for five years before returning to France in 1838 on the condition that he need not swear an oath of loyalty to the new king.

Prague was the home to BERNARD BOLZANO (October 5, 1781–December 18, 1848) who, like Cauchy, attempted to place analysis on a rigorous foundation. In 1817, Bolzano anticipated Cauchy's definition of a continuous function, and proved the Intermediate Value Theorem for the first time. The proof was not quite complete, since he relied on an assumption that the sum of a subsequence of a convergent geometric series had a limit, which

---

[1] Dirichlet *Werke*, I p. 13

would be validated when Weierstrass provided a rigorous theory of the real numbers. Unfortunately Bolzano's liberal tendencies (he believed in pacifism and economic justice) led to his dismissal from the Chair of Philosophy and Religion at the University of Prague in 1819, and his residence in Prague placed him far away from the main mathematical centers of Europe, so his actual influence is hard to gauge.

Back in France the radicals still sought to abolish the monarchy. In response, Louis-Philippe found himself resorting to repressive measures similar to those of Charles X. He banned the Artillery of the National Guard on December 31, 1830 and had some of its members arrested and tried for conspiracy. However, they were acquitted on May 9, 1831, and two hundred people attended a celebratory banquet that night.

Alexandre Dumas helped organize the banquet. Though known at the time for his plays, Dumas's enduring fame rests on his novels of historical fiction, a genre that had recently been popularized by the 1831 publication of *The Hunchback of Notre Dame* by Victor Hugo. Dumas's prolific output included *The Count of Monte Cristo* (1844–5), *Queen Margot* (1845), and, of course, *The Three Musketeers* (1844), which had two sequels, *Twenty Years After* (1845) and *Ten Years Later* (1848–1850). The last was a massive work whose last part is known to English readers as *The Man in the Iron Mask.*

Dumas generally began with actual events and persons and embellished them for storytelling purposes. For example, the fictional d'Artagnan was based on Charles de Batz d'Artagnan, the youngest son of a Gascon nobleman who became one of the king's musketeers in 1644, rising to high rank within the king's service. Dumas made d'Artagnan somewhat older so he could frustrate Richelieu (who died in 1642) and Buckingham (who in reality was an incompetent bungler).

In Chapter CCIV of his *Memoirs*, Dumas recounts that, towards the end of the dinner celebrating the acquittal of the National Guard members, EVARISTE GALOIS (October 25, 1811–May 31, 1832) raised a glass and a dagger in the same hand, and toasted, "To Louis-Philippe." As the crowd was noisy, Galois repeated the toast. It was a particularly threatening gesture, and Galois was arrested. According to Dumas, Galois later explained to the prosecutor that the knife was his dinner knife, and that he had in fact toasted, "To Louis-Philippe, if he defects," but the last phrase was drowned out by the noise of the crowd. Galois was acquitted, but was arrested on July 14, 1831, for wearing the uniform of the outlawed National Guard as well as being in possession of several loaded firearms. This time, his guilt was clear, and he was imprisoned until April 29, 1832.

Galois is known for creating group theory the night before his death in a duel, a story that is at least as accurate as Dumas's historical novels. In fact Galois's work on group theory began with a paper to Cauchy in 1830 on the impossibility of solving a fifth or higher degree equation by means of radicals. Cauchy found the ideas interesting, but pointed out to Galois that he had in fact been anticipated by Abel. Cauchy recommended that Galois rewrite the paper to include Abel's results. Galois revised it and (possibly at Cauchy's suggestion) submitted it in February 1830 for the Academy's Grand Prize in Mathematics. This time it went to Fourier. Unfortunately Fourier was in poor health, and died shortly afterwards; Galois's paper was never found, and as a result, the Grand Prize was awarded jointly to Abel (posthumously) and Sturm.

Meanwhile Galois continued to develop his theories and submitted a paper to Poisson outlining what is now called Galois theory. A modern mathematician who knew what

Galois was describing would find the paper difficult reading; Poisson found it incomprehensible. But while he rejected the paper as written, he encouraged Galois to rewrite it and elaborate the ideas contained within it. Galois began doing so (it was his fourth rewrite). Near one proposition, he noted, "There is something insufficient in this demonstration; I haven't the time to complete it." Galois had no suspicion of how very true this note would prove, due to events that began half a world away.

During the first half of the 18th century, France and Britain fought each other over the control of India. At the time, India was divided into a number of mutually hostile states, and the British and French supported one side or the other in an attempt to win trading concessions. The Seven Years' War removed France as a major rival, and after the American Revolutionary War, the British focused on India; Cornwallis would become Governor-General of India and establish a civil service renowned for its integrity through the simple expedient of paying bureaucrats a salary commensurate with their position.

Since the internecine wars in India provided opportunities for foreign powers (particularly France and Russia) to intervene, the British resolved to end them. This required a large army, most of which consisted of Indian troops (called sepoys) led by British officers. One of those officers was John De Morgan, a Lieutenant Colonel, stationed in Madras. Shortly after the birth of his son AUGUSTUS DE MORGAN (June 1806–March 18, 1871), the family returned to England.

De Morgan entered Cambridge and came to associate with a remarkable group: CHARLES BABBAGE (December 26, 1791–October 18, 1871), JOHN HERSCHEL (March 7, 1792–May 11, 1871), and GEORGE PEACOCK (April 9, 1791–November 8, 1858). In 1812, Babbage, Herschel, and Peacock founded the Analytical Society, which hoped to "promote the principles of d-ism as opposed to the *dot*-age of the university," a reference to Newton's dot notation (where $\dot{y}$ represents the derivative of y with respect to some variable) while continental mathematicians used Leibniz's differential notation (where $\frac{dy}{dx}$ represents the derivative of y with respect to x). Peacock became a tutor and lecturer at Cambridge in 1815, and in 1817 used his position as mathematical examiner to mandate the use of Leibnizian notation. De Morgan was impressed, and in 1836 wrote:

> Without discussing this point, we are inclined to consider the universality of the notation of Leibnitz throughout the whole of the civilized world, and the fact of most of the important discoveries made since the time of Newton, both in pure mathematics and physics, being expressed by means of it, as itself a sufficient reason for adopting it.[2]

De Morgan was also one of the first in England to introduce Cauchy's ideas of rigor. However, change was slow to come, and De Morgan came to be better known for his contributions to logic and probability.

The De Morgans' move from India back to Britain probably saved their lives. Europeans in India came into contact with cholera, a water-born bacterial infection endemic to southeast Asia. In 1817, the endemic disease turned into an epidemic. The number of native casualties is unknown, but the British Indian army suffered more than 10,000 casualties. The disease spread through the trade routes, becoming a pandemic. It reached

---

[2] A. De Morgan, *The Differential and Integral Calculus*, The Society for the Diffusion of Useful Knowledge, 1836, p. 34.

Moscow in 1830, Berlin in 1831, and Paris in 1832; it even crossed into the New World, reaching Quebec and New Orleans in 1832 and Mexico and Cuba by 1833. In Paris alone, 18,000 people would die from it, including the French prime minister Casimir-Périer and Sadi Carnot, the son of Lazare Carnot and discoverer of some of the laws of thermodynamics. Galois contracted the disease and was transferred to a hospital. Shortly thereafter he fell in love with the prison physician's daughter, Stéphanie-Felice du Motel.

For no clear reason (but definitely having to do with Stéphanie), Galois fought a duel with Perscheux d'Herbinville, a fellow Republican, on May 30, 1832. Galois was wounded and died the next day. The night before the duel, Galois wrote up an account of his work—his fifth version—and sent it to Auguste Chevalier for publication in the *Encyclopedia Review* (which happened), and that Jacobi and Gauss be asked *publicly* their opinions "not on the truth, but on the value" of the work (which did not happen).

9.1.2   Railroads and Computers

The industrial revolution made it necessary to transport goods and raw materials faster and in greater quantities than ever before. A new form of transportation had to be developed: the railroad. The first trains were pulled by horses along wooden rails, but in 1803, the first locomotive was built by the British engineer Richard Trevithick. In 1814, a Peruvian mining company placed a large order for Trevithick's steam locomotives. Trevithick went to Peru in 1816, only to be caught in the middle of the South American war for independence, and was unable to return to England until 1827.

By then, George Stephenson had begun building railroads, and on September 27, 1825, the *Active*, began making the first regular commercial passenger run, taking 450 passengers on the 27-mile trip between Darlington and Stockton at speeds of up to 15 miles per hour. The success of the Darlington and Stockton line prompted investors in Liverpool, the largest port city in England, to build a link with Manchester, 40 miles away. Stephenson's *Rocket* could attain the hitherto unprecedented speed of 36 miles per hour, and thanks to it, the Liverpool-Manchester line opened on September 15, 1830 with eight of Stephenson's locomotives making the trip.

Several important figures observed the opening of the rail line. One of them was Charles Babbage, who invented the pilot (sometimes called the "cowcatcher") for the railroad in 1837. However, Babbage is best known for his failed inventions. In 1822, he presented a mechanical device that could evaluate polynomials. Since "engine" is a general term for a mechanical device, and Babbage's computer used the theory of finite differences, it came to be known as the difference engine.

Babbage built a demonstration model, with which he found the successive values of $n^2 + n + 41$ (a remarkable polynomial that generates prime numbers for $n = 0$ through $n = 39$). To calculate these values, we note that the first differences form the sequence 2, 4, 6, ..., and the second differences are 2, 2, 2, ... Thus the difference engine would be given a set of three values: 2, 0, and 41, representing the second difference, initial first difference, and initial value. To find the successive values, the constant second difference would be added to the initial first difference to obtain a new first difference, namely $0 + 2 = 2$. This first difference would then be added to the initial value to obtain the next value of the function: $41 + 2 = 43$. The process would be repeated: the second difference would be added to the first difference to obtain a new first difference $2 + 2 = 4$, and the new first difference would be added to the current function value to obtain the next function value: $43 + 4 = 47$. The difference engine could calculate $n^2 + n + 41$ for all integers from $n = 0$ to $n = 39$ in about four minutes.

Babbage eventually obtained a grant of £1500 from the British government to develop a more sophisticated version. Unfortunately the mechanical precision necessary to make the more ambitious difference engine work properly did not exist, and the project floundered to an unsuccessful conclusion in 1834. By then, Babbage had designed a more sophisticated computer, the analytical engine, which (had it been built) would have been the first general purpose computer. The British government declined further funding.

However, others were interested. In August 1840 Babbage lectured at the Academy of Sciences in Turin to a group that included LUIGI FEDERICO MENABREA (September 4, 1809–May 24, 1896) and physicist Ottaviano Mossotti. Menabrea wrote up the lectures in French as *Elements of the Analytical Engine of Charles Babbage* (1842). AUGUSTA ADA KING (December 10, 1815–November 27, 1852), a longtime acquaintance and sometime student of Babbage, translated Menabrea's notes into English.

Ada King was born Ada Byron, the daughter of the English poet George Gordon Byron and Anna Isabella Millbanke. Lord and Lady Byron separated shortly after Ada's birth. Lord Byron left England in 1816, never to return in his lifetime, and settled briefly on the shores of Lake Geneva in Switzerland. There his household included his physician, John Polidori; the poet Percy Shelley (who had also separated from his wife); and Shelley's mistress and future wife, Mary Godwin. In late 1816, Byron suggested they all write a ghost story. Neither Byron nor Shelley finished their stories, but Mary wrote *Frankenstein* (1818) and Polidori wrote *The Vampyre* (1819), the first vampire story to be written in English.[3] In 1823 Byron went to Greece to fight for independence from the Ottoman Empire, and died of disease the next year.

Meanwhile, Lady Byron, whose mathematical abilities led Byron to call her his "princess of parallelograms" during their courtship, encouraged Ada to study mathematics. At the time, the education of women was considered to be unimportant, though *A Vindication of the Rights of Women* (1792) by Mary Wollstonecraft (the mother of Shelley's mistress) argued that women ought to receive the same educational opportunities as men. This treatise, addressed to Talleyrand (unofficial French ambassador to Britain and later the most important diplomat of the nineteenth century), expressed hope that the Constitution of the French Republic would be amended to include the rights of women and justice for "half of the human race." Wollstonecraft argued that the poor condition of women was largely due to their education, which was

> ...more anxious to make them alluring mistresses than affectionate wives and rational mothers; and the understanding of the sex has been so bubbled by this specious homage, that the civilized women of the present century, with a few exceptions, are only anxious to inspire love, when they ought to cherish a nobler ambition and by their abilities and virtues exact respect.[4]

---

[3] Byron and Polidori parted on bad terms. Much to Byron's annoyance, when *The Vampyre* appeared, it was attributed to him.

[4] M. Wollstonecraft, *A Vindication of the Rights of Woman*, J. Johnson 1796, p. 2.

Around the same time Wollstonecraft wrote *Vindication*, WILLIAM FREND (1757–1841) wrote *The Principles of Algebra* (1796). In the preface, Frend noted, "the greater part of mathematics now taught in the university of Cambridge may be made level to the capacities of boys and girls under seventeen years of age."[5] Frend, whose anti-religious views caused him to be dismissed from Cambridge for his *Peace and Union* (1793), became one of Ada's mathematics tutors, as did Lady Byron's neighbor, MARY FAIRFAX GRIEG SOMERVILLE (December 26, 1780–November 29, 1872).

Somerville's father, William Fairfax, took the traditional view of educating girls and forbade her to study mathematics; as might be expected, this simply meant she studied algebra and geometry in secret. Her situation improved slightly with her first husband Samuel Grieg, who was merely apathetic, so during their short marriage she studied French and advanced mathematics. Grieg died in 1807, so Mary, as a widow and inheritor of a small estate, could act more freely, and she began studying in earnest under the tutelage of John Playfair and Playfair's student WILLIAM WALLACE (September 23, 1768–April 28, 1843). In 1812, she married William Somerville, a doctor with the British army who became an Inspector to the Army Medical Board in 1816. This time, her husband supported her interests in mathematics and the sciences. At the request of Henry Brougham of the Society for the Diffusion of Useful Knowledge, which published biweekly pamphlets on a variety of mathematical and scientific subjects (including, some years later, De Morgan's calculus), Mary produced a translation and annotation of Laplace's *Celestial Mechanics*. Somerville's translation appeared in 1831 and made Laplace's monumental work comprehensible to a larger audience.

The Somerville's friends and acquaintances included De Morgan, Herschel, Peacock, and Babbage. It was through the Somervilles that Ada and Babbage met. He became the most influential of Ada's teachers, for he introduced her to the love of her life: the theory of computation. In particular, how could one reduce a mathematical process to one that could be replicated by machine? When Babbage learned of Ada's translation of Menabrea, he asked her why she had not written her own treatise. At Babbage's suggestion, she incorporated a number of her own notes into the translation, published in 1843 with its author's identity obscured by the use of the initials AAL.[6] The extensive annotations doubled the length of the treatise, and included a detailed description of how the analytical engine might be programmed to calculate the Bernoulli numbers; this arguably made her the first computer programmer. Ada might have gone on to make significant contributions to mathematics and computer science, but she died of cancer just short of her 37th birthday.

9.1.3   Fermat's Last Theorem

Another observer of the first run of the Liverpool-Manchester line was GABRIEL LAMÉ (July 22, 1795–May 1, 1870), a French engineer. In 1832 a friend of his, Benoit Paul Emile Clapeyron, proposed a rail line from Paris to St. Germain-en-Laye (a town just outside of Paris). Construction began in 1835 under the direction of Clapeyron and Lamé; the rail line

---

[5] W. Frend, *Principles of Algebra*, J. Davis 1796, p. xiv.

[6] In 1835, Ada married William King. King became the Earl of Lovelace in 1838, which made his wife Countess Lovelace.

opened in 1837. Clapeyron went on to develop the laws of thermodynamics, bringing Sadi Carnot's work to a larger audience.

Lamé was more mathematician than engineer; in fact, his teaching had been criticized for being "too theoretical" (while at the same time contemporaries criticized his mathematics as "too applied"). In 1839, Lamé proved Fermat's Last Conjecture for the $n = 7$ case. Then in April 1847, Lamé tried to prove Fermat's Last Conjecture for all cases. Lamé's proof relied on the factorization of $x^n + y^n$ as:

$$x^n + y^n = (x + y)(x + ry)\left(x + r^2 y\right)\left(x + r^3 y\right) \cdots \left(x + r^{n-1} y\right)$$

where r is a primitive root of $X^n + 1 = 0$; Lamé derived Fermat's Last Conjecture as a consequence of this factorization, and his countrymen hailed his great achievement.

Only one French mathematician was unsure of the validity of the proof: JOSEPH LI-OUVILLE (March 24, 1809–September 8, 1882). In 1836, Liouville began publishing the *Journal of Pure and Applied Mathematics*, universally referred to as "Liouville's journal," in response to a critical need: in 1831, both Gergonne's *Annals of Pure and Applied Mathematics* and the *Bulletin of the Mathematical, Astronomical, Physical, and Chemical Sciences* ceased publication, leaving French mathematicians with no French journal in which to publish their results. Liouville's journal published Lamé's proof of Fermat's Last Conjecture for $n = 7$ in 1840, and Galois's papers in 1846. When Lamé announced a proof of Fermat's Last Conjecture, Liouville was almost alone in suspecting that it was flawed.

A key problem was that Lamé's proof relied on the unique factorization of $x^n + y^n$ among the generalized Gaussian integers (numbers of the form $\sum_{i=0}^{n-1} \alpha_i r^i$, where r is a primitive nth root of unity). Unique factorization had been proved for the cases of $n = 2, 3$, and 4, but for no other cases. While these three examples made it plausible that unique factorization held for all n, Liouville found it easy to construct a counter-example for arbitrary complex numbers. For example, if we consider numbers of the form $a + b\sqrt{-17}$, then 13 and $4 \pm \sqrt{-17}$ are prime, but

$$169 = 13 \cdot 13 = \left(4 + 3\sqrt{-17}\right)\left(4 - 3\sqrt{-17}\right).$$

Hence, 169 can be written as the product of two different sets of primes, and unique factorization fails.

In fact, the failure of unique factorization for generalized Gaussian integers had already been proven by ERNST EDUARD KUMMER (January 29, 1810–May 14, 1893) in 1846. It shows how little read German works were, for Kummer's results appeared in the proceedings of the Berlin Academy of Sciences. Kummer's work was a portent of things to come: the primacy of French mathematics would begin to wane with the rise of the Berlin mathematical school.

Liouville edited his journal for nearly 40 years, and published many of his 400 papers there. His most important work began with a presentation to the French Academy on May 13, 1844. "On Very Extensive Classes of Quantities Whose Values are Neither Rational nor Reducible to Algebraic Irrationals" began the study of transcendental numbers, real numbers that are not the solution to a polynomial equation with integer coefficients. At the time, the only transcendental numbers Liouville could identify were highly artificial numbers like

$$\frac{1}{a} + \frac{1}{a^{1\cdot2}} + \frac{1}{a^{1\cdot2\cdot3}} + \frac{1}{a^{1\cdot2\cdot3\cdot4}} + \cdots,$$

though on June 8, 1845, Liouville showed that $\ln x$ generates transcendental numbers if x is rational (and not equal to 1).

9.1.4 The Second Empire

Louis Philippe did the best he could to steer France between the Ultras, who wanted a return to the pre-revolutionary days, and the radicals, who wanted an abolition of the monarchy. Unfortunately, he failed. In 1848, crop failures and the accumulation of eighteen years of grievances led to revolution. Faced with the choice between firing on the rebels and abdicating the throne, Louis-Philippe abdicated, leading to the formation of the Second Republic. One of the first acts of the interim government was to declare the principle of universal manhood suffrage, which increased the electorate from 200,000 to 9,000,000.

New elections were held on April 23, 1848, and moderate Republicans (including Liouville) won 500 of the 880 seats in the legislative assembly. The radicals won only 80 seats, and resented what they perceived as the theft of the revolution by the bourgeoisie. A new revolt against the government began on June 23. In contrast to Louis-Philippe's restraint, General Louis-Eugène Cavaignac bombarded rebel positions with artillery on June 26 (earning the nickname "The Butcher of June"). Fifteen hundred rebels were killed; 12,000 were later arrested and deported to Algeria.

Cavaignac stood for election to the Presidency in December, but lost to a candidate whose very name recalled the glory days of the Empire: Louis Napoleon, the nephew of Napoleon Bonaparte. Napoleon received about 5.3 million votes. The constitution prohibited the President from succeeding himself, so on December 2, 1851, three years into his four-year term, Napoleon orchestrated a coup, seized control of the government, and requested from the electorate the right to draw up a new constitution. In a plebiscite held on December 21, Napoleon obtained 7,500,000 votes for, compared to 640,000 votes against. Within a year, he rewrote the constitution and the Second Empire began with the reign of Napoleon III (Napoleon II, like Louis XVII, lived but did not reign).

At the start of Napoleon's reign, Paris had 47 acres of public parks, eighty miles of sewers, narrow medieval streets, and a water system that relied on the river Seine both as a source of water and as a sewage dump. Napoleon envisioned a Paris worthy of an empire, and appointed Georges Eugène Haussmann the Prefect of the Seine on July 29, 1853. By 1870, the city could boast nearly 5000 acres of parks, four hundred miles of sewers, broad boulevards, and a daily supply of five million gallons of fresh water pumped in from Châlons, a hundred miles away; fifteen thousand gas lamps turned Paris into the "City of Light."

The modern fashion industry began in Paris around this time. The English designer Charles Frederick Worth arrived in Paris in 1845 and achieved great prominence when the Empress Eugénie became one of his clients. Among Worth's innovations were clothing labels, seasonal lines, and live models. It was the start of haute couture, and this era is frequently referred to as the Age of Worth.

— Quand je vous dis que c'est une robe de chez Worth, je reconnais la touche.

Figure 9.2. Sketch by Bertall (Charles d'Arnoux): 'I told you the dress was from Worth's; I know the look.'

Worth's most important innovation was bringing the industrial revolution to the clothing industry. During the seventeenth century a new model for manufacturing consumer goods was developed: the domestic or workshop system. In this system, entrepreneurs would give raw materials to workers who would turn them into finished goods; these would be returned to the entrepreneurs, who would then sell them to the consumer. The workers owned their tools and chose the hours they worked; the system was especially convenient for farm families, who used it to supplement their income during the times of year (such as winter) when there was little farm work to be done.

The convenience of the domestic system for the worker had to be balanced against its initial cost for the worker, who had to purchase a complete set of tools. In many cases the worker bought the tools with money borrowed from the entrepreneur who supplied the raw materials. At some point a crucial realization was made: most of the tools were inactive most of the time. For example, a worker constructing clothing had to use a pair of scissors to cut the fabric into pieces. Five workers in five separate workshops had to have five pairs of scissors between them. On the other hand if those five workers did their work at a central location, a single pair of scissors might suffice. Actually, there would be times when two or more workers would need the scissors simultaneously, so there was a further change: workers began specializing, so (for example) one worker did nothing but cut fabric, and hence was the only worker who ever needed the scissors.

The industrial revolution, by harnessing the power of "prime movers" such as wind and water, meant that larger, more efficient machines could be used in place of a worker using a hand tool. But these machines were even more expensive, putting them even further out of the reach of the typical worker. Thus instead of hundreds of individual workshops with relatively inexpensive tools, there was a single factory with tools too expensive for an individual worker to own. Indeed, where once workers owned the tools they worked with, the new factory workers owned nothing and the capitalist owned everything. This trend was the central thesis of *The Communist Manifesto* (1848) of Karl Marx and Friedrich Engels, which foretold an uprising of the working-class proletariat against the capitalist-class bourgeoisie, and the replacement of capitalism with socialism (and ultimately communism).

Since it was inefficient to have workers starting and finishing at different times, the centralization of production also meant the development of regular work hours. Moreover, a single factory might replace hundreds of workshops, the "work" and the "shop" were separated, and the sole function of a shop was to store goods prior to sale. But stores that were already receiving goods of one type from one factory could just as easily be receiving goods of many types from many factories, leading to the opening of Aristide Boucicaut's *Bon Marché* ("Good Buy") in Paris in 1852, the world's first department store, named because the different types of goods were kept in different departments.

For all its faults, the industrial revolution made more consumer goods available to more people at lower prices than ever before. Worth's designs were sent to factories for mass production, allowing those with modest means to dress like the wealthy. The wealthy patronized Worth and other dressmakers directly. Worth and others realized that upper-class women, who were their main clients, spent most of their life in one of two activities: maternity and mourning. One woman reputedly told her dressmaker to be sure that she had a set of black: "I've got three sons at the Crimean front."

9.2 Great Britain

Russian foreign policy since Peter the Great included a quest for warm water ports. The Black Sea was the only reasonable choice, but its outflow is through the Dardanelles and Ottoman-controlled territory. Thus Russo-Turkish conflicts dominated Middle Eastern affairs during the nineteenth century. The result was one of the most pointless wars in history.

The ostensible cause of the Crimean War was the demand by Russia to act as protector of Orthodox Christians living in Ottoman lands. When Russia annexed the Ottoman's Danubian Principalities (modern Romania) in July 1853, France, Britain, and Turkey formed an alliance against Russia. On October 17, 1854, French and British forces began a year-long siege of Sevastapol, the main naval base of the Russian Black Sea fleet. Leo Tolstoy, an artillery officer serving in the city, would establish his reputation as one of the rising stars of Russian literature with three short stories published between 1855 and 1856, collected as the *Sevastapol Tales*.

It is significant that two of the British commanders at Sevastapol are remembered for their clothes and not their military prowess: Lord Raglan and Lord Cardigan. A third item of clothing got its name from the port of Balaclava, through which the British received their supplies. On October 25, 1854, the Russians attempted to take the port. They failed, but a confused order by Lord Raglan, who consistently referred to the enemy as "the French," led to the disastrous charge of the light brigade under the command of Lord Cardigan. Forty percent of the light brigade was wiped out, commemorated in Tennyson's poem, "The Charge of the Light Brigade" (1855); the French General Bosquet observed pithily "It was magnificent. . . but it has nothing to do with war."

Disease killed seven times as many soldiers as Russian guns and British officers combined. Often these diseases were contracted after arrival at a military hospital, and even in peacetime the British soldier was twice as likely to die from disease as the British civilian. Clearly something had to be done.

There were two problems. First was the immediate problem of giving the soldiers in the Crimea adequate medical care under sanitary conditions. The second was to reform the entire military medical establishment to ensure that soldiers in future conflicts would receive adequate medical care. Both problems would be attacked by FLORENCE NIGHTINGALE (May 12, 1820–August 13, 1910), named after her birthplace in Italy. Nightingale arrived in Scutari, Turkey on November 5, 1854, where she forced through massive reforms, such as demanding the army provide clean drinking water and firing nurses who drank on duty. Unfortunately this included most of the nursing staff, so she attended to most of the soldiers in Scutari by herself, earning her the nickname "Lady of the Lamp." In May 1855, she arrived in the Crimea, fought massive bureaucratic inertia, instituted many reforms, and greatly improved the chance of a British soldier surviving hospitalization.

The Crimean War dragged to an end in 1856, after Sardinia-Piedmont joined the alliance against Russia. When it became obvious that Russia was on the verge of surrender, Austria also joined the alliance, an obviously opportunistic move that alienated everyone. Nightingale returned to England, where she campaigned tirelessly for thorough reform of the medical practice within the British army. Her testimony before a Royal Commission on the Health of the Army, later compiled into *Notes on Matters Affecting the Health, Efficiency and Hospital Administration of the British Army* (1858), was instrumental in establishing the Army Medical School (1857).

We might compare Nightingale's success in the Crimea with the dismal career of Ignać Semmelweis, a physician in Vienna. In the 1840s, Semmelweis noted a dramatic difference in mortality rate from puerperal fever (a form of septicemia) between two divisions of a free clinic. The two divisions were identical in all important respects save one: the first division trained physicians, while the second division trained midwives. In particular, students in the first division would go from dissections to deliveries without washing their hands. Thus in 1847, Semmelweiss made the radical suggestion that doctors should wash their hands before examining patients. When this was done, the mortality rate in the first clinic dropped from 18% to 1%. The medical community drove Semmelweis from Vienna, and he settled in Pest, Hungary, where his unorthodox practices led to the survival of many women. Physicians in Vienna and elsewhere returned to their pre-Semmelweis practices, and accepted the high mortality rate as an inevitable consequence of the dangers of childbirth.

Nightingale's success can be attributed to several factors. First, she publicized her ideas in the press (whereas Semmelweis mainly presented his findings to professional societies). This brought the issue to the attention of the public in general, and to Isambard Kingdom Brunel in particular. Like his friend Babbage, Brunel was a brilliant engineer known primarily for a failure: the *Great Eastern*, launched in 1858. A technical marvel, it was the largest ship in the world at the time, and no larger ship would be built for a generation. Unfortunately, that was because no ship of its size would be needed for another generation. The owners were never able to fill the ship to capacity, and sold it at a sizable loss in 1864; the ship was scrapped a few years later.

Brunel's brother-in-law was Ben Hawes, the Permanent Under Secretary of the War Office. Anxious to avoid further criticism, Hawes approached Brunel on February 16, 1855, and requested that he design a portable hospital for the Crimea. Six days later, Brunel presented plans for a prefabricated 1000-bed hospital that incorporated many of Nightingale's suggestions. The parts arrived at Renkioi (on the shores of the Dardanelles) in May 1855, and the hospital opened on July 12.

Another component of Nightingale's success was her training as a statistician who could present data in ways that even politicians could understand. Before Nightingale, the use of statistical graphs was regarded as an interesting, but not terribly important, way to describe statistical data. Even as late as 1900, it was possible for authors of statistics texts to say:

> [graphs] have their use for popular lectures and handbooks, but do not add anything to the significance of the figures ... The use of statistical maps needs only a brief notice.[7]

Nightingale used a new type of graph, which she called a "coxcomb" (now called a Polar Area Diagram). In the coxcomb, the area of a region is proportional to the quantity it represents; since the area of a sector goes up as the square of its radius, then large variations can be more easily represented; moreover, the smaller quantities are not overwhelmed by the larger.

Nightingale studied mathematics in the 1840s under the tutelage of JAMES JOSEPH SYLVESTER (September 3, 1814–March 15, 1897), who had just returned from a brief stint teaching in the United States at the University of Virginia in Charlottesville. The University had been founded by Thomas Jefferson, who also played a large role in designing the Rotunda, drafting the curriculum, and selecting the first faculty. Curiously the University did not originally grant degrees: Jefferson viewed them as "artificial embellishments."

The Jewish Sylvester could not work at the more prominent English universities (all of which were Anglican and required professions of faith), so he accepted a position in Charlottesville in November 1841. By then, the university granted degrees in medicine and law (though it still did not grant undergraduate degrees), and its students were known for their "drunkenness and lawlessness." Indeed, just before Sylvester's arrival a student had murdered a faculty member (a professor of law). Sylvester, dismayed by the lack of respect shown by the students for the faculty, only stayed six months. The exact reason for his departure are unclear: there seems to have been some sort of student confrontation, and Sylvester did not receive the support from the university administration he felt that a faculty

---

[7]Bowley, A. L., *Elements of Statistics*, 1901, p. 156.

member was due. For the next two years he lived in New York City, unable to find another position in the United States; finally he returned to England in 1843 where he supported himself by working as an actuary for the Equity and Law Life Assurance Society.

Since legal training was then (and still is) valuable for actuaries, Sylvester entered the Inner Temple in 1846 and was called to the bar in 1850. During his studies he became friends with ARTHUR CAYLEY (August 16, 1821–January 26, 1895), another lawyer interested in mathematics. Sylvester, Cayley, and GEORGE SALMON (September 25, 1819–January 22, 1904) created the algebraic theory of invariants. Roughly speaking, an invariant is an expression of the coefficients of an equation that does not change under a linear transformation of the variables. For example, consider the quadratic equation $x^2 + px + q = 0$ and suppose we let x undergo the linear transformation $x = x' + b$ to form the new equation $x'^2 + p'x' + q' = 0$. It is relatively straightforward to show that $p^2 - 4q = p'^2 - 4q'$. Thus $p^2 - 4q$ is an invariant under the linear transformation $x = x' + b$. In 1846 Cayley conjectured that, given an arbitrary equation of the form $P(x, y) = 0$ and a linear transformation of the variables $x = \alpha x' + \beta y'$, $y = \gamma x' + \delta y'$, it was possible to find all invariants; this would be proved by Sylvester in 1877.

By then, Sylvester had returned to the United States. In 1876 he accepted an offer to found the mathematics department at the newly established Johns Hopkins University in Baltimore, Maryland. Sylvester realized that if mathematics was to have any future in the United States, it needed a reputable journal, so in 1878 he began editing the *American Journal of Mathematics*, the first successful mathematical journal published in the United States. But the United States would not retain Sylvester for long, and in 1883 he returned to England as the Savilian Professor of Geometry at Oxford.

Cayley played a crucial role in putting group theory in its modern form. One of his early papers was "On the Theory of Groups, as Depending on the Symbolic Equation $\theta^n = 1$," presented on November 2, 1853. In this paper, θ was an operator that acted on a finite set A, sending each element to an element in another finite set B. Thus θ^n simply meant that the operator θ was to be applied n times in succession, at which point the original set A was restored (the meaning of "symbolic equation $\theta^n = 1$"). In general:

A set of symbols

$$1, \alpha, \beta, \ldots$$

all of them different and such that the product of any two of them (no matter in what order) or the product of any one of them into itself, belongs to the set, is said to be a *group*. It follows that if the entire group is multiplied by any one of the symbols, either as further [left] or nearer [right] factor, the effect is simply to reproduce the group...[8]

Cayley recognized that if a group contained p elements, where p is prime, then the group is equivalent to the group $\alpha, \alpha^2, \alpha^3, \ldots, \alpha^p = 1$. On the other hand, if p is composite, then the different groups with p elements can be represented by their multiplication table. Thus Cayley began the classification of all finite groups.

The British made other contributions to algebra. One significant advance appeared in *The Mathematical Analysis of Logic* (1847) and *An Investigation into the Laws of Thought*

---

[8]Cayley, Arthur, "On the Theory of Groups, as Depending on the Symbolic Equation $\theta^n = 1$." *Philosophical Magazine* VII (fourth series), Jan–June 1854, p. 41.

(1854) by GEORGE BOOLE (November 2, 1815–December 8, 1864). In these books, he described what is now called an operator algebra, which made logic a branch of mathematics. In modern terms, let X, Y, Z, ... be sets and 1 designate the universe, and x, y, z be operators that select the elements of X, Y, Z, ... Then $x1$ (or simply x, since the 1 may be assumed) would indicate selecting the elements of X from the universe; yx would indicate first selecting the elements of X, and then selecting those that are also members of Y. There are several consequences of this notation. First, $xy = yx$ (the commutative law holds); second, $x(y + z) = xy + xz$ (the distributive law holds); finally, $x^n = x$ (this is known as the power law).

However, most mathematicians preferred to keep logic and mathematics distinct, so it would be some time before Boole's ideas were generally accepted. For example, chemist-turned-economist WILLIAM STANLEY JEVONS (September 1, 1835–August 13, 1882) praised Boole's work, but called his "mysterious mathematical forms" needlessly complex.

Like Newton, Jevons worked in the mint—in Australia (between 1854 and 1859). When he returned to Britain in 1859, he began putting together his ideas on economics (then called political economy). Among other things, he rediscovered Daniel Bernoulli's principle of decreasing marginal utility. In 1865, Jevons wrote *The Coal Question*, which concerned itself with the consequences of exhausting fossil fuel resources:

> And though others have been found to reassure the public, roundly asserting that all anticipations of exhaustion are groundless and absurd, and "may be deferred for an indefinite period," yet misgivings have constantly recurred to those really examining the question.[9]

Jevons was the first to approach resource depletion as a problem with serious economic effects, noting the disastrous consequences that would befall the nation when its supply of coal ran low. He derided the notion of using petroleum as an alternate source of fuel, pointing out that it was more limited than coal, and stressed conservation as the only viable method of preserving national prosperity.

Oddly enough, though Jevons estimated that the sun could supply a thousand times as much energy as the coal used in England (the actual figure is probably closer to a hundred thousand times), and discussed plans whereby wind, water, or tidal power could generate electricity to produce hydrogen, he dismissed these as impractical. What made his rejection of these sources peculiar is that in the same book, he described what is sometimes called Jevons's paradox: increasing efficiency of fuel usage is associated with a *rise* in consumption, as more ways are found to use the fuel. For example, the original steam engine, invented by Thomas Savery in 1698, used too much fuel to be practical. In 1718, Thomas Newcomen made several changes that improved the efficiency enough to make the steam engine useful in certain situations, particularly draining water from mines. By 1792, James Watt's steam engines were nearly ten times as efficient in their use of energy as the original Newcomen engine; coupled with the falling price of coal (made possible by the use of the steam engine), the operating cost of a steam engine dropped significantly, so many more uses were found. The industrial revolution was a direct result, as was a dramatic rise in coal consumption. By the same argument, increasing efficiency in the use of sun, wind,

---

[9] W. S. Jevons, *The Coal Question*, Macmillan 1865, p. viii-ix.

water, and tidal power would also make their usage more widespread; Jevons missed this implication of his own work.

Jevons proved ahead of his time in another way. He noted the existence of many types of mechanical calculators, including Babbage's analytical engine, which could in theory duplicate any work of a human calculator. However, analogous devices for logical inference did not exist. On January 20, 1870 Jevons presented a paper before the Royal Society describing a mechanical device, which he called a "logical piano," for evaluating the equivalence of logical statements.

For example, consider the three concepts A "iron," B "metal," and C "an element," along with their complements, a "not iron," b "not metal," and c "not an element." By choosing either A or a, B or b, and C or c, one has eight possible combinations, such as ABC, AbC, and so on. If we add two premises, namely "iron is a metal" (AB) and "metals are elements" (BC) some of these combinations are inconsistent and may be discarded. Given the premises, the logical piano deleted the inconsistent combinations automatically.

Jevons' work in logic led him to raise an early version of the computational complexity problem. In his *Principles of Science* (1874), he noted the difficulty of inverting some processes. For example, he observed that it would probably take a good computer (a person) many weeks to factor a number like 8,616,460,799, though he himself had produced the number in a manner of minutes. More importantly, it is easy to determine the statements that are logically consistent with a given premise. For example, the premise A implies B is consistent with the statements A and B; not A and B; and not A and not B. On the other hand, given a set of consistent statements, determining the premise is more difficult. With just two terms A and B, there are four possible combinations, and thus 2^4 possible sets of consistent statements. Jevons noted that the number of possible laws (each of which would generate a set of consistent statements) increased rapidly: for 4 terms, there are $2^4 = 16$ combinations, and thus $2^{16} = 65,536$ possible laws, while 6 terms would give rise to 18,446,744,073,709,551,616 possible laws!

9.3 Italy

The year 1848 is known in European history as "the year of revolutions," for it saw liberal uprisings in most European countries; generally speaking these uprisings sought to establish a constitutional monarchy. In foreign-ruled Italy the revolts took on a nationalistic tone and are collectively known as the Risorgimento ("Rising again"). The year began with a revolt in Sicily against the Spanish-descended Ferdinand II; on January 29, he was forced to agree to a liberal constitution. Other rulers forced to agree to constitutional monarchies included the Austrian Duke of Tuscany, Leopold II (February 17), the Italian King of Sardinia Charles Albert (March 4), and the Pope, Pius IX in Rome (March 14).

On March 23, Charles Albert declared war on Austria, hoping to capitalize on the nationalistic uprisings and establish a kingdom of Italy (with himself as king, of course). Unfortunately the Austrians sent an army under the command of Count Joseph Radetzky, a marshal of the Napoleonic Era. Radetzky crushed the revolt rapidly, and Johann Strauss the Elder honored him with the *Radetzky March*. The future "Waltz King" Johann Strauss the Younger did not share his father's reactionary tendencies and wrote the *Revolution March*, leading to detainment and interrogation by the Austrian secret police, though no case was brought against him.

On May 29, 1848, Radetzky marched to relieve the besieged city of Peschiera, and ran into a Tuscan army at Curtatone and Montanara. The Tuscan army included a contingent from the Universities of Pisa, Siena, and Florence led by several professors, including Leopoldo Pilla da Venafro, professor of geology, Ottaviano Mossotti, professor of physics, and ENRICO BETTI (October 21, 1823–August 11, 1892), who would play a key role in rebuilding a mathematical tradition in Italy. Tuscan losses were heavy, and the university contingent was especially hard hit (Venafro was killed at the beginning of the battle), but they delayed Radetzky's march long enough to allow Peschiera to be taken and for Charles Albert to win a string of victories. Radetzky reputedly chided his officers for taking six hours to defeat "a handful of boys."

Unfortunately the native Italian forces were too weak to expel the Austrians by force unaided. Radetzky and the Austrians regained lost ground, defeated Charles Albert, and moved on to the last holdout: Venice, whose defenders included another future mathematician, LUIGI CREMONA (December 7, 1830–June 10, 1903). During May 1849 the Austrian army hurled some 60,000 projectiles into Venice and on August 27, the city surrendered. Venice would remain in Austrian hands for another generation.

The events of 1848–1849 made it clear that the Italians needed outside assistance. In particular, they needed someone powerful enough to expel the foreigners, idealistic enough to support Italian independence, and foolish enough to be drawn into a conflict that would eventually pit them against Austria, Spain, and the Papacy. Fortunately, Napoleon III had just become the ruler of France.

The Prime Minister of Sardinia-Piedmont, Camillo Benso, the Count of Cavour, played a key role in Italian independence. At the peace conference that ended the Crimean War, Cavour pointed out that the disunity of the Italian peninsula posed a threat to European peace, for it was divided between three great powers: Spain in the south, the Papacy in the center, and Austria in the north. Bringing about Italian unification required expelling the foreign powers from the peninsula.

First, Cavour obtained an alliance with Napoleon against Austria. The Austrians then committed a colossal blunder and on April 23, 1859 demanded that the Sardinians demobilize or face invasion. This gave Napoleon the opportunity to intervene under the guise of protecting his ally. A series of bungled military campaigns in which the French proved slightly less incompetent than the Austrians culminated in the Battle of Solferino (June 24, 1859), in which Menabrea participated (by then he was a major general commanding an engineering corps). Napoleon was so appalled by the battle that he accepted an armistice offer from the Austrians.

Nearly 30,000 soldiers died at Solferino. During the battle a Swiss businessman, Henri Dunant organized medical services for wounded soldiers on both sides. In 1862, Dunant proposed the formation of an international organization to care for the casualties of war, regardless of whose side they were on; within a year the Geneva Society for Public Welfare became the International Committee for the Relief of the Wounded, and in 1875 the International Committee for the Red Cross. At a meeting in Geneva in 1864 the Committee (which included Dufour, the inventor of modern topographical mapping) helped write the Convention for the Amelioration of the Wounded in Time of War: the first of the Geneva Conventions.

Since Napoleon virtually abandoned the cause of Italian independence after Solferino, the southern Italian states were joined to the growing Kingdom by the work of Giuseppe Garibaldi. By then, Garibaldi had a long portfolio as a revolutionary, beginning with an 1834 revolt against Charles Albert, the King of Sardinia. The revolt proved unsuccessful, so Garibaldi spent the next fourteen years in exile, the last twelve in South America where he and a group of Italian mercenaries (the original Red Shirts) overthrew the Argentine dictator Juan Manuel de Rosas.

Garibaldi returned to Italy in 1848, established a Roman Republic (the Republic was subsequently invaded and the Pope restored by French troops), and went back into exile. By then his career had grown to epic proportions, and caught the attention of Alexandre Dumas, who further glamorized his deeds. By May 1860 Garibaldi was back in Italy, about to embark on his greatest adventure: the liberation of Naples and Sicily from the Spaniards. By year's end, all of southern Italy was in Garibaldi's hands; he turned it over to Sardinia's Victor Emmanuel (the son of Charles Albert), and on May 17, 1861, the Kingdom of Italy was declared, with Victor Emmanuel as king. Cavour lived just long enough to see its proclamation and died three weeks later.

A new nation was an opportunity to build up a new tradition in mathematics. One possibility involved a focus on abstract algebra. In 1852 Betti published "On the resolution of algebraic equations," which began with a commentary on the theory of substitutions and concluded with a development of Galois theory. It was the first important exposition on the subject, and though Betti's work was insufficiently rigorous by modern standards, and not as clear as it should have been, it was a major step in making Galois's work more accessible to mathematicians. Betti wrote up some notes on modern algebra, but never got around to teaching a course on the subject; one of his students, Giovanni Novi, intended to write a three-volume treatise on the subject based on Betti's notes, but only one volume appeared. Thus it would be another generation before group theory entered the mathematical mainstream, and Italian mathematicians would find their interests stimulated by developments in analysis.

In 1858 Betti, FRANCESCO BRIOSCHI (December 22, 1824–December 14, 1897), and FELICE CASORATI (December 17, 1835–September 11, 1890) visited the mathematical centers of Europe. At Göttingen, Betti met Riemann and began a lifelong friendship (this proved convenient when Riemann was sent to Italy to recover from a bout with pleurisy; he spent two years with Betti in Pisa). When Betti returned to Italy, he arranged for a translation of Riemann's inaugural lecture, which contained the ideas that Riemann was turning into modern complex analysis. Oddly enough Casorati, who would become one of the prominent names in complex analysis in Italy (he is known for the Casorati-Weierstrass Theorem), met Riemann neither in 1858, when he visited Göttingen, nor in the years Riemann was in Pisa, but instead visited him in Berlin in 1864.

Meanwhile Betti obtained the Chair of Higher Analysis in Pisa in 1859. Through his connections with Pisa's *Scuola Normale Superiore* (an institution founded by the original Napoleon in 1808), he founded a mathematical research community whose members included ULISSE DINI (November 14, 1845–October 28, 1918) and VITO VOLTERRA (May 3, 1860–October 11, 1940).

After unification several Italian mathematicians entered politics. Menabrea (whose political involvement predated unification) joined the ministry of Cavour's successor, Bettino Ricasoli; Menabrea was Prime Minister himself between 1867 and 1869. Betti became a member of parliament and Brioschi achieved high position in the Ministry of Education. Brioschi, who worked primarily on the algebraic theory of determinants and the theory of forms, took an active role in improving mathematics education in the newly formed kingdom. From 1863 he guided Milan's *Instituto Tecnico Superiore*, a version of the French Polytechnical School, and added Casorati and Luigi Cremona to its faculty. Betti and Brioschi later collaborated on a translation of the first six books of Euclid into Italian for use in secondary schools.

Modern Italy was nearly complete, but Rome was still firmly under the control of the Papacy, and Venice still firmly controlled by Austria. The king had another problem: Garibaldi, who had no love of the monarchy and who demanded the seizure of Rome, was beginning to become a political liability. Moreover, Dumas helped turn Garibaldi into "the hero of two worlds" (the Old and the New), and the king feared the revolutionary's popularity. Thus the king found it convenient to send Garibaldi on secret missions whose success would serve the cause of Italian unity, but whose failure could be pinned on a troublesome adventurer. The king hoped Garibaldi would attack the Austrians and recover northern Italy; unfortunately Garibaldi had his eyes set on Rome, and twice attempted to seize it. Only the intervention of the French prevented the Papal States from being annexed to the Kingdom of Italy, and for the next decade France would be the guardians of Rome.

9.4 Germany

The Congress of Vienna left the multinational Austrian Empire intact, but forced Prussia to cede its Polish territories to form the Congress Kingdom of Poland. Poland was independent in name only: the king of Poland was also the Tsar of Russia, and by 1863 even the nominal independence of Poland vanished, and it became a Russian province. In exchange for their Polish lands, Prussia received much of Saxony and other territories. Consequently the new Prussia was "all German," and the natural nucleus around which a German nation could be formed.

Prussia's defeat by Napoleon was followed by some much-needed internal reform. One of the main targets was the educational system, and in 1809, Wilhelm von Humboldt became the Prussian Minister of Education. Humboldt gave mathematics high priority at the *gymnasium* (secondary school) level; only Latin had more time devoted to it. Humboldt's vision of higher education included three institutions: the University of Berlin; a Polytechnical School, modeled after the French Polytechnical School; and a Normal School.

Unlike previous universities, which were professional schools to train lawyers, physicians, and theologians, the University of Berlin, which opened in 1810, was planned from the start to be a center for education and research as well. The actual researchers were to be trained at an institution modeled after the French Polytechnical School, and Humboldt offered Gauss the chance to be the director of the institution: his duties were to be primarily administrative and scientific, and above all he was not required to teach (a task that Gauss detested). Gauss was tempted by the offer, but ultimately declined. Meanwhile, teachers would be trained at the Normal School.

Humboldt's bold plan failed in part due to bad luck. For the Normal School, Humboldt recruited NIELS HENRIK ABEL (August 5, 1802–April 6, 1829), who was rapidly becoming one of the pre-eminent mathematicians in Europe. Abel's father, Søren Georg Abel, was part of a group of Norwegians who wished to open a university in Norway. In 1811, they secured a charter from King Frederick VI of Denmark; the Royal Frederick University opened in Christiana two years later in a rented building, with 17 students and 6 faculty, teaching in a rented building.[10]

The opening of the university was part of a broader initiative to improve the lives of Norwegians. A few years later, Søren Abel helped draft the Constitution of Eidsvoll on May 17, 1814. The Constitution of Eidsvoll turned Norway into a constitutional monarchy under the Danish crown prince, Christian Frederick, and gave the Norwegian parliament (the *stortig*) more authority than any other European constitutional body.

There was one problem: Norway was no longer part of the Kingdom of Denmark. The Treaty of Kiel (1814) gave Norway to Sweden, so it had no right to elect its own king; the drafters were technically rebels. Sweden's King Charles XIV John (formerly Napoleon's marshal Bernadotte) ended the rebellion in a mere fourteen days, forcing Christian to abdicate in July 1814. Charles John then took a drastic measure to prevent future revolts: he ratified the Eidsvoll constitution.

Niels Henrik Abel entered the Royal Frederick University in 1821, and later that year sent a paper to Ferdinand Degen, at the Royal Society of Copenhagen, describing a means of solving the general fifth-degree equation. Degen's enthusiasm was guarded, since the solution to the quintic had eluded mathematicians for centuries. Thus he wrote back to Abel, and made two comments that would profoundly affect Abel's life. The first was that, though he could find no obvious flaw in the paper, Abel could improve its readability by including an example. Degen also suggested that the field of algebraic solutions to equations was an overworked field that promised little in the way of new results, and that Abel ought to consider the new field of elliptic integrals.

Following Degen's suggestion, Abel tried to come up with an example, and found to his surprise that his method did not work. Indeed, he soon realized that he had the basis for a proof of the unsolvability of the general quintic, and published his results in 1824 in a pamphlet which, to save printing costs, had all unnecessary details omitted. This made it nearly impossible to read, but it helped to establish Abel's reputation, and in 1825 the Norwegian government granted him a scholarship to study abroad.

Abel intended to go to Paris, then the center of European mathematics, but immediately after graduation, he followed his physicist friends to Berlin. There he met the civil engineer AUGUST LEOPOLD CRELLE (March 11, 1780–October 6, 1855), who was about to begin editing one of the first German mathematical publications: *The Journal of Pure and Applied Mathematics*, popularly known as Crelle's journal.[11] In 1826, the first volume appeared; it contained seven papers by Abel, including his proof of the unsolvability of the quintic and his first papers on elliptic functions.

---

[10]Christiana changed its name to Oslo in 1938, and the next year the Royal Frederick University became the University of Oslo.

[11]Because its theoretical content far outweighed its applied content, it also became known as the Journal of Pure Unapplied Mathematics, an excellent pun in German.

Abel arrived in Paris, and on October 30, 1826 presented to the Paris Academy his main work, "Memoir on a General Property of a Very Extensive Class of Transcendental Functions." The paper contained some very fundamental results on integral equations, but Cauchy had little interest in it and the paper was not published until 1841. Abel was treated so indifferently by the Academy that he came back to Berlin at the end of 1826, and returned home to Norway in early 1827. Crelle, who realized that Norway was too poor to offer Abel a good position, sought Humboldt's help to obtain a position in Berlin. As a result, Humboldt offered him a position in the Normal School on April 8, 1829—two days after Abel's death at the age of 26.

Abel's death and Gauss's refusal to head the research institute meant that the functions of teacher and researcher training were absorbed by the University of Berlin. It was rapidly becoming the premier institute of higher education in Europe, with a faculty that included Dirichlet, who arrived there in 1828, and Georg Wilhelm Friedrich Hegel, who had been there since 1818 and espoused a philosophy of history later used by Marx: human history is a process whose ultimate goal is human freedom.

9.4.1   Jacobi

CARL GUSTAV JACOB JACOBI (December 10, 1804–February 18, 1851) entered the University of Berlin in 1821 to study mathematics. He displayed enough promise to be offered a position at one of the leading preparatory schools in Berlin. But like Sylvester, Jacobi was Jewish, and Jews could not receive university appointments. Thus in 1825 Jacobi renounced his religion and was offered an appointment at the University almost immediately thereafter. He stayed only a year before accepting a position at the University of Königsberg in May 1826. His rapid promotion over those with more seniority led to some objections until he established his mathematical reputation with *Foundations of a New Theory of Elliptic Functions* (1829).

Königsberg (now Kaliningrad, and different from Regiomontanus's birth town) is situated on the Baltic, and the harsh climate proved detrimental to Jacobi's health. At the request of Alexander von Humboldt (the brother of Wilhelm, architect of the Prussian educational system), Friedrich Wilhelm IV of Prussia sponsored a trip to Italy to help Jacobi recover, and a subsequent relocation to Berlin in June 1844. Past jealousies prevented him from being offered another position at the University of Berlin, but he was given a stipend by the government. His position in Berlin and his pension from the king would place him in the middle of the great political events of the late 1840s.

The revolts of 1848 came to Köln, part of Prussia's Rhineland territories, on March 3 when a mob stormed city hall and made several demands, including protection for laborers, a guaranteed minimum standard of living, and universal education. In Berlin, clashes between the army and the citizenry were becoming more and more violent. Friedrich Wilhelm IV vacillated, unwilling to order the army to fire upon his own citizens, but at the same time unwilling to agree to rebel demands. His hand was forced when, on March 13, Austria's Prime Minister, Prince Metternich, was forced to resign. The fall of Metternich, key architect of the Congress of Vienna and bulwark of the conservatives, energized the liberals, who increased the pressure on Wilhelm. Then on March 18, two shots were fired during a demonstration in front of the royal palace. No one was injured, but the incident sparked further street fighting. To prevent civil war Wilhelm agreed to virtually all the demands of the liberals, and took the opportunity to promote a unified Germany: "Henceforth Prussia shall be submerged in Germany."

It is interesting to note that German mathematicians, unlike Italian and French mathematicians, generally avoided any political activism. Part of the reason was self-protection, for the 1848 revolution showed that it was unwise for a German mathematician to become involved in politics. Indeed, even the suspicion of political activism would have unpleasant repercussions.

For example, during the streetfighting in Berlin, shots were fired on soldiers from a house on March 19. Those inside, including FERDINAND GOTTHOLD MAX EISENSTEIN (April 16, 1823–October 11, 1852), were arrested. Though he was released the next day, the arrest marked Eisenstein for the rest of his life as a supporter of liberal causes. This made it difficult for him to find a post in Germany, despite the support of Gauss (who considered him one of the most brilliant mathematicians in Europe), Dirichlet, and von Humboldt. Eisenstein's health, never good to begin with, gradually worsened and he died of tuberculosis a few years later.

Jacobi actually ventured into politics. In March 1848, liberal leaders from all parts of Germany called for a meeting in Frankfurt to draft a constitution for a united Germany. Delegates were to be elected from each of the German states. Jacobi, in Berlin, was put forward as a candidate for Prussia, and he accepted the nomination, but because he was receiving a pension from the king, the liberals rejected him as untrustworthy. Berliners sent Jacob Grimm instead.[12]

On May 18, 1848 the delegates met in Frankfurt in the Church of St. Paul. One of the first problems was deciding what constituted *Germany*. In particular, should the German-speaking Austrians be part of the united Germany, or should they be kept separate? Austria, with its large population and resources, would dominate such a Germany; moreover, a new Austrian constitution forbade the dismantling of the Austrian Empire, which meant that the new Germany either included Austria and its Empire (with its enormous non-German population), or no Austria at all.

The delegates chose to exclude Austria, and on March 28, 1849, adopted a constitution that included universal suffrage, a parliamentary government, and a hereditary monarchy. Friedrich Wilhelm's declaration for a united Germany made him the natural choice, but he rejected the offer, supposedly saying, "I would not stoop to pick up a crown from the gutter." Without a viable candidate for a monarch, the Frankfurt parliament found itself hamstrung.

Meanwhile Jacobi's involvement in the revolution made him suspect, and in 1849 Friedrich Wilhelm stopped his pension, forcing Jacobi to leave Berlin. Jacobi and family settled just outside of Erfurt in Gotha, a region known for its educational innovations. In 1642 the Duke of Gotha, Ernst the Pious, announced a purely secular school system which required compulsory education from the age of 5; the division of schools into lower, middle, and higher grades; and textbooks.

---

[12]Friedrich Wilhelm IV recruited Jacob and his brother Wilhelm for the University of Berlin after the brothers and five others, known as the Göttingen Seven, were dismissed from the University of Göttingen in 1837 because of their refusal to swear an oath of loyalty to the King of Hanover.

Ironically the unification movement followed Jacobi to distant Erfurt. At the end of May 1849, delegates from Saxony, Hanover, and Prussia met in Berlin to write a draft constitution for a new union. Twenty-eight states later agreed to join, and between March 20 and April 29, 1850, delegates met in Erfurt to finalize a constitution. Unfortunately Bavaria and Austria refused to participate, which led to the withdrawal of Saxony and Hanover; eventually the delegates gave up and went home.

Though Prussia failed to unify Germany at Erfurt, a later effort might be successful—and a Prussian-dominated Germany would be intolerable to Austria. Thus on May 16, 1850 the Prime Minister of Austria, Prince Felix Schwarzenberg organized a meeting in Frankfurt to reconstitute the old Germanic Confederation (dominated, of course, by Austria). Several states seceded from the Prussian Union to rejoin the Germanic Confederation. War between Prussia and Austria loomed, and a revolt in Hesse-Kassel against the reigning Duke pushed matters to the breaking point. The Duke appealed to the Frankfurt Diet for assistance, and Austria promised to intervene. But Hesse-Kassel was part of the Prussian Union, so Friedrich Wilhelm dispatched troops to Hesse-Kassel, though (hoping to avoid actual war) he called on the Tsar of Russia to mediate the dispute. To his dismay, the Tsar sided with Austria. Prussia was forced to abandon attempts to form a united Germany without Austria, an event known as the Punctation of Olmütz (a "punctation" is a statement of demands) or, more significantly, the Humiliation of Olmütz.

In the meantime Jacobi accepted a chair at the University of Vienna. The Prussian government, disinclined to allow any further humiliation at the hands of the Austrians, made up its mind not to lose Jacobi, so it restored his pension and gave him a minor position at the University of Berlin. Jacobi stayed in Prussia, though his family remained in Gotha, two hundred kilometers away; he would only see them on vacations. In January 1851, Jacobi contracted influenza, then smallpox, and died from complications.

9.4.2 Gottingen and Berlin

One of Eisenstein's students was GEORG FRIEDRICH BERNHARD RIEMANN (September 17, 1826–July 20, 1866). Eisenstein, whose interests included investigations into prime numbers, may have been a key inspiration of Riemann's later work, *On the Number of Primes Less than a Given Number* (1859), in which he introduced the ζ-function. To understand the implications of Riemann's work, a comparison might be made with the rational function $f(x) = 1/(1-x)$. If $-1 < x < 1$ then

$$\frac{1}{1-x} = 1 + x + x^2 + x^3 + \dots.$$

If $|x| > 1$, the series diverges; however, $f(x) = 1/(1-x)$ still has a value. In the same way, the ζ-function can be defined by the series

$$\zeta(s) = \sum_{n=1}^{\infty} \frac{1}{n^s}$$

where this series converges. This series, for $s > 1$, was first studied by Euler, but Riemann realized that even though the series diverged for $s \leq 1$, the ζ-function itself might have a value. In his 1859 paper, Riemann conjectured that all the complex zeroes of the ζ-function occur when the real part of s is equal to one-half.

Riemann began his studies at the University of Göttingen. But in spite of (or perhaps because of) Gauss's presence, Göttingen was a poor place to learn mathematics. Gauss hated teaching, and the only course he taught regularly was an introductory course on the method of least squares. Consequently Riemann transferred to the University of Berlin in 1847, where he studied with Eisenstein, Jacobi, and Dirichlet. However he returned to Göttingen for his doctorate: he had impressed Gauss sufficiently to be taken on as a student. In 1851, Riemann completed a thesis on complex function theory where, among other things, he introduced the notion of a Riemann surface.

A few years later, Riemann had the opportunity to become a *Privatdozent* at Göttingen. This meant he could give lectures and charge a fee. In order to do so he had to give a *Habilitationschrift*, an inaugural paper, as well as give a *Habilitationsvortrag*, an inaugural lecture. Riemann's inaugural paper was, "On the Representability of a Function by Means of Trigonometric Series." Since the coefficients of a Fourier series were determined by means of an integral, Riemann had to consider the meaning of $\int_a^b f(x)\, dx$. The result was the first description of the Riemann sum in connection with the definite integral.

Riemann delivered the inaugural paper at the beginning of December 1853, and began to prepare for his inaugural lecture. At that point he was to give three topics to the faculty, who would choose one of them. Riemann had prepared two of the topics beforehand, but Gauss chose the third, "On the Hypotheses that Underlie Geometry." Riemann delivered his lecture on June 10, 1854. The ideas were very far ahead of their time, and only Gauss understood them in their entirety, but Gauss's approval was enough, and Riemann became a *Privatdozent*.

In 1855, Riemann gave a course on Abelian functions. His ideas were very well developed, and a paper published in Crelle's journal contained so many groundbreaking results that Weierstrass, who had also been researching Abelian functions, withdrew a paper he had before the Berlin Academy. Riemann's course attracted three students, including a fellow faculty member at Göttingen: RICHARD DEDEKIND (October 6, 1831–February 12, 1916).

Dedekind was Gauss's last doctoral student, completing a dissertation on Eulerian integrals in 1854. Afterwards he joined the faculty at Göttingen, where in the winter of 1856–7 he taught the first course on Galois theory ever given; he offered the course again the next year. Dedekind also worked extensively with Dirichlet; he edited Dirichlet's lectures on number theory, which were published as *Introduction to Number Theory* in 1863. He would also help prepare the works of Gauss and Riemann for publication.

When Gauss died in 1855, Dirichlet received his appointment as Chair of Mathematics, and the trio of Dirichlet, Dedekind, and Riemann stood poised to make Göttingen into a major center of mathematical research. Unfortunately, fate intervened. Dirichlet died on May 5, 1859 at the age of 54. Riemann succeeded him, but in autumn 1862, Riemann contracted tuberculosis and spent the next few years traveling back and forth to Italy, trying to recover his health. He failed, and died on July 20, 1866 in Italy at the age of 39.

Meanwhile Dedekind accepted a position at the Polytechnic School in Zürich. While there he realized that no one had yet established an axiomatic basis for the real numbers, and he began developing a theory of arithmetic based on a "cut" (now a Dedekind cut). Consider the set of rational numbers Q. A cut is a division of Q into two sets, A_1 and A_2, where every element of A_1 is less than any element of A_2. Given any rational number q, we form the cut (A_1, A_2) by putting into A_1 all numbers less than q and A_2 containing all numbers greater than q; q itself can be in either set (Dedekind realized that there would be two different cuts, but considered them essentially equivalent). Hence every rational number corresponds to a cut. On the other hand, not every cut corresponds to a rational number. For example, let A_1 be the set of all real numbers x where $x^2 < 2$, while A_2 is the set of all real numbers x where $x^2 > 2$. In this case, Dedekind said, the cut creates a new, irrational number that is completely defined by the cut.

A key problem faced by previous authors was defining continuity; in particular, distinguishing it from denseness. Dedekind was the first to give a formal characterization of continuity in "Continuity and the Irrational Numbers" (1872), where he identified the continuity of the straight line with the property that every division of the line into two pieces (where the points of one can be said to be to the "left" of the other) is produced by a unique point. Carrying this over to cuts on sets of numbers, a set of numbers is continuous if every cut on the set is produced by a unique element of the set. Hence the real numbers form a continuous domain, since every cut corresponds to a unique real number, while the rational numbers do not, since there are some cuts that are not produced by elements of the set.

Dirichlet's departure for Göttingen in 1855 left the chair of mathematics open in Berlin, and Dirichlet recommended Kummer for the position. Dirichlet and Kummer knew one another well: an 1836 paper by Kummer on hypergeometric series led to a correspondence with Jacobi and later Dirichlet, and in 1840, Kummer married a cousin of Dirichlet's wife, though she died a few years later.

When Kummer accepted the position in Berlin, his chair in Breslau (for which Jacobi and Dirichlet recommended him in 1842) became open. The top two candidates for the position were KARL WEIERSTRASS (October 31, 1815–February 19, 1897) and FERDINAND JOACHIMSTHAL (March 9, 1818–April 5, 1861). Weierstrass was the better candidate, but Kummer threw his support behind Joachimsthal, his former student: Kummer knew that a position at Berlin would be opening up, and wanted Weierstrass, the better mathematician, with him at Berlin. Thus in October 1856, Weierstrass arrived in Berlin.

In 1861 Kummer and Weierstrass organized and ran the first German seminar on pure mathematics. The students who passed through the Berlin seminar read like a "Who's Who" of nineteenth century mathematics: Cantor, Klein, Netto, Lie, Minkowski, Mittag-Leffler, Schwarz, and others. Modern mathematics is very much a product of the Berlin seminar, and thus Weierstrass's ideas and methods continue to play a very important role.

The third key figure of the Berlin triumvirate was LEOPOLD KRONECKER (December 7, 1823–December 29, 1891). Kronecker, like Gauss, changed his interests from philology to mathematics, though in Kronecker's case it was *because* of one of his teachers: Kummer. Kronecker studied in Berlin under Dirichlet, obtaining his degree in 1845 after passing a comprehensive examination that included, besides questions on pure and applied mathematics, questions on Greek and the history of legal philosophy.

Kronecker's family was wealthy, so he had no need to find an academic position to support himself, nor had he any desire to put up with the restrictions imposed upon university teachers. He continued to do mathematical research while in Berlin, and Kummer proposed Kronecker for membership in the Berlin Academy, which was granted on January 23, 1861. This gave him the right to lecture at the University of Berlin, and Kummer encouraged Kronecker to do so. For the next decade Kronecker maintained good relationships with his colleagues at Berlin and elsewhere. But dramatic changes were in the air.

9.4.3 Bismarck and von Moltke

After the failure of the Frankfurt assembly in 1848 and the Erfurt parliament in 1850, the unification efforts in Germany centered around the *Zollverein*, a customs union (what we might now call a "free trade zone"). Originally established by Prussia in 1834, by the 1860s it included much of northern Germany, and a united Germany seemed inevitable. But would the united Germany be a loose confederation of independent states, or a centralized nation? The question was of more than theoretical interest: across the Atlantic Ocean, the United States was on the brink of a civil war over the question of how much authority the central government could exert over the individual states. Moreover Poland showed that European states that failed to establish a strong, central authority were easy prey for rapacious neighbors.

The architect of German unification was Otto von Bismarck, the son of minor nobility. Bismarck spent a few years at the Universities of Göttingen and Berlin studying to be a lawyer (his mother's wish), but entered the Prussian civil service after his mother died. He was elected to the Prussian Chamber of Deputies (the lower house) in 1849, was one of the delegates at Erfurt, and represented Prussia in Frankfurt after the Humiliation of Olmütz. He also served as ambassador to Russia and France, and spent time in Vienna and other capitals of Europe in special assignments for the new king, Wilhelm I. In 1862, Bismarck was the Prussian representative at the court of Napoleon III.

Wilhelm fought in the last of the Napoleonic Wars, and viewed the rise of Napoleon III as a threat to Prussia. Thus he sought to increase the size of the army and centralize its command structure. The Prussian Diet sought greater civilian control over the army, and since they controlled the purse strings, conflict between king and parliament seemed inevitable. Wilhelm let it be known that he would rather abdicate than allow the Diet to control the king, but was persuaded to recall Bismarck from Paris to fight a battle with the Diet. Thus Bismarck became prime minister in 1862. His early days in power were met with stiff opposition from the liberals, but Bismarck had the support of the king and the army. The Diet proved intractable, so Bismarck made them irrelevant by arguing that the state had the right to collect taxes even without parliamentary support to ensure that the civil service and other organs of government could continue to function; this brought the bureaucracy to his side as well. By sidestepping the Diet this way, Bismarck paved the way for another visionary: Count Helmuth von Moltke.

Moltke was born in Mecklenburg, but served for a while in the Danish army before joining the Prussian army in 1822. He was widely traveled, well-educated, and the author of a number of books including a novel (*The Two Friends*, 1827), some histories, travel journals, and—significantly—a treatise on how to choose routes for railways (1843).

The railway was new to Germany: the first Prussian rail line, between Berlin and Potsdam, opened on October 29, 1838 (Crelle supervised its construction). But Moltke realized

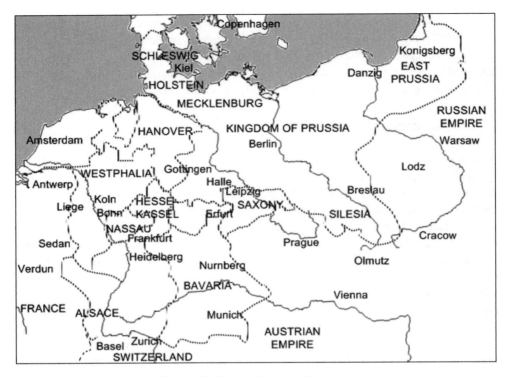

Figure 9.3. The Unification of Germany

that railroads could send soldiers and their equipment to an assembly area more quickly and in greater numbers than ever before, even in weather that turned the roads of Europe into impassable quagmires. Thus the railroad allowed the rapid assembly of an enormous army—a crucial factor in winning future wars.

Unfortunately, coordinating the mobilization and deployment of such a large army could not be done by a single person. Thus Moltke began reorganizing the Prussian General Staff when he joined it in September 1858. One of the greatest changes Moltke instituted involved *de*centralizing the command of the army: he believed that future battlefields would simply be too large for a single person to give detailed orders. Instead the field commanders were given General Directives, but allowed considerable discretion on how to achieve them. Moltke's ideas met with considerable resistance at first, since he suffered from one glaring defect: except for a brief period between 1836 and 1838, when he served as a military adviser to the Turkish army, von Moltke had never been in the field.

The first major step towards German unification concerned the Duchies of Schleswig and Holstein, which were in a peculiar political situation where they were members of the German Confederation but joined in personal union with the King of Denmark; in fact, Moltke was originally a Danish subject because his father had moved to Holstein. The Prusso-Danish War (1864) resulted in Schleswig being transferred to Prussia and Holstein to Austria. It also provided a small-scale test of Moltke's ideas.

Bismarck's experience in the Frankfurt Diet led him to mistrust Austrian motives, and for a long time he had believed that a showdown with Austria was inevitable. The Austrians also realized this, and sought to derail Prussian plans. The threat of war worked at Olmütz in 1850, but since then Russia had been defeated in the Crimea and Austria in Italy, so neither power was in a position to threaten effectively. This time, the Austrians reasoned, Prussia had to suffer an actual defeat. The Austro-Prussian War began on June 14, 1866, putting Moltke's theories to their first real test.

The railroads allowed the Prussian army to mobilize rapidly. Moreover, since the 1840s the army had been equipped with the Dreyse rifle, a breech-loading "needle gun" (a reference to the firing pin) that allowed Prussian soldiers to reload while prone; the Austrian army relied on the more archaic muzzle loaders, which could only be reloaded from a more vulnerable standing position. Prussia scored rapid victories against an alliance consisting of Austria, Bavaria, Saxony, and Hanover and the war ended so quickly it is best known as the Seven Weeks War. By the Treaty of Prague (August 23, 1866), Prussia annexed Holstein, as well as Austria's allies Hanover, Hesse-Kassel, Nassau, and Frankfurt. Austria later ceded Venice to Italy, one of Prussia's allies.

9.4.4 Klein and Lie

The Treaty of Prague left Austrian territory largely intact. It is often suggested that Bismarck engineered a gentle peace with Austria in preparation for his next war. In part, this is true: no sensible statesman generates enemies, and Bismarck believed that in national rivalries *both* nations were at fault. But the terms of the Treaty of Prague can be interpreted in terms of another, simpler objective. Since the time of Frederick the Great, Prussian territory consisted of Brandenburg, on the Rhine, and Prussia, near the Baltic, territories separated by Hanover, Hesse-Kassel, Nassau, and Frankfurt. After the Napoleonic Wars, Prussia's Brandenburg territories expanded to include much of the Rhineland over the objections of its inhabitants, who revolted in 1848; FELIX KLEIN (April 25, 1849–June 22, 1925) was born the next year during street fighting against the Prussians in Düsseldorf. Though the Prussians were able to maintain their hold on the Rhineland territories, transporting troops from Prussia to Brandenburg required crossing through the territories of other German states who might, for their own reasons, refuse passage. The Treaty of Prague created a contiguous Prussian state, the cornerstone of a new North German Confederation.

During the Austro-Prussian War, Klein studied with JULIUS PLÜCKER (June 16,1801–May 22, 1868) at the University of Bonn.[13] In 1830 Plücker described in Crelle's journal a new coordinate system, now called homogeneous coordinates, which described the position of a point in the plane with reference to the distance between the point and *three* lines. Since Cartesian coordinates in the plane only require the distance to two lines, the value of homogeneous coordinates might not be apparent. However they allow us to continue the process begun by Descartes: the algebraization of geometry.

We can easily describe lines and circles using algebraic equations. But how can we describe geometric operations, such as rotation, reflection, or translation? One possible way is through transformation of the coordinates. For example the transformation $(x, y) \rightarrow (-x, y)$ represents a reflection about the y-axis. In general, we can express a rotation, reflection, or translation of a point (x, y) to a new point (x', y') by a series of transformations

---

[13]The University of Bonn was founded in 1818, shortly after the region became Prussian, and grew rapidly. It has the distinction of being the last university founded in Germany before unification.

of the form:

$$x' = ax + by + c, \qquad\qquad y' = dx + ey + f$$

where $ae - bd = \pm 1$. If we describe the point (x, y) using three coordinates as $(x, y, 1)$, we can represent the transformation $(x, y, 1) \rightarrow (x', y', 1)$ as the matrix product AX, where

$$A = \begin{bmatrix} a & b & c \\ d & e & f \\ 0 & 0 & 1 \end{bmatrix}, X = \begin{bmatrix} x \\ y \\ 1 \end{bmatrix}$$

with det $A = \pm 1$. Thus the transformations of plane geometry can be described in terms of matrix multiplication. Since the set of matrices of this form is a group under matrix multiplication, then group theory can be applied to questions of plane geometry.

Unfortunately group theory had attracted little attention from continental mathematicians, despite the efforts of Liouville, Betti, Cayley, and Dedekind. The most recent attempt to bring group theory to the attention of mathematicians was by PETER LUDVIG MEJDELL SYLOW (December 12, 1832–September 7, 1918). Between 1862 and 1863 Sylow gave a series of very clear lectures on Galois theory to a Norwegian audience that included SOPHUS LIE (December 17, 1842–February 18, 1899).

Although Lie became one of Sylow's students, Sylow did not think highly of Lie's potential, and Lie himself had only a passing interest in mathematics. After graduation in 1865 he worked as a teacher of mathematics and as an astronomer's assistant, and it was not until 1868, when he read the works of Poncelet and Plücker, that Lie became inspired by mathematics. Lie was particularly interested in Plücker's use of geometric objects other than points as the basis for geometry, and used Plücker's ideas to develop some very interesting results, including a new representation of the complex plane. Like Abel, Lie chose to print his results in a pamphlet and like Abel, this led to a travel grant. He traveled to Berlin and became a participant in the Kummer-Weierstrass seminar alongside Klein. Since they were both converts to Plücker's ideas, it was natural that they became well-acquainted, and though they were temperamentally very different, they became good friends.

After Berlin, Lie went to Paris (by way of Göttingen) and met CAMILLE JORDAN (January 5, 1838–January 22, 1922). Jordan's "Memoir on the Algebraic Resolution of Equations" (1867) finally attracted enough attention from other mathematicians to make group theory an integral part of mathematics; this paper was followed by *Treatise on Substitutions and the Algebraic Resolution of Equations* (1870), the first text exclusively devoted to group theory. It was Jordan, not Sylow, who convinced Lie of the value of group theory; Lie in turn convinced Klein, who had also come to Paris to study with Jordan.

Klein soon realized that the different types of geometry might be unified using the group concept by studying what Klein called the "main group of spatial alterations" (now called transformation groups). For example, Euclidean geometry is the study of the spatial properties of figures that are invariant under rotation, translation, or reflection. This suggested to Lie a way of further generalizing geometry: rather than determine which group corresponded to a particular set of geometrical axioms, one could define the group first, then study the resulting geometry. Lie began the study of "finite continuous groups" (now called Lie groups), a peculiar name that needs some explanation: if a transformation is finite continuous, then any transformation can be reduced to a sequence of transformations that vary continuously from one another, while the group itself cannot be decomposed into discrete families. As a counterexample, Lie used Euclidean geometry: a reflection cannot be expressed as a sequence of continuous transformations. On the other hand translations and rotations can be expressed continuously; hence Euclidean geometry without reflections corresponds to a finite continuous group.

Within a few years, Klein would outline an ambitious program in the published version of his inaugural lecture at the University of Erlangen: the Erlangen Program hoped to unify the different types of geometry under the group concept. Unfortunately a different type of unification was about to occur, and it would force Klein, Jordan, and Lie apart.

9.4.5 The Birth of Three Nations

Events in Spain provided the final push for German unification. By 1833, Spanish liberals had staged a comeback, largely by being competent financiers and thus indispensable to the crown. This led to a complex three-way struggle between the reactionaries who sought a return to royal absolutism; the radicals who demanded a Republic; and the liberals who lobbied for a constitutional monarchy. The reactionaries suffered a key loss when Queen Isabella II left Spain for France in September 1868.

If no new monarch was appointed, the radicals might obtain a republic by default; thus the liberals had to find a replacement monarch quickly. One of the generals who led the revolution, Juan Prim y Prats, supported the candidacy of Prince Leopold of Hohenzollern-Sigmaringen: Wilhelm's cousin. However, this would surround France by German princes, so Napoleon sent Count Benedetti to meet with Wilhelm at Bad Ems to discuss the situation. On July 12, 1868, Benedetti persuaded Wilhelm to withdraw Leopold's name; the throne of Spain would then be offered to and accepted by Amadeus of Savoy, who arrived in Spain on December 30, 1870. Unfortunately Prim was assassinated on that day, and another civil war erupted (the Second Carlist War). The radicals forced Amadeus to abdicate on February 11, 1873, and announced the formation of a Republic. However the Republic proved so unstable that when Isabella's son Alfonso declared his support for a constitutional monarchy on November 30, 1874, disillusioned generals and others tired of the civil wars in Spain flocked to his side, and the republic simply vanished.

Meanwhile, Benedetti demanded a promise that no member of the House of Hohenzollern would seek the Spanish throne at any point in the future. The origin of the outrageous request is not clear, but Wilhelm refused to commit himself and all his descendants. Wilhelm sent a telegram to Bismarck detailing the events. Bismarck edited the Ems telegram for release to the press. Although he did not change the substance of the letter, he made Benedetti's demand more impolite and Wilhelm's refusal more angry, goading both the French and Prussians into a war frenzy.

The French had many reasons for confidence. French soldiers had proved victorious against the Russians and the Austrians, and derided Prussia's army of "lawyers and oculists." Moreover, the French army of 1870 was technologically superior to the Prussian army. The French *chassepot* (named after its inventor) had twice the range and firepower of the Dreyse rifle. Moreover the French had a "secret" weapon: the *mitrailleuse*.

In 1861 an American inventor, Richard Gatling, invented a multi-barrel gun with a crank mechanism that allowed each barrel to be loaded, rotated into place, and fired at a rate limited only by how fast a user could turn the crank. The Gatling gun saw very limited service during the American Civil War, and Britain and Prussia (among others), while noting the gun's tremendous rate of fire and its potential value, decided against adopting it, choosing instead to focus on cheaper, more portable rifles.

Meanwhile Napoleon III sponsored the production of a multi-barrel cannon invented by Auguste Verchère de Reffye and known as the *mitrailleuse* (anti-personnel loads for cannon were known as *mitraille*). The project was carried out in great secrecy: the parts were constructed in different locations, and even the budget came from Napoleon's own personal accounts and not the regular military budget. However it was a poorly kept secret, and articles about the weapon appeared in English, Austrian, Swiss, and Prussian publications.

Armed with technically superior weapons, including about 190 *mitrailleuse*, and a recent string of victories, the French were very confident when they declared war on July 19. The 22-year-old Klein, who was technically an enemy alien, left France and served as a medical orderly on the Prussian side. Lie was technically a foreign neutral, but decided to leave France by walking to Italy: in his university days, he was known for his athletic prowess, and a story is told that once he walked home to visit his parents one weekend, a distance of 60 kilometers, and when he found they were not home, he walked back!

En route, Lie was arrested by the French police, who accused him of being a German spy. He had in his possession letters from Klein, an enemy alien, and notebooks with entries written in a secret code. GASTON DARBOUX (August 14, 1842–February 23, 1917) intervened to obtain Lie's release, patiently explaining to the authorities that the code was, in fact, mathematics. Lie eventually returned to Norway and began applying group theory to differential equations; Lie groups have since become a major part of modern physics, being applied anywhere symmetries are of interest. As EMMY NOETHER (March 23, 1882–April 14, 1935) would point out in 1919, *every* conservation law in physics corresponds to a symmetry, so Lie groups are fundamental to theoretical physics.

Against the *mitrailleuse* and the *chassepot*, the Prussians had their antiquated rifles—and railroads. Within eighteen days, the Prussians sent an unprecedented 380,000 troops to the border with all their equipment. The French were still organizing their army for an invasion of Germany when the Prussians struck. The French armies disintegrated, and on September 1 at Sedan, the Emperor was captured. The *mitrailleuse* worked very well when used properly, causing great casualties among the Prussians at distances of up to 2000 meters (three times the range of the Dreyse); the *chassepot* performed equally well. Unfortunately the weapons were the only part of the French army that performed with distinction.

Meanwhile, news of the Emperor's capture led to a bloodless revolution in Paris that deposed the Emperor and continued the war. The siege of Paris began on September 19, and elsewhere French armies continued to fight against the Germans. The Army of the East was placed under the command of Charles Bourbaki, who failed to raise the siege of Belfort and was driven towards the Swiss border. With his position hopeless, Bourbaki attempted to commit suicide, but his pistol malfunctioned.

The terms of the Treaty of Frankfurt were unusually harsh in comparison with the Prusso-Danish and Austro-Prussian Wars: France ceded Alsace, and part of Lorraine (including Metz) to Germany, and agreed to pay a five billion franc indemnity. A German army would be stationed in northern France until the indemnity was paid (and the French were to pay for the cost of that army as well). Paris was still in chaos following the suppression of the Commune, so the French government signed the treaty at Versailles.

The Franco-Prussian War created three nations. The south German states, which Bismarck had dismissed with contempt earlier in his career, flocked to join the North German Confederation. Even before the peace treaty was announced, Wilhelm I was declared Emperor of Germany at Versailles, marking the formation of the Second Reich.

French troops withdrew from the Papal States at the beginning of the war, leaving Rome virtually defenseless. After a mostly symbolic resistance, the Papal army surrendered and the Papal States were annexed to the Kingdom of Italy on September 20, 1870; in October, Rome became the capital of a united Italy. On May 13, 1871 the Italian parliament passed the Law of Guarantees which conferred upon the Pope the dignities of a sovereign in the Vatican and the immediate region, as well as a pension to be paid from the Italian government, but Pope Pius IX refused to acknowledge the validity of the annexation of the Papal states and until his death declared himself "prisoner in the Vatican." It was not until 1929 that the Papacy recognized the state of Italy, then ruled by Benito Mussolini.

Finally in France itself, Napoleon III was deposed, establishing the Third Republic on the ruins of the Second Empire. The loss of Alsace and Lorraine and the humiliation of the Treaty of Versailles would poison relationships between France and the German Empire, and Bismarck of all people should have predicted its disastrous long-range effects. He is best known for saying that, "The great issues of today will not be settled by discussion and debate, but by blood and iron." He is less well known for saying that the next great war would begin over "some damned fool thing in the Balkans."

For Further Reading

For more details about this era, see [15, 53, 87, 99, 116].

10

The United States

10.1 From Colony to Country

In 1757, during the Seven Years' War, French forces captured Hanover. Rather than sending more troops to the continent to retake Hanover, William Pitt the Elder proposed an alternative strategy of attacking French colonial interests overseas: in effect, turning the periodic skirmishes between the British and French in the contested Ohio River Valley into a major war, now known as the French and Indian War. The British captured Fort Duquesne in 1758, renaming it Pittsburgh, and by 1760, all of French Canada was in British hands. By the terms of the Treaty of Paris (1763), Britain received all of French North America east of the Mississippi (with the exception of New Orleans), and all of French India acquired after 1749; Britain also received Florida from Spain. France withdrew its forces from Hanover, Hesse, and Brunswick, though in exchange she received a few small (but, since their main product was sugar, extremely valuable) islands in the West Indies.

The French and Indian War, though a resounding British victory, was a costly one. The British national debt nearly doubled, going from 75 million pounds in 1755 to 133 million pounds in 1763. Britain hoped the North American colonists (who benefited most from the war) would help pay for the peace. The proposed tax burden in America would be lower than in Britain itself, but archaic electoral practices meant that the American colonists had no representation in the British parliament. This led to protests against "taxation without representation." The protests and tax riots generally persuaded parliament to repeal or lower unpopular taxes, though invariably they imposed another tax to maintain the principle that the colonists could be taxed. Protests led to the American Revolutionary War (1775–1783) and the formation of the United States of America. Almost immediately the new country plunged into a depression.

Shipping to and from the British West Indies played a major role in the colonial economy, but independence cut off the colonists from this vital market. Trade within the colonies might have made up for this loss, but the thirteen colonies were governed by thirteen different sets of laws and most issued their own currency. Those that did not used the currency of England, Ireland, Spain, Canada, or France. The complexity of this trade is best illustrated by Nicolas Pike's *New and Complete System of Arithmetick, Composed for Citizens of the United States* (1788), where he included detailed instructions for converting between the different state currencies, as well as conversions between the weights and measures of the different countries in Europe with those of the United States.

A strong central government with the ability to make tariffs and currency uniform across the country was essential, but under the Articles of Confederation the central government had almost no effective power. Thus in May 1787, delegates from the colonies met in Philadelphia to revise the Articles of Confederation. After some debate, they decided to rewrite it entirely, and on September 17, 1787 they produced a new document, the Constitution of the United States of America. Tellingly the constitution gives Congress considerable power to regulate commerce and denies the states the right to set their own import duties or coin their own money.

Congress was also given the exclusive right to create a district, "not exceeding ten miles square" to house the seat of government of the United States. Land from Maryland (which included the city of Georgetown, named after George III) and Virginia (which included the city of Alexandria) were joined together to make a federal district. George Washington selected Pierre-Charles l'Enfant to plan what Washington called "Federal City" and l'Enfant called "Capital City," though it is now known as Washington D.C.

10.1.1 Benjamin Banneker

In 1791 l'Enfant hired Andrew Ellicott to survey the site for the new city. Ellicott in turn hired his 60-year-old neighbor, BENJAMIN BANNEKER (October 9, 1731–October 9, 1806), whose father was a freed slave. As slaves were considered property, they were generally not given last names; the name "Banneker" comes from his mother, Mary Banneky. At the age of 21, Banneker discovered watchmaking and constructed a wooden clock of remarkable accuracy. At the age of 57 Banneker taught himself mathematics so he could learn the astronomy necessary to construct an almanac.

Banneker completed his almanac by August 20, 1791, and sent a copy to James McHenry, signer of the Declaration of Independence and later a military surgeon who had served with George Washington. McHenry in turn forwarded the work to William Goddard and James Angell, two Baltimore abolitionists. They sponsored the printing of *Benjamin Banneker's Pennsylvania, Delaware, Maryland and Virginia Almanack and Ephemeris, for the Year of Our Lord 1792*. In the introduction, Goddard and Angell noted that:

> ...a sable Descendant of Africa, who, by this Specimen of Ingenuity, evinces, to Demonstration, that mental Powers and Endowments are not the *exclusive* Excellence of *white People*, but that the Rays of Science may alike illumine the Minds of Men of *every Clime*, (however they may differ in the *Colour* of *their Skin*) particularly those whom Tyrant-Custom hath too long taught us to depreciate as a Race inferior in intellectual Capacity...[1]

The almanac included an ephemeris, stories, information about the meetings of the federal courts and the members of the government, tables of depreciation and interest, and (since the state money had not yet disappeared) tables detailing the conversion of Federal dollars into the currencies of New Jersey, Pennsylvania, Delaware, and Maryland.

Banneker sent a copy to Thomas Jefferson (then Secretary of State) on August 17. In it he drew sharp contrast between Jefferson's words in the Declaration of Independence of

---

[1] *Benjamin Banneker's Pennsylvania, Delaware, Maryland and Virginia Almanack and Ephemeris, for the Year of Our Lord 1792*, William Goddard and James Angell, Baltimore, p. 2.

the right of all men to freedom, and the fact that Jefferson and others were guilty of "detaining by fraud and violence so numerous a part of my brethren under groaning captivity and cruel oppression." It probably took a considerable amount of effort for Jefferson to write a civil response to this accusation (however true) of hypocrisy; in a letter to longtime correspondent Joel Barlow on October 8, 1809, Jefferson wrote:

> ...we know he has spherical geometry enough to make almanacs, but not without the suspicion of aid from Elliot, who was his neighbor & friend, & never missed an opportunity of puffing him. I have a long letter from Banneker which shows him to have had a mind of very common stature indeed.

However, Jefferson sent a diplomatic response to Banneker on August 30, 1791:

> ...no body wishes more than I do to see such proofs as you exhibit, that nature has given to our black brethren, talents equal to those of the other colours of men, & that the appearance of a want of them is owing merely to the degraded condition of their existence both in Africa & America.

On the same day Jefferson wrote to Condorcet:

> I am happy to be able to inform you that we have now in the United States a negro, the son of a black man born in Africa, and of a black woman born in the United States, who is a very respectable mathematician. I procured him to be employed under one of our chief directors in laying out the new federal city on the Potomac, & in the intervals of his leisure, while on that work, he made an Almanac for the next year, which he sent to me in his own handwriting, & which I inclose to you. I have seen very elegant solutions of geometrical problems by him.

10.1.2 Nathaniel Bowditch

In 1791, an eighteen-year-old clerk named NATHANIEL BOWDITCH (March 26, 1773–March 16, 1838) (rhymes with "Now which") was permitted access to the scientific library of the Salem Philosophical Society. Subsequently Bowditch launched his scientific career by discovering an error in Newton's *Principia*.

The library came into the hands of the Philosophical Society as part of the spoils of war. By attacking British merchant ships, the colonists could keep the British navy occupied, making it easier for France and other countries to send money, men, and arms. But the colonists themselves had no navy, and if a private individual attacked a merchant ship, they were legally a pirate, subject to trial and summary execution. To avoid this unhappy fate, a ship captain might have a Letter of Marque and Reprisal from a government. This turned pirates into a privateers and gave them the status of combatants: they might be killed in the course of battle, but the rules of war (insofar as they exist) protected them in the event of their surrender or capture. On May 2, 1780, the Continental Congress issued a set of guidelines for privateers.

The guidelines issued by Congress were directed against those who took too many liberties with their right to attack shipping. For example in 1779 the *Pilgrim*, a privateer out of Beverly (a town just north of Boston) under the command of Captain Hugh Hill, seized a Danish vessel; ship and cargo were eventually returned, since Denmark was neutral. Later that year the *Pilgrim* seized a Spanish vessel that had taken on cargo in London. Again, Spain was neutral, though the seizure of the cargo was ruled legitimate. Finally, on September 5, 1780 the *Pilgrim* made a legitimate seizure: the *Duke of Gloucester*, sailing from Galway to London. The seized cargo included the library of Richard Kirwan, a prominent Irish chemist who was about to be elected to membership in the Royal Society of London (Kirwan himself had been living in London since 1777). The *Pilgrim* returned to Salem in 1781, and Joseph Willard of Beverly, who had just become the President of Harvard University, raised money to buy the books and donate them to the newly founded Salem Philosophical Society. Among the 115 books were the works of Newton, the Bernoullis, and the *Encyclopedia Britannica*. The books were eventually bought for $435.35, and kept in the home of the Reverend John Prince.

Bowditch's discovery of an error in Newton portended a brilliant scientific career, but before he made his mark on the world of science, Bowditch went to sea. He made four voyages between 1795 and 1799 before marrying Mary Ingersoll on October 28, 1800. This gave Bowditch reason to stay ashore, and over the next few years he began revising John Hamilton Moore's *Practical Navigator*.

Moore's work was a vast compilation of useful information for the navigator. Unfortunately, because of careless editing, it was also a vast compilation of errors as well—and in the open ocean, even a small error can be catastrophic. Bowditch corrected ten *thousand* errors in the texts (both in Moore and in the sources Moore used), and in June 1802 the first printing of his *New American Practical Navigator* appeared. It became the textbook from which generations of navigators would learn their trade and, since the ability to navigate was one of the prerequisites for becoming a ship's officer, a surprising number of sailors learned from it.

Bowditch might have stayed ashore, but the pepper trade lured him back for one final voyage. In 1602 the Dutch East India company (the *Vereenigde Oost-Indische Compagnie* or VOC) established a foothold in Indonesia and began a centuries-long domination of the region. By helping one side or the other in local conflicts, the VOC was able to establish control over most of Java and portions of Sumatra. Through a system of "contingencies" (taxes payable in kind) and forced deliveries, the VOC acquired spices at very low cost for resale on European markets at enormous profits. This gave the company the resources to gradually force other European powers out of the area. But by the end of the eighteenth century the Dutch hold on the region was beginning to slip.

In 1793 Captain Jonathan Carnes of Salem learned that northwest of Bengkulu in Sumatra, black pepper could be obtained directly from the producers at prices far below those charged by the VOC. Carnes voyaged to Sumatra in the *Rajah*, a small schooner, carrying a cargo of brandy, gin, iron, tobacco and dried fish worth about $18,000. Eighteen months later he returned to Salem and sold a cargo of black pepper at a profit of 700%. Soon Salem became a major center of the pepper trade, exporting up to eight million pounds a year (and incidentally generating as much as 5% of the *nation's* tax revenue); the connection with the East Indies would be memorialized in the city's seal, adopted 1839 and still used today, which depicts an Atjehnese man (from northwestern Sumatra) and a Latin inscription which translates as "To the farthest port of the rich east."

On November 21, 1802 Bowditch sailed as captain of the *Putnam*, a brand new ship.[2] It took five months to make the twelve thousand mile voyage; en route, Bowditch read the first volume of Laplace's monumental *Celestial Mechanics*. When the *Putnam* returned to Salem on December 25, 1803, it carried a cargo of 411,498 pounds of pepper and 27,625 pounds of coffee. Bowditch had paid the equivalent of $4000 for this cargo, which customs inspectors assessed at $26,071.13; as Captain, Bowditch's share was $1563.44. It would be Bowditch's last voyage. He had been lucky: a younger brother William died on a voyage to Trinidad in 1799, and an older brother Habakkuk drowned in Boston Harbor in 1800. The *Putnam* would be captured in Sumatra by Malay pirates in 1805 and its crew massacred.

Within a month of returning to Salem, Bowditch became president of the Essex Fire and Marine Insurance Company. He became involved in the bitter political fights of 1804 between the Federalists, led by Alexander Hamilton, and the Republicans, led by Thomas Jefferson. Roughly speaking the Federalists supported the development of commerce and industry (particularly shipping, which made them immensely popular in New England) and a strong national government, while the Republicans supported the development of agriculture and states' rights. However, events in Europe complicated the election in the Americas.

By the terms of the Treaty of Paris, France ceded the Louisiana Territory to Spain. By the Treaty of San Lorenzo (1795), Spain gave the United States the right to ship goods through the mouth of the Mississippi River without paying taxes, and the right to store goods temporarily (the "right of deposit") in the port of New Orleans. But Napoleon persuaded Charles IV of Spain to cede Louisiana to France on October 1, 1800, under the stipulation that it would not be given to a third power.

In 1802 Spain revoked the right of deposit granted by the Treaty of San Lorenzo. It was not clear whether Spain had the right to do so, since the territories were no longer Spanish possessions. Jefferson ordered Robert R. Livingston, the U.S. minister in Paris, to meet the French Foreign Minister Talleyrand and clarify the situation and, if possible, purchase the port of New Orleans. In 1803 James Monroe arrived in Paris to continue negotiations and, rather unexpectedly, Napoleon offered to sell the entire Louisiana Territory for a total of about $15 million (which included assuming $3,750,000 in claims against France). Thus on May 2, 1803 the United States acquired the Louisiana Territory, doubling the size of the country. Ironically, the purchase of the Louisiana Territory had the effect of extending the powers of the federal government, something the Republican Jefferson opposed on philosophical grounds.

Many New Englanders feared that their interests would be diluted by those living in the far west, and some even spoke of seceding from the union. However Hamilton threw his support behind the purchase, fatally dividing the Federalist party.

Hamilton himself did not have long to live. Jefferson's first-term Vice President Aaron Burr and Hamilton were longstanding political enemies: Hamilton helped organize Burr's defeat for governor of New York early in 1804, and subsequently George Clinton replaced Burr on the Republican ticket. The dispute turned deadly on July 11, 1804, when Burr

---

[2]The Putnam family was a prominent and notorious one in Salem history. In 1692 Ann Putnam and three other girls accused Tituba, a West Indian slave, of practicing witchcraft. The Salem witch hysteria resulted in the executions of 19 women by hanging and 1 man by pressing. Years later, Ann Putnam recanted, apologizing for her false accusations.

killed Hamilton in a duel over some disparaging remarks made by the latter. The Federalist party suffered an equally fatal blow that year. Massachusetts was won by Jefferson and the Republicans. Bowditch, William Gray and Jonathan Waldo all ran for representative as Federalists; all lost, and the Federalists declined in power, effectively disappearing by 1817.[3]

In a letter of October 26, 1818 the Republican Jefferson sent a three page letter to the Federalist Bowditch, thanking him for a pamphlet on Stewart's method of determining the distance of the sun, though he found the work "much above my mathematical stature."[4] Jefferson goes on to describe "an interest much nearer home, and on the subject of which I must make a long story."

Jefferson described the foundation of the Central College (the future University of Virginia at Charlottesville). "We conclude to employ no Professor who is not of the 1st-order of the science he professes, that where we can find such in our own country, we shall prefer them, and where we cannot, we will procure them wherever else to be found. The standing salary proposed is of 1000 to 1500 D with 25 D tuition fee from each student attending any professor, with house, garden &c. free of rent ... " A few paragraphs later Jefferson comes to the point: "We are satisfied that we can get from no country a professor of higher qualifications than yourself for our Mathematical Department, and we entertain the hope and with great anxiety that you will accept of it." Bowditch declined on November 4, 1818. In 1820 he rejected a similar offer for the Chair of Mathematics at West Point from Republican Secretary of War John C. Calhoun. In 1823 Bowditch accepted an offer to become the actuary for the Massachusetts Hospital Life Insurance Company and the president of the Commercial Insurance Company at a salary of $5000, nearly three times that he was being paid in Salem.

10.1.3 Robert Adrain

During the American Revolution, the British called for volunteers to help defend Ireland from a possible French invasion; by 1779 some 100,000 men marched under the banner of the Irish Volunteers. The fact that so many armed Irishmen could be united into an army concerned the British government, especially when the Volunteers marched in protest of the Navigation Acts. Parliament made many concessions that gave Ireland more autonomy. However, this liberalization of British rule was largely abandoned in the wake of the French Revolution, and in 1798 the Irish revolted. Unfortunately for Irish independence, some waited for French help, while others fought the British piecemeal, and the rebellion failed. ROBERT ADRAIN (September 30, 1775–August 10, 1843) was shot in the back by one of his own men and after recovery, Adrain fled to the United States, settling in Princeton, New Jersey.

Adrain was a self-trained mathematician who would make the first significant American contribution to mathematics. In 1804, George Baron of Philadelphia began publishing the first mathematics journal in the United States, the *Mathematical Correspondent*. Adrain was a frequent contributor, and became its editor in 1807, shortly before it ceased publica-

---

[3]Meanwhile the Republicans became known as the Democratic Republicans and, during the presidency of Andrew Jackson became known as the Democratic Party, their modern name.

[4]Matthew Stewart was Maclaurin's successor at Edinburgh.

tion. In 1808, Adrain began publishing *The Analyst*. The journal was to consist of problems, posed by correspondents, and solutions, sent in by readers. In the second issue Robert Patterson of Philadelphia posed the following problem:

Problem 10.1. *A surveyor walks around a five-sided parcel of land in the directions and for the distances indicated:*

1. *North by 45° east,* 40 *paces*

2. *South by 30° east,* 25 *paces*

3. *South by 5° west,* 36 *paces*

4. *Due west,* 29.6 *paces*

5. *North by 20° west,* 31 *paces, back to the starting point*

What is the most probable area of the field?

Since the path as recorded does not return to the starting point, there must have been errors in the recording of the directions and distances. Adrain made this the greater prize question for the issue, and promised a reward of ten dollars for the best solution; at the time, the salary of a United States senator was $1500 a year and the president earned $25,000 a year.

The only solution Adrain judged worthy of the prize was given by Bowditch, who began by supposing $ABCDE$ to represent the figure described by the survey, where the points A, E should coincide (note that Bowditch solved the problem for a four-sided plot of land). Since the points A, E do not in fact coincide, then AE is the total error of the measurements. Let AB' represent the most probable first side of the field. The first side was actually measured as AB, so the error in measuring this side was BB'. By substituting the most probable side, AB', for the measured side, AB, the points C, D, and E are moved, respectively, to C', D', and E' by amounts equal to BB' and in the same direction. (This is because the measured side BC no longer begins at B, but instead begins at B' and thus ends at C'; likewise, the measured side CD begins at C' and ends at D', and so on.) Again, let $B'C''$ represent the most probable second side of the field; since it was measured as BC (or $B'C'$, as the two are the same), then the error in the measurement is $C'C''$. Again, substituting the most probable side $B'C''$ for the measured side BC, the points D', and E' are moved to points D'', and E'', respectively. Continuing in this manner, the most probable vertices of the field are at A, B', C'', D''', and E'''', which should coincide with A.

The question then becomes determining the correction factors BB', $C'C''$, $D''D'''$, and $E'''E''''$ needed to make the final point, E'''', coincide with A. Bowditch's principle is that the corrections should all tend to decrease AE. In the simplest case, Bowditch assumed that all the errors should be in the same direction as AE, and the magnitude should be proportional to the measured lengths.

Adrain awarded the prize to Bowditch, and considered the problem further. By assuming that the magnitude of the most probable error is proportional to the quantity being measured, the probability X_m of obtaining an error of magnitude x in the measure of quantity s_m is dependent only on x, and that the errors x_1, x_2 actually obtained in making the measurement of s_1, s_2 correspond to the most probable errors, Adrain proved that the probability of obtaining an error of magnitude x is ae^{-mx^2}, where a, m are positive constants.

Consequently the probability of making errors of magnitude x_1, x_2 (assuming these probabilities independent) will be $ae^{-m(x_1^2+x_2^2)}$. This probability will be at a maximum when the $x_1^2 + x_2^2$ is at a minimum; thus the most probable errors are the ones that minimize the sum of the squares of the errors. This was the first proof of the validity of the method of least squares. It was the greatest mathematical contribution yet made by a mathematician in the United States. Unfortunately *The Analyst* failed to attract the necessary readers; indeed, Adrain and Bowditch were about the only two correspondents of mathematical significance, and within a year it ceased publication.

In 1813 Adrain accepted the position of Professor of Mathematics at Columbia College (now University) in New York City, where he stayed until 1826. While in New York, he wrote a number of textbooks for student use, and made a second attempt to found a mathematics journal, *The Mathematical Diary*, in 1825. Adrain introduced the first issue with:

> The principal object of the present work is to excite the genius and industry of those who have a taste for mathematical studies, by affording them an opportunity of laying their speculations before the public in an advantageous manner; and thus to spread the knowledge of mathematics in a way that is both effectual and agreeable.[5]

The Mathematical Diary fared little better than *The Analyst*, and a successful mathematical journal in the United States would not appear for more than half a century.

At the time, New York was the second largest city in the country (Philadelphia being slightly larger). But in 1811, a commission that included Stephen van Rensselaer, a wealthy landowner in the Albany region, suggested that a canal be built linking the Great Lakes to the Hudson River. Construction of the Erie Canal began on July 4, 1817, and was completed two years ahead of schedule, on October 26, 1825. The canal made a tremendous difference: before the canal, it took twenty days and $100 to ship a ton of freight from Buffalo to New York, while after the canal, it took eight days and $15. Because the bulk of federal revenues came from tariffs (taxes charged on goods as they cross borders), this made the Port of New York increasingly important to the nation, and within a generation New York City was the largest city in the country.

During the building of the canal, Amos Eaton surveyed portions of the route. His work brought him to the attention of Rensselaer, who recognized the lack of higher technical education in the United States. With Eaton's help, Rensselaer founded the first college in the United States to be devoted to the sciences. On January 3, 1825 the Rensselaer School (later Polytechnic Institute) opened, with a mandate for the students to learn not by hearing lectures and seeing experiments, but by doing these things under the supervision of their professors or a "competent assistant."

The state of mathematics education in the United States might be judged by the curriculum of its oldest schools. The will of John Harvard bequeathed 400 books and half his estate to New College in Cambridge; the next year the institution renamed itself in Harvard's honor. During the era of Cromwell, degrees granted at Harvard were given equal status to those given at Oxford or Cambridge, though this was more a matter of religious affiliation (as both England and New England were dominated by Puritans) than academic accreditation. At the time Harvard College was primarily an institution to train future cler-

---

[5] *The Mathematical Diary*, Vol. I, No. 1 (1825), p. iii.

gymen, though a candidate for the master's degree submitted a thesis on the topic "Is the quadrature of a circle possible?" in 1693 (the candidate answered "Yes"). The number of institutions of higher education grew rapidly in the United States after independence, but their quality is more difficult to judge. Certainly mathematical knowledge was not considered important. It was not until 1803 that Harvard (first mentioned as a University in the Massachusetts Constitution of 1780) made knowledge of the rule of three an entrance requirement; by 1819 some knowledge of algebra was required.

10.1.4 William Rowan Hamilton

The ability of some mathematicians to perform extraordinary feats of mental computation pales in comparison to that of Zerah Colburn. The otherwise-normal Colburn's remarkable ability came to the attention of the President of Dartmouth College, who offered to pay for Zerah's education, but his father declined, hoping instead to make money exhibiting his son's computational prowess. Thus in 1810 (when Zerah was six) they traveled, first to Boston, then on May 11, 1812, to England, where Zerah performed before audiences and answered questions like "Find the cube root of $268, 336, 125$" and "Factor $247, 483$ and $36, 083$." Because Britain and the United States were at war, Colburn was sometimes introduced as a Russian. Colburn spent the next ten years in Europe, returning only after the death of his overbearing father (who rejected not only the offer from Dartmouth, but withdrew Zerah from the Royal College after a year, despite a full scholarship).

One of those who saw Colburn perform was WILLIAM ROWAN HAMILTON (August 4, 1805–September 2, 1865), a child prodigy who could speak Latin, Greek, and Hebrew by the time he was five and at least 11 other languages by his ninth birthday. Colburn's performance interested Hamilton in mathematics, and he began to study it in earnest. In contrast, Colburn turned to the study of languages, eventually becoming professor of languages at Norwich University.

In 1833, Hamilton considered the algebra of what he called moment couples of the form (a_1, a_2), where a_1, a_2 were real numbers. Addition and subtraction were defined componentwise, which allowed Hamilton to define whole number multiplication as a repeated addition. The simplest way to preserve the associative law, the distributive law, and the uniqueness of quotients was to let the product $(0, 1) \times (0, 1) = (-1, 0)$. By identifying $(0, 1)$ with the complex unit i and $(1, 0)$ with the real number 1, the complex numbers could finally be put on a rigorous foundation.

Could the notion of complex numbers be extended? Hamilton began with two key properties of the complex numbers $a + bi$, where a and b are real numbers. The first property is that $i^2 = -1$, so a hypercomplex number might be of the form $a + bi + cj$, where a, b, and c are real numbers and $i^2 = j^2 = -1$. Second, given a complex number $a + bi$ and its modulus $\sqrt{a^2 + b^2}$, the modulus rule is that the modulus of the product of two complex numbers is the product of the moduli.

Suppose we have the hypercomplex number $a + bi + cj$ and define its modulus as $\sqrt{a^2 + b^2 + c^2}$. The modulus of $(a + bi + cj)^2$ ought to be $a^2 + b^2 + c^2$. However the expansion of $(a + bi + cj)^2$ seems to give:

$$(a + ib + jc)^2 = a^2 - b^2 - c^2 + 2iab + 2jac + 2ijbc$$

and satisfying the modulus rule requires that $ij = 0$. But this made no sense, since it meant that two non-zero quantities i and j had a product equal to zero. Hamilton wrestled with this problem for a long time then, while walking with his wife along the Royal Canal in Dublin on October 16, 1843, had a flash of inspiration: the modulus rule could be saved if ij and ji were not equal, but instead $ij = -ji$; in this case, the ij term of the above product vanishes, and the modulus rule can be saved—at the cost of the commutativity of multiplication.

What was ij itself equal to? If the modulus of a product is to be equal to the product of the moduli, then the product ij cannot be equal to either 1, i, or j. The only option was that $ij = k$, a third unit. Following the logic to the end it was necessary that the three units i, j, and k obey the multiplication rules $ij = k$, $jk = i$, $ki = j$, and, most surprisingly, $ji = -ij$, $jk = -kj$, and $ik = -ki$. Hamilton was so impressed with the unusual multiplication properties of the hypercomplex numbers that he carved the fundamental equation $i^2 = j^2 = k^2 = ijk = -1$ on a stone of Brougham Bridge along the Royal Canal.

The elements of this algebra with four units are of the form $a + bi + cj + dk$, and Hamilton, a classical scholar, named them quaternions, after the term used to designate a squadron of four Roman soldiers. It was fitting that Hamilton, who had been inspired to pursue mathematics by an American calculator, would in turn inspire the first major American contribution to mathematics: the study of linear associative algebras.

10.1.5 Benjamin Peirce

In 1824 student riots at Harvard University caused the state of Massachusetts to revoke an annual $10,000 subsidy. Harvard President John Kirkland neglected to adjust the school's budget to reflect this loss of revenue, and questions soon arose about the school's financial stability. Bowditch became one of Harvard's governors in 1826, and ordered a general audit. The man who found ten thousand errors in the *Practical Navigator* found Harvard's accounts equally flawed. Bowditch and Kirkland fought constantly, and eventually Kirkland resigned on May 28, 1828 (a year after having an incapacitating stroke).

Before his resignation Kirkland fought to save a professor of mathematics from dismissal. Bowditch attacked the professor's competence and pointed out that "Peirce of the sophomore class knows more pure mathematics than [the professor] does." Bowditch was referring to seventeen-year-old BENJAMIN PEIRCE (April 4, 1809–October 6, 1880). Like Bowditch, Peirce (pronounced "purse") came from Salem, and demonstrated his mathematical capabilities during the proofreading of Bowditch's translation and annotation of Laplace's *Celestial Mechanics*.

Bowditch began a translation and annotation of Laplace's *Celestial Mechanics* in 1814, incorporating recent developments and long derivations dismissed by Laplace with a casual "it is easy to see..." Although most of the work was completed by 1818, it was not published for several years afterwards both because of financial considerations (Bowditch would later spend a third of his own fortune, or $12,000, to publish it) and because Bowditch worked very hard to purge his translation of errors. Peirce learned a great deal of mathematics during the proofreading process, and not surprisingly, his early work focused on analytical and celestial mechanics. But soon he turned to the field of algebra, and when he learned of Hamilton's quaternions, he bemoaned the fact that he could not work on them as a young person could. Peirce was just thirty-four.

Hamilton limited the coefficients of the quaternions to real numbers and arrived at non-commutativity because the alternative was permitting divisors of zero (i.e., non-zero quantities a and b where $ab = 0$). Most mathematicians of the era, including Peirce's own son CHARLES SANDERS PEIRCE (September 10, 1839–April 19, 1914), believed algebraic structures with divisors of zero were useless, but Peirce investigated them systematically. He made one concession to the traditional rules of algebra: the associative law should hold. Since an expression of the form $a + bi + cj + dk$ is referred to as a linear combination, Peirce's objects of study are referred to as linear associative algebras.

By 1870 Peirce had analyzed over 150 algebras with anywhere between 1 and 6 units. His work was published by the National Academy of Sciences in Washington, DC by a somewhat unusual method: the lack of funds meant that the Academy could not publish Peirce's work (presented over the past few years) in a traditional work, so they hired a lady without mathematical training but excellent penmanship to transcribe Peirce's virtually unreadable notes onto lithograph stones, which could then be used to make a block print of the manuscript. One hundred copies were printed and distributed to mathematicians in the Americas and Europe.

After some brief introductory notes, Peirce's 1870 lithograph begins by defining mathematics as "the science which draws necessary conclusions." Peirce goes on to describe the varieties of linear associative algebras in detail, and introduced two key ideas of modern abstract algebra: an idempotent element, where $x^m = x$ for $m \geq 2$, and a nilpotent element, where $x^m = 0$ for some m.

Of equal if not greater importance was Peirce's influence in the teaching of mathematics. He became a tutor at Harvard in 1831, and by 1833 was the head of the mathematics department. Peirce introduced many innovations. In 1825, all Harvard students were expected to complete a three-year mathematics program, which included plane, solid, and analytic geometry; algebra; conic sections; trigonometry; and differential calculus. For obvious reasons, this rigorous course in mathematics proved unpopular with students and their parents, so Peirce instituted a new mathematics curriculum. A new, one-year mathematics course (on algebra, geometry, and plane and spherical trigonometry, and their applications to navigation and surveying) was required, but at the end of their first year, "at the written request of their parent," students could discontinue their study of mathematics. Those who continued their studies might take a one-year theoretical course for future teachers; or an advanced three-year course for research mathematicians; Peirce also eliminated arithmetic as a college level course in 1837.

The changes met with the approval of the President of Harvard and the parents of Harvard students. As might be expected, most students chose not to continue their study of mathematics: of the fifty-five students eligible to continue studying mathematics in 1839, only eight did so. Even then, the Committee on Studies noted that the freshman spent too much time in their mathematics courses, and mathematics continued to come under heavy criticism. One student reported that the best mathematics teacher he had while at Harvard was J. M. Child—better known for his work collecting and classifying folk ballads.

In 1848, Peirce taught the first graduate level mathematics courses in the United States at Harvard's Lawrence Scientific School. Peirce lectured on analytical and celestial mechanics, as well quaternions; Hamilton's first lectures on quaternions were also given in that year.

In addition Peirce wrote some of the first American textbooks on mathematics. This represented a break from a reliance on translated European texts (mostly those written in France for the Polytechnical School). Over the next few years Peirce wrote *Elementary Treatise on Plane Geometry* (1835), *Elementary Treatise on Spherical Geometry* (1836), *Elementary Treatise on Plane and Solid Geometry* (1837), and *Elementary Treatise on Algebra* (1837).

10.1.6 The Civil War

In 1839 the U.S. brig *Washington* intercepted the Spanish ship *Amistad*, adrift off the coast of Long Island, and brought it to New Haven. The *Amistad* carried over 40 men, women, and children of African descent, and two Cubans, Jose Ruiz and Pedro Montes. The Cubans claimed the Africans were slaves, bought in Havana for transport to another Cuban port; they accused the Africans of murder and piracy. Getting the Africans's story was hampered by a simple difficulty: no one knew their language. Fortunately, New Haven was the home of Yale University. Josiah Willard Gibbs, professor of linguistics at Yale, determined that the Africans spoke Mende, and soon native Mende speakers were located in the docks of New Haven and New York. Through them, a very different story emerged.

The Mende had been captured in Africa in April 1839, and were not born into slavery in Cuba as Ruiz and Montes claimed. Consequently on January 13, 1840 Judge Judson, the district court judge in New Haven, declared that the Africans were kidnap victims, not slaves, and ordered them returned to Africa, where they would be freed. However, this touched on a serious sectional issue: slavery. The economy of the southern states rested on cotton and its "peculiar institution" of slavery. In the north, the movement for the abolition of slavery was growing more politically powerful, and southern slaveholders feared the federal government might outlaw slavery. Even though the *Amistad* verdict recognized the legality of slavery, ruling that the cabin boy was legally the property of Ruiz and Montes, and freed the Africans because of existing law (importation of slaves from Africa was illegal), allowing the verdict to stand would mark President Martin van Buren as being "soft on slavery." Thus he appealed the case twice. Former President John Quincy Adams became involved in the case, arguing on their behalf before the Supreme Court of the United States. The Supreme Court upheld the lower courts's rulings in 1841, and the Mende were returned to Africa.

The fear that the government might abolish slavery tainted politics for the next generation, and matters reached a breaking point in the election of 1860. In that year the Democratic presidential candidate was Stephen A. Douglas, who suggested that a territory (*not* the federal government) could make slavery untenable through unfavorable legislation. Southern Democrats broke with the national party and threw their support behind John C. Breckenridge of Kentucky, a state that allowed slavery. The Constitutional Union Party put forward a third candidate, John Bell (his running mate was Edward Everett, a former governor of Massachusetts who helped bring the Prussian educational model to the United States), but with the south divided between Bell and Breckenridge, and the Democrats divided between Douglas and Breckenridge, the Republican candidate (and Douglas's former opponent for the senate seat) Abraham Lincoln won the election with less than 40% of the popular vote.

Lincoln believed that slavery would eventually disappear on its own, and pledged that he would not interfere with slavery where it existed. Despite this, South Carolina announced it was seceding from the union on December 20, 1860. Ten other states (Mississippi, Florida, Alabama, Georgia, Louisiana, Virginia, Arkansas, Tennessee, and North Carolina) followed, forming the Confederate States of America. Since secession would effectively neutralize the powers granted to the federal government under the constitution, it could not be permitted, and rather than four more years of political wrangling over the institution of slavery, South Carolina's actions precipitated four years of civil war.

Secession forced many to choose between their loyalty to the federal government and their loyalty to their state. More than 300 officers resigned their commissions with the United States Army, and about half as many resigned from the Navy. Replacements had to be found. The resignation of a professor of mathematics in the Navy led to Lincoln's nomination of SIMON NEWCOMB (March 12, 1835–July 11, 1909) as a replacement on December 17, 1861. Newcomb had been born in Canada and studied mathematics with Peirce at Harvard. After the war, Newcomb would write *A Critical Examination of Our Financial Policy During the Southern Rebellion* (1865), analyzing the Union's financial policy during the Civil War, and *Principles of Political Economy* (1885), one of the earliest to treat economics as a quantitative science. Unfortunately his distinguished career as an economist and a mathematician would be overshadowed by an article published in *McClure's Magazine* in September 1901, where he declared it extremely improbable that man would ever fly in a heavier-than-air craft.

Southerners hoped the British would offer them support, since an independent Confederacy could balance the territorial ambitions of the United States (which had invaded Canada during the War of 1812 and annexed large portions of Mexico in 1848). Moreover, the backbone of the British economy was the textile industry, and a victorious Confederacy might offer trade concessions to a supportive Britain. Thus two Confederate ambassadors, James Murray Mason and John Slidell, sailed for Britain to negotiate support. On November 8, 1861, Captain Charles Wilkes, commanding the Union warship *San Jacinto* boarded the British ship *Trent* and seized the two as they set sail from Havana; he then let the *Trent* continue en route to England. This led to the greatest diplomatic crisis of the Civil War.

Wilkes justified his interception of the ship of a neutral power on the grounds that the Confederate ambassadors were contraband. The British rejected this view, and held that Wilkes's seizure violated maritime law. In particular, while Wilkes might have had the authority to intercept the ship, he did not have the right to determine what was contraband and remove it on the high seas; such decisions could only be made at a Prize Court.

The *Trent* affair might have led to war with Britain, but Albert, the Prince Consort, intervened to defuse the situation. It would be one of Albert's last acts, for even then he was dying from typhoid fever. After his death on December 14, 1861, Queen Victoria mourned him for the rest of her life.[6] She wore black and retired from public view, earning her the nickname "Widow of Windsor."

---

[6]Victoria was George III's granddaughter, and thus she was the monarch of Great Britain. Albert, as her husband, had no claim to the throne, hence his title of "Prince Consort."

According to one apocryphal story, she enjoyed *Alice's Adventures in Wonderland* (1865) so much that one of her advisers thought to cheer her up by obtaining more books by the same author. Subsequently they presented her with *Elementary Treatise on Determinants* (1867) by Oxford mathematician CHARLES LUTWIDGE DODGSON (January 27, 1832–January 14, 1898).

Dodgson told the stories that became *Alice* to Alice Liddell and her sisters Lorina and Edith. At Alice's request, Dodgson wrote the stories down and published them. They appeared under the Latinized version of his name: Lewis Carroll.

Dodgson made some significant contributions to the theory of determinants, offering the first proof of:

Theorem 10.1. *If a system of n inhomogeneous equations in m unknowns is to be consistent, it is necessary and sufficient that the rank of the augmented and unaugmented matrices be equal.*

He also contributed to the theory of voting and apportionment in two pamphlets published in 1884, *Parliamentary Elections* and *The Principles of Parliamentary Representation*. One of his last problems concerned a problem posed by a New York correspondent: to find three rational right triangles with equal areas. On December 19, 1897, Dodgson noted in his diary that he had been up until 4 A.M. that morning trying to solve the problem; he found a pair with equal area ($20 - 21 - 29$ and $12 - 35 - 37$) but had no success in finding a triple.[7]

10.1.7 Vectors Victorious

Meanwhile, the Civil War dragged to an end. A perception that southern victory was tantamount to a perpetuation of slavery (reinforced by the Emancipation Proclamation of September 22, 1862), combined with the growing cotton industry of British-controlled Egypt and poor European harvests that made northern wheat more important than southern cotton, made British support of the Confederacy piecemeal at best. In April 1865, Ulysses S. Grant accepted the surrender of Robert E. Lee and the Army of Northern Virginia, and William Tecumseh Sherman accepted the surrender of the Army of Tennesse, led by Joseph E. Johnston.

The Civil War coincided with the first American doctorates. In 1862, Yale University became the first institution in the United States to grant a Ph.D. in mathematics, to John Hunter Worrall of Philadelphia. The next year JOSIAH WILLARD GIBBS (February 11, 1839–April 28, 1903), earned the first American Ph.D. in engineering for his dissertation, "On the Form of the Teeth of Wheels in Spur Gearing." Gibbs, son of the translator who identified the *Amistad* captives as Mende.

When describing the quaternions $a + bi + cj + dk$, Hamilton noted that the real part a could have any value on a scale from $-\infty$ to $+\infty$; hence he called it the scalar part. The imaginary part $bi + cj + dk$ could, by analogy with the complex numbers, be viewed as a line having a given direction and length, so Hamilton named it the vector part. Hamilton was convinced of the importance of quaternions, but was unable to find many converts besides PETER GUTHRIE TAIT (April 28, 1831–July 4, 1901), who agreed to

---

[7]The existence of infinitely many triples was not proven until 1996.

write a comprehensive treatise on the physical applications while Hamilton wrote a treatise on the mathematical applications. One consequence of Tait's involvement was the term *nabla* for the operator ∇: Tait's assistant, William Robertson Smith, was both physicist and Hebrew scholar, and noted the resemblance between ∇ and the Assyrian harp (called a nabla).

Tait and JAMES CLERK MAXWELL (June 13, 1831–November 5, 1879) had gone to school together, and between 1865 and 1871 Tait convinced Maxwell of the value of quaternions. One problem that had to be resolved was the orientation of the three axes (corresponding to i, j, and k): did they form a right-handed system (the "tendril of the vine," since vines coil counterclockwise) or a left-handed system (the "tendril of the hop," because hops coil clockwise). On May 11, 1871 Maxwell asked the London Mathematical Society to make an "official" ruling; with no reason to support the "tendril of the hop," they agreed to the right-handed convention. Maxwell incorporated the system in his *Treatise on Electricity and Magnetism* (1873), and consequently physicists (and thus mathematicians) use a right-handed coordinate system.

Maxwell's use of quaternions in his *Treatise* is somewhat contradictory. On the one hand, while describing the derivation of Ampère's laws, Maxwell notes appreciatively:

> The only experimental fact which we have made use of in this investigation is the fact established by Ampère that the action of a closed current on any portion of another current is perpendicular to the direction of the latter. Every other part of the investigation depends on purely mathematical considerations depending on the properties of lines in space. The reasoning therefore may be presented in a much more condensed and appropriate form by the use of the ideas and language of the mathematical method specially adapted to the expression of such geometrical relations—the *Quaternions* of Hamilton.[8]

On the other hand:

> In this treatise we have endeavoured to avoid any process demanding from the reader a knowledge of the Calculus of Quaternions. At the same time we have not scrupled to introduce the idea of a vector when it was necessary to do so.[9]

Thus while Maxwell agreed that quaternions were, indeed, valuable, he hesitated to inflict their use on the reader, while he recognized the value of the vector part of the quaternion. Others had come to the same realization; independently of Tait and Hamilton, HERMANN GÜNTHER GRASSMANN (April 15, 1809–September 26, 1877) developed his own form of vector algebra, but like Hamilton failed to attract many converts. Like Hamilton, Grassmann had one important convert: Gibbs.

In 1881 and 1884, Gibbs published a pamphlet for his students that introduced vector algebra in essentially its modern form. Vectors and scalars were no longer parts of a quaternion: they were entities in their own right. It is a measure of the growing relevance of American mathematics that Tait accused Gibbs of retarding the development of quaternions with his "hermaphrodite monster" formed from the ideas of Hamilton and Grassmann. Gibbs responded with restraint and compared the two systems, noting in particular that vectors

---

[8] J. C. Maxwell, *Treatise on Electricity and Magnetism*, Clarendon Press 1873, Vol. II p. 159.
[9] J. C. Maxwell, *Treatise on Electricity and Magnetism*, Clarendon Press 1873, Vol. II p. 236.

could be easily extended to more than three dimensions, while quaternions could not. Tait dismissed the extensibility of vectors by asking why physicists should care about more than three spatial dimensions. The debate between supporters of vectors and quaternions went on for three years, and in the end vectors won out. Since then, the importance of vectors in physics has continued to grow, while quaternions have been relegated to specialized applications (primarily the analysis of rotations, though recently they have gained a place in the mathematics of computer-generated imagery).

10.1.8   The Howland Will

A little over two months after the end of the Civil War, mathematics appeared in one of most contentious court cases of the era. On July 2, 1865 Sylvia Ann Howland died, leaving an estate worth slightly over two million dollars. About half the estate was to be given to her niece Henrietta ("Hetty") Howland Robinson. But Robinson produced a will, dated January 11, 1862, which named her the sole beneficiary of her aunt's estate, and included directions that no later will should be honored. This led to a legal battle over the estate. Howland's executor, Thomas Mandell, claimed that two of the three signatures on the Robinson will were copies of existing signatures: forgeries, in other words.

In the subsequent lawsuit Robinson and Mandell both sought expert witnesses, and for the first time in history, statistical analysis played a role in the courtroom. Oliver Wendell Holmes (senior), professor of anatomy and physiology at Harvard, and Louis Agassiz, professor of zoology at Harvard, testified for Robinson, saying they could find no physical evidence of forgery. Benjamin Peirce and his son Charles testified for Mandell. Howland's signature included 30 downstrokes (for example, the downstroke that completes a lower case "d"). By an exhaustive empirical analysis of 42 genuine Howland signatures, Charles Peirce concluded there was approximately one chance in five of any particular downstroke matching any corresponding downstroke. Since all thirty downstrokes in two of the signatures on the disputed will matched, Benjamin Peirce concluded the probability that the signature was in fact that of Howland was $1/5^{30}$. The alternative was that the signature was a forgery.

In 1871 the court ruled against Robinson, though the decision was based on invalidating the direction in the 1862 will to disregard any later will, and not on Peirce's arguments (which are mathematically dubious, though this fact escaped the opposing lawyer). Robinson, who had married in the meantime and became Hetty Green, would still inherit half of her aunt's estate. Through a combination of excellent management and miserliness, she earned the nickname "the witch of Wall Street" and when she died in 1916, left her two children with an estate valued at over $100,000,000.

10.2   The Gilded Age

In 1876 JAMES ABRAM GARFIELD (November 19, 1831–September 19, 1881) published a proof of the Pythagorean Theorem, which appeared in *The New England Journal of Education* (Volume 3, No. 161). The proof is essentially the following: let $EDBC$ be a trapezoid with ED parallel to BC, and angles C and E right angles; moreover, let the triangles be constructed so that $EA = BC = a$ and $AC = ED = b$; consequently $DA = AB = c$.

By comparing the area for the trapezoid $EDBC$ and the three constituent triangles ABC, AED, and ABD, we have:

$$\frac{1}{2}(a+b)(a+b) = ab + \frac{1}{2}c^2.$$

Consequently $a^2 + b^2 = c^2$. Garfield, who resigned his post as Major General during the Civil War to serve as a representative from Ohio, noted the proof evolved out of discussions with other members of Congress: "We think it something on which the members of both houses can unite without distinction of the party." Garfield did not know it at the time, but within a year he would be called upon to help decide the most controversial election in U.S. history.

In 1876 Rutherford Birchard Hayes, the Republican governor of Ohio, faced Samuel Jones Tilden, the Democratic governor of New York. On November 8, the Republican National Chairman Zachariah Chandler announced that Hayes had the 185 votes necessary to win. Actually, Hayes had 165 votes to Tilden's 184, with 20 votes still uncertain. A single vote would give Tilden the presidency, while Hayes could only win if he took the electoral votes of four contested states: Oregon, Louisiana, South Carolina, and Florida. Accusations of voter intimidation (by Democrats to keep blacks from voting Republican) and electoral irregularities (in Florida, the Republican-controlled election board overturned the Tilden majority and gave the state's electoral votes to Hayes) led to the formation of a congressional Electoral Commission in January 1877. The commission consisted of seven Democrats, seven Republicans (including Garfield), and one supreme court justice, Joseph P. Bradley, ostensibly non-partisan. When the disputed votes came up for review, the commission consistently split along party lines, with Bradley invariably supporting the Republicans. Thus Hayes won the "Stolen Election" of 1876 to become President.

From the start Hayes intended to be a one-term President, and guaranteed this would happen by his vigorous attacks on the spoils system of public office, whereby the newly elected administration distributed most of the available government jobs to its supporters. At the time, there was nothing technically illegal about this practice. By the 1870s this practice had become so entrenched in American society that the era was known as The Gilded Age (1873), after the novel by the same name written by Charles Dudley Warner and his neighbor, the rising star of American literature: Mark Twain. The novel, which describes the corrupt political atmosphere in Washington D.C., is arguably Twain's worst, but the satirical critique of American society and government corruption sold forty thousand copies in a few months.

In 1880, Garfield became President. He continued Hayes's attack on the spoils system, with fatal consequences. On July 2, 1881 Charles Guiteau, a mentally disturbed lawyer who believed his unsolicited work for the Republican party entitled him to an ambassadorial post, shot President Garfield "several times as I wished him to go as easily as possible." He failed: one shot struck Garfield in the arm, but the other lodged somewhere in his body and could not be found. Alexander Graham Bell offered the use of a new device, a prototype metal detector, but it proved ineffective (later it was determined that the metal bedframe interfered with the working of the device, but at the time no one realized this). Incompetent medical care, including exploratory surgery to find the bullet, led to his death on September 19, 1881, a martyr for the cause of civil service reform. Subsequently Senator George Hunt

Pendleton (like Garfield, from Ohio) sponsored the Pendleton Civil Service Reform Act (January 16, 1883), which established competitive examinations for most government jobs. Today, while some positions (such as the ambassadorial ones sought by Guiteau) are still given out as rewards to party supporters, most are not subject to political interference.

10.2.1 The Rise of American Mathematics

The Baltimore businessman and philanthropist Johns Hopkins was named after his grandfather, who was named after his mother Margaret Johns and father Gerard Hopkins. When he died in 1873, Hopkins left $7 million to fund a hospital and a university. The first President of Johns Hopkins University was Daniel Coit Gilman (yet another graduate of Yale, in 1852); at the advice of Joseph Henry of the Smithsonian Institution and Benjamin Peirce, Gilman offered Sylvester the chance to build a mathematics department. Thus in 1876 Sylvester returned to the United States.

Sylvester realized that a journal was critical for the development of a mathematical tradition in the United States. To this end Sylvester, William Story and Simon Newcomb began editing the *American Journal of Mathematics* in 1878. The first issue included articles by Newcomb, Cayley, and Sylvester, as well as the first part of "Researches in the Lunar Theory" by the reclusive New Yorker GEORGE WILLIAM HILL (March 3, 1838–April 16, 1914). In this three-part paper, Hill turned celestial mechanics from a study in practical mathematics whose goal was to provide ever more accurate predictions of celestial motions, into a theoretical study of orbital possibilities. For example, Hill introduced the concept of a surface of zero velocity which a body moving under gravitational influences alone could not pass through (since, by definition, if it was on the surface its velocity was zero). One consequence was that the orbit of the Moon could be proven stable, for it was limited to a specific region; Hill computed that the Moon could recede no farther than 109.964 Earth-radii from the center of the Earth (its mean distance is somewhat more than half that amount). Hill's achievements in celestial mechanics were among the most significant yet to be produced in the United States, and Poincaré went so far as to say that all subsequent progress in celestial mechanics could be traced back to ideas presented in Hill's papers.

The success of the *American Journal of Mathematics* implied a thriving mathematical culture in the United States. The next logical step was taken by THOMAS SCOTT FISKE (May 12, 1865–January 10, 1944) of Columbia College (now University). In the spring of 1887, Fiske (a second year graduate student at the time) went to Cambridge University, where he met Cayley and other English mathematicians, and attended lectures. Fiske also met JAMES WHITBREAD LEE GLAISHER (November 5, 1848–December 7, 1928), who took him to meetings of the London Mathematical Society and entertained him with gossip about contemporary and historic mathematicians. When Fiske returned to Columbia, he and two other students, Edward Lincoln Stabler and Harold Jacoby, issued a call to organize a mathematical society. On Thanksgiving Day at 10:00 AM (November 24, 1888) six mathematicians met to form the New York Mathematical Society, with Amringe as President and Fiske as Secretary.

In January 1891 the New York Mathematical Society elected KARL AUGUST RUDOLF STEINMETZ (April 9, 1865–October 26, 1923) as a member. Steinmetz completed a dissertation in 1888 at the University of Breslau. The dissertation was well-received and he would have received his degree, but he had become involved with a socialist club that was subsequently banned; to avoid arrest, Steinmetz fled, first to Switzerland, and then to New York in 1889.

Fiske read Steinmetz's doctoral dissertation (published in the *Zeitschrift für Mathematik und Physik*), which indicated that the author was then living in New York. Fiske contacted him, suggested he write papers in English, and offered to help him with any linguistic issues. Steinmetz did so, publishing a few papers in the *American Journal of Mathematics*. By then Steinmetz renamed himself Charles Proteus Steinmetz, after the Anglicized version of Karl and his nickname while at university.[10] Steinmetz took a job as an electrical engineer with a small electrical company owned by Rudolf Eickenmeyer of Yonkers (just north of New York City).

Steinmetz distinguished himself by applying mathematics to the analysis of electrical circuits, particularly to the theory of alternating current (AC); he developed a method of using complex numbers to greatly simplify computations. In 1893 Steinmetz presented his methods to the International Electrical Conference, but few members of the audience members understood them. At the time, mathematics played little part in the training of electrical engineers. To remedy this he and Ernst J. Berg co-authored *Theory and Calculation of Alternating Current Phenomena* (1897); Steinmetz later expanded this book into three volumes.

Eickenmeyer's company was purchased in 1893 by General Electric, and Steinmetz was one of the company's major assets. He was promoted to head of the calculating department and taught his electrical engineers the mathematics necessary to understand his theory of alternating current. Between Steinmetz's mathematics, Croatian-born New Yorker Nikola Tesla's engineering, and George Westinghouse's business acumen, the United States became AC-electrified.

10.2.2   Oil

During the Civil War, the Confederacy commissioned several commerce raiders from English shipyards to attack Union vessels on the high seas. Commerce raiders are most effective when they attack ships carrying vital commodities. The union was self-sufficient in food and munitions, but there was one vital commodity that came from overseas: oil. Specifically, oil from whales hunted in the oceans of the world, which was used to light lamps. A remarkable 80% of the world's whaling fleet sailed from the ports of New England; about 15,000 whales were killed each year to satisfy the need for oil.

It was a dangerous industry. On November 20, 1820 the whaleship *Essex*, sailing out of Nantucket, Massachusetts, was rammed and sunk by a sperm whale in the South Pacific. Seven of its 20 man crew would be rescued from various locations the next year. *Narrative of the Wreck of the Whaleship Essex* (1821) by the first mate Owen Chase was a national sensation. The public's interest in whalers was further piqued in 1839, when J. N. Reynolds published "Mocha Dick, or the White Whale of the Pacific" in the *Knickerbocker* magazine. The whale, "white as wool," had been seen in the waters around Mocha (off the Chilean

---

[10]Proteus is a figure in Greek mythology who knows the answers to all questions, but only answers those who can seize and hold him while he tries to escape by undergoing a series of shape changes.

coast) as early as 1810, and survived over a hundred whalers; when it was finally killed, twenty harpoons from previous encounters were taken from its back. From these two accounts and his own experience aboard whaling ships, Herman Melville crafted the novel *Moby Dick* (1851).

The whalers were easy prey for the commerce raiders. In a single week, Captain James I. Waddell of the *Shenandoah* virtually destroyed the North Pacific whaling fleet, sinking 24 whalers in the Sea of Okhotsk near Alaska, recently purchased by the United States from the Russian Empire.[11] Unbeknownst to Waddell, the war had ended two months before.

With the whaling fleet largely destroyed, substitutes had to be found. For millennia, the existence of a flammable liquid coming from underground had been known; even before the Civil War, petroleum (literally "rock oil") was refined into kerosene for lamps. The problem was obtaining it. In some areas, notably northwestern Pennsylvania, it appeared as a contaminant in salt wells (dug for brine to be boiled down for salt, a valuable commodity inland). A positive report on the potential of Pennsylvania oil by Yale professor of Chemistry Benjamin Silliman Jr. (like Gibbs, the son of a Yale professor of the same name) led to the foundation of the Pennsylvania Rock Oil Company of Connecticut on September 18, 1855. Edwin L. Drake was assigned to supervise production, and on August 27, 1859, Drake's well in Titusville struck oil.

While many sought their fortune drilling for oil, John Davison Rockefeller realized that the bottleneck would be refining the oil into a usable product. In 1863 Rockefeller and chemist Samuel Andrews invested in a Cleveland oil refinery. It was the birth of what would become Standard Oil, the greatest monopoly ever created in the United States. In 1882 Rockefeller began using the device of having the majority shareholders in different companies allow a small group of trustees to manage their holdings, forming a trust. The net result is that different companies, which appear to be owned by different people, are in fact operated as if they were a single entity. By 1904 the Standard Oil Trust controlled 90% of the production and distribution of refined petroleum in the United States. Rockefeller's practices led to the closure of inefficient businesses, and his own economies of scale meant that the price of kerosene dropped significantly; the consumers benefited (as did 15,000 whales each year). But fear of what a monopoly might do without competition led to a federal investigation, which would eventually result in the breakup of Standard Oil into 34 separate companies, including Standard Oil Company of New Jersey (which become Exxon), Standard Oil Company of New York (which became Mobil), Conoco, Standard Oil of California (which became Chevron), Atlantic Refining Company (now part of ARCO), and others.

Rockefeller pursued efficiency as relentlessly in his philanthropic efforts as his business pursuits: rather than dole out small amounts of money to many institutions that might or might not produce a useful return, Rockefeller preferred to give substantial amounts to a few, so that they would thrive and be assured of producing returns year after year. One problem came to the attention of the Baptist Rockefeller. Frederick T. Gates, Secretary of the American Baptist Education Society (ABES), pointed out that because Baptists had not founded many schools in the Midwest, it frequently happened that Baptist ministers studied theology at non-Baptist institutions. The ABES hoped to found an institution in

---

[11] Securing access to the marine life off Alaskan waters was the main impetus for acquiring what was then known as "Seward's Folly."

the Midwest, in what was rapidly becoming the second largest city in the United States: Chicago.

10.2.3 Chicago

Shekakwa is an Algonquian Indian word of uncertain meaning, but with the connotation of something strong smelling (skunks or onions). When the city of Chicago incorporated in 1833, it had a population of 350 persons, but like New York City, the town's population and importance exploded following the building of a canal. In Chicago's case, it was the completion of the Illinois and Michigan canal in 1848, which linked the Great Lakes to the Mississippi River. The arrival of the railroad in 1852 meant that Chicago became the center of a great transportation web that spanned the Midwest, Northeast, and South.

By 1870 the city's population was 300,000. The rapid expansion meant that houses were built close together out of the cheapest and most readily available material: wood, which was used not only for houses but for sidewalks as well. On October 8, 1871, a fire broke out in the barn of Patrick and Catherine O'Leary. For two days the great fire raged, destroying 18, 000 homes, leaving 90, 000 homeless, and about 300 dead. Ironically the prevailing winds caused the fire spread *away* from the O'Leary home, which emerged unscathed. An international relief effort helped Chicago rebuild rapidly; within a year, most of the evidence of the fire had gone and Chicago continued its rapid growth.

Two remarkable figures came to Chicago in 1889. Rudyard Kipling passed through the city on his way to London from India. In *American Notes* (1891), Kipling called San Francisco a "pleasure-resort as well as a city", and Salt Lake City "a phenomenon." Chicago was the "first American city I have encountered . . . Having seen it, I urgently desire never to see it again." Kipling's *American Notes* would be eclipsed by later works such as *The Jungle Book* (1894), *Kim* (1901), and *Just So Stories* (1902).

Whereas Kipling merely passed through and did his great work elsewhere, ELIAKIM HASTINGS MOORE (January 26, 1862–December 30, 1932) arrived to stay. Moore studied at Yale under HUBERT ANSON NEWTON (March 19, 1830–August 12, 1896), (also a Yale graduate). Newton arranged for Moore to study abroad, first in Göttingen in the summer of 1885, then in Berlin with Weierstrass and Kronecker between 1885 and 1886. This overseas "finishing" would be significant: American mathematical contributions had largely been in applied fields, such as celestial mechanics (Hill, Newcomb, and Newton) and mathematical physics (Gibbs). Moore returned from Berlin with an appreciation and aptitude for theoretical mathematics, which he would convey to all his students.

Moore arrived in Chicago to teach at Northwestern University, founded in 1855 by John Evans.[12] But in 1891 the University of Chicago opened with Yale graduate William Rainey Harper as its President. Originally Rockefeller offered the ABES $600,000 to start an undergraduate college, but Harper had bold plans that included a graduate school, with senior professors freed from teaching responsibility to do research; a printing press; an extension school; and integration of adult education into the university curriculum. Harper persuaded Rockefeller to donate another $1,000,000, and in time, Rockefeller would eventually do-

---

[12]Evans would become governor of the Colorado Territory and, with the help of Moore's father, David Hastings Moore, would found the University of Denver in 1864.

nate more than \$35 million to the University. To achieve these ambitious goals, Harper collected the best faculty he could, and gained notoriety for his raids on other universities. In 1892 Harper recruited Moore from Northwestern by offering him a full professorship and position as acting chair of the mathematics department.

Harper also raided Clark University in Worcester, Massachusetts; the newly founded university (it opened in 1887) lost half its faculty to Chicago. Among those who left Clark was OSKAR BOLZA (May 12, 1857–July 5, 1942). Bolza entered the University of Berlin in 1875, intending to study physics (which he chose over philology). But after a few years, he realized that experimental work was not for him, so he switched to mathematics. Consequently it took him eleven years, under Klein's supervision, to earn his Ph.D. By then Klein had moved to Göttingen, and Bolza and his friend HEINRICH MASCHKE (October 24, 1853–March 1, 1908) followed him there.

Bolza found working with Klein discouraging: he realized he could never hope to match Klein's brilliance, and felt that securing a university position in Germany would be impossible. Thus in 1888 Bolza emigrated to the United States, where, after spending the winter of 1888-1889 at Johns Hopkins, he accepted a position at Clark University. The death of his father and the unwillingness of his mother to relocate to the United States made Bolza consider returning to Germany, but Harper was persuasive. Bolza agreed to join the faculty at Chicago on the condition that Maschke, also in the United States, be offered a position as well. The indefatigable Harper secured funds to hire another mathematician, and Moore, Bolza, and Maschke became the triumvirate that would shape American mathematics. Between 1896 and 1910, the University of Chicago awarded 39 doctorates in mathematics. The very first, in 1896, went to LEONARD EUGENE DICKSON (January 22, 1874–January 17, 1954), who would join the department in 1900.

Chicago would host one of the earliest international mathematical conferences. The year 1892 was the 400th anniversary of Columbus's arrival in the New World, and several cities vied for the honor of hosting an international quadricentennial exposition. Chicago won the bid, much to the dismay of New Yorkers who felt the city too unsophisticated to be able to handle an international event. They may have been right: although the exposition officially opened on October 21, 1892, the fairgrounds themselves did not open until May 1, 1893.[13]

The organizers designated August 21–28, 1893 as a week for congresses of science and philosophy, so in early 1893 invitations were sent to mathematicians in the United States and Europe. Forty-five mathematicians attended, including CHARLOTTE CYNTHIA BARNUM (May 17, 1860–March 27, 1934), one of the first American women to receive a Ph.D. in mathematics (from Yale in 1895), and Sylvester's student GEORGE BRUCE HALSTEAD (November 23, 1853–March 16, 1922) of the University of Texas. Four European mathematicians attended, including Felix Klein, designated by the Prussian Ministry of Culture as its official representative. In addition, mathematicians from Germany, Austria, France, Switzerland, Italy, and Russia sent papers to be presented.

Klein stayed after the Congress to give a two week colloquium at Northwestern University, and returned the next year to the United States to give a talk to the New York Math-

---

[13] It is sometimes claimed that Chicago's nickname, "The Windy City," stems from the acrimonious debate over who would host the fair, but the term seems to have originated in the 1870s over a baseball rivalry between the Chicago White Stockings and the Cincinnati Red Stockings.

ematical Society. The Society was rapidly gaining prominence, with membership spread throughout the country. Thus in August 1894, the society presented itself at the Brooklyn meeting of the American Association for the Advancement of Science with a new name: the American Mathematical Society. Hill would be the first elected President of the AMS.

A second mathematical society in the United States came about through the efforts of a graduate of the University of Chicago, HERBERT ELLSWORTH SLAUGHT (July 21, 1861–May 21, 1937). In 1892, the Mathematics Department at the University of Chicago began giving out two-year fellowships to promising students; Slaught was one of three to receive a fellowship that year. When his fellowship ran out, he joined the faculty, though he did not receive his doctorate (supervised by Moore) until 1898. He remained at the University of Chicago for the rest of his life.

In 1913 Slaught, at Dickson's suggestion, became managing editor of the *American Mathematical Monthly*, a journal founded in 1894 by Missouri schoolteacher Benjamin F. Finkel. The *Monthly* had been founded because "Most of our existing Journals deal almost exclusively with subjects beyond the reach of the average student or teacher of Mathematics...No pains will be spared on the part of the Editors to make this the most interesting and most popular journal published in America."[14]

Slaught petitioned the American Mathematical Society to take over the journal, but they rejected this proposal in April 1915. However the AMS Council encouraged the creation of a new society. Slaught sent out query letters, and 104 people attended an organizational meeting held on December 30–31, 1915 to form the Mathematical Association of America.

10.2.4 Progressivists and Imperialists

Chicago played a central role in the history of American labor, and Moore arrived between two key events: the Haymarket Riot, which occurred shortly before Moore's arrival, and the Pullman strike. Most of George Pullman's workers, who manufactured the sleeping cars of the same name, lived, worked, shopped, and prayed in buildings rented from Pullman; one worker quipped, "We are born in a Pullman house, fed from the Pullman shops, taught in the Pullman school, catechized in the Pullman Church, and when we die we shall go to the Pullman Hell."

In the wake of the depression of 1893, Pullman reduced wages but not rents. In response Pullman workers began a strike on May 11, 1894, and the American Railway Union (organized a few years earlier in Chicago by Eugene V. Debs) announced a sympathy strike: as of June 26, 1894 they would not work trains that included Pullman cars. In Chicago, whose politicians had learned long ago the voting power of organized labor, the police were given orders to be neutral and the governor of Illinois, John Altgeld, declined to call in the state militia to break up the strike. Over the protests of Governor Altgeld, President Grover Cleveland sent more than 2000 troops into Chicago to operate the trains and break up the strike. What had been a peaceful strike turned violent, and within a few days rioters destroyed 700 railcars and burned seven of the buildings at the Colombian Exposition; confrontations with federal troops left at least 12 dead. Debs and others were arrested, and the striking workers returned to work on August 2, 1894.

---

[14]*American Mathematical Monthly*, Vol. 1, No. 1 (January 1894), p. 1–2.

By alienating labor, Cleveland became the only serving President to lose support of his own party (the Democrats), who chose William Jennings Bryan to head the ticket in 1896. But Mark Hanna, an Ohio Republican notorious for putting the interests of the wealthy ahead of those of the working class, used $100,000 of his own money to secure the nomination and eventual election of William McKinley. McKinley's first term left the United States in possession of an empire.

Since 1895 Spain had been fighting an insurrection in Cuba and the Philippines. The newspaper syndicate of William Randolph Hearst promoted intervention in Cuba on the side of the rebels, and in 1898 war was declared. On May 1, 1898 Commodore George Dewey led an American naval squadron into Manila Bay and destroyed the antiquated Spanish fleet at the cost of seven U.S. sailors injured. On July 3, a similar fate befell the Spanish navy in Santiago Harbor, Cuba, and the war was effectively over by July 17. By the terms of the latest Treaty of Paris (December 10, 1898), Cuba became independent; Spain received $20,000,000 in compensation; and the United States acquired Guam, Puerto Rico, and the Philippines.

Industrialist Andrew Carnegie offered to pay the US government $20 million in exchange for Philippine independence, but the government declined and under Emilio Aginaldo, the Filipinos continued their insurrection, this time against the United States. The capture of Aguinaldo in 1901 and the surrender of Miguel Malvar and the remnants of the Philippine Army the next year led to a declaration that the war was over on July 4, 1902. The insurgency continued for another eleven years, particularly in provinces held by the Islamic Moros. The Moros attacked US troops with little regard for their own lives; soldiers found their standard issue .38 had insufficient "stopping power," and as a result the US Army adopted the Colt .45 (officially the M1911) in 1911. With very few changes, it would remain the army's standard issue sidearm for nearly seventy-five years.

Meanwhile, McKinley won a second term in 1900. During the nominations, the Republicans found themselves in a quandary: Theodore Roosevelt, McKinley's Secretary of the Navy, gained fame during the Spanish-American War. This was used to make him governor of New York, where (to the dismay of the Republican party), he proved all too effective at rooting out corruption in New York politics. There is a traditional remedy in American politics for troublemakers within a party: make them Vice President, where their political power would be reduced to casting a tie-breaking vote in the Senate. Hanna was horrified at the proposal, and reputedly said, "You've put that damned cowboy a heartbeat away from the Presidency!" Over Hanna's objections, Roosevelt became the Republican choice for Vice President. McKinley won the election but on September 6, 1901, he was shot by the anarchist Leon Czolsgosz and died eight days later. Theodore Roosevelt (the "damned cowboy" who spent the first twenty-six years of his life in New York City and Boston) became the President of the United States.

Roosevelt went on to be one of the most colorful and important Presidents in American history. It was during his term in office that the Presidential mansion came to be known as the White House. In 1902 Roosevelt, an avid hunter, chose not to shoot a bear cub in front of a group of reporters; the incident became a cartoon, published November 16, 1902, and Brooklyn toy manufacturers Morris and Rose Michtom seized on the marketing possibilities with the first of "Teddy's bears." Though Roosevelt's policy of "speak softly, but carry a big stick" was a throwback to medieval notions of government, he proved extremely progressive in social matters. He designated 194 million acres (five times more land than all his predecessors combined) as national forests, preventing their commercial exploitation. His greatest successes came in breaking up monopolies.

The Sherman Act (1890) made illegal any agreement or conspiracy to restrain trade. The Act was used against union organizers, since strikes did, indeed, restrain trade; however, its use against companies that engaged in monopolistic practices was haphazard at best. This began to change on February 19, 1902, when Roosevelt laid suit against the Northern Securities Company, a railroad combine formed the previous year by a group that included J. P. Morgan and Rockefeller. In 1904 the Supreme Court ruled against the company, which was forced to dissolve. Roosevelt would go on to break up more than forty major trusts; ironically the companies owned by the Northern Securities Company, along with several others, are all now part of BNSF Railway.

In 1906 Roosevelt secured the passage of the Pure Food and Drug Act and the Meat Inspection Act, in 1906. The passage of the latter was helped by Upton Sinclair's novel *The Jungle* (1906), where he presented, in fictionalized form, how the dismal lives of the working class could be made better by socialism. Sinclair chose to portray a family of Lithuanian immigrants working in Chicago's meat packing industry. The public missed the point, and focused on the nauseatingly unsanitary conditions in the factories described by Sinclair, who later quipped that "I aimed for the public's heart, and by accident I hit it in the stomach."

10.2.5   Princeton and the Presidency

In 1900 Chicago native GILBERT AMES BLISS (May 9, 1876–May 8, 1961), received his Ph.D. under Bolza's supervision. Bliss and Moore's student OSWALD VEBLEN (June 24, 1880–August 10, 1960), one of the founders of topology, would go to Princeton in 1905.[15]

Princeton alumnus and professor Thomas Woodrow Wilson became President of Princeton in 1902 and introduced several key innovations. For example, in 1904 he introduced a system whereby undergraduates would spend their first two years in a general education program, followed by two more years of advanced study in a single subject area: the academic major.[16]

In June 1905 Wilson appointed 45 preceptors, who had the rank of assistant professor and met with small groups of students in what was effectively a group tutorial for all the courses the students were taking in a department; Bliss and Veblen were among the first preceptors in mathematics. Veblen remained at Princeton for the rest of his career.

Bliss returned to Chicago in 1908 under tragic circumstances: in February 1908 Maschke entered a hospital for emergency surgery, and died on March 1. With Maschke dead, Bolza had no ties to the United States, and returned to Germany in 1910. Bliss succeeded Maschke, and ERNST JULIUS WILCZYNSKI (November 13, 1876–September 14, 1932), a German immigrant to the United States who returned to study at the University of Berlin, joined the faculty as Bolza's replacement.

---

[15] Veblen was the nephew of Chicago faculty member Thorstein Veblen, author of *Theory of the Leisure Class* (1899), which introduced such terms as "conspicuous consumption."

[16] In contrast, we might note that the top forty of Harvard's Class of 1880, only a quarter took more than five classes in their primary subject area.

Wilson also left Princeton. In 1910, he became the governor of New Jersey. Then in 1912, he received the Democratic party's nomination for President, running against William Howard Taft. Taft's administration had seen more than 80 trusts busted, a strengthened Interstate Commerce Commission, and the passage of the 16th amendment (which permitted the imposition of income taxes) and 17th amendment (which provided for the direct election of senators) through congress. Unfortunately Taft was an inept politician and alienated many key groups. When Taft won the Republican nomination in 1912, Roosevelt and his supporters formed the Progressive Party, known as the Bull Moose Party after Roosevelt's declaration that he felt as "fit as a bull moose" for the office of the Presidency. The phrase gained notoriety after a deranged would-be assassin John Shrank shot Roosevelt on October 14, 1912 just before a campaign speech in Milwaukee. The wound did not deter Roosevelt from giving his speech (the bullet actually passed through the speech and his metal glasses case), saying "It takes more than that to kill a bull moose."

In fact it took the former President of Princeton. With the Republican party split between Roosevelt and Taft, Wilson became the twenty-eighth President of the United States, the only President to date to have earned a doctorate (from Johns Hopkins University, on ways of making the American government more efficient and more answerable to public opinion). In his first term Wilson oversaw the passing of the Clayton Act (which clarified the Sherman Act and specifically exempted unions from prosecution under the Sherman Act), the passage of the 16th and 17th amendments to the constitution, the eight-hour day for railroad workers, and a law prohibiting child labor. But the great triumphs and tragedies of Wilson's Presidency would be determined by events abroad.

For Further Reading

For history of the United States, see [5, 8, 74, 100, 136].

11

The Modern World

11.1 Before the Great War

Between 1830 and 1870, total enrollment in German universities remained steady at around twelve to thirteen thousand, but after unification the student population grew dramatically, nearly tripling by 1900. Many of the new students came from non-aristocratic backgrounds. Dueling clubs became very popular, as the right to carry a sword was traditionally reserved for the aristocracy. A dueling scar identified the bearer as sympathetic to (and deserving of sympathy from) the professionals and military officers who ran Germany.

The dueling societies at the University of Heidelberg are legendary, and in 1885 a student at the Massachusetts Institute of Technology wrote a report describing the first-year experience of a typical German student:

> He arrives at the University, and, as a general rule, has no inclination for study; so, for the first year or two, devotes his time to the amusements which the place affords . . . He very soon gets acquainted with other students, and is soon initiated into their habits. The first thing which occupies his mind is a mania for dueling, and, if he has plenty of money, and the inclination, he joins a corps; if not, generally some private dueling society.[1]

A few years later another student noted that Heidelberg banned football as too dangerous an activity!

11.1.1 Kovalevskaya

Because of archaic practices, the University of Heidelberg missed the opportunity of being known for something more productive than dueling: it could have been the school that graduated SOFIA (OR SONYA) KOVALEVSKAYA (January 15, 1850–February 10, 1891), born Vasilyevna ("Vasily's daughter"). A descendant of Matthias Corvinus, she developed an interest in mathematics at a young age (supposedly learning off her wallpaper, which consisted of pages from her father's school text on differential and integral calculus). In 1867 she took a course on analysis from Aleksandr N. Stannolyubsky, who recognized her potential as a mathematician and encouraged her to study the subject further.

---

[1] *The Tech*, Vol. V, No. 2 (November 4, 1885) p. 17

The two giants of Russian mathematics in the nineteenth century were NIKOLAI IVANOVICH LOBACHEVSKY (December 2, 1792–February 24, 1856), who had done much to develop non-Euclidean geometry, and PAFNUTI LVOVICH CHEBYSHEV (May 16, 1821–December 8, 1894), who helped prepare two volumes of Euler's unpublished papers on number theory (1849) prior to joining the faculty of the University of St. Petersburg. Since Lobachevsky spent most of his life in distant Kazan (about 500 miles east of Moscow, slightly greater than the distance between Paris and Berlin), Chebyshev proved more important in the development of mathematics.

Chebyshev's main interest was probability, which subsequently became the mainstay of the St. Petersburg school of mathematics. An important feature of his work is a focus on discrete probabilities. His proof of the law of large numbers, for example, relies on sums and requires only simple algebra to understand. Chebyshev's greatest student, ANDREI ANDREYEVICH MARKOV (June 14, 1856–July 20, 1922), would investigate the probability of event sequences. Markov's inspiration came from studying the distribution of letters in Russian poetry. For example, *e* is the most common English letter while *u* is relatively rare; however, if the letter *q* appears in a word, it will almost certainly be followed by *u*.

Vasilyevna could not attend classes at the University of St. Petersburg, so in 1868 she entered into a marriage of convenience with Victor Kovalevsky (hence her married name Kovalevskaya). They went to the University of Heidelberg to study, where she could attend classes, provided the lecturers did not mind. Two of those who permitted her to sit in were the chemist Gustav Robert Kirchoff and the physicist Hermann von Helmholtz, but Kovalevskaya (much to their dismay) chose to study mathematics.[2] Thus she went to Berlin in 1871.

Berlin was no more progressive than Heidelberg, so Weierstrass tutored her privately, and set her to solve a problem he had been working on. Previously he had proven the existence and uniqueness theorem for ordinary differential equations, but was unable to extend the theorem for partial differential equations. He gave the topic to Kovalevskaya, and by 1873 she proved that, under a very broad range of conditions, existence and uniqueness held. Had Kovalevskaya been a man (and thus a regular student at the University), her doctorate would have been assured. But since she could not be a regular student, she could not obtain a degree from the University of Berlin. Fortunately a German university student could study at several different universities and take his final examinations at a different one (generally one far away from his old associates, who might distract him from his studies). Hence Weierstrass was able to arrange for her to receive her doctorate from the University of Göttingen in 1874.

Shortly thereafter the Kovalevskys returned to Russia where Vladimir became involved in a financial scandal, and resolved the problem by committing suicide. GOSTA MITTAG-LEFFLER (March 16, 1846–July 7, 1927), who studied with Weierstrass after Kovalevskaya returned to Russia, recruited her for the University of Stockholm, then known as the Högskola.

Kovalevaskaya began to lecture at the University in 1884, and carried out some of her most important research there, particularly in mathematical physics. In 1886 she won a prize from the French Academy of Science for an investigation of the rotation of a solid

---

[2] Kirchoff and Helmholtz had another unusual student: Josiah Willard Gibbs, the first person to receive a Ph.D. in engineering from an American school, who toured Europe's academic centers after the Civil War.

body around a fixed point; the Academy was sufficiently impressed with her entry that they raised the value of the prize from 3000 francs to 5000. Unfortunately, like Descartes, she would die from pneumonia in the midst of a Swedish winter.

11.1.2   The Stockholm Prizes

In 1881, Lie met Mittag-Leffler in Stockholm and suggested that he begin publishing a Scandinavian journal of higher mathematics. It was a bold suggestion; even Lie doubted that Mittag-Leffler could find a sufficient quantity of publishable material. But Mittag-Leffler threw himself wholeheartedly into the project, and tapped into a unique source: JULES HENRI POINCARÉ (April 29, 1854–July 17, 1912), recommended to him by their common teacher CHARLES HERMITE (December 24, 1822–January 14, 1901). Mittag-Leffler hoped that Poincaré's papers would make the reputation of *Acta Mathematica*, just as Abel's papers had made that of Crelle's Journal. To this end, he wrote to Poincaré directly on March 29, 1882, requesting that he submit some of his work to the new journal. Poincaré agreed.

Next, Mittag-Leffler sought funding. King Oscar II of Sweden donated 1500 crowns (Mittag-Leffler's annual professorial salary was, by comparison, 7000 crowns); the governments of Denmark, Sweden, and Norway each promised 1000 crowns annually after 1883. Even Hermite sent a donation of about 720 crowns. On December 12, 1882, Mittag-Leffler presented Oscar II with the first copy of the first issue of *Acta Mathematica*, which contained articles by Hermite, Poincaré, PAUL APPEL (September 27, 1855–October 24, 1930), EDOUARD GOURSAT (May 21, 1858–November 25, 1936), CHARLES EMILE PICARD (July 24, 1856–December 11, 1941), LAZARUS FUCHS (May 5, 1833–April 26, 1902), and EUGEN NETTO (June 30, 1848–May 13, 1919). The second volume of *Acta* contained translations into French of seven articles by GEORG FERDINAND LUDWIG PHILIP CANTOR (March 3, 1845–January 6, 1918), introducing Cantor's theory of infinite quantities to a broad audience.[3]

Very few German mathematicians published in *Acta*. Part of the reason had to do with the existence of several prominent German mathematical journals, including Crelle's. Indeed, since 1880 Crelle's journal had been edited by Weierstrass and Kronecker, so German mathematicians may have hesitated to submit work to a foreign journal when they might have impressed Weierstrass and Kronecker instead.

However, Mittag-Leffler knew that German contributions would be essential for the success of *Acta*. Thus he persuaded Oscar II to offer a prize of 2500 crowns, to be given on January 21, 1889 (the king's sixtieth birthday) for the best paper on one of four topics. The judging committee would consist of Mittag-Leffler, Weierstrass, and Hermite. Kronecker, who was not on the judging committee, objected to the contest, claiming it was merely a device for attracting publicity. Indeed, Kronecker might have gone farther: every one of the questions related to one of Poincaré's research interests.

The question Poincaré examined concerned the stability problem raised by Newton and tentatively solved by Lagrange and Laplace. The differential equations describing the motions of the planets could be solved formally, but as Weierstrass noted in a letter to

---

[3]Unfortunately, after publishing many of Cantor's papers, Mittag-Leffler would persuade Cantor to withdraw one of his articles, noting it was about a hundred years ahead of its time. After that, Cantor would have nothing further to do with *Acta*.

Kovalevskaya on August 15, 1878, he could not prove their convergence. He suggested it as a dissertation topic, though by then Kovalevskaya was already working on other topics.

Poincaré chose to focus on what is now called the first return map (inspired in part by Hill's work). Imagine a plane through the center of the sun and not containing the orbit of a planet, and let P_0, P_1, P_2, etc. be the points on the plane through which the planet passes on successive orbits (we now call the set of points $\{P_0, P_1, P_2, \ldots\}$ the orbit of P_0). These points either form a compact set, or the set is unbounded. In the former case, the orbit of the planet may be said to be stable; in the latter, unstable.

The solution was prepared for the October 1889 edition of *Acta*, but between printing and distribution, Poincaré was dismayed to discover a significant error in it, and agreed to pay for the cost of reprinting the entire issue (which took up most of the prize money). While doing so, he made a crucial discovery. Imagine some starting point P_0 that produced a stable orbit. Consider any point P_0' close to P_0, and consider the distance between P_n and P_n'. Continuity would imply that these distances would stay small for a while, but we would not be surprised if, after a period of time, the distances grew greater than any assigned limit. Poincaré discovered something even more remarkable: arbitrarily close to any point P_0 is a point P_0' whose orbit would retreat indefinitely far away from the orbit of P_0—but then would return arbitrarily closely to the orbit. Poincaré called these points doubly asymptotic points, though they are now known as homoclinic points.

The more famous of the Stockholm prizes had its origin around the same time the Oscar II prize competition was announced. In 1888, a French newspaper reported that Alfred Nobel, "the merchant of death," had died in Cannes. Nobel, born in Stockholm, made his fortune from the manufacture of nitroglyercin, which had replaced black powder as the explosive of choice in mining and tunneling. However, nitroglycerin is extremely unstable, so its manufacture is hazardous. In 1864, one of Nobel's factories was destroyed in an explosion, killing his younger brother Emil and several other people. In 1867 Nobel made a key discovery: if nitroglycerin was mixed with a substance called *kieselguhr* (diatomaceous earth), the result would be an explosive that had nearly as much power as pure nitroglycerin, with none of the latter's instability. Nobel named the new product dynamite, from the Greek *dynamos* or "power." Within limits, dynamite could be burned, crushed, or even set afire without danger; detonation required the use of a blasting cap of mercury fulminate (another Nobel invention).

It was actually Nobel's brother Ludvig who died, but Nobel may have been dismayed at the prospect of being remembered as the "merchant of death." Thus the bulk of his fortune went to establishing the Nobel Foundation, which was to give annual prizes in recognition of those who have "conferred the greatest benefit on mankind." There were to be five categories: physics, chemistry, medicine and physiology, literature, and peace. The fact that three of the five prizes are for sciences has raised the question of why Nobel did not establish a mathematics prize. One might as well ask why there is no prize for music, painting, or sculpture.

There is a persistent story that Nobel detested mathematicians because his mistress, Sophie Hess, left him for a mathematician (Mittag-Leffler is the usual culprit). This myth may be a distorted version of actual events. Mittag-Leffler had been trying, before Nobel's death, to secure a bequest for the Högskola. When Nobel's will gave nothing to the University, some members of the faculty (including the chemist Svante Arrhenius, who would later win the 1903 Nobel Prize in Chemistry) blamed Mittag-Leffler, and later in life Mittag-Leffler himself seems to have contributed to the myth by expressing a belief that he was responsible for Nobel's dislike of mathematics.

11.1.3 Mathematicians and Antisemitism

By the time Kovalevskaya received her degree from Göttingen, its position as the preeminent center of German mathematics and science had been eclipsed by the University of Berlin. Göttingen's revival would be largely due to the work of Felix Klein, who arrived there in 1885 hoping to revive the fruitful connection between mathematics and the other branches of science, embodied by Gauss's work, in contrast to the more abstract work that was representative of the University of Berlin.

At the time, most German universities filled vacancies by offering ("calling") a prominent individual a position in the hopes of enticing them away from their current school. Between 1882 and 1907, these calls were reviewed by Friedrich Althoff, the undersecretary responsible for higher education in the Ministry of Culture. Althoff threw himself so assiduously into his work that when he left in 1907 after a change in the Ministry, he had to be replaced by four people. Klein, with Althoff's support, made Göttingen once again the mathematical and scientific center of Germany. But the road to success would have to overcome a significant obstacle: anti-Semitism.

Between 1870 and 1873, 857 new companies sprang into existence, more than had been formed in the twenty years preceding unification. These "Foundation Years" came to an abrupt end in 1873, when a worldwide economic depression devastated the economy; about a fifth of the newly formed companies went bankrupt, and Germany entered a long recession. Bankers (and by association, Jews) were blamed. Anti-Semitism at the universities was particularly acute: the rising costs of a university education, as well as the influx of non-traditional students, coincided with an increase in the number of Jewish academics, and it was easy to believe the two events were related.

The government's response to the recession was to help big companies become even larger by sanctioning cartels and other monopolistic practices: between 1882 and 1895, the total number of companies grew by a mere 4.6 percent, but the number of companies employing more than fifty workers grew by 90 percent. German industrial output grew rapidly. Iron and steel production went from 169,000 metric tons annually in 1870 (compared to England's 286,000 metric tons), to more than 2 million tons in 1890, and nearly 14 million tons in 1910, when England's production was slightly over 6 million tons. By the 1890s the recession had largely ended, but the anti-Semitism remained.

Thus in 1892 Klein faced a difficult choice: he wanted to recruit ADOLF HURWITZ (March 26, 1859–November 18, 1919), his student from Munich, for Göttingen. But Göttingen already had one Jewish faculty member, ARTHUR SCHÖNFLIES (April 17, 1853–May 27, 1928). For years, Klein had struggled unsuccessfully to make Schönflies a salaried extraordinary professor. But Hurwitz was a better mathematician, so in a letter to Althoff, Klein offered to sacrifice Schönflies in order to have Hurwitz appointed. When Althoff declined to appoint Hurwitz, who eventually accepted a post in Zurich, Klein wrote a furious letter to Althoff on April 10, 1892, saying that if he failed both Hurwitz and Schönflies, he would be marked as irrelevant in higher education, and he "would then be forced to advise young mathematicians not to turn to me, if they hope to make further advancement in Prussia" [102, p. 435]. Shortly afterwards, Schönflies obtained his salaried position, though he would leave Göttingen for Königsberg in 1899.

Klein's first major success came in 1895 when DAVID HILBERT (January 23, 1862–February 14, 1943) came to Göttingen. Meanwhile the mathematical group at the University of Berlin lost its founders: Kronecker died in 1891, Kummer in 1893, and Weierstrass in 1897. When Berlin attempted to recruit Hilbert in 1902, Althoff, at Klein's urging, created a position at Göttingen for Hilbert's friend HERMANN MINKOWSKI (June 22, 1864–January 12, 1909). Hilbert became the first mathematician to ever turn down an offer from the University of Berlin, and remained at Göttingen for the rest of his career, despite a second offer from Berlin and numerous offers from other institutions.

Minkowski's unorthodox appointment would have serious and lasting consequences. Althoff's intervention threatened the autonomy of the Göttingen faculty, and broke the unspoken rule that limited the number of Jews at any single institution (Göttingen already had one Jew: the astrophysicist Karl Schwarzschild, who joined the faculty in 1901). But for Klein, it was a great victory: it meant that he was free to recruit talented mathematicians regardless of their religion. This would give a dramatic boost to Göttingen's preeminence in the mathematical world—then, later, bring it to an equally dramatic end.

Klein, Hilbert, and Minkowski would teach a remarkable group of students at Göttingen that included OTTO BLUMENTHAL (July 20, 1876–November 12, 1944), FELIX BERNSTEIN (February 24, 1878–December 3, 1956), MAX BORN (December 11, 1882–January 5, 1970), RICHARD COURANT (January 8, 1888–January 27, 1972), and EMMY NOETHER (March 23, 1882–April 14, 1935).

Perhaps the greatest products of the new Göttingen came from the mathematical investigation of the theory of relativity. The special theory of relativity, developed by ALBERT EINSTEIN (March 14, 1879–April 18, 1955) in a landmark 1905 paper, is based on a straightforward principle: all observers measure the velocity of light in a vacuum to be the same (roughly $300,000$ km/sec), regardless of the relative motions of the source of the light or the observers. This seemingly simple principle leads to some remarkable results.

For example, consider an observer A moving at half the speed of light who passes a second observer B who is stationary. At the instant they pass, A turns on a flashlight. B measures the velocity of light to be $300,000$ km/sec, and A's velocity to be $150,000$ km/sec, so according to B, A *should* measure the velocity of light to be $300,000$ km/sec $-$ $150,000$ km/sec. But A *also* measures the light to be moving at $300,000$ km/sec, not $150,000$ km/sec. The only way to reconcile these two observations is to conclude that distance and time are measured differently by the two observers.

In his 1905 paper, Einstein considered the transformations that took the coordinates of an event in a "stationary" system (x, y, z, t) into the coordinates of the same event in a moving coordinate system (ξ, η, ζ, τ). Under the assumption of invariance of the speed of light, Einstein derived what is now known as the Lorentz transformation (which had already been derived by Voigt, FitzGerald, and Lorentz in different contexts); the Lorentz transformations, which are continuously variable functions of velocity, form a Lie group, and consequently correspond to a geometry. As Minkowski pointed out in 1908, this geometry is one with three spatial coordinates and one time coordinate, now called Minkowski space (and refuting Tait's assertion that space of more than three dimensions was irrelevant to physicists). The next year Poincaré lectured at Göttingen on special relativity. It is interesting to note that Poincaré, who had been developing ideas similar to Einstein's in the years prior to 1905, conspicuously avoided any reference to Einstein's work.

The special theory of relativity concerns unaccelerated motions only, so Hilbert, Minkowski, and Einstein began working on a general theory. Unfortunately Minkowski suffered an untimely death from appendicitis in 1909. Göttingen faculty recommended Blumenthal, Hurwitz, and Berlin *privatdozent* EDMUND LANDAU (February 14, 1877–February 19, 1938). All three happened to be Jewish; Althoff offered the position to Landau, who accepted. It should be noted that even though Landau's work (in number theory) had little obvious applicability, Klein was pleased nonetheless at Landau's appointment.

In 1911 Bernstein, another mathematician who happened to be Jewish, joined the faculty as extraordinary (untenured) professor of statistics and actuarial sciences; he would go on to found the Institute of Mathematical Statistics at Göttingen in 1921. Klein retired in 1913; by overcoming anti-Semitism and hiring talent regardless of background, he had set Göttingen well on its way towards being the center of the mathematical world.

French mathematicians would combat anti-Semitism in their own way, and become involved in the most notorious scandal of the Third Republic: the Dreyfus Affair. French intelligence paid a cleaning woman at the German embassy to bring in waste papers for analysis. In 1892, a paper scrap from the trash of military attaché Max von Schwarzkoppen referred to "That scoundrel D.," apparently a well-placed source within the military. Then in 1894, the trash produced a schedule (*bordereau*) which described a number of classified documents, with a note that, if desired, they could be made available—a clear offer to sell state secrets.

Suspicion fell on Captain Alfred Dreyfus, an artillery captain assigned to the war office. Assuming Schwarzkoppen was stupid enough to use the actual initial of his intelligence source, Dreyfus might indeed be referred to as "that scoundrel D." Dreyfus had one other useful characteristic: he was an Alsatian Jew, and would be easy to convict in the court of public opinion. On October 15, 1894 Dreyfus was arrested for treason; he was tried, convicted, and sentenced to life imprisonment, to be served at Devil's Island in French Guiana.

Almost immediately suspicions arose about the military's case. Dreyfus's brother Mathieu proclaimed his brother's innocence, and in 1896, Colonel Georges Picquart discovered evidence that suggested Major Ferdinand Walsin Esterhazy was the actual author of the *bordereau*. Picquart was subsequently reassigned to Tunisia, while Esterhazy was acquitted after a cursory trial. Belief in Dreyfus's innocence continued to grow, and on January 13, 1898 the Paris newspaper *L'Aurore* (The Dawn) printed a letter from the writer Emile Zola to the French President Felix Faure about the affair.[4] Zola accused several members of the government (but *not* the President) of improprieties; because of this the editor (and future Prime Minister) Georges Clemenceau titled the letter "*J'accuse*" and put it on the front page.

Zola's letter brought international attention to the case, and in 1899 a new trial began. Attention focused on the *bordereau*. Preceding the original trial, a handwriting analyst from

---

[4]The French President is the head of state, while the Prime Minister, appointed by the President, is the head of government. During the Third Republic, Prime Ministers were dismissed and new ones appointed on a regular basis, but only rarely was the President forced from office.

the Bank of France compared Dreyfus's handwriting to that on the *bordereau*, and concluded that the latter might have been written by someone else. Thus Alphonse Bertillon, a police analyst, was called in to analyze the letters; Bertillon was also told that Dreyfus's guilt was already established by other evidence. Not surprisingly, Bertillon declared that the *bordereau* was almost certainly written by Dreyfus; Bertillon's testimony played a key role in the original conviction.

By 1899 Dreyfus could count several mathematicians among his supporters. His wife Lucie Hadamard was a second cousin to JACQUES HADAMARD (December 8, 1865–October 17, 1963). Hadamard persuaded PAUL PAINLEVÉ (December 5, 1863–October 29, 1933) that there was something dubious about the case against Dreyfus. At Painlevé's request, Poincaré wrote a letter to the court declaring Bertillon's testimony worthless.

Nevertheless on September 9, 1899, the new court convicted Dreyfus again, though it declared the existence of extenuating circumstances and shortened his sentence to ten years. This was a mockery of justice, for prisoners at Devil's Island had to remain in Guiana as colonists for a period of time equal to their original sentence, or for life if their original sentence was longer than eight years. Domestic and international dissatisfaction with the verdict was profound, and ten days later the President, Emile Loubet (Faure had died in office on February 16, 1899 under scandalous circumstances) pardoned Dreyfus and ordered him freed, though it took until July 12, 1906 before he was exonerated, in part due to the testimony of Poincaré, Appel, and Darboux.

It was clear to many that Dreyfus was convicted primarily because he was Jewish. One of the reporters who covered the case was Theodor Herzl, a Viennese Jew. Herzl concluded that anti-Semitism flourished because the Jews had no country of their own. On August 29–31, 1897 the first of six Zionist Congresses met in Basel, Switzerland, to organize Jews internationally and campaign for the foundation of a Jewish state. To this end, Herzl visited Kaiser Wilhelm of Germany; the Sultan of the Ottoman Empire; and Joseph Chamberlain, the British Colonial Secretary. Only the British offered to help: they offered Uganda to the Jews.

11.1.4   Cape Colony

Chamberlain, as Colonial Secretary, played an important role managing Britain's far-flung empire. One outpost of the Empire was Cape Colony in South Africa, which had been settled by the Dutch East India company but acquired by the British in the aftermath of the Napoleonic Wars. In the 1830s about 12,000 inhabitants, mostly farmers (*Boers*) of Dutch descent began what is known as the Great Trek to escape British rule, and established the Transvaal Republic and the Orange Free state. The Transvaal was annexed by the British in 1877, but regained its independence four years later.

Cape Colony attracted many British settlers, including FRANCIS GUTHRIE (January 22, 1831–October 19, 1899), a student of Augustus de Morgan. In 1852, while still in England, Guthrie advanced the conjecture that every plane map could be colored with at most four colors; however, AUGUST FERDINAND MÖBIUS (November 17, 1790–September 26, 1868) raised a similar question in 1840. Guthrie was unable to prove the four color conjecture. In 1861 he moved to South Africa, eventually becoming professor of mathematics at South African College (later the University of Cape town) in 1879.

In 1892, the mathematician Edward Thornton Littlewood arrived in Cape Town with his family to take up the position of Headmaster at the Wynberg Interdenominational School. Littlewood turned Wynberg, which could trace its roots to 1822, into a first-class institution.

A few years before the Littlewoods arrived, gold was discovered in the Transvaal at Witwatersrand (a chain of mountains that forms part of the continental divide). Thousands swarmed into the region, looking to make their fortune, and in time *Uitlanders* ("foreigners," mostly of British descent) came to outnumber the Boers, who passed laws limiting their rights. On December 29, 1895 Leander Starr Jameson, a physician and local administrator, led a group of 494 men armed with rifles, Maxim guns, and artillery pieces into the Transvaal Republic, intending to lead an *Uitlander* uprising and overthrow Transvaal President Paul Kruger.

The Jameson Raid turned into a major embarrassment for the British. Jameson and his men were captured and eventually turned over to the British for trial, and the Prime Minister of Cape Colony, Cecil Rhodes, was forced to resign. Kruger accused Chamberlain of staging the raid. An official parliamentary investigation concluded that Chamberlain had no prior knowledge of the raid, but Chamberlain himself admitted, "I did not want to know too much."

Meanwhile another crisis struck South Africa. In 1894 a bill presented before the Natal Parliament would have disenfranchised Indians (who were British subjects). The lack of organized opposition to the bill caught the attention of the lawyer Mohandas Karamchand Gandhi, who resolved to stay and fight the bill. He intended to stay a month or so before returning to his family in India, but the bill passed and Gandhi would spend nearly twenty years in South Africa fighting for the rights of the Indian community there and earning the title *Mahatma*, "Great Soul." Though Chamberlain expressed sympathy for Gandhi's cause, he refused to override the Natal parliament.

Tensions between Cape Colony and the Boers continued to escalate, and in 1899 Chamberlain demanded that Kruger give British citizens living in Transvaal full political equality; Kruger in turn demanded the British remove the forces they had assembled along the Transvaal border. On October 11, 1899 the Boers struck first, invading Natal and Cape Colony. The Boers won several early battles and captured many British prisoners, though one proved hard to keep: the war correspondent Winston Churchill. Churchill escaped from captivity on December 12, 1899, and made his way back to Durban in Natal, earning international fame for his exploits.

By September 1900 the British were in control of the cities of both the Orange Free State and the Transvaal Republic. Bringing the rest of the country under British control proved much more difficult, and the British resorted to destroying farms and croplands that might be used to support the guerillas. To support those whose livelihoods had been destroyed, the British relocated them to centralized locations where they could be fed and housed. Such locations were called concentration camps, as they concentrated the civilians in one place. Originally intended as a humanitarian measure, the number of refugees overwhelmed the facilities available, and thousands died from neglect. By 1902 these scorched-earth policies forced the remaining guerrillas to agree to a cease-fire, and the Treaty of Vereeniging (May 31, 1902) lay the groundwork for the founding of the Republic of South Africa eight years later.

In the midst of the Boer War, Littlewood sent his son, JOHN EDENSOR LITTLEWOOD (June 9, 1885–September 6, 1977), to England to complete his education. Littlewood arrived at Cambridge in 1900, and while there ERNEST WILLIAM BARNES (April 1, 1874–November 29, 1953) suggested he investigate the Riemann hypothesis. Around 1911 Littlewood began a 35-year collaboration with GODFREY HAROLD HARDY (February 7, 1877–December 1, 1947) that would result in the publication of more than a hundred joint papers. The collaboration proved remarkably successful, perhaps in part due to four rules they adopted as early as 1912:

1. When one wrote to the other, it was completely indifferent whether what they wrote was right or wrong.

2. When one received a letter from the other, he was under no obligation to read it, let alone answer it.

3. Although it did not really matter if they both simultaneously thought about the same detail, still it was preferable that they should not do so.

4. It was quite indifferent if one of them had not contributed the least bit to the contents of a paper under their common name.

Much of their joint work concerned the Riemann ζ-function. It was known that $\zeta(s) = 0$ for all negative even numbers. The Riemann hypothesis holds that all the other zeroes have $\text{Re } s = 1/2$. In 1914 Hardy proved that an infinite number of complex zeroes appear on the critical line $\text{Re } s = 1/2$.

Later that year Hardy and Littlewood proved a remarkable result based on the Riemann hypothesis. If $\pi(n)$ is the number of primes less than n, and $\text{Li}(n) = \int_2^n \frac{1}{\ln x}\, dx$, then for all calculated values $\pi(n) < \text{Li}(n)$, and mathematicians of the caliber of Gauss and Riemann believed that this held true for all n. Hardy and Littlewood proved that in fact the inequality must reverse infinitely often. In 1937, SAMUEL SKEWES (who, like Littlewood, came from South Africa) proved that if the Riemann hypothesis was true, then the inequality must switch for some $n < 10^{10^{10^{34}}}$; Skewes later showed that if the Riemann hypothesis is false, the inequality would still switch, this time for some $n < 10^{10^{10^{10^{10^{1.46}}}}}$. The two Skewes numbers are some of the largest numbers ever to appear in a mathematical proof.

In 1916 Hardy received a letter from SRINIVASA RAMANUJAN (December 22, 1887–April 26, 1920), who had achieved fame in India for contributions to the *Journal of the Indian Mathematical Society*. Some of Ramanujan's results were sent by C. L. T. Griffith to M. J. M. Hill, a professor of mathematics at University College in London (where Griffith studied); Hill thought the results interesting, but evidence of poor mathematical training, since among the things Ramanujan claimed were results on divergent series like:

$$1 + 2 + 3 + \ldots + \ldots = -\frac{1}{12},$$
$$1^2 + 2^2 + 3^2 + \ldots = 0.$$

Since the series are divergent, the sums are meaningless. However, when Ramanujan sent these same results to Hardy in a letter dated January 16, 1913, along with over a hundred others, Hardy was impressed since Ramanujan, in his own way, had found the correct values for $\zeta(-1)$, $\zeta(-2)$, and others. Hardy realized that Ramanujan had the makings of a first class mathematician, and arranged for a scholarship to be given so Ramanujan could study at Trinity College in Cambridge. Ramanujan arrived in England on April 14, 1914.

11.2 The Era of the Great War

On June 28, 1914 Serbian nationalist Gavrilo Princip assassinated Archduke Francis Ferdinand and his wife Sophie in Sarajevo. But what should have been a local crime turned into a great conflagration through what can only be described as a tragedy of errors. Germany assured Austria of its full support against Serbia, so Austria issued several demands to Serbia, including a free rein to investigate the assassination. Serbia agreed to most of the terms, but refused this last, so Austria declared war on Serbia on July 28. Serbia's ally Russia began mobilizing for war against Austria, so Germany declared war on Russia on August 1, an action that would bring France into the war. The von Schlieffen plan had been to knock France out of the war by trapping most of its army between the Franco-German border and a massive invading army coming through neutral Belgium and Luxembourg. Thus Germany invaded Luxembourg on August 2 to demand passage through Belgium en route to France, and on August 3 Germany declared war on France.

The only major European country that had no treaty obligations to any of the combatants was Britain. But an invasion of Belgium was imminent, and since the Concert of Vienna, Britain had been the guarantor of Belgian neutrality. On August 3, 1914 the British foreign secretary Edward Grey noted wryly, "The lamps are going out all over Europe; we shall not see them lit again in our lifetime." That night Germany invaded Belgium, and the next day Britain reluctantly declared war on Germany.

To keep Russia preoccupied Germany began negotiations with the Ottoman Empire. In 1908, growing discontent with the medieval structure of the Ottoman Empire culminated in a series of military rebellions led by a group of reformers known as the Young Turks. They established a constitutional government, secularized the legal system, and improved educational opportunities for everyone (including women). In addition, they sought to promote the industrialization of the Empire and a growing sense of Turkish nationalism. Many had lived in exile in Paris, where they saw firsthand the military power of Germany; consequently they viewed an alliance with Germany favorably. On October 29, 1914 the Ottoman navy launched an attack on Russia's Black Sea ports. The Ottoman, Habsburg, and German Empires formed the Central Powers (because of their location), while Serbia, France, Britain, Russia, and their allies formed the Allied Powers.

On February 21, 1916 the Germans advanced towards Verdun, a strategic point that the French had to hold at all costs. But the French Commander-in-Chief, Joseph-Jacques-Césaire Joffre believed that the German movement was a feint, and the real attack would occur elsewhere. It took four days before an army under the command of General Philippe Pétain arrived and halted the German advance. By July 400,000 Germans had been killed, wounded, or lost in action. The French suffered a comparable number of casualties, and among the dead were two of Hadamard's sons. Fighting might have continued at Verdun but the British launched the Battle of the Somme on July 1, 1916, drawing German attention elsewhere. British troops went "over the top" at Albert (about 100 miles north of Paris), beginning the battle of the Somme and forcing the Germans to halt the advance on Verdun.

On September 15, the British unveiled a new weapon, developed in great secrecy under the code name tank (referring to a water container).

The British plan at the Somme had been to destroy the German defensive positions beforehand with a week-long artillery bombardment. Unfortunately the barrage failed: not only did it warn the Germans of the impending attack, but many of the shells failed to detonate, leaving future generations of farmers to deal with the "iron harvest" of unexploded munitions. Finally the time between the barrage and the arrival of the advancing infantry gave the Germans time to emerge from protected bunkers and reform their defensive positions. British infantry, burdened by 70 pounds of equipment, walked towards German positions defended with machine guns. After four months, the British had advanced five miles, at the cost of 420,000 casualties. The French lost about 195,000 and the Germans 650,000.

In the meantime Robert-Georges Nivelle, a graduate of the Polytechnical School, made a reputation for himself at Verdun by retaking nearly all of the ground lost to the Germans since the beginning of the year. He was the first to successfully employ a new artillery technique known as the rolling barrage (also called a creeping barrage). In a rolling barrage, artillery targeted positions just ahead of the advancing infantry. Obviously this required great coordination of the infantry and the artillery: if the infantry advanced too quickly, they would fall into the zone of fire, while if the artillery advanced too quickly, the defenders would have time to regroup before the infantry arrived. Most importantly, it required that artillery officers knew precisely where their shells would land, one of the main results of the mathematical science of ballistics.

In December 1916, Nivelle replaced Joffre and presented a bold plan to attack the Germans in Champagne. Meanwhile Erich Ludendorff, the newly appointed (with Paul Hindenburg) co-Commander in Chief of the German army, ordered a new, stronger defensive line built in Champagne, and at the end of February the German troops withdrew behind it. Nivelle insisted on carrying out his planned offensive. Pétain disapproved, but the final decision was in the hands of Paul Painlevé.

Painlevé credited the Dreyfus affair with piquing his interest in politics. His importance to the administration came through his connections to the rapidly developing science of aviation. On October 28, 1908 French aviation pioneer Henri Farman flew about two kilometers in a plane, carrying Painlevé as a passenger. Farman's planes included many innovations, such as wheels for takeoff and landing and ailerons for control surfaces (the Wright flyer, in contrast, used skids for landing and rudders for steering). Painlevé became professor at the École Supérieure d'Aéronautique in 1909, and chaired naval and aeronautical commissions after his election as a representative for Paris in 1910. He became the Minister of War in April 1917, and had to evaluate Nivelle's plan almost immediately. With some hesitation, he approved it, and Nivelle launched his attack on April 16.

The much-publicized plans for the first day called for an advance of six miles; in fact, French troops advanced 600 yards at a cost of 120,000 casualties. Although Champagne was not a complete disaster (the French captured 28,000 German troops), it was enough to end Nivelle's career. Painlevé replaced him with Pétain on May 15, 1917. Pétain, whose career before the war had been hampered by his insistence on the importance of defensive strategy, counseled patience: "We must wait for the Americans and the tanks."

The United States entered World War One on April 6, 1917. With a population of 92 million (compared to Germany's 70 million) and an untouchable industrial base, her entry would guarantee allied victory *provided* men and materiel could make their way across the Atlantic. Debate raged over whether it was safer for single ships to speed across the ocean, minimizing the time they spent in submarine infested waters, or whether ships should sail in convoys protected by destroyers, which would have to travel at the speed of their slowest member and give the submarines more targets. The grisly experiment would be carried out in the North Atlantic.

In April, German submarines sank 852,000 tons of merchant ships (more than 400 vessels, or about a quarter of those attempting to cross the Atlantic to Britain), with a concomitant loss of war materiel.[5] Convoys, protected by destroyers, proved their value in the next few months: losses dropped to 500,500 tons in May, and 200,000 tons by November.

The entry of the United States into the war balanced the loss of Russia. After three years of war, most of it badly managed and all of it badly equipped, Russia lost more than nine million killed, wounded, or missing. Between March 8 and 12, 1917 (February 24–28, O.S.), protests and a mutiny of Russian troops in Petrograd (St. Petersburg) led to the abdication of Tsar Nicholas II in favor of a provisional government that promised to uphold the commitments made to her allies. Thus Germany allowed Vladimir Ilich Ulyanov, better known by his pen name "Lenin," to travel from neutral Switzerland, through Germany, and into Russia. The Bolsheviks ("majority," though in fact they represented a minority viewpoint) under Lenin helped overthrow the provisional government on November 6–7, 1917 (October 24–25, hence this is known in Russian as the "October Revolution") and established *soviets* (worker's councils) as the primary instrument of local government. The Treaty of Brest-Litovsk (March 3, 1918) officially ended Russian participation in World War One by ceding to Germany large portions of the Russian Empire, including Finland, the Ukraine, the Caucasus, the Baltic states, and Poland.

11.2.1 Ballistics

Techniques like Nivelle's rolling barrage required an ability to accurately predict the path of an artillery shell through the air. The problem could be solved using partial differential equations, but required a parameter known as the ballistics coefficient which took into account the shell's size, shape, and mass. The ballistics coefficient was usually determined empirically.

In the United States the existing testing range at Sandy Hook was too close to New York City and the busy harbor, so testing was moved to a new site near the town of Aberdeen, on the western shore of the Chesapeake Bay. On January 2, 1918, during a blinding snowstorm, the first munitions tests occurred at the Aberdeen Proving Grounds. Two days later Veblen, now a Captain in the United States Army, arrived to take up his post as Head of the Division of Experimental Ballistics.

In February Veblen began to construct a range table for the army's 2.95 inch cannon (which could be carried by horseback). Unfortunately February 1918 was one of the coldest on record, and it took four weeks for Veblen to gather the necessary data. Veblen originally

---

[5]The tonnage of a merchant ship is a measure of how much cargo it can carry; the tonnage of a warship is a measure of its actual weight.

planned to have the tedious computations done by the mathematics instructors 150 miles away at Sandy Hook (Lieutenants Alger, Schwartz, and Cobb), but long distance collaboration proved impractical. Veblen made four visits to Sandy Hook in March, completing the ballistics table after a marathon session ending around 3 in the morning of March 26. Because of this experience, Veblen brought Schwartz and Cobb to Aberdeen, and recruited more mathematicians to help with the growing workload.

The army also established an office of mathematical ballistics, headed by FOREST RAY MOULTON (April 29, 1872–December 7, 1952), named because his mother likened his birth to a ray of sunlight shining through a forest. Moulton, a mathematical astronomer, had a career not too unusual by the standards of late 19th century American academics. He entered Albion University with almost no formal education, joined the faculty of the University of Chicago in 1896, became Associate Professor in 1898 and Director of the Department of Astronomy, and then earned his Ph.D. in 1899.

With Veblen's help, Moulton recruited Bliss from Chicago, where he was teaching a navigation course to about 100 students, and NORBERT WIENER (November 26, 1894–March 18, 1964), a child prodigy who graduated with a degree in mathematics from Tufts University (near Boston) at the age of 14 and a Ph.D. in mathematical logic from Harvard at 18. Afterwards Wiener went to Cambridge, where he took courses from Hardy, and Göttingen, where he studied with Hilbert on the eve of World War One, but when he returned to the United States he pursued philosophy until recruited by Veblen. Wiener was inspired to return to mathematics, and after the war ended, accepted a position at the Massachusetts Institute of Technology. There he would become a famously bad lecturer and one of the founders of communications theory.

Meanwhile American President Woodrow Wilson presented a framework for a "peace without victory" in January 1918. Two ideas were central: national self-determination (in particular, the breakup of multinational states like the Ottoman Empire and the Habsburg Empire, as well as the reconstitution of Poland), and the call for "A general association of nations ... for the purpose of affording mutual guarantees of political independence and territorial integrity to great and small states alike." The Germans rejected Wilson's overtures, hoping that submarine warfare could prevent the United States from effective intervention. This hope proved fleeting. During the first six months of 1918 destroyers sank 40 U-boats (up from an average of 2 per month before 1918). American shipbuilding capacity expanded to replace the ships, and by October 1918, more than 500,000 tons were being launched each month. By then British and Arab forces (some under the control of archaeologist-turned-soldier Thomas Edward Lawrence) swept through the Ottoman Empire, taking control of Palestine, Syria, Mesopotamia, and Gaza. The Habsburg Empire disintegrated with the secession of Hungary (October 24), Czechoslovakia (October 28), Poland and Yugoslavia (October 29), and finally Austria itself (October 30). This left Germany fighting alone, and the entry of the United States made it clear to the German leaders that the best they could hope for was Wilson's offer of "peace without victory." On November 11, 1918 they accepted an armistice. After four years of war, more than eight million soldiers had died, along with more than thirteen million civilians.

Unfortunately Wilson's hope for "peace without victory" would not be realized. The Treaty of Versailles incorporated many of the fourteen points—when they happened to favor the allies or lead towards the dismemberment of the Central Powers. The allied representatives, particularly French Prime Minister Georges Clemenceau, wanted to ensure that Germany could never again wage general war in Europe. Germany was blamed for the war, and had to promise to pay reparations, whatever their eventual cost (the so-called "blank check" clause). In addition Germany had to agree to sharp limitations on the size and power of its military. There was no negotiation with Germany: the delegates were given the choice between signing or facing a renewal of the war. They signed.

Even more disastrously, the United States rejected the Versailles Treaty because of the clause establishing the League of Nations. The "League fight" pitted two Ph.D.s against each other: Wilson, supporter of the League, and Henry Cabot Lodge, Senator from Massachusetts and recipient of Harvard's first Ph.D. in political science. Lodge disingenuously argued that the League would require the United States to intervene to maintain colonial empires, particularly that of Britain (never a popular country in Massachusetts, both with its colonial history and large Irish population). Wilson took the case to the American people. In September 1919 he toured the country, making 39 speeches in three weeks in support of the League. He might have made more, but his doctor, noting the signs of overwork, canceled the rest of the tour on September 25; Wilson had a stroke on October 2, 1919 that left him largely incapable of handling the physical tasks of President. The victory of Warren Gamaliel Harding in 1920 and a separate peace with Germany ended any possibility of United States involvement in the League. According to some pundits, Harding was elected for one reason only: he was more handsome than the haggard Wilson, and women in America now had the right to vote.

11.2.2 Suffrage

One problem faced by Moulton is that the partial differential equations of projectile motion cannot be solved analytically. Mathematical algorithms exist for finding numerical solutions to differential equations, and by breaking the algorithms down into discrete steps, the actual computations could be performed by computers: enlisted men with computational skills. For obvious reasons, the war made this type of computer hard to find. Thus Moulton hired eight women, including Elizabeth Webb Wilson, a graduate of George Washington University (1917). Moulton was not the first to hire female calculators; that distinction belongs to Edward Pickering of the Harvard Observatory. Pickering's motive: he could pay them half what he could pay a man for the same job. For this and other reasons, unions opposed the entry of women into the workforce.

Wartime needs, combined with pressure from business and government, overrode union objections, and women began filling traditionally male jobs like bank tellers, stenographers, ticket sellers, and switchboard operators. During the war, women began working in munitions plants, airplane factories, steel mills, and railroad switchyards, as well as some military positions (for example, in the signal corps, where French-speakers were particularly valued). Since many of these women had families, they in turn hired others for domestic tasks like cleaning and cooking, causing even more women to enter the workforce. The sudden presence of so many women in the workplace gave new impetus to the suffrage movement.

In Britain in 1897, Millicent Fawcett became the President of the National Union of Women's Suffrage Societies, which slowly converted many politicians to the cause of women's suffrage. But Fawcett's progress was too slow for some, so in 1903 Emmeline Pankhurst and her daughters Christabel and Sylvia founded the more confrontational Women's Social and Political Union, later known as the Suffragettes. Some suffragettes resorted to arson and vandalism, but declared a moratorium on violence when the war began, and in 1918 Parliament passed the Representation of the People Act, which gave limited franchise to women over thirty who owned property, or whose husbands owned property.

In the United States the Suffragists avoided the violence of the Suffragettes, possibly because the early leaders of the Suffragists learned their organizational and oratorical skills through their involvement with the Quaker-dominated Abolitionist movement; some of them were Quakers themselves. In 1840 at an international anti-slavery convention in London, seats were refused to Elizabeth Cady Stanton and Lucretia Mott, a Quaker. In response, they organized a meeting on women's rights in Seneca Falls, New York on July 19–20, 1848. The meeting attracted about 300 people. Frederick Douglass, an ex-slave and anti-slavery activist spoke at the event, but noted wryly in his *Rochester North Star* newspaper that a discussion of the rights of animals would have been more acceptable to the general populace than the discussion on the rights of women. Stanton met temperance advocate Susan Brownell Anthony, another Quaker, at an 1851 anti-slavery convention in Syracuse, New York. Lucy Stone, another prominent suffragist, also campaigned for abolition and organized a Women's Rights conference in Worcester, Massachusetts in 1850, and worked closely with abolitionists Julia Ward Howe and Josephine Ruffin, whose father was born in Martinique.

Despite the work of Stanton, Mott, and Anthony in New York, and Stone, Howe, and Ruffin in New England, the right of women to vote in all elections would first be achieved out west. In 1869 Wyoming Territory not only gave women the right to vote, but also the right to hold political office. When disaster did not follow, other territories followed suit. In 1916 Wilson would win re-election in part by his strong support among female voters in San Francisco, which helped him take California and its 13 electoral votes. Momentum was gathering for the passage of the Nineteenth Amendment.

The suffragists overlapped with abolitionists of another kind: the temperance leagues. The Women's Christian Temperance Union (founded 1874) supported the suffragists because they believed women were more likely than men to support a legal ban on alcohol. Unfortunately, lobbyists used this same argument to limit the franchise. In any case, female voters did not influence the 1919 passage of the 18th amendment, prohibiting the sale, manufacture, or transport of alcoholic beverages, since women did not have a nationally recognized right to vote until the 1920 passage of the 19th amendment.

11.2.3 Painlevé and Borel

On April 28, 1921, an Allied commission presented the reparations bill to Germany: $33,000,000,000, 52% of which would go to France to help rebuild the devastated northeastern part of the country. By mid-1922, France had already spent $7.5 billion for reconstruction and benefits for veterans, expecting German reparations to pay for them, and when German payments were not forthcoming, French Prime Minister Raymond Poincaré (Henri's cousin) sent troops to occupy the Ruhr on January 10, 1923. This was a violation of the League Charter, but no action was taken against France, reducing the credibility of the League as a whole.

The German government called for a campaign of passive resistance by the citizens of the Ruhr (about 10% of the total population of Germany), encouraged by promises of financial aid to those who lost their means of support, and enforced by penalties for those who cooperated with the French. Factories and mines did not operate, railways did not run, customs duties were not paid. In retaliation the French and Belgian authorities blocked exports from the Ruhr. Since about 80% of Germany's coal, steel, and iron came from the Ruhr, the German economy collapsed and the Mark fell rapidly in value: by August 8, 1923 the Mark was trading at 5 million to the dollar. The government fell four days later, and a new government announced it would cooperate with the invaders, and began negotiating a new payment plan. The occupation troops began to withdraw in late 1924. The high cost of the occupation forced the Poincaré government to call for a 20% tax increase on all classes; as they had in the past, the privileged classes in France rejected the idea that they pay a fair share of the cost of government, and brought down ministries that dared suggest they do so.

Paul Painlevé, a member of the Republican-Socialist party, became Prime Minister in April 1925, and oversaw the final withdrawal of the French troops. Painlevé's cabinet included Pierre Laval, who would turn Hitler's invasion of France into a political opportunity, and Aristide Briand, who shared the Nobel Prize for Peace in 1926 with German Foreign Minister Gustav Stresemann for the Locarno Treaty, guaranteeing peace between Belgium, France, and Germany.

Painlevé appointed fellow mathematician EMILE BOREL (January 7, 1871–February 3, 1956) Minister of Marine (which controlled the Navy). Unlike Laplace, whose political career consisted mainly of receiving awards and titles, or Painlevé, whose mathematical career largely ended after he entered politics, Borel continued his mathematical researches while in politics.

In 1913 ERNST ZERMELO (July 27, 1871–May 21, 1953) proved that in chess (or in general any symmetric two-player game) either side could force a win or a draw; this was the first paper on what would become the theory of games. Borel delivered the first of four papers on game theory to the French Academy of Sciences on December 19, 1921, and his last on January 4, 1927. In the last paper Borel developed the idea of a *tactic* (now called a strategy), which assigned a probability to each move a player could make that would maximize his payoff regardless of what information was known to the other player. For example, in the game rock-paper-scissors, a strategy is to pick rock one-third of the time, paper one-third of the time, and scissors one-third of the time. Even if the other player knew you were using this strategy, he could not construct a counter-strategy that would allow him to win more than one-third of the time.

Painlevé proved no better than his predecessors at stabilizing French finances, and when in November 1925 he proposed a 1 percent tax on capital (essentially an excise tax), his government fell. Briand became the next Prime Minister; Borel returned to the Chamber of Deputies, and Painlevé became Briand's Minister of War (a post he would keep through several changes of government).

In 1926 Raymond Poincaré once again became Prime Minister; he made Briand the Minister of Foreign Affairs and kept Painlevé as Minister of War. The next year Briand and Frank Kellogg of the United States authored a multinational treaty whose signatories agreed to "condemn recourse to war for the solution of international controversies, and renounce it, as an instrument of national policy in their relations with one another." The

Kellogg-Briand Pact would be signed by sixty-two nations, including Germany, Italy, and Japan. For his efforts, Kellogg would win the Nobel Prize for Peace in 1929.

The Briand-Kellogg Pact provided a legal basis for prosecuting world leaders for waging war. But criminalizing war and preventing it are two entirely different things; the French still feared an invasion by Germany. Because France was a signatory to the Briand-Kellog and Locarno Pacts, and consequently bound herself not to wage aggressive war, French strategy centered around building a purely defensive line of forts along the border with Germany and Belgium. In 1929 Painlevé, still Minister of War, gave the final approval to build the Maginot Line (named after André Maginot who, like Painlevé, served as Minister of War multiple times in the 1920s).

11.2.4   Noether and Courant

The economic disasters of the 1920s affected the lives of Emmy Noether and Richard Courant, and through them, the mathematical community of Germany and the United States.

Klein established the precedent of recruiting talent without regard to ancestry; Hilbert took the next step and invited Noether to join the Göttingen faculty in 1915. Noether studied at Göttingen during the winter of 1903–4, taking courses from Hilbert, Klein, Minkowski, and Blumenthal (who had been appointed *privatdozent* at Göttingen after taking his degree). At the time, women could only audit courses with the instructor's permission, but in 1904 the University of Erlangen permitted women to enroll as regular students; consequently, Noether transferred there and earned her degree in December 1907 (finding invariants of ternary biquadratic forms, work which she later disdained).

Shortly after Noether's return in 1915, Einstein presented six lectures at Göttingen that gave a preliminary (and, as it turned out, somewhat flawed) version of general relativity. One of the main problems he could not resolve had to do with the conservation laws, the cornerstones of physics. The problem is that the general theory of relativity appeared to violate the law of conservation of energy. At Hilbert's suggestion, Noether began to work on the problem, and by 1919 resolved the difficulty; she presented "Invariant Variation Problems" as part of a *habilitation* lecture. One of the key ideas she introduced is now known as Noether's theorem: every symmetry in the laws of nature corresponds to a conservation law, and vice versa. In particular, symmetry of the laws of nature under a translation in time would imply conservation of energy. But in general relativity the laws of nature are *not* symmetric under a translation in time; consequently, the conservation of energy does not hold.

Hilbert tried to have Noether appointed to the faculty, but the history and philology faculty objected, since as a regular faculty member, she would be eligible for a seat on the university senate. Hilbert retorted that the sex of a candidate should be irrelevant: "Göttingen University is an academic institution, not a bathhouse." The best Hilbert could do was get her a position as Adjunct Extraordinary Professor, an expansive title that concealed an insignificant position ("ordinary" professors were roughly equivalent to tenured faculty, while "extra" ordinary professors were junior faculty) and non-existent salary. It took until April 22, 1923 before she obtained a paid position at the University. The hyperinflation following the French invasion of the Ruhr made her salary worthless.

The hyperinflation also affected the Mathematics Institute of Göttingen. In 1922, Hilbert's student Courant founded the institute, but with the German mark worthless, the Institute existed on paper only. Fortunately the United States was in the midst of an economic boom, and in 1927, a donation from the Rockefeller Foundation permitted the Institute to construct a building to house its members. The institute opened on December 2, 1929.

The Rockefeller Foundation also gave grants that allowed promising scholars to study at prominent research centers like Paris, Cambridge, and Göttingen. JOHN VON NEUMANN (December 28, 1903–February 8, 1957) received such a grant that allowed him to study at Göttingen between 1926 and 1927. During this period he presented his first work on the theory of games to the Göttingen Mathematical Society on December 7, 1926, around the time Borel delivered his last paper on the subject. Although a latecomer to the theory of games, von Neumann immediately had a profound impact; the paper, published in 1927, proved the minimax theorem.

11.2.5 The Black Chamber

In 1921 an international conference held in Washington D.C. sought to limit the size of navies in the hopes of preventing future conflicts. Japan hoped for parity with the United States or Britain, but a coded message to its ambassador instructed him to accept a ratio as little as 6 to 10. The message was decoded by Herbert Osborne Yardley, and given to the American negotiator, who pressed for a settlement that set the tonnage limits at 10 : 10 : 6 : 3 : 3 for Britain, the United States, Japan, France, and Italy respectively. It was the greatest success of what was officially known as MI-8, but unofficially as the Black Chamber, a government agency whose sole purpose was to obtain and decipher coded messages.

Both codes and ciphers work by substituting one symbol for another, changing an uncoded message (the plaintext) into a coded one (the ciphertext); the main difference is the basic unit of encryption. Codes generally encrypt words or concepts, while ciphers encrypt letters. For example, a code might take the word "GREEN" (the concept) and encrypt it as the word "RIVER," while a cipher might take the letters G, R, E, and N (the symbols) and encrypt them as H, S, F, and O, turning GREEN (the plaintext) into HSFFO (the ciphertext). This last is an example of what is called a rot-n (for rotation) cipher: If we assign each letter in the alphabet a unique number (for example, $A = 0$, $B = 1$, and so on up to $Z = 25$; note that this identification is itself a cipher!), then the letter corresponding to c is replaced with the letter corresponding to $d \equiv (c + n) \mod 26$. The preceding is thus a rot-1 cipher.

Since every occurrence of a given letter in the plaintext is replaced by the same letter in the ciphertext, this type of code is known as a single substitution (sometimes monoalphabetic) cipher: in a rot-1 cipher, G is always encoded as H. Single substitution ciphers are notoriously easy to break because not all letters appear equally often. For example, E is by far the most common letter in written English, so if a symbol appears more frequently than any other, the likelihood is that it represents the letter E. Thus one standard technique ("attack") used to break this type of cipher is a statistical analysis.[6]

---

[6] Sherlock Holmes fans may recall "The Case of the Dancing Men," whereby Holmes breaks a single substitution cipher using this method.

The security of substitution ciphers can be improved in several ways. The simplest is a polyalphabetic cipher. Mathematically, we can form a polyalphabetic cipher by choosing an arbitrary set of k numbers $\{n_0, n_1, n_2, \ldots, n_{k-1}, \}$. If c is the ith letter of the message to be encrypted, it will be replaced with the letter

$$d \equiv (c + n_j) \mod 26 \quad \text{where} \quad j \equiv i \mod k.$$

k is referred to as the key length; if $k = 1$, the result is a single substitution cipher. In the years preceding World War II, several nations produced mechanical devices that would allow an operator to type in a plaintext message, which would be encoded using a polyalphabetic cipher with an enormous key length; the German Enigma machine was of this type. For example, we might form the polyalphabetic cipher with key $\{1, 5, 7, 3\}$. To encode the word GREEN, the first letter (G) would be replaced by the letter 1 space after it (H); the second letter (R) would be replaced by the letter 5 spaces after it (W); the third letter (E) by the letter 7 spaces after it (L); the fourth letter (also an E) by the letter 3 spaces after it (H). The pattern would repeat for the fifth through eighth letter, so N would be replaced by O, and the plaintext GREEN would be transformed into the ciphertext HWLHO.

Even greater security can be obtained using a polygraphic cipher, which breaks the message into groups of two or more letters (called "blocks" or polygraphs) for encryption. Because polygraphs have a more even statistical distribution than individual letters, polygraphic ciphers are more difficult to break. The difficulty is finding a way to use polygraphic ciphers easily. For a digraph cipher, the traditional method is to use a prearranged square table of the letters of the alphabet: the relative location of the two letters in the digraph is used to find the corresponding encryption.[7] But this method is infeasible for trigraphs or polygraphic ciphers in general; it might seem the only conceivable way to use a polygraphic cipher is to construct a massive code book, listing every possible n-graph and its encrypted equivalent. For a tetragraph cipher, one would require a code book with $26^4 = 456,976$ entries, from AAAA to ZZZZ.[8]

Although mathematicians like Vietè and Wallis had been involved in breaking ciphers, mathematics played no substantial role in making ciphers until August 1929, when the American mathematician LESTER SANDERS HILL used matrix algebra to construct a generalized polygraphic cipher. The basis of Hill's method is the following. Let x be an $n \times m$ matrix corresponding to a block of nm letters in the plaintext message. Let A be an invertible $n \times n$ matrix; then $y = Ax$ will correspond to a block of letters in the ciphertext. The plaintext is easy to recover as $x = A^{-1}y$. In its simplest form, x is an $n \times 1$ column vector, though Hill considered several variations. The Hill cipher could be used to form polygraphic ciphers of any block size, and matrix multiplication is incorporated in many modern encryption schemes. Unfortunately it was impractical to apply with the technology of the 1930s, and if we know y (the ciphertext) and x (the plaintext), we can generally find a good deal of information about A (a "plaintext attack").[9]

---

[7]This is known as a Playfair cipher, named after Lyon Playfair of the British Foreign Office and actually invented by physicist Charles Wheatstone.

[8]The polygraphs do not have to follow the standard rules of spelling, since one ignores spaces and the letters may stretch across two or more words.

[9]Plaintext attacks might seem impractical: after all, if one knows the plaintext, there is no need to break the code. Actually, they are one of the more common ways of breaking codes; Vietè effectively employed plaintext attacks when breaking the Spanish codes, because he had some idea of the contents of the encrypted messages.

Despite the crucial role the Black Chamber played in the Washington Conference, it was disbanded a few years later after Secretary of State Henry Stimson declared fatuously that "Gentlemen do not read each other's mail." After decoding more than ten thousand messages, the Black Chamber shut down on October 31, 1929. It was a bad time to become unemployed.

11.3 The Depression Era

During the 1920s the American stock market had seen record gains (a "bull market," because bulls attack by thrusting their horns upwards). But the tulip craze, South Seas Bubble, and Mississippi Scheme proved that, sooner or later, overpriced items will drop to their true value. On Black Monday (October 28), the Dow Jones Industrial Average (DJIA) lost a record 13% of its value.[10] The next day, Black Tuesday (October 29) was the worst yet: the DJIA lost an additional 12% of its value, and the decline in stock values had spread throughout the market. The bull market of the twenties was followed by the bear market of the thirties (because bears attack by swinging their claws downwards) and a general economic collapse known as the Great Depression. It should be pointed out that 1929 was the start of the Great Depression in the United States; most of Europe had been in an economic crisis since the end of the Great War.

By itself, the Great Crash is not considered one of the main causes of the depression. Indeed, there have been greater crashes since 1929 (the largest percentage drop in the history of the DJIA occurred on October 19, 1987, where the DJIA dropped 22%) that have not led to an economic depression. In rough terms, the bull market of the 1920s masked a failing global economy; when the stock market crashed, the weakness of the economy became apparent, and the loss of confidence led to disaster.

Between 1929 and 1932, the stock market had lost 80% of its value; more than 40% of the nation's banks had failed; manufacturing output had been cut in half; and more than 12 million people had lost their jobs, giving rise to an unemployment rate around 25%. Some members of the Republican party, notably the Secretary of the Treasury Andrew Mellon, believed that the "boom-bust" cycle was an inevitable feature of capitalism, and that the government should not interfere. However, President Herbert Hoover broke with his own party and established the Reconstruction Finance Corporation (RFC) on January 22, 1932, to extend loans to businesses to help stimulate the economy.

Of course, money given to businessmen does not always translate into employment for workers. In Britain, the Liberal Party launched public works projects to provide jobs for the unemployed. However, this required running a budget deficit, traditionally regarded as anathema to national stability. The English economist John Maynard Keynes, who had been lured away from mathematics by his economics teacher Alfred Marshall at Cambridge, argued that the benefits of employment outweigh the costs of a government deficit; Keynes summarized his results in *The General Theory of Employment, Interest, and Money* (1936).

---

[10]The DJIA is a measure of the "average" value of 30 stocks, though in this case "average" is even more misleading than usual. The divisor is not equal to the number of stocks, but instead is chosen to make the value invariant under a variety of "allowed permutations." For example, the companies that make up the DJIA changed periodically; if a company whose stock has a price of $25 is replaced by a company whose stock has a price of $50, the mean price of the shares of the companies would increase, but the divisor would change so the DJIA would remain the same. Currently the DJIA divisor is around 0.125.

Promising a "new deal" for the working class, Franklin Delano Roosevelt became President of the United States in 1932. In his first hundred days in office, he established the Civilian Construction Corps and the Tennessee Valley Authority, which managed large projects that provided employment for hundreds of thousands of people, and built much-needed infrastructure.

In 1935, Roosevelt added the Social Security Act and the Works Progress (later Projects) Administration to the New Deal. At its peak, the WPA employed more than 3.3 million people (in November 1938) working at a variety of tasks, including the construction of buildings, bridges, and roads. Seven percent of the WPA budget was allocated to support creative artists: writers, actors, artists, and musicians.

11.3.1   Gertrude Blanch

In January 1938, the WPA established several projects in New York City, in part to help the re-election campaign of Mayor Fiorello LaGuardia, a personal friend of Roosevelt (even though LaGuardia was a Republican). One of the more remarkable WPA programs was the Mathematical Tables Project, directed by physicist Arnold N. Lowan (a part-time instructor at Brooklyn College, among other places).

The technical director was GERTRUDE BLANCH (February 2, 1897–January 1, 1996), born in Kolno.[11] Her family emigrated to the United States in 1907, where she graduated from a Brooklyn High School in 1914. To earn money to go to college, she worked as an office clerk for the next fourteen years, graduating from New York University in 1932 and earning a Ph.D. from Cornell in 1936. Unfortunately her job prospects were dismal. Falling enrollment and endowments made worthless by the market crash meant few positions existed for newly graduated mathematicians. She worked as a substitute teacher at Hunter College before joining the Mathematical Tables Project.

Blanch's fourteen years in the business world gave her a unique insight. In the 1880s, Frederick Winslow Taylor began the first time-and-motion studies of industrial workers to optimize labor efficiency. Key to the Taylor system were worksheets that described, in detail, precisely how individual tasks were to be performed. By World War One, Taylor's ideas had been adapted for clerks working in business offices. Blanch, who had managed two different offices before returning to the academic world, was very familiar with the Taylor system and incorporated many of its key ideas in the design of the project.

Unlike the computers at Aberdeen, most of those at the Mathematical Tables project had no training in advanced mathematics. Most were unemployed office clerks, and many lacked even basic arithmetic skills. Blanch divided the workers into groups; each group would specialize in a *single* type of arithmetic calculation (addition, subtraction, multiplication, or—for the most advanced workers—division). By April 1938, the project published a table of powers of the integers. While the content of the table was mathematically uninteresting, it was enough to convince the WPA to continue funding the project. By 1939 a more sophisticated table (exponential function values) was produced, and interest in the project began to grow. Tables of logarithms and trigonometric function values were produced as well, but both Lowan and Blanch realized that in order to make the project viable, they had to tackle unsolved problems.

---

[11] At the time of her birth, Kolno was part of Russia, though it is now part of Poland.

Lowan solicited problems from his colleagues. The first significant request came from the astrophysicist Hans Bethe, who had just solved the problem of how the sun produces energy using fusion reactions. Bethe suggested another problem that required solving a fourth order non-linear differential equation by numerical integration, and was sufficiently impressed with the results that he recommended the project to his colleagues.

11.3.2   Poland

The thirteenth of Wilson's fourteen points called for a re-establishment of the nation of Poland, and it was easy enough for the victorious allies to create a nation from the territory of the Central Powers.

Polish intellectual life suffered from decades of neglect as well as outright suppression. For example, the University of Warsaw, established in 1815, was reorganized as the Imperial University in 1870, with a largely Russian faculty and with all instruction conducted in Russian. The Germans re-established the University of Warsaw when they occupied the city in 1915, and after the war, rebuilding a Polish intellectual community became a priority of the new nation. Mathematics benefited: it is relatively inexpensive to support mathematicians, and a thriving mathematical community could in turn promote more practical areas of science and engineering.

But what areas of mathematics? It would be difficult to make groundbreaking contributions in well-researched fields such as the theory of functions or differential equations. Set theory, on the other hand, offered considerable promise. Not only was this a relatively new branch of mathematics but Kronecker's opposition, particularly to any notion of infinite sets or processes, meant that few besides Cantor worked in the field.

In the 1890s Cantor published a series of groundbreaking papers that put the theory of the transfinite ordinals on an axiomatic basis. However, Cantor's work raised an important question. The cardinality of the natural numbers is designated by $\aleph_0$, and the next higher transfinite number in Cantor's hierarchy is designated by $\aleph_1$. The cardinality of the real numbers is equivalent to the cardinality of the power set of the natural numbers, and can be designated by $2^{\aleph_0}$. It is tempting to suppose that $\aleph_1 = 2^{\aleph_0}$, but Cantor could not prove this. The continuum hypothesis was the very first of twenty-three problems posed by Hilbert in a speech to an international conference of mathematicians in Paris in 1900.

In 1907 WACŁAW SIERPIŃSKI (March 14, 1882–March 21, 1969) learned of Cantor's result that a one-to-one map existed between the plane and the real numbers. Consequently he began to study set theory, and in 1909 taught the first course ever given on set theory at the University of Lviv. Another convert to Cantor's set theory was Sierpiński's colleague at Lviv, ZYGMUNT JANISZEWSKI (June 12, 1888–January 3, 1920), and Sierpiński's student STEFAN MAZURKIEWICZ (September 25, 1888–June 19, 1945).

In 1917 Janiszewski made several recommendations for creating a strong Polish tradition in mathematics. One of his recommendations was pedestrian: a center for mathematical research should be developed. Another recommendation was somewhat more radical: Polish mathematicians ought to focus on areas where Poles had already established an international reputation (set theory was the obvious choice). The most radical suggestion was the publication of an international journal specifically devoted to a single field (again, set theory was the obvious choice).

After the war, Sierpiński, Janiszewski, and Mazurkiewicz accepted positions at the University of Warsaw, where they joined a faculty that also included STANISLAW LESNIEWSKI (March 18, 1886–May 13, 1939) and JAN LUKASIEWICZ (December 21, 1878–February 13, 1956). In 1916 Lukasiewicz began a debate over the place of logic in mathematics. One view held that logic is a tool of mathematics but not, properly speaking, part of mathematics. Lukasiewicz and his students, including KAZIMIERZ KURATOWSKI (February 2, 1896–June 18, 1980), held that logic was itself a branch of mathematics. The view of Lukasiewicz and his students prevailed, and mathematical logic joined set theory as a central field of study for Polish mathematicians.

Lesniewski played a crucial role in the history of Polish mathematics by convincing seventeen-year-old ALFRED TARSKI (January 14, 1902–October 26, 1983), born Tajtelbaum, to study mathematics instead of biology. Tajtelbaum's papers began appearing in 1919, but in a display of nationalistic fervor, he converted from Judaism to Catholicism and changed his name to the more Polish-sounding "Tarski" in 1924; since Polish law required the two names be used simultaneously during the transitional period, a few papers appeared as "Tajtelbaum-Tarski" before Tajtelbaum disappeared forever.[12]

Meanwhile, the first issue of *Fundamenta Mathematica*, a journal specifically devoted to set theory and related topics, appeared in 1920. Polish mathematics was well on its way to preeminence. Unfortunately a chain of events begun thousands of miles away led to Janiszewski's premature death.

On March 11, 1918, an army private at Fort Riley, Kansas reported to the camp hospital with a case of flu. By noon, over one hundred other soldiers had been admitted with the same condition; over 500 would be hospitalized by the end of the week. They were the first victims of a great influenza pandemic that spread rapidly around the world. Soldiers from the United States carried the disease to Europe, where the close quarters of trench warfare and limited sanitation helped spread the disease. After the war, jubilant populations held mass celebrations, giving the disease further opportunities to spread from person to person, and returning soldiers brought the disease to Australia and India, and started a new outbreak in the United States. The major phase of the pandemic ended in 1919, but there were periodic recurrences over the next few years. Exact numbers are difficult to determine, but more than twenty million people (including Janiszewski) are believed to have died from the disease between 1918 and 1920—more than had died in all four years of World War One.

Poland faced other difficulties. The Treaty of Versailles provided for the existence of Poland, but failed to give the new nation clearly defined boundaries. Even the Poles were divided over what should constitute the new nation. Some, like Jozef Piłsudski, hoped to build a federation of allied states that included Poles, Lithuanians, Belarusians, and Ukrainians, and stretched from the Baltic to the Black Sea. Others, like Roman Dmowski, believed that the inclusion of other ethnic groups would only weaken the Polish state. The rift between the two threatened to turn into civil war, and was only healed through the intervention of the composer Jan Paderewski, the former Prime Minister.

However the border issue was unresolved, and made more complex by the fact that Lithuania, Belarus, and Ukraine declared their independence of Russia after the Bolshe-

---

[12] He and his brother Wacław evidently picked the name Tarski because it sounded Polish but was unused.

vik revolution. The situation in Lviv (Lwow in Polish, Lvov in Russian, and Lemberg in German) was typical. Lviv was the capital of the region known as Galicia, part of Austrian-occupied Poland. Most of the inhabitants of Galicia are Ukrainian, though Lviv itself was largely Polish, and included an important university which, in the years before Polish independence, could boast of a faculty including Janiszewski, Sierpiński, and Lukasiewicz, and students of the caliber of Kuratowski, Lesniewski, and STEFAN BANACH (March 30, 1892–August 31, 1945).

In late October 1918, Ukrainians in Lviv declared it the capital of the West Ukrainian People's Republic. The Polish majority in the city revolted and drove the small Ukrainian garrison from the city, and a Polish army invaded Galicia. Meanwhile Poland drifted into war against Bolshevik Russia, which was trying to reconquer the Ukraine, Belarus, and the Baltic States. In 1920 Ukraine gave Galicia to Poland in exchange for an alliance against the Bolsheviks. At the Battle of Warsaw (August 13–25, 1920), Piłsudski led the Poles to victory over a Soviet army led by Mikhail Tukhachevsky. The Treaty of Riga (1921) ended the Polish-Bolshevik War. To the dismay of Pilsudski, the agreement partitioned the Ukraine and Belarus between Russia and Poland, giving Russia a border with Poland. The Bolsheviks formed the Union of Soviet Socialist Republics (usually referred to as the Soviet Union) out of Russia and the briefly independent Belarus, Ukraine, and other parts of the Russian Empire.

Banach joined the faculty at Lviv after annexation to Poland, and helped make it a second center of mathematical research in Poland. One of Banach's key contributions was applying measure theoretic ideas to probability, which led to "New Foundations of the Theory of Probability (Definition of Probability Based on the Theory of Sets)," by ANTONI MARIAN ŁOMNICKI (January 17, 1881–July 4, 1941) of the University of Lviv. "New Foundations" appeared in the 1923 volume of *Fundamenta*, and began by noting that probability occupied a "rather infamous" position in mathematics, mainly because it had yet to be built up from an axiomatic basis. In 1929 Banach, along with HUGO DYONIZY STEINHAUS (January 14, 1887–February 25, 1972) established *Studia Mathematica*, a journal specifically devoted to functional analysis and related topics.

Banach and Tarski proved a remarkable result in 1924:

> In a Euclidean space with $n \geq 3$ dimensions, two arbitrary regions, bounded and containing their interior points (for example, two spheres of different radii), are equivalent by finite decomposition. An analogous theorem holds for regions on the surface of a sphere; however the corresponding claim for a 1- or 2-dimensional Euclidean space is false.[13]

This result is known as the Banach-Tarski Paradox, since it implies that one can take a sphere of one radius and, through a finite set of dissections and compositions, construct a sphere of a different radius.

11.3.3 Austria and Germany

The breakup of the Habsburg Empire after World War One separated industrialized Austria from agricultural Hungary and mineral-rich Czechoslovakia. After the war all three nations

---

[13] *Fundamenta Mathematica*, Vol. 6, p. 244.

plunged into an economic depression. Vienna itself was an imperial capital without an empire; moreover, wartime refugees swelled the population and strained the limited resources of the city. Tuberculosis plagued the population so badly it was known as the Viennese disease, while runaway inflation wiped out savings and pensions. Politicians with socialist programs flourished in this environment. Red Vienna became a forerunner of the modern welfare state, with state-subsidized education, health care, and housing.

One of the key innovations of Red Vienna was the construction of public housing designed to maximize community spaces. The greatest of these was the Karl Marx House, designed by the architect Karl Ehn and completed in 1930. The kilometer-long building was designed to include nearly 1400 apartments, as well as medical offices, kindergartens, libraries, recreational areas, open air gardens, laundry facilities, and meeting rooms.

Ehn was a student of Otto Wagner, a proponent of the use of modern construction materials and techniques. The style of Art Nouveau was permeating Europe in the late 1800s, and Wagner began using flowing, organic motifs along exterior surfaces. In 1894 Wagner won a contract to design the stations along Vienna's new Stadtbahn (a light rail system), which would link the suburbs with the city center; later that year he became a professor at Vienna's prestigious Academy of Fine Arts, and with Gustav Klimt, would found the Secessionist School (the Austrian version of Art Nouveau).

In 1907 an eighteen-year-old high school dropout applied for entrance to the Academy of Fine Arts. While his sketches showed promise, they were primarily of landscapes and buildings, so the examiner suggested that the applicant enter the School of Architecture instead. After giving it serious consideration, Adolf Hitler realized that his lack of education would make becoming an architect nearly impossible.[14] After spending a few years as a struggling artist, he joined the Bavarian Army when World War One broke out.

After the war, Hitler joined the German Worker's Party to report on its activities to the government. Instead he became an active supporter. The party became the National Socialist German Worker's Party (the first two words, in German, are *National-sozialistiche*, hence Nazi). In 1923 Hitler showed his grasp of history by attempting to overthrow the Republic by force; the monarchists tried this tactic in 1920 and the communists in 1923, and both failed. The Nazis did no better, and for his role in the Munich *putsch*, Hitler was sentenced to five years in prison.

Meanwhile, the Viennese government ended postwar inflation in 1923 by declaring 13,000 crowns equal to 1 shilling. By law, property owners could not charge more than the shilling equivalent of rents in preinflationary crowns: effectively, this reduced rents to 1/13,000 of what they were before the war. Luxurious apartments could be had for next to nothing, though there were risks: owners had no incentive to maintain their property. A half-ton of falling stucco nearly ended the life of KARL MENGER (January 13, 1902–October 5, 1985).

Menger, at the invitation of HANS HAHN (September 27, 1879–July 24, 1934), joined a remarkable group of philosophers, scientists, and mathematicians known in later years as the Vienna Circle. Other group members included Menger's doctoral advisor and Circle founder MORITZ SCHLICK (April 14, 1882–June 22, 1936); RUDOLF CARNAP (May 18, 1891–September 14, 1970); and OTTO NEURATH (December 10, 1882–December 22, 1949).

---

[14] The fact that Hitler drew buildings led newspapers to deride him as an "Austrian house painter."

The Circle promoted the doctrine of logical positivism, which grew out of the philosophy of University of Vienna professor Ernst Mach. Mach, a pioneer in the study of supersonic projectiles, took the scientific method to its ultimate conclusion: statements that can neither be proven nor disproven have no place in science, a doctrine known as verificationism. This seems to be axiomatic, but consider a statement like "The diameter of the Earth is approximately 13,000 kilometers." We can verify this statement, of course, by measuring the diameter of the Earth. But how do we do that? The strictest application of Mach's principles requires precise definitions of previously accepted terms like velocity, time and distance; Einstein would later identify Mach as one of the key sources of inspiration for the theory of relativity.

In March 1921 Hahn began a weekly two-hour seminar entitled "News about the concept of curves." At the first meeting, Hahn noted that all previous definitions of a curve were unsatisfactory, and that to date no mathematician had given a proper definition. Menger considered the problem over the next week, and a few minutes before the next session suggested a new definition to Hahn. Let C be a continuum, defined by Hahn (following Jordan) as a closed set that cannot be expressed as the union of two disjoint closed sets, and take any point p in C. If C is not a curve, then the intersection of any neighborhood of p and C forms a continuum. Hence C is a curve if for any sufficiently small neighborhood of p, the intersection of the neighborhood and C does not form a continuum. Menger's definition intrigued Hahn, who asked him where he had learned so much about topology; Menger replied that he was a physics student, and his only prior experience had been Hahn's lecture the previous week.

Menger's work on the theory of dimensionality brought him into contact with Polish mathematicians. Although Poland had traditionally aligned herself with France politically and intellectually, most French mathematicians of the postwar era found set theory unpalatable; notable exceptions included HENRI LÉON LEBESGUE (June 28, 1875–July 26, 1941), who applied measure theoretic ideas to integration, and MAURICE FRÉCHET (September 2, 1878–June 4, 1973), whose 1906 thesis introduced the concept of a metric space. Some German mathematicians, particularly ABRAHAM FRAENKEL (February 17, 1891–October 15, 1965), FELIX HAUSDORFF (November 8, 1868–January 26, 1942), and CONSTANTIN CARATHÉODORY (September 13, 1873–February 2, 1950) achieved prominence in set theory and developed what we now call point-set topology, but animosity tainted the relationship between Poland and Germany, making it difficult for intellectual cross-fertilization.

With France uninterested and Germany hostile, Polish academicians turned to their traditional supporters: Austria. Despite the partitions of Poland, relationships between Austria and Poland were reasonably cordial, and Vienna supported Polish intellectual life in many ways. Instruction in primary and secondary schools in Galicia was in Polish, and the University of Cracow flourished; the Austrians even established the University of Lviv and the Cracow Academy of Sciences.

What was lacking was awareness of the Polish school of logic; consequently Menger invited Tarski to Vienna in 1930. Unfortunately Tarski's topics were of little interest to circle members, and the first lecture, on February 19, 1930, was delivered to an audience of three: Hahn, Menger, and Carnap. Personal phone calls by Menger helped ensure full attendance at Tarski's third lecture, on February 21.

After the talks, Menger was approached by KURT GÖDEL (April 28, 1906–January 14, 1978), who had been his student in a 1927 course on dimension theory and was a regular attendee of Circle meetings. Gödel asked if he could arrange a meeting with Tarski to discuss Gödel's thesis on the completeness of first order logic. Tarski and Gödel met, and Tarski was impressed with the work. Gödel delivered his most surprising result on January 21, 1931. In his landmark "On Completeness and Consistency," Gödel proved the existence of theorems about arithmetic that were true but unprovable. At the time, Menger was in the United States, visiting the Rice Institute in Houston, Texas. When he received word of Gödel's remarkable result, he informed his colleagues, and consequently they became the first outside of Vienna to learn of Gödel's Theorem.

11.3.4   The Soviet Union

In Marx's ideal communism, the workers own the means of production, and keep all the fruits of their labor; Marx envisioned the existence of a "dictatorship of the proletariat" during the transitional period between capitalism and communism. During the transitional period the state would own all the means of production, receiving from the workers their labor and returning to them the necessities of life. Unfortunately Lenin's attempt to put this theory into practice resulted in economic disaster, leading to the "New Economic Program" (NEP) which tolerated capitalistic practices in smaller factories and farms. The *nepmen* formed the backbone of the Soviet economy in the years immediately following the Revolution.

Lenin's premature death at the beginning of 1924 caused a power struggle between two Bolshevik factions. The faction led by Leon Trotsky wished to export the revolution to other countries. But Nikolai Bukharin advocated "socialism in one country:" communists had to consolidate power in the Soviet Union first, and world revolution would come after. This faction, led by Josef Visserionovich Dzugashvili, son of a Georgian shoemaker, eventually prevailed. Like Lenin, Dzugashvili was also known by a pseudonym: "Stalin," man of steel (from his great physical strength).

Bukharin and Stalin both originally supported the *nepmen*, but Stalin, desiring a rapid industrialization of Russia, announced the first Five Year Plan on October 1, 1928. Although the first Five Year Plan failed to achieve most of its industrial objectives, leaving the majority of the Soviet people worse off in 1933 than in 1928, the educational initiatives proved successful. The number of elementary school students tripled (to $22,000,000$) and 80% of the children between eight and fourteen were enrolled in government schools. Government leaders realized that success of the Five Year Plan (and the Soviet Union in general) relied on training a generation of engineers, skilled workers, and technicians. Consequently technical education flourished, and the best students could hope to participate in the Mathematical Olympiads (begun in 1935), where winners would be exempted from university entrance examinations.

Key to the Five Year Plan's rapid industrialization of Russia was the annihilation of the *nepmen*. The most brutal measures were taken against the *kulaks*, peasants who (thanks to the original revolution) owned their land; the Five Year Plan took the land away and reduced the peasants to workers on state-owned collective farms (*kolkhozy*). In 1929 Stalin appointed 21 prominent communists to oversee this process; an estimated five million *kulaks* resisted collectivization—and vanished. But reaction against collectivization threatened to turn into open revolt, so in March 1930 Stalin published an article, "Dizzy With Success," where he castigated the collectivizers for their excessive enthusiasm and stressed that collectivization was to be a voluntary process; nineteen of the twenty-one members of the commission were subsequently arrested by the GPU (secret police) and executed. Collectivization continued, and when peasants in the Ukraine, Kazakhstan, and the Caucasus revolted, authorities punished the region in 1932–3 by confiscating all grain supplies (even seed grain for future planting). Seven million people died in the ensuing famine.

In addition to murdering the *kulaks* and taking their land, the Soviets also established enormous collective farms in hitherto uncultivated regions. But generally speaking, any place that crops *can* be raised is already under cultivation; consequently, the new collective farms were generally located on arid or otherwise barren land. Consequently Stalin exulted in a discovery, made by obscure agriculturalist Trofim Denisovich Lysenko, that crops could be grown on these lands without fertilizers. Lysenko would periodically announce other improbable discoveries, and is particularly associated with promulgating the doctrine of inheritance of acquired characteristics, which (if valid) would mean that plants could "learn" to grow in unsuitable climates and pass on this knowledge to their offspring.

Lysenko helped Stalin with another problem. The peasants who survived collectivization had no incentive to do more work than what was absolutely necessary (in contrast, as *nepmen* they had been able to sell their surpluses on the open market). Because Lysenko came from a peasant, not academic, background, he could rally the peasants behind him, particularly by attacking the academic scientists whose researches, in his view, did nothing for the farmers. Stalin, who voiced his support for "practice" over "theory," supported Lysenko wholeheartedly. Since any competent scientist could discredit Lysenko's "practice," and thereby endanger the collectivization movement as a whole, geneticists and others were ruthlessly persecuted, and Soviet biology would suffer for a generation.

Mathematics was both more and less fortunate. On the one hand, it could neither help nor hinder any political agenda, so it was relatively immune to government interference. On the other hand, because mathematics is fundamentally about ideas, it is subject to criticism based on Marxist principles. Key to such attacks is the doctrine of dialectical materialism. Materialism holds that all knowledge is ultimately derived from the senses: if the tree falls in the forest and no one hears it, there is no sound. However, materialism is tempered in Marxism by the dialectic, the apparent contradiction between thesis and antithesis whose resolution produces a synthesis. If a tree is observed to fall and a sound is heard, the thesis is that the tree made the sound, while the antithesis is that the tree did not make the sound. Ordinary logic requires that only one of two contradictory statements be true. Dialectical materialism holds that both contain portions of the truth, and the synthesis (in this case, perhaps that the falling tree set air into motion which is what we perceive as sound) is what is true.

Since one cannot build a mathematical structure by reconciling contradictory axioms, materialist mathematicians focused on the source of those axioms. This is a question whose origins can be traced back to Aristotle, but Cantor's work brought the issue to the forefront of mathematical philosophy. The opposing philosophies came to be known as formalism and intuitionism, after LUITZEN EGBERTUS JAN BROUWER (February 27, 1881–

December 2, 1966) used these terms in his inaugural lecture at the University of Amsterdam.

The difference between the two points of view may best be illustrated by looking at geometry. Hilbert (a key supporter of formalism) held that all the statements of a properly constructed geometry would hold true if we replaced "points," "lines," and "planes" with "tables," "chairs," and "beer mugs." In other words: mathematics is independent of the meaning of its symbols. In contrast the intuitionist would argue that "point," "line," and "plane" are abstractions made from our experience; consequently, they have properties that are not amenable to logical derivation. It should be stressed that both sides held that mathematical statements must be logically derived; they differed only on the application of logic.

One key issue concerns the existence of mathematical objects. Consider Cantor's proof of the uncountability of the real numbers, which begins by assuming the existence of a one-to-one correspondence between the whole numbers and the real numbers, then proving the existence of a real number that corresponded to no whole number. Intuitionists reject this proof because the real number in question cannot be constructed in a finite number of steps. This objection reflects a deeper and more fundamental difference between the two schools of thought. Formalists accept indirect proofs, because they hold that if a statement is not true, it must be false: this is the so-called law of the excluded middle. Extreme Intuitionists reject the law of the excluded middle: they would argue that, at best, Cantor proved the falsity of the statement "The real numbers are countable," and not the truth of the statement "The real numbers are not countable."

Because of its rejection of the law of the excluded middle (which thus allows for the possibility of reconciling contradictory statements) and by its insistence on sensory impressions as the ultimate source of knowledge, dialectical materialists embraced Intuitionism. Since labeling someone a formalist was tantamount to labeling them anti-Communist, mathematical philosophy became a powerful tool to enforce ideology.

The first All Union Mathematical Conference was held in Kharkov in June 1930. The organizers of the conference, DMITRI FEDOROVICH EGOROV (December 22, 1869–September 10, 1931), the President of the Moscow Mathematical Society, and SERGEI NATANOVICH BERNSTEIN (March 5, 1880–October 26, 1968), invited a number of foreign mathematicians to attend, and many did; Hadamard, Cartan, and several others gave plenary lectures. Bernstein assured the foreign mathematicians that politics would not be discussed, but a group of activists, led by O. YU. SCHMIDT (1891–1946) circulated a telegram to be signed and sent to the Party Congress then being held in Moscow, where Stalin had just announced a campaign for the "advance of socialism on all fronts." Despite the refusal of Egorov to sign and Bernstein's protests that the foreign mathematicians had been promised an apolitical conference, the telegram was sent, and the authorities took note of Egorov and Bernstein's opposition.

Egorov was arrested in September by the GPU, and after a sham trial, was given a long prison term, dying after going on a hunger strike in 1931. Meanwhile a group of prominent mathematicians signed a petition congratulating the GPU for uncovering counterrevolutionaries within the scientific community. The petitioners included one of Egorov's students, LEV SHNIRELMAN (January 2, 1905–September 24, 1938), and ALEXKSANDR OSIPOVICH GELFOND (October 24, 1906–November 7, 1968). Egorov's immediate successor as President of the Moscow Society was Arnosht Kol'man. But Kol'man was a political appointee, and was soon replaced by PAVEL SERGEEVICH ALEKSANDROV (May 7, 1896–November 16, 1982).

Gelfond and Aleksandrov were, at least, competent mathematicians who happened to support the government. Gelfond solved Hilbert's seventh problem in 1934 by proving that a^x is transcendental if a is algebraic and x is irrational, while Aleksandrov proved several of Cantor's conjectures regarding what are now called Borel sets, including Cantor's conjecture that a Borel set is either countable or has the power of the continuum. Aleksandrov also arranged for Emmy Noether to lecture at the University of Moscow during the winter of 1928–9.

However, most mathematicians merely tolerated the government, so in an effort to ensure the ideological purity of academia, the government inflicted many communists on the mathematical community. Most proved better at polemics than proofs. G. G. Lorentz, a student at Leningrad University, observed that one of the materialist mathematicians, V. V. Lyush, built an entire theory of numbers of the form $x + iy + jz$, despite the fact that Hamilton had already proven such a system could not be consistent, while the leader and spokesman of the materialists, L. A. Leifert ("a zero of a mathematician" according to Lorentz) began one class by listing the values of several useful integrals, then polling the students to see if they agreed with the formulas.

In February 1931, Leifert attacked the Leningrad Mathematical Society for its acceptance of "bourgeois" mathematics; in consequence, the Society issued a declaration of its new goals: the first three involved purging mathematics of its bourgeois concepts, and only the last two items concerned the promotion of mathematics. Leifert himself fell from power in February 1932 after being accused of deviationism; after some years of exile to the University of Rostov-on-Don (in the Caucasus), he was arrested and shot by the NKVD (state security) sometime in 1938.

Leifert's execution was part of the Great Purge. On December 1, 1934 Sergei Kirov, Stalin's hand-picked leader of the Leningrad Communist Party, was murdered. Stalin accused the assassin, Leonid Nikolaev, of being part of an international conspiracy headed by Trotsky to overthrow the Soviet Union, and empowered NKVD Director Genrikh Yagoda, one of the two survivors of the collectivization committee, to investigate. Over the next four years, some eight million people would be arrested, and more than 300,000 would be executed (not counting those who died after being sent to forced labor camps), including Bukharin and, eventually, Yagoda himself.

Though primarily directed at high ranking members of the Communist Party who might oppose or replace Stalin (there is some evidence that Yagoda arranged Kirov's assassination), the Great Purge reached every aspect of Soviet culture. On January 26, 1936 Stalin and a group of high officials attended a performance of the opera *Katerina Ismailova* (later known as *Lady MacBeth of Mtsensk*), the latest creation of leading Soviet composer Dmitri Shostakovich. But when Stalin and the other officials walked out before the fourth act, Shostakovich collapsed in panic. Two days later, an unsigned editorial titled "Muddle Instead of Music" appeared lambasting the work, which attempted to humanize a greedy and brutal merchant's wife. Shostakovich survived, but took great care to avoid giving offense in future works; most Soviet composers followed suit. By threatening Shostakovich, the authorities were able to bring the entire musical establishment into line.

Shostakovich's counterpart as a leading figure in a discipline was NIKOLAI NIKO-LAEVICH LUZIN (December 9, 1883–February 28, 1950), Egorov's student and close col-laborator. Egorov recognized Luzin's potential and arranged for him to visit Paris, where he met Borel and Lebesgue, and Göttingen, where he met Hilbert. In 1917 Luzin joined Egorov at Moscow University, where they formed a remarkable research group known as *Luzitania*. Its students included Aleksandrov, Shnirelman, and ANDREI NIKOLAEVICH KOLMOGOROV (April 25, 1903–October 20, 1987).

Luzin's work in set and measure theory automatically placed him at odds with the ma-terialists. Formalism suffered a potentially serious blow in 1904, when Zermelo proved a remarkable and counterintuitive result: the continuum can be well-ordered. In particular, *any* set of real numbers has a least element (according to some ordering scheme). The proof relied on the axiom of choice: given any number of disjoint sets, it is possible to construct a new "representative" set consisting of exactly one member from each of the sets. The seemingly paradoxical result lent weight to the arguments of the intuitionists that only finite processes could be considered mathematically rigorous. The Banach-Tarski paradox also relies on the axiom of choice. But few mathematicians could abandon the law of the excluded middle and non-constructive proofs. Some mathematicians (notably Borel, Lebesgue and Baire) suggested a compromise position known as Effectivism, which rejected the axiom of choice but kept the other basic axioms of set theory. When Lebesgue praised Luzin's work, Luzin was marked as an Effectivist, and Luzin's international repu-tation meant that the best-known Soviet mathematician was not materialist.

On June 27, 1936 Luzin wrote an editorial in *Izvestia*, praising the mathematical achieve-ments of a middle school's students. The Director of the School, G. I. Shuliapin, attacked Luzin in a July 2 rebuttal in *Pravda*: Luzin's praise was "unpatriotic," since the students were weak and their instruction poor. A relative nobody like Shuliapin could hardly be ex-pected to attack Luzin without prompting; suspicion focuses on Kol'man, by then head of the science department of the Moscow Soviet.

The Shuliapin letter was the first of many articles detailing Luzin's "crimes." Dialecti-cal materialists might be able to reconcile the contradictory accusations hurled at him: pub-lishing first-rate articles in foreign journals discredited Soviet science; praising the achieve-ments of middle school students was seditious; obtaining a preeminent position in Soviet mathematics because of his research record was unfair; and in any case his research was largely plagiarized from his students. These unpatriotic activities came to be referred to as the crime of *Luzinshchina*.

On July 7, 1936 the Soviet Academy of the Sciences (recently relocated from Leningrad to Moscow) accused Luzin of behavior detrimental to Soviet science and the Soviet Union, and convened a session to investigate the charges. If Luzin was convicted in the Academy, the case would be referred to the NKVD—which had already forced one of Luzin's friends to confess to the existence of a conspiracy to overthrow Stalin and install mathematical physicist SERGEI ALEKSEEVICH CHAPLYGIN (April 5, 1869–October 8, 1942) as Prime Minister and Luzin as Foreign Minister. Referral to the NKVD would be tantamount to a death sentence.

The trial was overseen by Aleksandrov, who managed to have Kol'man excluded from the proceedings and declared that Luzin had never made any anti-Soviet statement, thereby depoliticizing the proceedings. Indeed, of the mathematicians on the investigating commission, only SERGEI LVOVICH SOBOLEV (October 6, 1908–January 3, 1989) and Schmidt were members of the Communist Party. But Aleksandrov had collected a daunting array of mathematicians to testify against Luzin, including Gelfond and Shnirelman. Meanwhile Luzin's sole supporters on the commission were Bernstein; IVAN MATVEEVICH VINO-GRADOV (September 14, 1891–March 20, 1983), whose work in number theory would lead him to a 1937 proof that every sufficiently large odd number is the sum of three primes (consequently every sufficiently large even number is the sum of four primes); and mathematical physicist ALEKSEI KRYLOV (August 15, 1863–October 26, 1945). Kolmogorov testified briefly, noting Luzin's enormous contributions to Soviet mathematics, and admitting that publishing in foreign journals was a standard practice among mathematicians. Better advising and guidance, Kolmogorov said, would have spared Luzin from indiscretions.

Luzin obtained help from a surprising source: Stalin sent a letter to the commission, advising them to continue the vilification of Luzin, but to rephrase the accusations in "academic language" and change them to "behavior unworthy of a Soviet scientists and incompatible with the dignity of a Soviet citizen" [75, p. 206]. The charges of plagiarism should be dropped. Since no one was willing to reject Stalin's "advice," Luzin's punishment was mild. He was dismissed from most of his administrative positions, and the *Great Soviet Encyclopedia* (1938) nowhere mentions him. But he was permitted to continue his mathematical research (and even, a few years later, return to the positions he had been dismissed from).[15] Indeed, the exclusion of Kol'man and Stalin's disinclination to refer the case to the NKVD meant that for the most part, mathematicians who avoided politics could continue their work without interference from dialectical materialists or the government. The main lasting effect concerned where Soviet mathematicians published their work: since one of Luzin's "crimes" was publishing in foreign journals, Soviet publications in foreign journals declined by about a third after 1938.

Meanwhile Kolmogorov rapidly became the star of Soviet mathematics, helping to lay the foundations of the mathematical theory of probability; his standing was so great that he could even criticize Lysenko openly. His role in the Luzin trial endeared him to the authorities, and he established himself as a politically reliable academic by stating, in the *Great Soviet Encyclopedia*, that dialectics played a crucial role in the development of mathematical algorithms. Others reexamined the history of mathematics in an attempt to prove that mathematics was indeed the product of dialectical materialism, but overall, the victory of the materialist mathematicians was largely symbolic.

11.4 World War Two

While in prison for his involvement in the Munich revolt, Hitler wrote the two volumes of *Mein Kampf* (1925 and 1927), which outlined every salient feature of his policies for the next twenty years. He had to write quickly: although sentenced to five years, he served only nine months in the relative comfort of Landsberg castle. Key to Hitler's worldview was a belief in the existence of an international Jewish conspiracy to destroy western civilization.

In *Mein Kampf*, Hitler noted that only one man in America resisted the "controlling masters" of Judaism. He was probably referring to Henry Ford. Though Ford revolutionized

---

[15] Shnirelman, for his part, was arrested by the NKVD in 1938, and committed suicide.

the car industry by paying his workers an unprecedented $5 per day (as well as reducing the workday from 9 hours to 8), he also organized a department to monitor the personal lives of his employees to ensure they lived according to standards he set, and took aggressive action against union organizers. Between 1920 and 1922 Ford's newspaper, the *Dearborn Independent*, published a series of anti-Semitic articles portraying international Judaism as the greatest threat facing western civilization. Much of Ford's evidence was drawn from a pamphlet called the *Protocols of the Elders of Zion*, first printed in 1903 by a Russian newspaper; Ford himself sponsored an American printing in 1920.

The *Protocols* contain detailed information about a Jewish-Masonic plot to take over the world through such nefarious schemes as progressive taxation and temperance laws. Fortunately, Ford and others renewed interest in the pamphlet, prompting *Times of London* reporter Philip Graves to investigate it. On August 16–18, 1921 Graves published an exposé showing that the *Protocols* were mostly plagiarized from Maurice Joly's 1864 satire, *Dialogues in Hell between Machiavelli and Montesquieu*. In Joly's satire Jesuits, not Jews, were the prime conspirators. Claiming he had been duped, Ford retracted his support of the *Protocols*. Despite the fact that the *Protocols* are badly plagiarized from a mediocre satire, belief in their veracity continues to this day.

The failure of the Munich revolt caused Hitler to reconsider his tactics: his next attempt to take control of the government would be through legal means. Over the next few years the Nazi party grew in popularity, and in a frightening display of how easily the democratic process can be subverted, Hitler was given dictatorial powers by the German parliament on March 23, 1933. He declared the formation of the Third Reich (the first and second being the Holy Roman Empire and the German Empire, respectively) and triumphantly declared it would last a thousand years.

Hitler believed in the existence of an Aryan race (related to the Vedic Aryans in name only), responsible for all human progress. This became the Nazi doctrine of the existence and superiority of Aryan science and mathematics: Aryan mathematics was intuitive, geometrical, and based on nature, instead of logical, algebraic, and based on axioms. For example, consider the problem of subtracting $1/2$ from $7\ 1/4$: the logical procedure converts both into fractions with the same denominator and subtracts $29/4 - 2/4$, while the intuitive procedure subtracts $1/4$ from each and performs the subtraction $7 - 1/4$. Presumably Nazi-style Aryans made decisions based on intuition, while non-Aryans used logic.

Logic would say that mathematics and the sciences, where the ultimate test is whether a theory works or fails, are immune to racial distinctions. Nazi intuition dictated otherwise, so on April 7, 1933 the Law for the Reestablishment of the Civil Service allowed for the dismissal of any member of the civil service (which included all university professors) for the crime of political unreliability or being of "non-Aryan descent," defined four days later as anyone with at least one Jewish grandparent. There were exemptions for those who had served in World War One, but these were swept away with the Nuremberg Code of September 15, 1935. Between 1933 and 1937, about half of the 523,000 Jews in Germany left for neighboring European countries, and some of the more prescient fled Europe altogether.

LUDWIG BIEBERBACH (December 4, 1886–September 1, 1982), whose Bieberbach conjecture (1916) concerning the coefficients of the power series expansion of a type of complex function was proved in stages over the next eighty years, became notorious for his persecution of Jewish mathematicians. Since his teacher Klein had brought a number of

Jewish mathematicians to Göttingen, Bieberbach probably felt it important to establish that Klein (who died in 1925) anticipated the existence of Aryan mathematics. Thus Bieberbach pointed to one of Klein's lectures at Northwestern University following the World Fair in Chicago. On September 2, 1893 Klein observed:

> It would seem as if a strong naive space-intuition were an attribute pre-eminently of the Teutonic race, while the critical, purely logical sense is more fully developed in the Latin and Hebrew races. A full investigation of this subject, somewhat on the lines suggested by Francis Galton in his researches on heredity, might be interesting.[16]

However much Bieberbach may have wanted to portray Klein as a proto-Nazi, the facts suggest otherwise. Indeed, Klein's statement (which simply noted the existence of racial differences, rather than implying the superiority of one race over the others) came after a discussion of the role of intuition in geometrical discovery. Klein contrasted the view of MORITZ PASCH (November 8, 1843–September 20, 1930), who believed that intuition could be discarded entirely, with his own belief that intuition played a role in mathematical discovery. This is essentially an early appearance of the debate between Formalism and Intuitionism. Moreover, there is Klein's indisputable record of recruiting talent regardless of ethnic background. Finally, while Klein might have speculated on the existence of racial differences, it is evident from his other writings that he felt that these differences were complementary. For example, he pointed to Jacobi as the forefront of a wave of new mathematicians, of Jewish or French origin, who helped reinvigorate German mathematics.

In 1933, the Mathematical Institute of Göttingen dismissed Emmy Noether. Director and founder Richard Courant was also dismissed, despite the fact that he qualified for an exemption based on his service in World War One (he would have been dismissed in 1935 in any case). He was replaced by his student OTTO NEUGEBAUER (May 26, 1899–February 19, 1990). However Neugebauer lasted only one day: he was not Jewish, and was a veteran besides (enlisting as an artillery lieutenant in World War One to avoid final examinations!), but had opposed the rise of the Nazis and was thus not politically reliable. HERMANN WEYL (November 9, 1885–December 8, 1955) became the new head of the Institute, but was even then planning to leave Germany with his Jewish wife. Altogether eighteen mathematicians left or were dismissed. A year later, the Nazi Minister of Culture Bernhard Rust asked the aged Hilbert if the Mathematics Institute had suffered from the removal of Jews and the politically unreliable; Hilbert replied that it hadn't suffered—it had ceased to exist.

The Rockefeller Foundation, whose 1927 donation gave the Mathematical Institute its own building, was appalled at the dismissal of so many of its key members. At the urging of Veblen and Roland G. D. Richardson of Brown University, the Foundation joined forces with the Institute for International Education and in May 1933 organized the Emergency Committee for Displaced Foreign Scholars. The Emergency Committee included Edward R. Murrow, who left the committee to join CBS in 1935; Murrow's coverage of World War Two would catapult him to international fame.

One difficulty that the Committee had to overcome was the Johnson-Reed Act (1924), which limited the number of immigrants from a country to 2% of the number already in the

---

[16] F. Klein, *The Evanston Colloquium: Lectures on Mathematics*, American Mathematical Society 1911, p. 46

United States in 1890, and prohibited entirely the immigration of people from Asia (adding further tension between the United States and Japan). Since 1890 predated the arrival of significant numbers of people from southern and eastern Europe, it favored northern and central European immigrants: 57,000 Germans could enter each year, but only 4000 Italians. There was a specific exemption for academics, provided they had already secured employment, but this was problematic: Murrow estimated that the Depression had eliminated an estimated 2000 of the 27,000 academic positions. Placing the emigrés in academic jobs would inevitably cause friction between them and have a chilling effect on the future of American mathematics.

Fortunately there was a solution: The Rockefeller Foundation could create endowed research positions for the emigrés. Since only about one-sixth of the thousand or or so mathematics Ph.D.s in the country were actively engaged in research, and since most new Ph.D.s sought teaching or instructorship positions, placing the emigrés in newly created research positions would reduce direct competition. Even so, Wiener, Veblen, Richardson, Birkhoff, and others expressed concern that the new arrivals would take away opportunities for young American mathematicians, causing them to become (in Birkhoff's words, quoting the Bible) "hewers of wood and drawers of water."

Because of the necessity of providing both money and a willing host, the Committee only managed to place 26 mathematicians. Courant went to New York University; Noether to Bryn Mawr college; Weyl to the Institute for Advanced Study at Princeton; Neugebauer went to Denmark and then Brown University in Providence (where he would found *Mathematical Reviews*).

Others who fled Europe (or declined to return after visiting the United States) include Hadamard, Gödel, Tarski, von Neumann, and Menger. Gödel, who was neither Jewish nor politically active, fled across Siberia, to join Weyl and Einstein at Princeton. Perhaps the most significant group of emigrés included Lise Meitner, who discovered fission in 1938 (who emigrated from Austria to England); Enrico Fermi, who built the first self-sustaining nuclear reaction underneath the squash field at the University of Chicago in 1942; Leo Szilard, Eugene Wigner, and Albert Einstein, who sent a letter to Roosevelt describing the dangers of the atomic bomb, should Germany get one first. Funds were voted to begin the Manhattan Project to build the atomic bomb on December 6, 1941, and work on the bomb was materially aided by emigrés like Hans Bethe, Edward Teller, Stanislaw Ulam, and other physicists.

11.4.1 Applied Mathematics

Mathematics in the United States had taken on a decidedly theoretical bent since 1900. SOLOMON LEFSCHETZ (September 3, 1884–October 5, 1972), a Russian who emigrated to the United States in 1901 and lost both hands in an industrial accident in 1907, made important contributions to the fixed point theorem and helped found and name the field of topology (previous authors called it *analysis situs*). Veblen helped promote differential geometry, and his student ROBERT LEE MOORE (November 14, 1882–October 4, 1974), no relation to Eliakim Hastings Moore, helped found point-set topology.

The shift is remarkable because mathematical tradition in the United States began with the work of Bowditch, Peirce, Newcomb, and Hill, all in mathematical physics. GEORGE DAVID BIRKHOFF (March 21, 1884–November 12, 1944) revitalized American interest in mathematical physics, taking inspiration from Poincaré (even while a student of Moore at the University of Chicago); in 1931, Birkhoff proved the ergodic theorem, a crucial part of modern statistical mechanics. Birkhoff's student MARSTON MORSE (March 24, 1892–1977) would develop the eponymous Morse theory. In 1932, Veblen helped organize Princeton's Institute for Advanced Study with Einstein, von Neumann, Weyl, all mathematical physicists, and JAMES WADDELL ALEXANDER (September 19, 1888–September 23, 1971), Veblen's student and a topologist, as founding members.

Interestingly enough, in 1929 Veblen opposed the founding of an AMS journal for applied mathematics, noting in effect that physicists, engineers, chemists, and economists were applied mathematicians. However, the mathematical training of physicists and engineers in the United States was often haphazard. In 1930, THEODORE VON KÁRMÁN (May 11, 1881–May 7, 1963) arrived at the California Institute of Technology (Cal Tech) and campaigned for an improvement in the mathematical education of engineers.

Von Kármán was a Hungarian-born engineer who had studied mathematics under Hilbert and Klein in Göttingen. The rising tide of anti-Semitism and the growing strength of the Nazi party in Germany caused him to accept a position from Cal Tech (he had been there in 1926, supervising the design of a wind tunnel to support the rapidly growing aviation industry in Southern California).

After World War One, engineers in the United States began a construction competition similar to the cathedral building frenzy that struck Europe during the Middle Ages. In 1930, 40 Wall Street (now the Trump Building) and the Chrysler Building were completed, vying for the title of world's tallest building at 952 and 950 feet respectively (discounting purely cosmetic spires that add nothing to usable space). Both would be outdone by the Empire State Building at 1250 feet, completed in 1931. The Depression put a damper on constructing office skyscrapers, but government funds helped build the Hoover Dam (1936) and the Grand Coulee Dam (1941), and promoted the building of bridges.

Unlike buildings or dams, where "size" can be interpreted in various ways, the unequivocal measure of the length of a bridge is the main span (the greatest distance between two of its support structures). The longest bridges in the world are suspension bridges, where the roadway is suspended from cables strung between two towers. The Bear Mountain Bridge, across the Hudson River near Peekskill, opened on Thanksgiving Day 1924; with a main span of 1632 feet, it was longest suspension bridge in the world. It would not hold the title for long, and in 1931, the George Washington Bridge opened in New York City, with a main span of 3500 feet. The "Bridge Wars" of the thirties ended when the Golden Gate Bridge opened in 1937 with a main span of 4200 feet.

The problem is that structures of this size had never before been built, and engineers had little training in physics or mathematics to guide them. Instead, they relied on the extrapolation of tables giving the dimensions necessary to construct safe structures. Moreover a designer might modify the basic structure of a bridge without fully understanding the consequences. For example, Leon Moisseiff, one of the co-designers of the Golden Gate Bridge, promoted the deflection theory of bridge design, which held that the cables and roadway had sufficient mass to resist the force of the wind; consequently, strengthening the deck was unnecessary.

After the Golden Gate Bridge, Moisseiff went on to produce what he regarded as his most elegant design: the Tacoma Narrows bridge over Puget Sound, Washington. With a main span of 2800 feet, only the George Washington and the Golden Gate were longer. Moisseiff's design resulted in a thinner and more elegant bridge than ever before (it is instructive to compare the massive towers of the Brooklyn Bridge with the towers of the Golden Gate), but even during construction, prevailing winds caused undulations of the roadway, earning the bridge the nickname "Galloping Gertie." The bridge opened on July 1, 1940, but on November 7, 1940, undamped oscillations led to the collapse of the bridge around 11 A.M., killing one dog and Moisseiff's career.

The precise reason for the collapse of the six million dollar bridge remains a subject of debate to this day, but von Kármán pointed to a phenomenon he had described in 1911: vortex shedding, which alternately exerts upward and downward forces on a surface. This was an important factor in the design of airplane wings but, von Kármán pointed out, bridge designers did not understand that the same forces affected both airplane wings and bridge roadways. More generally, there was a general lack of mathematical training for engineers. Shortly afterward, cracks began to appear in the Grand Coulee Dam, and von Kármán once again pointed to a lack of understanding of the mathematical and physical theory that led to a flawed design. Fortunately he was able to suggest, and construction crews able to implement, a solution before any disasters occurred.

It was apparent that a new type of mathematician was needed: one who was well-versed in mathematical techniques, and capable of applying them to problems in physics and engineering. Von Kármán once compared working as an applied mathematician to shopping in a warehouse of "mathematical knowledge." The applied mathematician browsed the shelves, seeing how they might make use of the items thereon; alternatively, they might go to the warehouse looking for the tools to solve a particular problem.

Rebuilding the applied mathematics tradition in the United States was made considerably easier by the emigrés. Richardson (of the Emergency Committee) established the first program in applied mathematics in the United States at Brown University in 1941, using emigré faculty members WILLIAM PRAGER (May 23, 1903–March 16, 1980) who arrived via Turkey; WILLIAM FELLER (July 7, 1906–January 14, 1970) who arrived via Sweden; and STEFAN WARSCHAWSKI (April 18, 1904–May 5, 1989) who arrived via the Netherlands. HILDA GEIRINGER (September 28, 1893–March 22, 1973), an Austrian who was about to be appointed as an extraordinary professor in Berlin in 1933, taught at Brown during the summer of 1942 (she had a permanent position at Bryn Mawr college); like Prager, she too arrived in the United States via Turkey. STEFAN BERGMAN (May 5, 1895–June 6, 1977), who fled Germany to Russia before coming to the United States, joined the Brown University group during World War Two.

At New York University Courant founded a research group (later named the Courant Institute) with former student KURT FRIEDRICHS (September 28, 1901–December 31, 1982), who fled Germany because his fiancé was Jewish, and JAMES J. STOKER (1905–1992), another emigré: between the three of them they would supervise the dissertations of nearly a hundred students. RICHARD VON MISES (April 19, 1883–July 14, 1953), who had set up the Institute for Applied Mathematics at the University of Berlin, would arrive in the United States in 1939 also via Turkey and join Harvard's faculty, recruiting Bergman from Brown (Bergman left shortly thereafter for Stanford University), and marrying Geiringer in 1943 (the two had known each other since 1921, when both were at Berlin).

11.4.2   Science Fiction

Von Kármán's efforts to promote applied mathematics notwithstanding, many American mathematicians still felt the draw to theoretical mathematics. Cal Tech colleague ERIC TEMPLE BELL (February 7, 1883–December 21, 1960), proudly noted in 1938 that American algebraists showed a "tendency to abstractness" earlier than anyone else. Bell wrote more than 200 mathematical papers and became President of the MAA in 1930, though he is best-known as an influential historian of mathematics (albeit one with a tendency towards hagiography).[17] Bell shares with Charles Dodgson the distinction of being a popular writer under another name: John Taine. His first novel, *The Purple Sapphire*, appeared in 1924.

Taine is sometimes called a science fiction writer, though properly speaking most of his works are adventure stories with some scientific content (like contemporary Edgar Rice Burroughs, whose John Carter of Mars series was extremely popular in the teens and twenties). Science fiction itself is a relatively new branch of literature, one where science and technology plays a central role in the storyline. Mary Shelley's *Frankenstein* (1818) dealt with the reanimation of a corpse by electricity, and lawyer Jules Verne wrote adventure novels inspired by scientific progress (hence an early term for the field was "scientific romance"). But biologist Herbert George Wells (who studied with Darwin defender Thomas Huxley) is arguably the first writer of modern science fiction, exploring the effects of scientific and technological advances on society.

For example, Wells's first novel, *The Time Machine* (1895) was inspired by an observation by Benjamin Disraeli in his novel *Sybil, or The Two Nations* (1845): the industrial revolution increased the divide between the rich and poor to the point where there were "two nations" in England that had very little to do with one another. Wells turned the sociological division into a biological one: the rich, whose pursuit of pleasure, comfort, and beauty cause them to decay towards the "feeble prettiness" of the surface-dwelling Eloi, while the workers maintained a mechanical civilization even as they evolved into the subterranean Morlocks. Wells's works include "The Stolen Bacillus," (1895) a short story concerning biological terrorism; and *The War of the Worlds* (1898), his reaction towards the colonization of Africa by Europeans, where human natives could do little against the superior technology of the invaders, who succumbed in far greater numbers to diseases.

In 1926 Hugo Gernsback, a Luxembourg-born engineer who emigrated to the United States, began publishing *Amazing Stories*, the first magazine exclusively devoted to "scientifiction." As with mathematics, the existence of a journal caused the field to expand rapidly. It was the era of "pulp fiction," where cheaply printed magazines and books brought entertainment to the masses.

Most of the stories published in the pulp era were of low quality; science fiction was no exception. But by its very nature, the field drew the interest of engineers and scientists, both as writers and readers. Thus a generation of readers was exposed to the possibilities and perils of scientific advance, while a generation of scientists was exposed to the necessity of writing prose. John Wood Campbell published his first story "When Atoms Failed" (1930) while studying physics at MIT, which includes one of the earliest descriptions of electronic computers. "The Man Who Awoke" (1933) by Lawrence Manning noted that our descen-

---

[17] It is worth noting that Bell's *Men of Mathematics* (1937) does *not* specify the arithmetic sequence summed by the young Gauss.

dants may revile us for our reckless consumption of irreplaceable natural resources like coal and oil, while the title character of "A Logic Named Joe" (1946) by Murray Leinster is an electronic service linking the world's libraries and universities that allowed one to find information on any subject, including educational and informational materials, as well as personal information, pornography, and detailed instructions on how to commit crimes.

The more gullible believe that writers who predict the future have some supernatural insight, but "all" that is needed is careful observation of the world around us. For example, the last years of the nineteenth century saw an international competition to create the largest or fastest or most luxurious passenger liner, but no law required these passenger liners to carry enough lifeboats for all passengers. *Futility* (1898), by the American writer Morgan Robertson, described a fictional vessel, the *Titan*: the largest, most luxurious, and fastest ship in the world. The *Titan* sank after striking an iceberg in the North Atlantic, with the loss of nearly all aboard.

On the night of April 14–15, 1912, the *Titanic* sank after striking an iceberg in the North Atlantic, resulting in about 1500 deaths because there were too few lifeboats for all the passengers. The similarities between the fictional *Titan* and the real *Titanic* are often exaggerated: the fictional *Titan* capsized on its third voyage, returning from America, and sank so rapidly that the number of lifeboats was moot, while the *Titanic* sank on its maiden voyage to America, sinking so slowly that it could have been evacuated had there been enough lifeboats. Moreover, Robertson reissued the novel after the *Titanic* disaster with some changes to increase the similarity between the two vessels. Robertson's foresight is only remarkable because it points to the lack of foresight of others.

In 1912 Robertson wrote another short story, based on events in the Far East. With both the United States and Japan seeking to expand their influence in China, conflict of one sort or another seemed inevitable. Robertson's short story, "Beyond the Spectrum," described a future conflict between the United States and Japan. The title refers to a new weapon invented in the United States but used by the Japanese: searchlights using ultraviolet light (hence "beyond the spectrum") to blind ship crews. The Japanese had shown their predilection for surprise attacks during the Russo-Japanese War, so in the story the hero singlehandedly (using the weapon from a captured Japanese submarine) stops a Japanese fleet preparing to launch a surprise attack on the United States.

11.4.3 Appeasement and War

Events in Europe moved rapidly after Hitler's ascent to power. On March 12, 1938 Germany annexed Austria and no one objected. Hitler's demand for the (mostly German) Sudetenland led to the dismemberment of Czechoslovakia at Munich on September 30, 1938. Poland and Hungary took the opportunity to seize some disputed territories, leaving Czechoslovakia in a state of chaos, and on March 15, 1939 the German army entered Czechoslovakia and annexed the rest of the country. After perfunctory demands for the city of Gdansk (assigned to Poland by the Treaty of Versailles, despite its largely German population), the army of the Third Reich entered Poland from the west on September 1, 1939, beginning World War Two. On September 17, the army of the Soviet Union entered Poland via the eastern border so shortsightedly created by the Treaty of Riga; by September 29, Poland had ceased to exist as a nation. The battlefront moved so rapidly that reporters used the term *blitzkrieg* ("lightning war") to describe the situation, though the term was not used by the military.

In contrast to the blitzkrieg that conquered Poland, so little happened in the fall of 1939 and winter of 1940 that these few months were called the "Phony War," or *sitzkrieg* ("sitting war," a parody of *blitzkrieg*). The Soviet Union invaded Finland and the Nazis regrouped and rearmed. The British dropped leaflets and the French waited behind the Maginot Line. *Sitzkrieg* ended with blitzkrieg as Denmark and Norway were invaded on April 9, 1940. The Danes offered only token resistance: Hitler's prewar rhetoric led them to believe that they exemplified the Aryan ideal so cherished by the Nazis. Norwegian resistance was thwarted by Vidkun Quisling. During the invasion Quisling, leader of the National Unity party, declared a new government, and canceled mobilization orders sent to the Norwegian army. By April 15, his name had entered the English language as a synonym for traitors and collaborationists.

Because of the Nazi invasion of Norway, MAX DEHN (November 13, 1878–June 27, 1952), who solved Hilbert's third problem in 1904 by showing the volume of a tetrahedron could *not* be found using finite decomposition, fled across the Soviet Union via the Trans-Siberian Railroad. After arriving in San Francisco on January 1, 1941, Dehn accepted short term positions at a number of institutions around the United States (including St. John's College, in Annapolis, Maryland), but settled at Black Mountain College, an experimental school a few miles northeast of Asheville, North Carolina. Although Black Mountain (which closed in 1956) offered no degrees and had no mathematics program, Dehn apparently enjoyed his time there, since the college's loose structure allowed him to teach not only mathematics, but philosophy, Latin, Greek, and other subjects of interest.

On May 10, Germany invaded Belgium, Luxembourg, and the Netherlands, which fell quickly to the Nazis. Another casualty was the government of Neville Chamberlain, who was succeeded on May 13 as Prime Minister by Winston Churchill. On May 17, the Germans entered France, bypassing the Maginot Line by cutting through the Ardennes forest (believed impassable by tanks). A heroic effort at the evacuation of Dunkirk (May 26–June 4, 1940) saved 340,000 British and French soldiers, at the cost of nearly all their equipment. Italy declared war against France and Britain on June 10.

Far too many Frenchmen preferred Nazi occupation to war. For the liberals, the horrors of the First World War had to be avoided at all costs. For the conservatives, the totalitarian state was the only feasible solution to France's problems. Laval (Painlevé's Minister of Public Works) orchestrated the resignation of Prime Minister Paul Reynaud and his replacement with Pétain. Anticipating that Pétain would surrender to the Germans, Charles de Gaulle (Reynaud's Minister of War) fled to England; on June 17, broadcasting over the BBC, de Gaulle called Frenchmen to resist the invaders. De Gaulle's fears came true on June 22 when Pétain signed an armistice with the Nazis. The cost of peace was three-fifths of France (mostly in the north and west, including Paris). The remainder, with its capital at Vichy, was left in the hands of a puppet French government. Laval, who had long schemed to make Pétain a figurehead Prime Minister while retaining control of the government, convinced the Chamber of Deputies to give Pétain nearly dictatorial power. The Vichy regime began by arresting its political opponents, including Borel, who served some time in prison before being released. Borel would eventually join the French Resistance.

11.4.4   The Two Enigmas

The invasion of Poland surprised no one who had read *Mein Kampf*, where Hitler outlined the need of the Germans to expand eastward for land and raw materials. But no natural boundary separates Poland from the Soviet Union, and it was only a matter of time before Hitler turned on his ally. Hence Churchill noted wryly in a radio address on October 1, 1939 that Soviet foreign policy (such as its support of Hitler) was "a riddle, wrapped in a mystery, inside an enigma."

A second enigma played a crucial role in World War Two: the Enigma cipher machine, a commercial device subsequently modified for use by the German military. In the years preceding World War Two, Polish Military Intelligence had gained an understanding of the encryption methods used by the Germans. In 1928 at the University of Poznan, a group of students took a special course in cryptology. The students included MARIAN REJEWSKI (August 16, 1905–February 13, 1980).

Previous codebreaking efforts relied on statistical analysis, but the Enigma machines used a polyalphabetic cipher with an enormous key length that made statistical analysis nearly impossible. Instead Rejewski applied the theory of permutation groups, though the exact techniques remained classified until the 1970s, and Rejewski's "An application of the theory of permutations in breaking the Enigma cipher" did not appear in *Applicationes Mathematicae* until 1980.

Mathematically, a polyalphabetic cipher can be represented by a sequence of permutations A, B, C, ..., where the permutation A is applied to the first letter, B to the second, and so on. Because of the construction of the Enigma machine, each permutation can be derived from the preceding one according to a fixed rule which can itself be described as a permutation G. Thus $B = GA$, $C = G^2 A$, and so on.[18] Since the plaintext is unknown, this knowledge is of no use. However, for technical reasons Enigma operators began each message with a set of three letters, repeated twice; the triplet would be different for different messages and for different operators. If enough messages were intercepted, some of them would contain the same letters. For example, one message might begin *cks dkl*, the second *dqx tfp*, and the third *teg eoq*. Because the first and fourth letters of the plaintext are the same, then (from the first message) the encrypted c and d must correspond to the same letter. Likewise, d and t in the second message correspond to the same letter, and t and e.

Suppose the first letter of the first message is x. The first message tells us two things: $Ax = c$, and $G^3 Ax = d$. Putting these two together we find $G^3 c = d$, which we can determine even if we have no idea what x actually is. Suppose the first letter of the second message is y. The second message tells us $Ay = d$, and $G^3 Ay = t$, and putting these two together we find $G^3 d = t$. Likewise the third message tells us $G^3 t = e$. Hence we know that G^3 contains a cycle that begins $(cdte\,...)$. In this way we can reduce the number of possibilities of G^3 from 26! down to a more manageable number (though still in the hundreds). This type of analysis then produced several hundred possible keys, one

---

[18]This is a drastic oversimplification, though it does not affect the basic approach. As Rejewski described it, the actual Enigma cipher is equivalent to the permutations of the form $A = SPNP^{-1}QPN^{-1}P^{-1}S^{-1}$, $B = SP^2NP^{-2}QP^2N^{-1}P^{-2}S^{-1}$, and in general the kth letter of the message is encrypted by the permutation $SP^k NP^{-k}QP^k N^{-1}P^{-k}S^{-1}$, where S, N, Q are permutations that correspond to the initial settings of the Enigma machine, and P is the rot-1 permutation.

of which (hopefully) would serve to decrypt the message. To test these keys, Rejewski designed and built a device called (for no clear reason) the "Bomb."

After the fall of Poland, Rejewski and his colleagues eventually joined British cryptanalytic efforts. The task of breaking the Enigma codes was given to a group at Bletchley Park which, for bureaucratic reasons, excluded the Polish cryptographers. Despite this, they made remarkable progress, at least in part due to the presence of MAXWELL HERMAN ALEXANDER NEWMAN (February 7, 1897–February 22, 1984) and ALAN MATHISON TURING (June 23, 1912–June 7, 1954). While at Princeton, Turing applied Newman's and Gödel's ideas to questions about the real numbers. "On Computable Numbers" (1937) described what is now called a universal Turing machine which could be fed a set of instructions and output a real number. By consideration of how the instructions could be described, Turing showed that the number of computable numbers was countable.

At Bletchley Park, Turing modified the Bomb to produce the Bombe, while Newman helped design the Colossus, one of the earliest computers.[19] By breaking Enigma, the British gained considerable foreknowledge of German military plans. Unfortunately, even this information might not be enough: the British could break the Enigma *cipher*, but the German *codes* proved elusive. Thus the British knew that the Germans planned an attack against a target in late 1940, but the target's identity was concealed under the code name *Korn*. On the night of November 14, 1940, *Korn*'s identity became apparent when the city of Coventry (a major industrial center) was hit with a massive bombing raid, killing about 600 civilians and destroying sixty thousand buildings, including the historic cathedral and 75% of the city's factories.

Although devastating, bombing civilians and industrial centers proved less effective than Hitler would have hoped, and the losses of Luftwaffe aircrew could not be sustained. As in World War One, Britain could hold out as long as she maintained control of the seas. In July, the British neutralized the French fleet, seizing ships that had fled to Britain and sinking or disabling those in North Africa.

The fate of the Italian navy was more portentous. On November 11, 1940 twenty-one antiquated British planes flying off the aircraft carrier HMS *Illustrious* attacked the Italian fleet at Taranto and sank one battleship, damaged two others and a light cruiser at the cost of two planes and four crew (two captured and two lost). Taranto not only proved the possibility of shallow water torpedo attacks, hitherto believed impossible, but it also showed that airplanes could inflict decisive damage on an unprepared opponent.

Meanwhile, Hitler invaded the Soviet Union on June 22, 1941. Within five days, some units had advanced as far as Minsk, 300 kilometers inside the border. Just behind the advancing *Wehrmacht* were members of the *Abwehr* (German Intelligence) and the 3000 members of the *Einsatzgruppen* ("Deployment groups"), who reported directly to Reinhard Heydrich, the notorious Nazi at the head of German State Security (including the Gestapo). The sole purpose of the *Einsatzgruppen* was the extermination of Jews, Gypsies, and Commissars who had the misfortune to be in German-occupied territory. Bletchley

---

[19]The question of "Who designed the first computer?" is very much like asking where mathematics originated: it depends on your definition. A computer processes information. If that processing is done without moving parts, the computer is said to be electronic; if that processing requires moving parts, the computer is mechanical. Thus the abacus is a mechanical computer, while a modern computer—even though it has moving parts (the disk drive)—is classified as electronic, since the information processing requires no moving parts. The Bomb and Bombe were electromechanical computers: they had moving parts, powered by electricity.

Park cryptanalysts decoded some *Einsatzgruppen* reports about these killings, but with Britain fighting for her survival, nothing could be done.

Since dictatorship flourishes best among an ill-educated population, the Nazis launched a systematic program to annihilate Polish academia. Events in Lviv were typical. On July 2, 1941 the Nazis arrested Polish mathematician and former Prime Minister KAZIMIERZ BARTEL (March 3, 1882–July 26, 1941). Over the next month the *Abwehr* and German army massacred Polish professors and their families, including many of the department chairs and institute heads at the University of Lviv and the Lviv Polytechnic School. Łomnicki and about fifty others died on July 4; Stanislaw Ruziewicz, one of Sierpiński's students, was murdered on July 12; Bartel was shot on July 26 after rejecting a German offer to head a collaborationist government.

The massacres of the *Einsatzgruppen* and *Abwehr* pale in comparison to the systematic program of deportation, forced labor, and extermination outlined by Heydrich at an eighty minute conference at Wannsee, on the outskirts of Berlin, on January 20, 1942. The Final Solution (to the "problem" of the existence of Jews in Europe) involved an initial relocation of Jews to ghettoes, prior to their removal to slave labor camps or (in the cases of those too unhealthy to be worked to death) extermination camps.

Some Jews, like Hausdorff, took the path of Condorcet: Hausdorff, his wife, and his sister-in-law were interned at Endenich, a suburb of Berlin. After writing a note to a friend that included the grim play on words "Endenich is not the end" (*Endenich ist Ende nicht*), Hausdorff, his wife, and his sister-in-law committed suicide. ALFRED TAUBER (November 5, 1866–late 1942), of Tauberian Integral fame, died at the concentration camp in Theresienstadt, as did Otto Blumenthal. ALFRED RÉNYI (March 30, 1921–February 1, 1970), who would make substantial progress towards proving Goldbach's Conjecture (eventually proving every even number could be written as the sum of a prime and the product of *a* primes) was more fortunate: he was interned at the Budapest ghetto in 1944 but managed to escape. To rescue his parents, he obtained a soldiers uniform, walked into the ghetto, and demanded they be released into his custody. The ruse worked: in a totalitarian state, one dare not question the orders of authority.[20]

11.4.5 The United States

The closure of the Black Chamber by Stimson simply meant the State Department had no codebreakers. Army and Navy codebreakers continued their work and later, when the State Department decided that they needed to have the extra insight provided by decoded transmissions, they turned to the military. Since extra responsibility implies extra budget and extra prestige, neither service was willing to let the other do all the codebreaking, and the state department began receiving decoded intercepts from the army on even numbered days and the navy on odd numbered days. In 1940 two of Dickson's students, OLIVE CLIO HAZLETT (October 27, 1890–March 8, 1974) and ABRAHAM ADRIAN ALBERT (November 5, 1905–June 6, 1972) joined the Cryptanalysis Committee of the AMS, assisting the Army's Signal Intelligence Service (SIS); other mathematicians also joined the code breaking efforts (though many of the details remain classified).

---

[20] Rényi is credited with defining a mathematician as a device to turn coffee into theorems.

On September 24, 1940 the SIS completed a machine that could decode Japanese diplomatic messages encrypted using the new Purple diplomatic code. It was a technical marvel (the code name for the intercepts was "Magic"): in modern terms, SIS reverse engineered the machine entirely from intercepted messages. From that point on, the United States had nearly complete information about Japanese intentions, in some cases decrypting messages before their intended recipients.

In the early morning hours of December 7, 1941 the Navy intercepted a short message to the Japanese ambassador in Washington instructing him to break diplomatic relationships with the United States at 1:00 PM, Eastern time; along with orders to destroy all remaining ciphering equipment and codebooks, it signaled that Japan was about to go to war with the United States. Though the Navy decoded the message by 5 A.M., it was not until 11 A.M. that it was read by the Army Chief of Staff, General George C. Marshall. Marshall considered the most likely target to be in the Philippines, or possibly Panama, and ordered warnings sent to United States military installations around the world.

Unfortunately weather conditions prevented a radio message from being sent to Hawaii; instead, a message was sent by Western Union. It was delivered to the naval base at Pearl Harbor at 2:40 PM, Hawaiian time. By then, 350 planes from the Japanese navy had attacked Pearl Harbor. Eighteen ships were either sunk or badly damaged, and more than two thousand military personnel had been killed (as well as 68 civilians).

The architect of Pearl Harbor, Admiral Isoroku Yamamoto, was very familiar with the United States, having spent two years (1919–1921) studying the American oil industry at Harvard University (during Morse's time as Benjamin Peirce Lecturer at Harvard, though it is unlikely that their paths ever crossed), and hitchhiking from Boston to Mexico during one vacation. This firsthand experience with the country left him with one unshakable opinion: if it came to war, he might be able to "run wild for six months, maybe a year," but beyond that he could make no guarantees of success. Exactly six months after Pearl Harbor, the Japanese Navy was effectively destroyed as a fighting force at the Battle of Midway (June 4–7, 1942), in large part due to the work of cryptanalysts.

In the aftermath of Pearl Harbor, the means by which the United States acquired and decrypted coded communications underwent a complete reorganization at the request of the Secretary of War, the very same Henry Stimson who shut down the Black Chamber in 1929. The newly reorganized intelligence service would eventually become the National Security Agency (NSA), an extremely secretive branch of the United States government that is said to employ more mathematicians than any other organization in the world.

Other wartime research projects were overseen by Vannevar Bush at the Office of Scientific Research and Development (OSRD). In Fall 1942, Bush reorganized the National Defense Research Committee (NDRC), the largest division within the OSRD, to include an Applied Mathematics Panel (AMP). Warren Weaver, the Chief of the AMP, and MINA REES (August 2, 1902–October 25, 1997), on leave from Hunter College, helped turn the problems posed to the OSRD into mathematical problems, and identified the researchers and institutions most likely to be able to solve them using the AMP's connections with eleven universities and the Mathematical Tables Project, taken over from the WPA, though Blanch and Lowan remained in charge.

At its peak the Mathematical Tables Project employed 450 computers, working on calculations applicable to a wide range of problems: the soundproofing of aircraft; the scattering of radar waves; values for Bessel, Struve, hypergeometric, and binomial distribution functions; and others. In the fall of 1943 an optimization problem arrived from emigré JERZY NEYMAN (April 16, 1894–August 5, 1981), one of the founders of modern theoretical statistics; the computers solved it by mid-December 1943 without knowing precisely *what* was being optimized. The greatest invasion in history was being planned, and the army needed a bombing pattern that would maximize the number of mines destroyed on the beaches of Normandy.

Although some mathematicians later said they did no mathematics during the war, others used the opportunity to improve the state of applied mathematics in the United States. Hilda Geiringer at Bryn Mawr and the applied mathematics group at Brown University worked on classified problems for the AMP (mainly connected with the mathematical theory of deformable materials, pioneered by Geiringer during her time in Turkey). Courant, Friedrichs, and others at New York University worked on fluid dynamics; in 1948 they would publish *Supersonic Flows and Shock Waves*, the first modern textbook on the subject. At Harvard, David Birkhoff's son GARRETT BIRKHOFF (January 19, 1911–November 22, 1996) worked on problems in underwater ballistics.

Mathematical physics was not the only beneficiary of war research. An important advance in statistical analysis originated from a problem in munitions testing where batches are accepted or rejected based on the results of test firings of large samples. But this is tedious, costly, and occasionally dangerous work, and in some cases it might be possible to accept or reject the batch based on the first few test firings. For example, if the failure rate of a batch is supposed to be less than 1 in 1000, but the first two items tested fail, might one be justified in rejecting the batch? Experienced ordnance testers like Garrett Schuyler, a Captain in the Navy, had rules of thumb they used to accept or reject production runs based on the results of the first few tests, but Schuyler asked the AMP if there was some mechanical rule that would allow someone with less experience to make an equally valid decision. Future Nobel prize winning economist Milton Friedman, working as a statistician for the AMP, and mathematician Allen Wallis began to examine the problem. Friedman came up with an example that suggested there might be some practical way to answer Schuyler's question, but they realized that the mathematics were beyond what they were trained for. They passed the problem to emigé ABRAHAM WALD (October 31, 1902–December 13, 1950), who would develop the theory of sequential analysis from this preliminary work.

After Germany declared war on the United States in the aftermath of Pearl Harbor, the war in Europe became a contest between the ability of the United States to supply Britain and the Soviet Union, and the ability of the Germans to wear them down by attrition. History repeated itself. Napoleon, unable to conquer Britain, invaded Egypt, then Russia, and suffered disastrous losses. Hitler, unable to conquer Britain, invaded Egypt, then the Soviet Union, and suffered disastrous losses. Germany, in World War One, failed to keep the United States out of the war, and could not sink convoys faster than they could be built; Germany, in World War Two, failed to keep the United States out of the war, and could not sink convoys faster than they could be built. On April 30, 1945 Hitler committed suicide, and on May 8 his successor, Admiral Karl Doenitz, arranged for the unconditional surrender of all German forces.

The war in the Pacific continued, but once again the virtually untouchable industrial base of the United States proved decisive and the eventual defeat of Japan was inevitable.

However, what might have been a long struggle against Japan at the side of the Soviet Union ended abruptly following the dropping of two atomic bombs on Hiroshima (August 6) and Nagasaki (August 9). In an unprecedented step, Emperor Hirohito of Japan addressed his people over the radio on August 15 and announced the end of the war. An estimated 62 million people died during the war; large sections of Europe, Asia, and Africa were destroyed; and the existence of the atomic bomb cast a shadow over the postwar era.

For Further Reading

Additional material on twentieth century history can be found in [5, 6, 38, 75, 80, 83, 102, 103, 126, 127].

Epilog

In a very real sense, the history of the world prior to the world wars could be treated as a collection of individual regional histories: the history of Asia, the history of the Americas, the history of Europe, and so on. Occasionally one region might influence another: the Mongol invasion of Europe; the scientific accomplishments of the Islamic states; European imperialism in the Far East. But the effects were either brief (the Mongol Empire virtually vanished within a generation), limited (European colonies in China were mainly limited to the coastal regions), or episodic (Islamic mathematics played a formative role in European mathematics, but almost no role after 1600). After World War II, it becomes impossible to treat each region of the world separately. The Soviet invasion of Afghanistan in 1979 led to the creation of the terrorist group that staged the attacks on the World Trade Center in New York City on September 11, 2001. Changes in the lending laws in the United States during the 1980s led to a world-wide financial crisis in 2008. Uncertainty in Middle Eastern affairs causes renewed interest in the resources of sub-Saharan Africa. We can no longer disentangle the history of one part of the globe from another.

Additionally, mathematics has become ever more specialized and in turn requires even more specialized knowledge to understand its applications. The most obvious example is the growth of computerization after World War II, which has led to new fields of mathematical research and new applications of old ideas: fractals, information theory, compression algorithms, operations research, and encryption schemes just to name a few.

Perhaps the most unexpected example is the importance of mathematics in economics. When von Neumann and OSCAR MORGENSTERN (January 24, 1902–July 26, 1977) published *Theory of Games and Economic Behavior* (1944), their application of mathematics to economics was unprecedented. Papers on economics rarely included graphs, let alone equations. Twenty-five years later, the first Nobel Prize in Economics was awarded to RAGNAR FRISCH (March 3, 1895–January 31, 1973) and JAN TINBERGEN (April 12, 1903–June 9, 1994) for their work on developing mathematical models of economic activity. Other prize winners have included KENNETH JOSEPH ARROW (b. August 23, 1961) who shared the 1972 prize for his research on the mathematics of social welfare; JOHN FORBES NASH (b. June 13, 1928), REINHARD SELTEN (b. October 5, 1930), and JOHN HARSANYI (May 29, 1920–August 9, 2000), who shared the 1994 prize for their work on game theory; and ROBERT C. MERTON (b. July 31, 1944) and MYRON S. SCHOLES (b. July 1, 1941), who shared the 1997 prize for their work on the mathematics of stock options. Merton, Scholes, and FISCHER BLACK (January 11, 1938–August 30, 1995), who would have received a share of the 1997 prize had he lived, are the main persons responsible for the explosive growth of financial mathematics over the past twenty years.

To tell the history of the last fifty years would take more space than the history of the previous fifty centuries. Thus, we choose to end the book here, in the aftermath of the destruction of World War II. Those who survived the war likely asked the same question we might ask today: What next?

Bibliography

[1] C. N. Anderson, *The Fertile Crescent*, Sylvester Press, 1968.

[2] E. Balazs, *Chinese Civilization and Bureaucracy*, Yale University Press, 1964.

[3] W. C. Bark, *Origins of the Medieval World*, Stanford University Press, 1958.

[4] M. E. Baron, *The Origins of the Infinitesimal Calculus*, Dover, 1969.

[5] J. L. Bates, *The United States, 1898–1928*, McGraw-Hill, 1976.

[6] F. L. Benns, M. E. Seldon, *Europe: 1914–1939*, Meredith Publishing, 1965.

[7] J. L. Berggren, *Episodes in the Mathematics of Medieval Islam*, Springer-Verlag, 1986.

[8] R. E. Berry, *Yankee Stargazer*, McGraw-Hill, 1941.

[9] N. H. Bingham, P. Holgate, Studies in the History of Probability and Statistics XLV. The Late Philip Holgate's Paper 'Independent Functions: Probability and Analysis in Poland Between the Wars', *Biometrika*, Vol. 84, No. 1 (Mar. 1997), pp. 159–173.

[10] B. Birkeland, Ludvig Sylows Lectures on Algebraic Equations and Substitutions, Christiania (Oslo), 1862: An Introduction and a Summary, *Historia Mathematica*, Vol. 23 (1996), 182–199.

[11] F. C. Bourne, *A History of the Romans*, D. C. Heath, 1966.

[12] C. Boyer and U. C. Merzbach, *A History of Mathematics*, John Wiley and Sons, 1991.

[13] F. Braudel, *Capitalism and Material Life*, Harper, 1973.

[14] R. Briggs, *Early Modern France, 1560–1715*, Oxford University Press, 1998.

[15] S. C. Burchell, *Imperial Masquerade: The Paris of Napoleon III*, Atheneum, 1971.

[16] P. Burke, *Popular Culture in Early Modern Europe*, Harper and Row, 1978.

[17] F. Cajori, *The Teaching and History of Mathematics in the United States*, Government Printing Office, 1890.

[18] V. G. Childe, *New Light on the Most Ancient East*, Grove Press, 1952.

[19] C. M. Cipolla, *Before the Industrial Revolution*, W. W. Norton, 1976.

[20] C. M. Cipolla, *Guns, Sails, & Empires*, Pantheon, 1965.

[21] M. Clagett, *The science of mechanics in the Middle Ages*, University of Wisconsin Press, 1961.

[22] M. Clagett, *Ancient Egyptian Science: A Source Book*, Diane Publishing, 1989.

[23] W. E. Clark (trans.), *The Āryabhaṭīya of Āryabhaṭa*, University of Chicago Press, 1930.

[24] H. F. Cohen, *Quantifying Music*, Kluwer Academic Publishers, 1984.

[25] V. Cronin, *The Florentine Renaissance*, E. P. Dutton, 1967.

[26] B. Datta, *The Science of the Sulba*, University of Calcutta, 1932.

[27] B. Datta, On the Relation of Mahavira to Sridhara, *Isis*, Vol. 17, No. 1 (1932), 25–33.

[28] E. S. Duckett, *Death and Life in the Tenth Century*, University of Michigan Press, 1988.

[29] R. S. Dunn, *The age of religious wars, 1559–1715*, Norton, 1979.

[30] C. Dyer, *Standards of living in the later Middle Ages*, Cambridge University Press, 1989.

[31] J. Elkins, Piero della Francesca and the Renaissance Proof of Linear Perspective, *The Art Bulletin*, Vol. 69, No. 2. (Jun., 1987), pp. 220–230.

[32] J. Fauvel and J. Gray, ed., *The History of mathematics: a reader*, Macmillan Press, 1987.

[33] C. P. Fitzgerald, *A Concise History of East Asia*, Praeger, 1966.

[34] K. Freeman, *Greek City-States*, Norton, 1963.

[35] D. J. Geanakoplos, *Medieval Western civilization and the Byzantine and Islamic worlds: interaction of three cultures*, D. C. Heath, 1979.

[36] J. Gernet, *Daily Life in China on the Eve of the Mongol Invasion, 1250–1276*, Stanford University Press, 1970.

[37] F. Gies, J. Gies, *Cathedral, Forge, and Waterwheel*, Harper Collins, 1994.

[38] F. Gilbert, *The End of the European Era, 1890 to the Present*, W. W. Norton, 1984.

[39] J. Gimpel, *The Medieval Machine*, Penguin Books, 1976.

[40] C. C. Gillespie, *Dictionary of Scientific Biography*, Scribner, 1970–1980.

[41] R. J. Gillings, *Mathematics in the Time of the Pharaohs*, MIT Press, 1972.

[42] L. C. Goodrich, *A Short History of the Chinese People*, Harper, 1958.

[43] A. Goodwin, *The French Revolution*, Harper, 1962.

[44] J. R. Hale, *Florence and the Medici*, Thames and Hudson, 1977.

[45] A. Harding, *England in the Thirteenth Century*, Cambridge University Press, 1993.

[46] T. L. Heath, *Elements*, E. P. Dutton and Co., 1933.

[47] T. L. Heath, *Diophantus of Alexandria*, Cambridge University Press, 1885.

[48] T. L. Heath, *A History of Greek Mathematics*, Clarendon Press, 1921.

[49] T. L. Heath, *The Works of Archimedes*, Cambridge University Press, 1912.

[50] H. Heller, *Labour, science, and technology in France, 1500–1620*, Cambridge University Press, 1996.

[51] J. C. Herold, *The Age of Napoleon*, American Heritage Publishing, 1963.

[52] H. V. Hillprecht, *The Babylonian Expedition of the University of Pennsylvania: Series A, Cuneiform Texts*, University of Pennsylvania, 1906.

[53] E. Hobsbawm, *The Age of Revolution, 1789–1848*, Random House, 1962.

[54] M. G. S. Hodgson, *The Venture of Islam (3 vol.)*, University of Chicago, 1974.

[55] C. Warren Hollister, *Medieval Europe: A Short History*, Alfred A. Knopf, 1982.

[56] A. Hourani, *A History of the Arab Peoples*, Belknap Press, 1991.

[57] S. K. Jayyusi, ed., *The Legacy of Muslim Spain*, E. J. Brill, 1992.

[58] D. Jensen, *Renaissance Europe*, D. C. Heath, 1981.

[59] N. Jolley, ed., *The Cambridge companion to Leibniz*, Cambridge University Press, 1995.

[60] T. B. Jones, *From the Tigris to the Tiber*, Dorsey Press, 1978.

[61] D. Kahn, *The Codebreakers*, Macmillan, 1967.

[62] S. Katz, *The Decline of Rome*, Cornell University Press, 1955.

[63] V. Katz, *A History of Mathematics*, Addison-Wesley, 1998.

[64] V. Katz, *The Mathematics of Egypt, Mesopotamia, China, India, and Islam*, Princeton University Press, 2007.

[65] P. M. Kendall, *Louis XI*, Cardinal, 1974.

[66] E. Keuls, The Apulian 'Xylophone': A Mysterious Musical Instrument Identified, *American Journal of Archaeology*, Vol. 83, No. 4, (Oct., 1979), pp. 476–477

[67] G. E. Kirk, *A Short History of the Middle East*, Praeger, 1964.

[68] S. N. Kramer, *The Sumerians*, the University of Chicago Press, 1963.

[69] G. W. Leibniz, *Mathematische Schriften*, ed. C. I . Gerhardt, 1855, Olms, 1971.

[70] M. Levey, The Encyclopedia of Abraham Savasorda: A Departure in Mathematical Methodology, *Isis* Vol. 43, No. 3. (Sep., 1952), pp. 257–264.

[71] M. Levey, Abraham Savasorda and His Algorism: A Study in Early European Logistic, *Osiris*, Vol. 11 (1954), p. 50–64.

[72] A. E. Levin, Anatomy of a Public Campaign: "Academician Luzin's Case" in Soviet Political History, *Slavic Review*, Vol. 49, No. 1 (Spring 1990), pp. 90–108.

[73] J. Lindsay, *The troubadours & their world of the twelfth and thirteenth centuries*, F. Muller, 1976.

[74] A. S. Link, *Woodrow Wilson and the Progressive Era*, Harper & Row, 1954.

[75] G. G. Lorentz, Mathematics and Politics in the Soviet Union from 1928 to 1953, *Journal of Approximation Theory*, Vol. 116 (2002), pp. 169–223.

[76] C. Manceron, *Age of the French Revolution* (5 vols.), Simon and Schuster, 1972–1989.

[77] M. Masi, ed., *Boethius and the Liberal Arts*, Peter Lang, 1981.

[78] G. Mattingly, *The Armada*, Houghton Mifflin, 1959.

[79] W. H. McNeill, *Venice, the hinge of Europe, 1081–1797*, University of Chicago Press, 1974.

[80] K. Menger, *Reminiscences of the Vienna Circle and the Mathematical Colloquium*, ed. L. Golland, B. McGuinness, and A. Sklar, Kluwer, 1994.

[81] Y. Mikami, *The Development of Mathematics in China and Japan*, Chelsea, 1913.

[82] D. Morgan *The Mongols*, Blackwell Publishers, 1986.

[83] R. Murawski, Philosophical reflection on mathematics in Poland in the interwar period, *Annals of Pure and Applied Logic*, Vol. 127 (2004), 325–337.

[84] O. Neugebauer, *The Exact Sciences in Antiquity*, Dover, 1962.

[85] O. Neugebauer, *Mathematical Cuneiform Texts*. American Oriental Society, 1945.

[86] I. Newton, *Mathematical Works. Assembled with an Introduction by Derek T. Whiteside*, Johnson Reprint Corporation, 1964–7.

[87] H. Nicolson, *The Congress of Vienna*, Viking Press, 1946.

[88] A. Özdural, A Mathematical Sonata for Architecture: Omar Khayyam and the Friday Mosque of Isfahan. *Technology and Culture*, Vol. 39, No. 4. (Oct., 1998), pp. 699–715.

[89] A. Özdural, On Interlocking Similar or Corresponding Figures and Ornamental Patterns of Cubic Equations. *Muqarnas*, Vol. 13 (1996), pp. 191–211.

[90] S. R. Packard, *12th century Europe; an interpretive essay*, University of Massachusetts Press, 1973.

[91] G. Parker, *The military revolution : military innovation and the rise of the West, 1500–1800*, Cambridge University Press, 1996.

[92] K. Pearson, A. de Moivre, R. C. Archibald. A Rare Pamphlet of de Moivre and Some of His Discoveries. *Isis*, Vol. 8, No. 4 (Oct. 1926), pp. 671–683.

[93] L. M. Paterson, *The world of the troubadours: medieval Occitan society, c. 1100–c. 1300*, Cambridge University Press, 1993.

[94] P. Pesic, Secrets, Symbols, and Systems: Parallels between Cryptanalysis and Algebra, 1580–1700. *Isis*, Vol. 88, No. 4. (Dec., 1997), pp. 674–692.

[95] R. L. Pounds, *The Development of Education in Western Culture*, Meredith Corporation, 1968.

[96] N. V. Riasanovsky, *A History of Russia*, Oxford University Press, 1963.

[97] P. Riché, *Daily Life in the World of Charlemagne*, trans. JoAnn McNamara, University of Pennsylvania Press, 1978.

[98] G. Rodis-Lewis, trans. J. M. Todd, *Descartes: his life and thought*, Cornell University Press, 1998.

[99] J. E. Rodes, *The Quest for Unity: Modern Germany 1848–1970*, Holt, Rinehart, Winston 1971.

[100] E. H. Roseboom, *A History of Presidential Elections*, Macmillan, 1970.

[101] F. Rosen, *The algebra of Mohammed ben Musa*, J. Murray, 1831.

[102] D. E. Rowe, "Jewish Mathematics" at Göttingen in the Era of Felix Klein, *Isis*, Vol. 77, No. 3 (Sep. 1986), pp. 422–449.

[103] D. E. Rowe, Klein, Hilbert, and the Göttingen Mathematical Tradition, *Osiris* (2nd Series), Vol. 5 (1989), pp. 186–213.

[104] Sacrobosco, *Algorismus*.

[105] A. S. Saidan, The Earliest Extant Arabic Arithmetic: Kitab al-Fusul fi al Hisab al-Hindi of Abu al-Hasan, Ahmad ibn Ibrahim al-Uqlidisi, *Isis*, Vol. 57, No. 4. (Winter, 1966), pp. 475–490.

[106] G. Sarton, *A History of Science*, Science Editions, 1952.

[107] N. Sastri, *A History of South India*, Oxford University Press, 1958.

[108] O. G. von Simson, *The Gothic cathedral: origins of Gothic architecture and the medieval concept of order*, Princeton University Press, 1988.

[109] D. E. Smith, J. Ginsburg, *A History of Mathematics in America Before 1900*, MAA Press, 1934.

[110] D. E. Smith, J. Ginsburg, Rabbi Ben Ezra and the Hindu-Arabic Problem, *The American Mathematical Monthly*, Vol. 25, No. 3. (Mar., 1918), pp. 99–108.

[111] D. E. Smith and M. Latham, *The Geometry of Rene Descartes*, Dover, 1925.

[112] R. L. Spang, *The Invention of the Restaurant*, Harvard University Press, 2000.

[113] C. N. Srinivasiengar, *The History of Ancient Indian Mathematics*, World Press, 1967.

[114] W. H. Stahl, R. Johnson, E. L. Burge (trans.), *Martianus Capella and the Seven Liberal Arts*, Columbia University Press, 1977.

[115] R. Steele, *The Earliest Arithmetics in English*, Oxford University Press, 1922.

[116] J. H. Stewart, *The Restoration Era in France, 1814–1830*, D. van Nostrand Company, 1968.

[117] S. M. Stigler, *The History of Statistics*, Harvard University Press, 1986.

[118] D. Struik, *A Concise History of Mathematics*, Dover 1987.

[119] D. Struik, *A Source Book in Mathematics 1200–1800*, Harvard University Press, 1969.

[120] J. Suzuki, *A History of Mathematics*, Prentice-Hall, 2002.

[121] F. J. Swetz, *The Sea Island Mathematical Manual*, Pennsylvania State University Press, 1992.

[122] K. Thomas, Numeracy in early modern England, *Transactions of the Royal Historical Society*, 5th Ser., Vol. 37. (1987), pp. 103–132.

[123] L. Thorndike, *University Records and Life in the Middle Ages*, Columbia University Press, 1944.

[124] I. Todhunter, *A History of the Mathematical Theory of Probability, from the time of Pascal to that of Laplace*, Macmillan, 1865.

[125] F. L. Utley, *The forward movement of the fourteenth century*, Ohio State University Press, 1961.

[126] A. Vucinich, Mathematics and Dialectics in the Soviet Union: The Pre-Stalin Period, *Historia Mathematica*, Vol. 26 (1999), 107–124.

[127] A. Vucinich, Mathematics and Dialectics in the Stalin Era, *Historia Mathematica*, Vol. 27 (2000), 54–76.

[128] J. Welu, Vermeer's Astronomer: Observations on an Open Book, *The Art Bulletin*, Vol. 68, No. 2, (Jun., 1986), pp. 263–267

[129] H. Wieruszowski, *The medieval university*, van Nostrand Reinhold Company, 1966.

[130] J. E. M. White, *Ancient Egypt*, Dover Publications, 1970.

[131] L. T. White, *Medieval technology and social change*, Clarendon Press, 1962.

[132] G. A. Williams *Artisans and Sans-Culottes*, W. W. Norton, 1969.

[133] J. B. Wolf, *Louis XIV*, W. W. Norton, 1968.

[134] S. Wolpert, *A New History of India*, 5th ed., Oxford University Press, 1997.

[135] W. H. Woodward, *Studies in Education during the Age of the Renaissance, 1400–1600*, Cambridge University Press, 1924.

[136] L. B. Wright, *The Cultural Life of the American Colonies*, Harper & Row, 1957.

[137] L. L. Yong, The Jih yung suan fa: An Elementary Arithmetic Textbook of the Thirteenth century, *Isis*, Vol. 63, No. 3 (Sep. 1972), 370–383.

[138] C. Zaslavsky, *Africa Counts*, Prindle, Weber & Schmidt, 1973.

Figure Citations

Fig. 1.1, p. 1: Photo courtesy of the Science Museum of Brussels.
http://en.wikipedia.org/wiki/File:Ishango_bone.jpg.

Fig. 1.3, p. 4: from *The Rosetta Stone in Hieroglypics and Greek* by Samuel Sharpe. Published by J. R. Smith, 1871. Original from Harvard University.
http://books.google.com/books?id=Q5Ow9Ub_BIgC&dq=The%20Rosetta%20
Stone%20in%20Hieroglyphics%20and%20Greek&pg=PP23.

Fig. 1.4, p. 9: from *The Babylonian Expedition of the University of Pennsylvania*. Published by Pub. by the Dept. of Archaeology, University of Pennsylvania, 1906. Item notes: v. 20, pt. 1. Original from the New York Public Library.
http://books.google.com/books?id=Xw4ZAAAAYAAJ&pg=RA1-PA1.

Fig. 5.2, p. 127: from *Summa de arithmetica, geometria, proportioni et proportionalita* by Luca Pacioli, 1494.
http://archimedes.mpiwg-berlin.mpg.de/cgi-bin/toc/toc.cgi?step=thumb&dir=pacio_
summa_504_it_1494.

Fig. 5.3, p. 129: from Lectionnaire de Luxeuil. Initiale T. Luxeuil. Fin du VIIe sicle. Paris, Bibliothèque Nationale.
http://commons.wikimedia.org/wiki/File:Lect_Luxeuil_144.jpg

Fig. 5.5, p. 140: Page from the Sketchbook of Villard de Honnecourt,(about 1230). Paris, Bibliothèque Nationale.
http://commons.wikimedia.org/wiki/File:Villard_de_Honnecourt_-_Sketchbook_-_
64.jpg

Fig. 5.7, p. 143: from *Elementa Arithmetica* by Jordanus Nemorarius, 1496.
http://gallica.bnf.fr/ark:/12148/bpt6k52595c.zoom.f144.langEN

Fig. 6.2, p. 160: Fresco from a house in Pompeii.
http://commons.wikimedia.org/wiki/File:Pompeii_Fresco_002.jpg

Fig. 6.3, p. 161: *The Flagellation of Christ* by Piero della Francesca, 1444. Currently located at Palazzo Ducale in Urbino.
http://commons.wikimedia.org/wiki/File:Piero_della_Francesca_042.jpg

Fig. 6.4, p. 162: Letter A as designed and published by Fra Luca Pacioli: *Divina Proportione*, Paganino dei Paganini, Venice 1509.
http://commons.wikimedia.org/wiki/File:Fra_Luca_Pacioli_Letter_A_1509.jpg

Fig. 6.4, p. 162: Letter A as designed and published by Albrecht Dürer: *Unterweysung der Messung*, 1528.
http://commons.wikimedia.org/wiki/File:Albrecht_D%C3%BCrer_Letter_A_1528.png

Fig. 6.6, p. 176: *Melancholia* by Albrecht Dürer, 1514.
http://commons.wikimedia.org/wiki/File:D%C3%BCrer_Melancholia_I.jpg

Fig. 7.2, p. 203: *The Night Watch* by Rembrandt van Rijn, 1642.
http://commons.wikimedia.org/wiki/File:The_Nightwatch_by_Rembrandt.jpg

Fig. 7.3, p. 204: *The Astronomer* by Johannes Vermeer, 1668.
http://commons.wikimedia.org/wiki/File:JohannesVermeer-TheAstronomer(1668).jpg

Fig. 7.4, p. 205: *The Geographer* by Johannes Vermeer, 1669.
http://commons.wikimedia.org/wiki/File:Jan_Vermeer_van_Delft_009.jpg

Fig. 8.1, p. 240: *Gin Lane* from *Beer Street and Gin Lane* by William Hogarth, originally 1751. This print comes from an re-engraving circa 1806-09 by Samuel Davenport from Hogarth's originals.
http://commons.wikimedia.org/wiki/File:William_Hogarth_-_Gin_Lane.jpg

Fig. 8.2, p. 241: *Beer Street* from *Beer Street and Gin Lane* by William Hogarth, originally 1751. This print comes from an re-engraving circa 1806-09 by Samuel Davenport from Hogarth's originals.
http://commons.wikimedia.org/wiki/File:William_Hogarth_-_Beer_Street.jpg

Fig. 8.3, p. 273: *The Third of May* by Francisco de Goya, 1814.
http://commons.wikimedia.org/wiki/File:Francisco_de_Goya_y_Lucientes_023.jpg

Fig. 9.2, p. 287: Cartoon by French illustrator Bertall from *La Comédie de notre temps : études au crayon et à la plume*, Paris, vol. 2, 1875.
http://commons.wikimedia.org/wiki/File:Bertall_-_Une_robe_de_chez_Worth.jpg

Index

About the Author

Jeff Suzuki was born in California, and received his B.A. in mathematics and history from California State University at Fullerton, and M.A. and Ph.D. in mathematics from Boston University. His previous publications include *A History of Mathematics* (Prentice-Hall, 2002); *A Brief History of Impossibility* (Mathematics Magazine 81, 27–38); *Lagrange's Proof of the Fundamental Theorem of Algebra* (American Mathematical Monthly, 113, 705–714); and *The Lost Calculus: Tangency and Optimization Without Limits* (Mathematics Magazine, 78, 339–353), for which he won the Carl B. Allendoerfer Award from the MAA for an article of expository excellence. He is an Associate Professor of Mathematics at Brooklyn College.